Introductory Physics

Second Edition

Robert Karplus

UNIVERSITY OF CALIFORNIA
BERKELEY

Introductory
Physics

Edited by Fernand Brunschwig
EMPIRE STATE COLLEGE
NEW YORK

a model approach

Second Edition

Introductory Physics: A Model Approach, Second Edition
Copyright © 2011 by Fernand Brunschwig
(Previous copyright © 2003 by Elizabeth F. Karplus, licensed to Fernand Brunschwig and published by Captains Engineering Services. First Edition Copyright © 1969 by W. A. Benjamin, Inc., assigned to Robert Karplus in 1980, Library of Congress Catalog Card Number of First Edition: 70-78614.)

Editor of Second Edition: Fernand Brunschwig

ISBN-10: 0-982-72580-9
ISBN-13: 978-0-9827258-0-1
Printed in the United States of America, 2011, by Lightning Source, Inc.

Publisher:
Fernand Brunschwig
New York, NY
fbrunsch@gmail.com

To Betty

Foreword

Robert Karplus was my graduate advisor and mentor from 1968 to 1972 at the University of Calfornia, Berkeley. *Introductory Physics: A Model Approach* was just being published when I first met Karplus, and I assisted in teaching several courses based on the book. *Introductory Physics* made a very strong impression: there was Karplus' unique approach to physics, his brilliant insight, his clear, direct language, and his powerful teaching methods, all expressions of his joyful personality, creativity, and love of humanity. However, soon after I finished my Ph. D. and left Berkeley, the book went out of print. Karplus and his wife Elizabeth bought the rights back from the publisher in 1980, and he hoped to publish a second edition. But this was not to be, and he passed away in 1990.

Recently, Robert G. Fuller edited and published a wonderful collection of Karplus' work on science education, along with essays about him by many of his closest collaborators: *The Love of Discovery* (Kluwer Academic/Plenum Press, 2002). Reading the essays, especially the ones by Alan Friedman, Rita Peterson and Fuller, reminded me about how inspiring and influential Karplus was for me and for so many others. More pointedly, re-reading the two chapters of *Introductory Physics* included in the collection made me realize, even more than when I was in Berkeley, how unusual the book was and how valuable it still could be. I resolved to try to bring *Introductory Physics* back to life.

My initial thought was to reprint the 1969 edition of the book. But this would not really have been in Karplus' spirit. He would have wanted to fix errors and clarify the text wherever possible. At first, I hesitated to change anything, but I found that I could indeed distinguish what really needed change from the many valuable features that had to be preserved. This required a substantial effort – more than a year – but it has been a labor of love.

With the permission and encouragement of Elizabeth Karplus, I have edited *Introductory Physics*, and this second edition is published in memory of Robert Karplus.

Robert Karplus was a well-known physicist, physics teacher and science educator: Professor at UC Berkeley, Director of the Science Curriculum Improvement Study (SCIS), Director of Berkeley's Lawrence Hall of Science, President of the American Association of Physics Teachers, and a creative, prolific researcher in physics and science education. In addition, Karplus was a warm, generous person, a wonderful teacher, a terrific speaker, a passionate advocate for good science teaching, and a tireless campaigner for innovation and improvement in science education at all levels.

After more than ten years at the top echelon of theoretical physics research, Karplus became interested and then immersed in science education. His subsequent career spanned a unique period – the Post-Sputnik effort to improve understanding of science.

In this long effort, Karplus was a pioneer: he was one of the very first (possibly the first) leading scientists to focus on elementary science education, and he was instrumental in extending the national priorities, and funding, which focused first on the college level then on the secondary level and finally, after Karplus, on the elementary level. He realized the importance of Piaget's work very early, became a vocal expert on Piaget's ideas, and did significant original research on the reasoning process.

Karplus originated the Science Curriculum Improvement Study in 1963, and over the next 15 years, in collaboration with Herbert Thier and many others, he carried through a major national elementary-level curriculum reform effort. Karplus, Thier and SCIS brought good science (including hands-on investigations) into schools throughout the country and beyond. With a deep-seated belief in the values of public education and democracy, and a self-evident spirit of cooperation and goodwill, SCIS made a very significant contribution to elementary education in the US.

In fact, Karplus was one of the most effective and charismatic leaders in the many-faceted effort of the 1960s and 70s to develop science curricula, train teachers, carry out research, educate voters and political leaders, and generally upgrade the understanding and teaching of science throughout the US. While he is best known for his work on SCIS at the elementary level, Karplus taught physics regularly at Berkeley, he was active in the American Association of Physics Teachers, and he made many significant contributions to physics teaching at the college level. *Introductory Physics* is the fruit of these efforts.

A longer biography of Robert Karplus appears at the end of the book. I would like to express my gratitude to Elizabeth Karplus for her permission to revise and publish the book, and especially for her strong and nurturing encouragement along the way. I would also like to thank Empire State College (where I have taught since 1972), Donald S. Cook of the Bank Street College of Education, Alan Friedman of the New York Hall of Science, my wife Jennifer Herring, my publisher Timothy Johnson (a distinguished graduate of Empire State College), Harold Ohrbach of the New York Academy of Sciences, and the various individuals, publishers and organizations listed in the Acknowledgments for their permission to include illustrations, diagrams or photographs in the Second Edition.

Fernand Brunschwig

fbrunsch@esc.edu
New York, New York
May, 2003

Introduction to the Second Edition

Robert Karplus wrote this innovative textbook on physics for non-science students between 1965 and 1969. The book first appeared in 1966 as a preliminary edition, and in 1969 W. A. Benjamin, Inc. published the first edition in hardback. For reasons not connected with the quality of the book, it never got to a second edition and has been out of print for many years.

As Karplus says in the Author's Preface below, ". . . this book is addressed particularly to readers with little scientific or mathematical background: the only requirements are common sense, experience, and reasoning ability." Writing an understandable physics textbook for readers with little background is a laudable goal – often announced, but seldom achieved. In fact, a large variety of magazines, popular books, and textbooks have been published for this audience since 1969. So, you ask, why republish *Introductory Physics: A Model Approach* now?

The reason is that this is a genuinely outstanding physics textbook - understandable to students and rewarding to teach. Whether you are a physics student, a teacher, a prospective teacher, or a general reader, you can benefit from the many insights and innovations that Karplus poured into *Introductory Physics*. As in all his work, you will find here an original attack on the subject at the most fundamental level expressed in clear, direct language with an extraordinary sense of the joy of discovery.

The feature of *Introductory Physics* that is most valuable to students and teachers, and which sets it off from most other physics texts, is that Karplus really does start at a beginner's level. In fact, the first four chapters (Part One, a full quarter of the book) are a carefully graduated "on ramp," a *tour de force* of physics teaching.

Karplus and his team at SCIS had worked hard and long to create, test and revise an entire curriculum (including a sequence of topics and activities exemplifying Piaget's ideas, plus an imaginative set of hands-on investigations grounded in children's experiences). Karplus took full advantage of all this; Part One of *Introductory Physics* focuses on the topics he knew to be essential: reference frames, relative motion, working and mathematical models, systems and subsystems, interaction, radiation, and energy. In his unique way, Karplus develops these ideas in Part One, drawing on many illustrations and concrete examples from SCIS plus his mastery of Piaget's theory, to build the reader's concrete and abstract reasoning skills.

In Chapters 3 and 4, Karplus' extraordinary capacity to see physics as a whole, and his skills in simplifying and synthesizing, come into play: he focuses on the concepts of interaction, fields, radiation and energy, drawing upon gravity, electricity, magnetism, heat, and many other phenomena as examples to provide an accessible, powerful, integrated and satisfying overview of the field. By the end of Chapter 4, the student has some real experience with the process of understanding the natural world and is familiar with what physics involves, how it is done, the key concepts and terminology, and its rewards and shortcomings.

Beyond its value as a textbook, *Introductory Physics* provides a window through which we can see how a master synthesized his expertise in physics, psychology, and education, and how he built on the innovations generated by the dynamic curriculum and teaching reform projects of the time. Naturally, as described above, he drew upon his 15 years of work with elementary school children and elementary science curricula to create Part One of the book. Another very important influence on Karplus' thinking about how to make physics understandable to a beginner was Francis Friedman, the convenor of the Physical Science Study Committee (PSSC) at MIT in 1956. By 1969, the PSSC, now under the direction of Jerrold Zacharias and with ongoing support from the National Science Foundation, had generated the first and second editions (1960 and 1965) of the innovative *PSSC Physics Course* for secondary schools.

PSSC Physics was based on Friedman's willingness to face the fact that velocity, acceleration, force and Newton's laws of motion, though the starting point for all physics courses of the time, actually demanded a level of abstract thinking for which most students were not prepared. Friedman identified other topics, particularly the study of light (ray and wave optics), as requiring substantially less abstract thinking. Therefore, *PSSC Physics* reversed the conventional order of topics, using the study of light, especially a critical comparison of the wave and particle models, plus other topics, to build up the reasoning skills needed to understand Newtonian mechanics.

Karplus enthusiastically adopted this approach, which fitted well with his own view of physics as well as his research on the development of reasoning. Karplus had also spent more than 10 years on the forefront of theoretical quantum field theory; he was intimately familiar with the difficulties of trying to apply the classical Newtonian approach in modern physics, and he was very aware of the on-going value of concepts such as interaction, radiation, field, momentum, and energy. This conceptual background very likely also made Karplus especially receptive to Friedman's idea.

Karplus also built on the PSSC comparison of the wave and particle models. But, in Karplus' hands the study of scientific models became the main theme. In Part One, he explains many trenchant examples illustrating how models are built, adapted and discarded. In Part Two (Chaps. 5-8), he compares the wave and particle models for light and (in distinction with PSSC) also for sound. Chapter 8 ("Models for Atoms") represents the climax of the first half of the book. Karplus uses the wave model (rather than force and Newton's laws) as his primary tool to elucidate modern physics. He succeeds brilliantly here: allowing the concepts to grow naturally from their context and explaining ideas such as the Bohr model, electron diffraction, wave mechanics, the wave-particle duality and the uncertainty principle in a way that makes sense to beginners.

Karplus' teaching approach was also strongly influenced by Gerald Holton, a former colleague in graduate school at Harvard. Holton, a physicist and a historian of science at Harvard, had a deep understanding of the relationships between science and other human concerns. In collaboration with

Fletcher Watson and F. James Rutherford, Holton developed the *Project Physics Course*, which took a substantially different approach to secondary school physics than PSSC. Karplus was very appreciative of *Project Physics*, which identified ways to teach physics as part of the human experience and integrated the development of scientific concepts within a historical and cultural context. Karplus borrowed some specific ideas from *Project Physics*, for example the attempt to isolate a single light ray in Section 7.2, and he drew upon Holton's wonderful textbook *Development of Concepts and Theories in Physical Science* (1952) as well as Thomas Kuhn's seminal work, *The Structure of Scientific Revolutions*, which came out in 1962. Karplus also clearly did substantial reading of his own in the scientific and historical literature.

There are many notes in the margins about the historical context. More important, there are innumerable points in the course of an explanation of the physics content where Karplus calls up the historical context or the thinking of the discoverer in just the way needed to help a beginner gain a deeper understanding or to relate the concepts together in a memorable way.

This edition of Karplus' book is still quite close to its original form. While I edited the text in many ways – to clarify key points, correct typos, and modernize the most glaringly out-of-date items – I kept the page layout as close to its original form as possible. This imposed a rather demanding, yet salutary, requirement: modifications had to fit within the same number of pages as the original. Matching words in this way with the master gave me a workout at a level that I had not experienced since I completed my dissertation under his direction over 30 years ago.

This edition of Karplus' text could well be updated and supplemented in many ways. In particular, the references could be brought more up to date, many valuable innovations in physics teaching of the last 30 years could be incorporated, and additional hands-on experiments could be included. I would hope that the value of Karplus' work will not be dimmed by any shortcomings in the editing or by the fact that it has been out of print for so long.

Is understanding of the fundamental concepts of science necessary for non-scientists today? Is the capacity to apply scientific principles, think logically, and solve problems important for national and world citizenship? To Karplus the answers were obvious. His book provided and, I believe, still provides a positive and realistic response to these on-going challenges. I hope that a new generation of teachers and students find the book as valuable, provocative, and on-target as I, and many others, did during its brief life in print.

Fernand Brunschwig

New York, New York
June, 2003
email: fbrunsch@esc.edu

Author's Preface

What is physics? Every introductory physics text tries to answer this question in its own way, and this text is no different. However, this book is addressed particularly to readers with little scientific or mathematical background: the only requirements are common sense, experience, and reasoning ability. To make the subject meaningful to this audience, it was necessary to alter the customary approach of most introductory texts.

The usual sequence of topics in introductory physics courses for liberal arts students begins with a study of Newtonian mechanics and its roots in astronomy. There is good reason for starting with mechanics, because Isaac Newton's brilliant achievement three hundred years ago marked the beginning of the interplay of experiment, imagination, and deductive logic that has characterized physics ever since. Unfortunately for the novice, Newton's imaginative leap from observation to formulation of axioms was so great that few can follow it. Introducing Newtonian mechanics at the outset of the course presents an extremely difficult problem for many students. This problem often acts as a deterrent to their full appreciation of physics since many students never realize that other subject areas (for example, heat and temperature, sound, light, elasticity) do not make the same demands for abstract reasoning when they are treated at an elementary level.

To remove this impediment, I have rearranged the subject matter and have placed Newtonian physics at the end of the text. The result is an emphasis on the "interaction" and "energy" concepts rather than on the more abstruse "force" concept; thus, physics is related more effectively to the common sense views of most students entering a beginner's course. The instructor accustomed to the more usual approach will have to be careful to remain within the conceptual sequence of this text.

The book is primarily intended to meet the requirements of a one-semester course for non-science students. It consists of four parts, an epilogue, and an appendix reviewing the mathematical background for the course. Part One is an overview of physical phenomena, which are taken up in more detail in the later parts. A number of fundamental concepts are introduced: scientific models, reference frames, systems, interaction, matter, energy, equilibrium, steady state, and feedback. This initial part lays a foundation for the remainder of the text, and it will be helpful to any beginning student of science whether he continues the study of physics or moves into other areas.

Part Two is concerned with light, sound, radiation, and the structure of atoms. This part introduces the student to twentieth century physics and will enable him to appreciate the many excellent books on relativity theory and quantum physics listed in the bibliographies. In Part Three, operational definitions of energy are used to construct simple mathematical models for thermal energy (specific heat), elastic energy, gravitational field energy, and electric power. Finally, Part Four deals with the Newtonian theory of particle motion relating force to acceleration. Immediate applications of the theory lead to a mathematical model for kinetic energy, to a description of the motion of falling objects, and to an understanding of the action of rocket engines. More extensive investigations include the study of periodic motion (for example, the solar system and the pendulum), gas pressure and heat engines, and the kinetic theory of gases. In this final part, the extensive chapter-end bibliographies will be particularly helpful in guiding the student to further study of topics that interest him most.

It is clear from this survey of the text that its overall organization does not follow the historical development of physics. As previously mentioned, I have been motivated instead by a desire to construct a bridge between the preconceptions of the nonscientific reader and the concepts and theories of physics. In order to provide a helpful and interesting historical context for some aspects of the subject, however, the margins include historical background, biographical notes, and quotations from the works of prominent scientists. Much of this material illustrates the transient nature of most scientific theories and the extensive modifications that scientific models have undergone. What is esteemed by one generation may be discarded by the next!

Although the present sequence of topics is the one I consider most fruitful, the organization of the text into four separate parts (the latter three of which represent fairly independent entities) does allow some flexibility in the order in which these topics may be treated. It is possible, for example, to omit Part Two or to exchange the order of Parts Two and Three. Through such a modification the needs of particular courses or the preferences of the instructor may be met.

The text may also serve as a full year's course. Parts One and Two could be studied in the first semester, and Parts Three and Four in the second semester. To supplement the text material and to expand the course to fit the requirements of a full year's program, selections from the extensive bibliographies could be the focus of many assignments and discussions. These bibliographies also enable the instructor to adapt the course to encompass the diverse interests of the students.

As a final thought, I suggest that the reader begin with a reading of the Epilogue and conclude with Chapter 1. I do not mean that the book should be read backwards. After the Epilogue, the book may be read in the proper sequence, but the reader should return to Chapter 1 after he has read the Epilogue for a second time. By comparing his reaction to these two sections of the book on the first and second reading, he will be able to gauge his progress in scientific literacy.

My work on this book has benefited from the assistance and encouragement of many colleagues, friends, and students, and it is a pleasure to acknowledge them. Abraham Fischler, now at Nova University, teased me into teaching an "impossible" physics course for liberal arts students and thereby indirectly initiated the project. Thomas Kuhn, Gerald Holton, and the late Francis Friedman of the Physical Science Study Committee strongly influenced my view of what is important in physics. My associates in the Science Curriculum Improvement Study, especially Carl Berger, Joseph Davis, Chester A. Lawson, Marshall Montgomery, Luke E. Steiner, Laurence Strong, and Herbert D. Thier, helped me in selecting and refining my approach to the subject matter. Stephen Williams and Michael von Herzen assisted in the preparation and evaluation of a preliminary edition of this text. Diane Bramwell and my wife Elizabeth contributed to the production of the present edition.

I wish to acknowledge also the assistance of many publishers, organizations, and individuals who have made figures and photographs available to me. A detailed listing appears below in the Acknowledgments.

Two large groups of young people have contributed greatly to the preparation of this work. One of these consists of my physics students at Berkeley, who responded sometimes with enthusiasm and sometimes with despair to my teaching and to a preliminary edition. The second of these consists of the many elementary school children (including Beverly, Peggy, Richard, Barbara, Andy, David, and Peter Karplus) in whose classes I have had the privilege to teach. The knowledge gained from the feedback from both these groups has been of immeasurable value.

Finally, the current spirit of innovation in education, which has been fostered by the National Science Foundation during the past decade, has provided the indispensable background for my undertaking.

Robert Karplus

Berkeley, California
January, 1969

Acknowledgments

The sources of figures or margin illustrations are listed below.

Figure   1.1(b) – United Press International
Figure   1.1(c) – Lookout Mountain Laboratory, USAF
Figure   1.1(d) – Kaiser Steel Corporation
Figure   1.2(b) – Robert Goss
Figure   1.2(c) – Kaiser Steel Corporation
Figure   1.2(d) – Barbara Costanzo
Figure   1.9 – National Institute of Standards and Technology
Figures   2.12 and 2.13 – Copyright © Harold and Esther
          Edgerton Foundation, 2003, courtesy of Palm Press, Inc.
Figures   2.14 and 2.15 – Science Curriculum Improvement
          Study, Systems and Subsystems, D. C. Heath, Boston,
          Massachusetts, 1968, now published by Delta
          Education, Nashua, New Hampshire
Figure   3.4 – Ealing Corporation, Cambridge, Massachusetts
Figure   3.15 – Science Curriculum Improvement Study, Systems
          and Subsystems, D. C. Heath, Boston, Massachusetts,
          1968, now published by Delta Education, Nashua, New
          Hampshire
Figures   3.16 and 3.17 – *PSSC Physics*, D. C. Heath, Boston, Massa-
          chusetts, 1965*
Figure   5.5 – *PSSC Physics*, D. C. Heath, Boston, Massachusetts,
          1965*
Figure   5.6 – Sperry Rand Corporation
Figures   5.7, 5.8, 5.9, 5.10, 5.17, 5.18(a) – *PSSC Physics*, D. C.
          Heath, Boston, Massachusetts, 1965*
Figures   6.18, 6.19, 6.20, 6.21, 6.26(a) – *PSSC Physics*, D. C.
          Heath, Boston, Massachusetts, 1965*
Figure   6.29 – *PSSC Physics*, D. C. Heath, Boston,
          Massachusetts, 1965*
Figure   7.1 – U.S. Army Ballistic Research Laboratory, Aberdeen
          Proving Ground, Maryland
Figure   7.4 – *PSSC Physics*, D. C. Heath, Boston, Massachusetts,
          1965*
Figure   7.7 – *Harvard Project Physics*, Harvard University
Figures   8.3 and 8.4 – adapted from *Chemistry: Experiments and
          Principles* by P. O'Connor, J. E. Davis, E. L. Haenisch.
          W. K. MacNab, and A. L. McClellan. by Jesus Garcia,
          Donna M. Ogle, C. Frederick Risinger, Joyce Stevos ,
          and Winthrop D. Jordan. Copyright © 1982 by D.C.
          Heath and Company. All rights reserved. Reprinted by
          permission of Mcdougal Littell Inc.

Figure   8.5 – Radio Corporation of America

Figures   8.7 and 8.9 – adapted from *Chemistry: Experiments and Principles* by P. O'Connor, J. E. Davis, E. L. Haenisch. W. K. MacNab, and A. L. McClellan. by Jesus Garcia, Donna M. Ogle, C. Frederick Risinger, Joyce Stevos , and Winthrop D. Jordan. Copyright © 1982 by D.C. Heath and Company. All rights reserved. Reprinted by permission of Mcdougal Littell Inc.

Figure   8.20(a) – Chas. Pfizer and Co.

Figure   8.20(b) and (c) – Perry R. Stout

Figure   8.21 – Lawrence Berkeley National Laboratory, University of California, Berkeley

Figure   8.24 – U.S. Air Force

Figure   8.25 –General Atomics

Figure   10.7 – American Science and Engineering, Inc.

Figures   13.4, 13.6, 14.6, 14.7, 14.8, 14.9, 14.10, 14.11(a), 14.12(a), 14.13, 14.14, 14.16 – *PSSC Physics*, D. C. Heath, Boston, Massachusetts, 1965*

Figure   15.8 –History of Science Collections, University of Oklahoma

Margin, page 401 – Old Sturbridge, Inc.

Figure   16.1 – Pratt and Whitney, A United Technologies Company

Figure   16.9 – Wallach&Tiernan GmbH, Germany

---

*The Physical Science Study Committee (PSSC) developed the PSSC Physics Course beginning in 1956 at the Massachusetts Institute of Technology, with primary support from the National Science Foundation. In 1958, Educational Services Incorporated (ESI) was established to continue the course development, and the first and second editions of *PSSC Physics* were published, in 1960 and 1965, by D. C. Heath and Company. The photographs and illustrations attributed above to *PSSC Physics* are from the second (1965) edition of *PSSC Physics* and, as such, are published pursuant to the royalty-free licensing agreement which applies to the first (1960) and second (1965) editions. ESI was later reorganized as the Education Development Center, Inc. (EDC), which has custody of the original negatives. The current (seventh) edition of *PSSC Physics* (1991) by Haber-Schaim, Dodge, Gardner, and Shore is published by Kendall/Hunt, Dubuque, Iowa.

Brief Contents

Detailed Contents

Part two   *Waves and atoms*

Part four   *Motion*

GALILEO GALILEI
(1564–1642)

Part one: a first view

Science is a good piece
of furniture for a man to have
in an upper chamber provided
he has common sense on
the ground floor.

OLIVER WENDELL HOLMES
1870

The nature of science

Have you ever sorted the books in your library according to their subject matter, only to find a few remaining that "didn't fit"? In a way, this problem is similar to problems that face a scientist. For example, a scientist collects data on crystals or atomic particles or orbiting planets and must face the fact that some of the data does not fit expectations. Such an experience can be unsettling, but it can also lead to new understanding and insight.

One of the primary objectives of this text is to introduce you to a few of the powerful interpretations of natural phenomena used by the physicist to help organize experience. The text discusses some of these phenomena and the patterns of behavior they exhibit. You, in turn, are asked to examine your own experience for additional data to support or contradict these ideas. Occasionally, an unexpected outcome may compel you to reorganize your thinking. A critical approach to all aspects of the text is in order.

Unfortunately, modern culture has become fragmented into specialties. Science was once a branch of philosophy. In modern times, however, science, especially physics, is no longer an intellectual discipline with which every educated person is familiar. There are many reasons for this state of affairs (Fig. 1.1). Probably the most important is that many individuals do not feel a need for a formal study of nature. They develop a commonsense "natural philosophy" as a result of their everyday experiences with hot and cold objects, moving objects, electrical equipment, and so on. For most people, this seems quite adequate.

A second reason is that many of the questions with which modern physicists are concerned seem remote from everyday life. Physicists now study sub-nuclear particles, matter at ultra-low or extremely high temperatures, cosmic-sized objects such as galaxies, the beginning of the universe, and other extraordinary phenomena. The physics that is accessible to the beginning student has a cut-and-dried aspect that lacks the excitement of a quest into the unknown. Therefore, many students tend to think of physics as a finished story that must be memorized and imitated, rather than as a challenge to the creative imagination.

A third reason is the frequently indirect nature of the evidence on which physicists base their conclusions. As a result of this indirect evidence, experimental observations are related to theoretical predictions only through long and complicated chains of reasoning, often of a highly mathematical kind.

A fourth reason, of relatively recent origin, is that science has become identified with the invention of destructive weapons (the atomic bomb and biological warfare) and technological advances whose byproducts (smog, detergents) threaten our natural environment. Many individuals reject science, and especially physics, as alien to sensitive, imaginative, and compassionate human beings.

In this text we will try to overcome those difficulties. We will limit the diversity of topics treated, make frequent reference to the phenomena of everyday experience, and examine carefully the ways in which

"Why does this magnificent ... science, which saves work and makes life easier, bring us so little happiness? The simple answer runs – because we have not yet learned to make a sensible use of it."

Albert Einstein

observations can be interpreted as evidence to support various scientific theories. The goal is to develop your understanding of how physical concepts are interrelated, how they can be used to analyze experience, and that they are employed only as long as there are no better, more powerful alternatives.

The reasons why an educated person should have some understanding of physics have been stated many times (Fig. 1.2). Physics is a part of our culture and has had an enormous impact on technological developments. Many issues of public concern, such as air and water pollution, industrial energy sources, disarmament, nuclear power plants, and space exploration, involve physical principles and require an acquaintance with the nature of scientific evidence. Only a wider public understanding of science will ensure that its potential is developed for our benefit rather than devoted to the destruction of civilization. More personally, your life as an individual can be enriched by greater familiarity with your natural environment and by your ability to recognize the operation of general principles of physics everyday, such as in children swinging and hot coffee getting cold.

1.1 The scientific process

The present formulation of science consists of concepts and relationships that humankind has abstracted from the observation of natural phenomena over the centuries. Throughout this overall evolutionary process occasional major and minor "scientific revolutions" (or,

*"Few things are more be-
nighting than the condescen-
sion of one age for another."*

Woodrow Wilson

*What happens to an object
released in space, far from
the earth or another body?*

possibly more accurately, "transformations") have reoriented entire
fields of endeavor. Examples are the Copernican revolution in astronomy, the Newtonian revolution in the study of moving objects, and the
introduction of quantum theory into atomic physics by Bohr. The net
result has been the development of the conceptual structure and point of
view with which modern scientists approach their work.

An investigation. Let us briefly and in an oversimplified way look at
the way a scientist might proceed with an investigation. For instance,
consider a ball that falls to the ground when you release it. After additional similar observations (other objects, such as pieces of wood, a
feather, and a glass bowl, all fall to the ground when released), we are
ready to formulate a hypothesis: all objects fall to the ground when released. We continue to experiment. Eventually, we release a helium-
filled balloon and find that instead of falling, it rises. That is the end of
the original hypothesis. Can we modify it successfully? We could say,
"All objects fall to the ground when released in a vacuum." This statement is more widely applicable, but it is still limited to regions near the
earth or another large heavenly body where there is a "ground." In
space, far from the earth, "falling to the ground" is meaningless because
there is no ground.

This simple description has skipped over two important decisions that
we made. First was the judgment as to what constituted "similar"

(a)

Figure 1·2
Is physics relevant?
*(a) Do you base your actions on a crystal ball or on
scientific evidence and reasoning?*
(b) What clues enable you to identify the vertical direction?
*(c) The waste gases from a steel mill are cleaned by the
action of "precipitators," which make use of electric fields.
Compare with Fig. 1·1 (d).*
*(d) Medical x-ray photograph after an unlucky
fall on a skiing trip.*

(d)

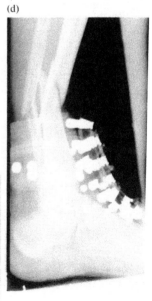

(b)

(c)

"... from my observations, ... often repeated, I have been led to that opinion which I have expressed, namely, that I feel sure that the surface of the Moon is not perfectly smooth, free from inequalities and exactly spherical, as a large school of philosophers consider with regard to the Moon and the other heavenly bodies, but that, on the contrary, it is full of inequalities, uneven, full of hollows and protuberances, just like the surface of the Earth itself, which is varied everywhere by lofty mountains and deep valleys."
 Galileo Galilei
 Sidereus Nuncius, 1610

"Matter" includes all solid, liquid, and gaseous materials in the universe. In this text, we will not define "matter" more precisely; we will treat "matter" as an undefined term, with a meaning that must be grasped intuitively. Properties of matter, to be described later in this text, include mass, extent in space, permanence over time, ability to store energy, elasticity, and so on.

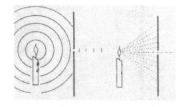

The work of Christian Huygens (1624-1695) and Isaac Newton (1642-1727) on the nature of light will he discussed in Chapters 5, 6, and 7.

observations. For instance, in the example, we included the balloon along with the ball, weed, feather, and so on. Yet we might have considered the balloon to be very different from the other objects observed. Then the balloon rising rather than falling would not have been considered pertinent to the hypothesis of falling objects. Even for some time after Galileo's telescopic observations of the moon more than 300 years ago, there was controversy as to whether it and other heavenly bodies were material objects to which the hypothesis of falling objects should apply.

The second decision was the judgment about what aspects of the observations were to be compared. We decided to compare the motion of the bodies after they were released. Aristotle, a Greek philosopher who also thought about falling bodies, was more concerned with such objects' ultimate state of rest on the ground, and therefore he reached conclusions very different from those we found above.

The scientific point of view. Usually the answers to these two kinds of questions are tacitly agreed upon by the members of the scientific community and constitute what we may call the "scientific point of view." One aspect of this point of view is that a real physical universe composed of matter exists, that we are a part of this universe, and that matter participates in natural phenomena. A second is the assumption that natural phenomena are reproducible: that is, under the same set of conditions the same behavior will occur. A third aspect is that while we ourselves are part of the physical world, we are also able to observe the natural world and to think about our observations. Other aspects of the point of view have to do with the form of an acceptable explanation of a phenomenon. This scientific point of view provides a context for scientific knowledge and for what is (and is not) accepted as scientific knowledge. Occasionally, however, it is very difficult to interpret new observations in a way that is consistent with the accepted scientific way of thinking. Then there is the need for bold and imaginative thinking to develop a new point of view. Hopefully, this new approach will be better able to explain the new observations and the known phenomena. Eventually it may become the accepted scientific point of view. The key idea here is that the scientific point of view (that is, the criteria for what is scientific knowledge) has gradually changed and is certain to continue to change.

The theory of light. A fascinating story in the history of physics that illustrates these remarks deals with the nature and interactions of light. Two competing ideas were advanced in the seventeenth century. Isaac Newton thought that light consisted of a stream of corpuscles, while Christian Huygens believed that light was a wave motion (see illustration to left). Up to that time, experiments and observations on light rays had apparently been made without questioning further the nature of the rays.

In spite of contradictory evidence, Newton's corpuscular theory of light was preferred by the scientific community, largely because of the success of Newton's laws of the motion of material bodies subject to forces. Small bodies (corpuscles) probably provided a more acceptable explanation to Newton's contemporaries and followers than did the waves proposed by Huygens. During the nineteenth century, however, new experimental data on the passage of light near obstacles and through transparent materials contradicted Newton's corpuscular theory conclusively and supported the wave theory. Waves and their motion became the accepted way to explain the observed properties of light.

This point of view flourished until the beginning of the twentieth century, when results of further experiments on the absorption and emission of light by matter conflicted with the wave theory and led to the presently accepted quantum theory of light. Already, however, there are contradictions within this theory, so that it, too, will have to be modified. This is one field of currently active research, and several proposals for new theories are being studied intensively to determine which holds the most promise.

Scientific "truth." Science is, therefore, never complete; there are always some unanswered questions, some unexpected phenomena. These may eventually be resolved within the accepted structure of science, or they may force a revision of the fundamental viewpoint from which the phenomena were interpreted. Progress in science comes from two sources: the discovery of new phenomena and the invention of novel interpretations that illuminate both the new and the well-known phenomena in a new way. Scientific truth is therefore not absolute and permanent: rather, it means agreement with the facts as currently known. Without this qualification, the statement that scientists seek the truth is misleading. It is better to say that scientists seek understanding.

1.2 Domains of magnitude

When and how does a person's experience of space and time originate? Probably the foundations are laid before birth, but the most rapid and important development takes place during an infant's early exploration of the environment. By crawling around, touching objects, looking at objects, throwing objects, hiding behind objects, and so on, an infant forms simple notions of space. By getting hungry and feeling lonely, by enjoying entertainment and playing, by watching things move and by moving himself, he forms notions of time. Even though an adult commands more effective skills with which to estimate, discriminate, and record space-time relations, our need to relate the environment to ourselves is never really outgrown.

Size. As you look about and observe nature, you first recognize objects, such as other people, trees, insects, furniture, and houses that are very roughly your own size. We will call the domain of magnitude of these objects the *macro domain*. It is very broadly defined and spans living creatures from tiny mites to giant whales. All objects to which you relate easily are in this domain.

All other natural phenomena can be divided into two additional domains, depending on whether their scale is much larger or much smaller than the macro domain. The former includes astronomical objects and happenings, such as the planet earth, the solar system, and galaxies. We will call this the *cosmic domain.* Much smaller in scale than the macro domain is the one that includes bacteria, molecules, atoms, and subatomic units of matter; we will call it the *micro domain.*

The phrase "geologic times" is sometimes used to denote very long time intervals because geologic processes (such as changes in the shape of the Earth) are extremely slow.

Time. It is useful to introduce the concept of domains into time scales as well as into physical size. Thus times from seconds or minutes up to years are *macro times* in the sense that they correspond to the life spans of human beings and other organisms. Beyond centuries and millennia are *cosmic times,* whereas *micro times* are very small fractions of a second. As with physical sizes, the mental images you make for processes of change always represent in seconds or minutes what really may require cosmic times or only micro times to occur.

Applications. In order of size, then, the three domains are the micro, macro, and cosmic. The division is a very broad one, in that the earth and a galaxy, both in the cosmic domain, are themselves vastly different in scale. Likewise, bacteria and atomic nuclei are vastly different. Nevertheless, the division is useful because the mental images you make of physical systems are always in the macro domain, where your sense experience was acquired. You therefore have to remember that your mental image of a cosmic system, such as the solar system, is very much smaller than the real system. Similarly, your mental image of a micro system is very much larger than the real system. As you make mental images of these systems, you will find yourself endowing them with physical properties of macro-sized objects, such as marbles, ball bearings, and rubber balls. This device can be very misleading because, of course, your images are in a different domain from the objects themselves.

When we pointed out in the introductory section to this chapter that physicists frequently must interpret indirect evidence, we had in mind, among other things, the three domains of magnitude. Since our sense organs limit us to observations in the macro domain, all interpretations concerning the other domains require extended chains of reasoning. An illustration relating the domains of magnitude to units of space and time measurement is presented at the end of this chapter in Fig. 1.11.

1.3 Theories and models in science

In the preceding section we contrasted the roles played in science by observation and interpretation. Observations of experimental outcomes provide the raw data of science. Interpretations of the data relate them to one another in a logical fashion, fit them into larger patterns, raise new questions for investigation, and lead to predictions that can be tested.

Scientific theories are systematically organized interpretations. Examples are Dalton's atomic theory of chemical reactions, Newton's theory of universal gravitation, Einstein's theory of relativity, and Piaget's theory of intellectual development. Within the framework of a scientific theory, observations can be interpreted in much more far-reaching ways than are possible without a theory. In Newton's theory of gravitation, for instance, data on the orbital motion of the moon lead to a numerical value for the total mass of the earth! In Dalton's theory, the volumes of chemically reacting gases lead to the chemical formulas for the compounds produced. All theories interrelate and extend the significance of the facts that fall within their compass.

Working models. Theories frequently make use of simplified mental images for physical systems. These images are called *working models* for the system. One example is the sphere model for the earth, in which the planet is represented as a uniform spherical body and its topographic and structural complexities are neglected. Another example is the particle model for the sun and planets in the solar system; in this model each of these bodies is represented as a simple massive point in space, and its size as well as its structure is ignored. Still another example is the "rigid body model" for any solid object (a table, a chair) that has a definite shape but may bend or break under a great stress.

Unlike other kinds of models (Fig. 1.3), a working model is an abstraction from reality. Our thoughts can never comprehend the full complexity of all the details of an actual system. Working models are always simplified or idealized representations, as we have already pointed out. Working models, therefore, and the theories of which they are a part, have limitations that must be remembered when their theoretical predictions fail to agree with observations.

Figure 1.3 The word "model" has many connotations in the English language, and most of them are not applicable to the scientific meaning of the word. A scientific "working model" has very little in common with a scale model (model airplane, left), a sample for examination (model home, below left), a visual replica (architectural model, below center), or a person (artist's or fashion model, below right).

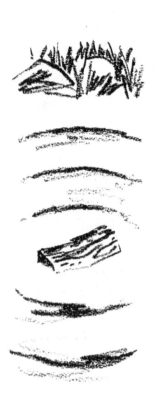

The scientist's relationship to the models he constructs is ambivalent. On the one hand, the invention of a model engages his creative talent and his desire to represent the operation of the system he has studied. On the other hand, once the model is made, he seeks to uncover its limitations and weaknesses, because it is from the model's failures that he gains new understanding and the stimulus to construct more effective models. Both creative and critical faculties are involved in the scientist's work with models.

One feature of working models is frequently disturbing to nonscientists: no model perfectly matches reality, and you never know whether a particular model is "right." In fact, the concepts "right" and "wrong" do not really apply to models. Instead, a model may be more or less adequate, depending on how well it represents the functioning of the system it is supposed to represent. Even an inadequate model is better than none at all, and even a very adequate model is often replaced by a still more adequate one. The investigator has to determine whether a particular model is good enough for his purposes or whether it is necessary to seek a better one.

Analogue models*.* Before a scientist constructs a theory, he often realizes that the system he is studying operates in a way similar to another system with which he is more familiar, or on which he can conduct experiments more easily. This other system is called an analogue model for the first system. You may, for instance, liken the spreading out of sound from a violin to the spreading out of ripples from a piece of wood bobbing on a water surface.

The analogue model for one physical System A is another, more familiar, System B, whose parts and functions can be put into a simple correspondence with the parts and functions of System A. For example, an analogy may be drawn between the human circulatory system and a residential hot water heating system (Table 1.1, below). It is clear that

TABLE 1.1 ANALOGUE MODEL FOR THE HUMAN CIRCULATORY SYSTEM

System A: Human circulatory system	System B: Residential hot water heating system
veins, arteries	pipes
blood	water
oxygen	thermal energy
heart	pump
lungs	furnace
capillaries	radiators
hormones	thermostat
(model fails) (or dilation of veins & arteries)	overflow tank
blood pressure	water pressure
white blood cells	(model fails)
carbon dioxide	(model fails)
kidneys	(model fails)
intestine	(model fails)

the human circulatory system fulfills several functions, whereas the heating system fulfills only one. The analogue model is, therefore, not complete, but it is nevertheless instructive.

The virtue of an analogue model is that System B is more familiar than System A. This familiarity can have several advantages:

1. Features of the analogue model can call attention to overlooked features of the original system. (Had you overlooked the role of hormones in the circulatory system, the room thermostat would have reminded you.)
2. Relationships in the analogue model suggest similar relationships in the original system. (Furnace capacity must be adequate to heat the house on a cold day; lung capacity must be adequate to supply oxygen needs during heavy exercise.)
3. Predictions about the original system can be made from known properties of the more familiar analogue model. (Water pressure is high at the inflow to the radiators, low at the outflow; therefore, blood pressure is high in the arteries, low in the veins.)

The limitations of the analogue model can lead to erroneous conclusions, however. On a cold day, for instance, the water temperature is higher in the radiators; therefore, you might predict that the oxygen concentration in the blood will be higher during heavy exercise. Actually, the heartbeat and the rate of blood flow increase to supply more oxygen - the oxygen concentration does not change greatly.

"There are two methods in which we acquire knowledge - argument and experiment."
Roger Bacon (1214-1294)

Thought experiments. In a thought experiment, a model is operated mentally, and the consequences of its operation are deduced from the properties of the model. A thought experiment differs from a laboratory experiment in that the latter serves to provide new information about what really happens in nature, whereas the former seeks new deductions from previous knowledge or from assumptions. By comparing the deductions with observations in real experiments, you can find evidence to support or contradict the properties or assumptions of the model.

A simple example of a mystery system (Fig. 1.4) can be used to illustrate these ideas. Two working models for what might be under the cover in Fig. 1.4 (a) are shown in Figs. 1.4 (b) and (c). If you conduct simple thought experiments with these models, you quickly find out how satisfactory they are. In the first thought experiment, you imagine turning handle A clockwise. In model G, handle B will turn somewhat faster, because the second gear is smaller than the first, but it will turn counterclockwise. This prediction is in disagreement with the properties of the mystery system. In the second thought experiment, you turn handle A in model S. What can you infer from this second experiment? Can you suggest a satisfactory working model?

Thought experiments are important tools of the theoretical scientist because they enable him to make deductions from a working model or a theory. These deductions can then be compared with observation. The usefulness of a theory or model is determined by the agreement between the deduction and observation. Some very general theories,

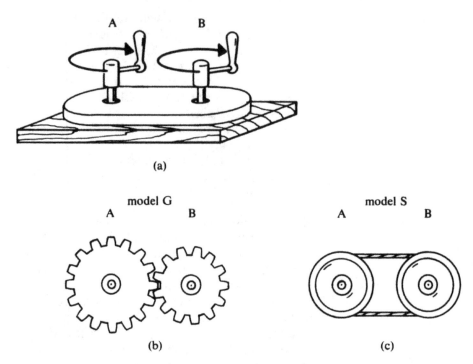

Figure 1·4 A mystery system. (a) When handle A is turned one revolution clockwise, handle B makes 2½ revolutions clockwise. Make models for what is under the cover. (b) Large and small gear model. (c) Two pulley and string model.

Equation 1.1

Mathematical model (algebraic form):
number of turns of
    handle A = N_A
number of turns of
    handle B = N_B

$$N_A = N_B$$

Equation 1.2

Mathematical model (algebraic form):
    distance   = s
    speed   = v
    time   = t
        $s = vt$

such as the theory of relativity, lead to consequences that appear to apply universally. Some models, such as the corpuscular model for light, are useful only in a very limited domain of phenomena.

Mathematical models and variable factors. Scientific theories are especially valuable if they lead to successful quantitative predictions. Working models G and S for the mystery system in Fig. 1.4 both lead to quantitative predictions for the relationship between the number of turns of handles A and B. The relationship deduced from model S (that the handles turn equally) can be represented by the formula in Equation 1·1. We will call such relationships *mathematical models;* the formula in Equation 1.1 is an algebraic way of describing the relationship, which we have also described in words, and which can be described by means of a graph (Fig. 1.5).

A familiar example of a mathematical model, applicable to an automobile trip, is the relation of the distance traveled, time on the road, and speed of the car (Equation 1.2). The distance is equal to the speed times the time. At 50 miles per hour, for example, the car covers 125 miles in 2 ½ hours (Fig. 1.6).

The physical quantities related by a mathematical model are called *variable factors* or *variables*. The numbers of turns of handles A and B are two variable factors in Equation 1.1 and Fig. 1.5. The distance and elapsed time are two variable factors in Equation 1.2 and Fig. 1.6, The speed in this

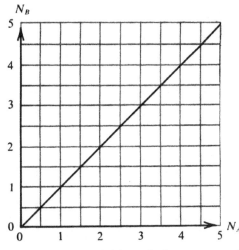

*Figure 1.5 Mathe-
matical model
(graphical form).*

*Number of turns of
handle A = N_A;*

*Number of turns of
handle B = N_B;*

mathematical model is called a constant, because it does not vary. Under different conditions, as in heavy traffic, the speed might be a variable factor.

Like the working model for a system, the mathematical model for a relationship is not an exact reproduction of a real happening. No real car, for instance, should be expected to travel at the perfectly steady speed of 50 miles per hour for 2½ hours. The actual speed would fluctuate above and below the 50-mile figure. The actual distances covered at various elapsed times, therefore, might be a little more or a little less than those predicted by the model in Eq. 1.2 and Fig. 1.6. Nevertheless, the model gives a very good idea of the car's progress on its trip, and it is very simple to apply. For these reasons, the model is extremely useful, but you must remember its limitations.

working model

thought
experiment

mathematical model

Scientific theories. The making of a physical theory often includes the selection of a working model, the carrying out of thought experiments, and the construction of a mathematical model. All physical theories have limitations imposed by the inadequacies of the working model and the conditions of the thought experiments. Occasionally a theory has to be

*Figure 1.6 Mathematical
model of relationship be-
tween distance and time
(graphical form):*

*Distance   = s (miles),
time   = t (hours),
speed   = 50 miles
per hour.*

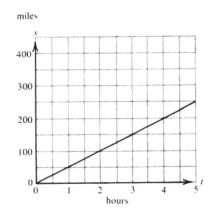

abandoned because it ceases to be in satisfactory agreement with observations. Nevertheless, physical theories are extremely useful. It is probably the power of the theory-building process we have described that lies behind the rapid progress of science and technology in the last 150 years.

1.4 Definitions

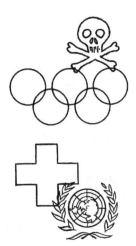

The primary function of language is to communicate information from one individual to others. Human language consists of signs, gestures, spoken sounds, and marks on paper that function as symbols of some sensed experience. Communication by means of human language is possible so long as the communicants have a common understanding of the meaning of the symbols, that is, so long as all persons relate a given symbol to a particular common experience and to none other.

In learning a language, you must first learn to recognize the symbol, then to relate that symbol to a particular experience. A symbol may refer to a material object, the relation or state of material objects, other symbols, or relations of symbols. The normal device for conveying the meaning of a symbol is the definition, of which we will distinguish two types. These are formal definitions, which use words, and operational definitions, which use operations.

Formal definitions. The familiar dictionary definitions, which identify the meaning of a symbol by the use of words or other symbols, are included in the category of formal definitions. Synonyms, paraphrases, lists of properties, and names of examples are the usual techniques of formal definition.

An example of the use of synonym is "bottle = jar." Synonyms may have exactly the same meaning, but they usually have slightly different meanings. For example, both "bottle" and "jar" are "containers made of glass" (paraphrase) but usually connote different shapes.

bottle = jar   ?

An example of definition by paraphrase is "photosynthesis = the conversion of light energy into chemical energy in green plants." Another example is "velocity of an object = the distance traveled divided by the time taken." The paraphrase definition is similar to the definition by synonym, except that the paraphrase contains more words. The paraphrase definition leads to efficiency in communication (or thought, which is self-communication) in that you can substitute the shorter term for the longer phrase. We will occasionally use paraphrases based on a mathematical process to define physical terms, as we did in the velocity example just given.

Operational definitions. The use of real objects and operations (not merely words) to produce, measure, or recognize an instance of a term is the essence of the operational definition. For example, the operational definition of color words, such as red, yellow, mauve, and lime, may be based on a set of color chips that have sample colors on one side and their names on the back. The objects used in this definition are the color chips. The operation is that of comparison of the hue of an unidentified color with those of the color chips. This operational definition is in general use in paint stores.

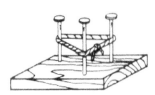

Words can be used to describe an operational definition but the definition itself consists of operations on real objects and not of words. For example, you can construct a triangle by driving three nails into a board and connecting the nails with a stretched string. A figure that matches the figure constructed in this way is also a triangle. These objects and operations define a triangle.

An example of an operational definition that leads to measurement is as follows: the number of seats in an auditorium is the auditorium's capacity. Here the actual seats in the auditorium and the counting operation are combined in an operational definition of the auditorium capacity.

We will soon introduce operational definitions for measuring basic physical quantities, such as length, time, and mass. Each of these definitions makes reference to a standard object that serves as the unit of measurement (in the definition of auditorium capacity, the chair served as the unit of measurement) and a comparison operation that allows the unit to be compared with other objects.

For science, the significance of operational definitions is that their use keeps the description of models and the statement of theories meaningful and testable in the physical world. In contrast to the scientist's operationally defined language, that of the poet rests mainly on terms (for example "beauty," "love," and "grace") that are not defined operationally. However, it is also worth pointing out that the language of a poem generally *does* have a close relationship (or multiple relationships) with the significance, sound, and/or meaning of the words as they are used in the language at large. We also must realize that while scientific concepts must always be somehow logically tied to operational definitions, many scientists use concepts that are only tied to an operational definition through a series of formal definitions. Therefore, scientists often use language that appears just as distant from the real world as the poet's! Finally, poets have anticipated key scientific developments, for example, in ancient times, Lucretius speculated about atoms in his poem, *On the Nature of Things*.

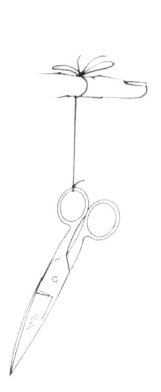

Comparison of formal and operational definitions. In science, formal definitions are frequently used to define one concept in terms of other concepts. For instance, the term "triangle" could have been defined by paraphrase as "a plane figure bounded by three nonparallel straight lines." This definition uses concepts, such as "plane," "nonparallel," "three," and "straight line," for which definitions have to be provided or that may properly remain undefined.

Let us consider another term, "vertical," that can be defined operationally or formally. In the operational definition, a freely hanging plumb line is allowed to come to rest; vertical is the direction indicated by the plumb line. The formal definition is "vertical = the direction toward the center of the earth." The latter definition is a paraphrase that is useful for theoretical purposes, but impossible to apply in practice, as when a house's walls are to be built.

The difference between formal and operational definitions is illustrated especially clearly by their application to "intelligence" and "IQ."

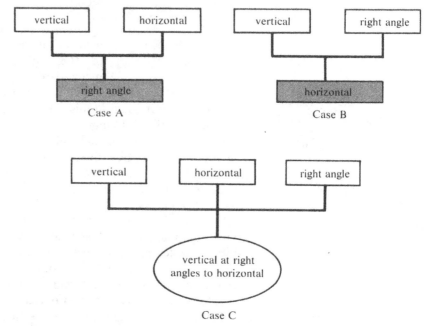

Figure 1·7 Definitions of vertical, horizontal, and right angle. Open box: operationally defined; shaded box: formally defined; oval: experimentally discovered. The definitions are described in Table 1.2

The dictionary defines intelligence as "the ability to apprehend the interrelationships of presented facts in such a way as to guide action toward a desired goal." The value of this formal definition as a positive personal trait seems obvious. It is very difficult, however, to rank individuals according to their intelligence, because this requires applying the definition operationally to specific cases. The intelligence

---

TABLE 1.2 THREE ALTERNATIVE DEFINITIONS OF VERTICAL,
HORIZONTAL AND RIGHT ANGLE

Case A. Define vertical: direction of a free plumb line at rest.
   Define horizontal: direction of a free water surface at rest.
   Define right angle: the angle between the vertical and horizontal.

Case B. Define vertical: direction of a free plumb line at rest.
   Define equal angles: angles that match when superposed.
   Define straight line: matches a stretched string.
   Define right angle: draw two intersecting straight lines on a given
      (flat) board so that four equal angles are produced. Each angle is
      a right angle.
   Define horizontal: the surface at right angles to the vertical.

Case C. Define vertical: direction of a free plumb line at rest.
   Define horizontal: direction of a free water surface at rest.
   Define right angle: draw two intersecting straight lines on a given
      (flat) board so that four equal angles are produced. Each angle is
      a right angle.
   Experimental relation: vertical and horizontal make a right angle.

quotient (IQ) can be defined operationally by a standard score on a specific test combined with a person's age. However, the *meaning* of the IQ as a personality trait and its functional value (that is, the relationship between an operationally defined IQ and its more generally accepted formal definition) are subjects of controversy that are far from being resolved.

Formal definitions and operational definitions each have their advantages and disadvantages. Operational definitions, as we have already stressed, make direct reference to the physical world and to human perception. This property gives them the advantage of being concrete. At the same time, their dependence on specific objects (such as auditorium seats) limits their scope of application. The definition of "capacity" given for an auditorium, for instance, could not be applied to the gasoline tank of a car. A definition of temperature using an ordinary thermometer would not be applicable in the interior of the sun. Operational definitions tend to be cumbersome in that they demand the availability of certain equipment.

Formal definitions, by contrast, are more concise and efficient. They relate concepts to one another directly. The definitions are much more generally valid. The price that is paid for these advantages is that the language becomes very abstract, because direct connections with reality are buried in the foundations on which the system of formal definitions rests.

In this text we will place more reliance on operational definitions than is customary, because we believe that concrete ties to reality are more valuable to you than efficiency and generality. Our approach, therefore, will be somewhat different from that of other texts. However, the physical world that is being described is the same; the differences are in the logical development and not in the content itself. To illustrate the diversity of possible approaches to the logical development of ideas, Fig. 1.7 and Table 1.2 show how the concepts "vertical," "horizontal," and "right angle" may be defined and related to one another in three different ways.

1.5 Length, time, and mass

That we relate most easily to the macro domain of magnitudes is reflected in the fact that units for measuring length have, since ancient times, been derived from our bodies (Fig. 1.8). The ready availability of the human body made the foot and the inch convenient units, but there was a great deal of local variation, depending on whose foot or thumb was used. With the growth of an international scientific community, it became necessary to adopt standard units of measurement that would be accepted by scientists everywhere. The French Academy of Sciences in 1791 suggested a new unit of length, the meter, which was to be one ten-millionth of the distance from the pole to the equator of the earth. Accordingly, a platinum-iridium bar with two marks separated by the "standard meter" was prepared after seven years of surveying the earth in Spain and France. The original is kept in the Bureau of Weights and Measures near Paris and accurate copies are kept by the National Bureau of Standards near Washington (Fig. 1.9) and by similar agencies

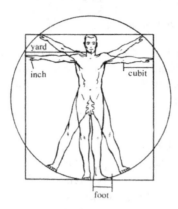

Figure 1-8 Units of measurement related to the human body.

elsewhere. The marks on the rulers you use are derived, through a long chain of copying, from the original standard meter in France.

Widely accepted units of measurement are essential to our technological culture. The story of weights and measures and the continuing search for improved standard units will never end.

We turn now to the operational definitions of the basic quantities of length, time, and mass. Since the most primitive measurement operation is that of counting, the definitions involve procedures for comparing the quantity to be measured with accepted standard units and counting the number of standard units that are required.

Length and distance. Length and distance are defined by a matching procedure in which the length of any object can be used as the unit. The generally accepted standard unit of length is the meter, described above. After the meter had been established, it was found that the earlier measurements of the earth had been inaccurate, so the geographical definition was abandoned, but the platinum-iridium bar was kept. However, duplicating the standard length was cumbersome and tended to introduce additional errors. As a result, the current definition of the meter in terms of the wavelength (see Chapter 7) of a specially designed light source was adopted. This definition allows the standard meter to

Figure 1.9 Replicas of the international standards of length and mass. (a) The standard meter bar. (b) The standard kilogram cylinder, whose size is close to that of a small egg.

(a)

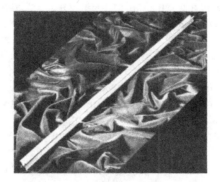

(b)

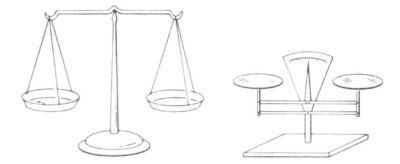

Figure 1.10 Equal-arm balances.

be replicated conveniently as needed: you simply measure the wavelength of the standard source to whatever accuracy is required.

Units of length associated with the meter are the centimeter (one hundredth of a meter), millimeter (one thousandth of a meter), and kilometer (1000 meters).

Time. Time intervals are defined by a matching procedure in which the unit of time may be the swing of a pendulum, the emptying of an hourglass, or the completion of some other repeated pattern of motion. The generally accepted standard unit of time is based on the repeating (periodic) motion of the earth around the sun (year) and the rotation of the earth on its axis (day). By means of a pendulum or other such system with a short time of repeating its motion, the second has been defined as 1/86,400 of a mean solar day, which is 1/365.2 ... of a year. As in the case of length, a standard unit of time associated with atomic vibrations has been substituted for the astronomical definition.

Mass. Mass is defined by a matching procedure with an equal-arm balance. The unit of mass could be any object, a stone, or a nail, for example. The accepted unit of mass since 1889 is the kilogram, the mass of a metal cylinder kept under carefully controlled conditions near Paris (Fig. 1.9). The kilogram was intended to be the mass of 1000 cubic centimeters of water at 4° Celsius. Later, more accurate measurements showed that the original determination was slightly in error, so that the reference to water was abandoned. The operational definition of mass makes use of the equal-arm balance (Fig. 1.10), which responds to the downward pull of the earth (commonly known as the weight). Therefore, mass, as we are referring to it here, is called the *gravitational mass*. This idea of mass as intimately connected with the gravitational attraction exerted by the earth will come up again in Section 3.4 where we will explain the related, but distinct, concept of *inertial mass*.

Units of mass derived from the kilogram are the gram (one thousandth of a kilogram), very closely equal to the mass of 1 cubic centimeter of water, and the metric ton (1000 kilograms), very closely equal to the mass of 1 cubic meter of water.

Size in meters		Time in seconds	
10^{+24}		10^{+24}	
10^{+21}	Milky Way (our galaxy)	10^{+21}	
10^{+18}		10^{+18}	age of Universe
			life on earth
10^{+15}		10^{+15}	first mammals
	Solar System		first humans
10^{+12}	orbit of Jupiter	10^{+12}	
	orbit of Earth		recorded history
10^{+9}	Sun diameter	10^{+9}	human lifetime
	Earth diameter		year
10^{+6}	Chicago to NY	10^{+6}	month
			day
10^{+3}	1 mile	10^{+3}	hour
			minute
10^{0} (1 m.)	Human stride	10^{0} (1 sec.)	heartbeat
10^{-3}	millimeter	10^{-3}	1 vibration (audible sound wave)
	razor edge		
10^{-6}	single cells	10^{-6}	1 vibration (AM radio wave)
	DNA, proteins		
10^{-9}	molecules	10^{-9}	1 cycle (1 GHz computer)
	atoms		
10^{-12}		10^{-12}	1 cycle (1000 GHz computer)
	atomic nuclei		
10^{-15}		10^{-15}	1 vibration (visible light)
10^{-18}		10^{-18}	1 vibration (X-ray)
10^{-21}		10^{-21}	

Cosmic domain

Macro domain

Micro domain

Figure 1.11 Time and size scale of cosmic, macro, and micro domains.

Other variable factors. It is possible to define units for all other physical variables through definitions based on mass, length, time, and temperature (to be defined in Chapter 10). We will, however, take a different approach, in which we introduce operational definitions for several concepts, such as energy and force, because such an operational procedure makes the physical meaning of the concepts clearer. You will have to accept one disadvantage of this procedure: operational definitions are limited by the technique or operation used and thus will not be the most general ones possible.

Domains of magnitude. We now briefly return to the three domains of magnitude introduced in Section 1.3: the cosmic domain, the macro domain of the everyday world, and the micro domain. Using the definitions of the standard units of measure that we have described, we can approximately characterize the domains by their relationship to these units. Figure 1.11 illustrates this relationship and shows the time and size scale of the various domains.

Summary

The phenomena studied by the physical scientist are highly diverse, ranging from the orbital motion of satellites to the propagation of light, from the turbulent motion of gases in the sun to the structure of the atomic nucleus. The space and time dimensions of phenomena are conveniently divided into three domains: the macro domain, roughly comparable to the human body; the cosmic domain of the very large or very enduring phenomena; and the micro domain of the very small or highly transient phenomena.

In the growth of science, the discovery of new facts and the formulation of new theories go hand in hand. New theories encompass the new facts and may reorganize previously established fields. Working models, thought experiments, and mathematical models are the components of a theory. The terms used to describe models and experiments are related to the real world through operational definitions or to concepts through formal definitions. Measurement (quantitative observation) is introduced through the counting of standard units in the operational definitions of length (distance), time intervals, and gravitational mass.

"Go, wondrous creature!
Mount where Science
guides;
Go measure earth, weigh
air, and state the tides;
Instruct the planets in what
orbs to run,
Correct old Time, and
regulate the Sun."

Alexander Pope
Essay on Man, 1732

List of new terms

scientific point of view	mathematical model	standard object
scientific "truth"	thought experiment	length: meter
domains of magnitude:	variable factor	time: year, second
micro, macro, cosmic	constant	mass: kilogram
theory	formal definition	equal-arm balance
working model	paraphrase	
analogue model	operational definition	

Problems

1. Give two examples from your own life where you had to revise your expectations (or prejudices) in the light of experience.

2. Describe your feelings toward the study of physics.

3. Describe the values of studying physics as part of a liberal education. Comment on these values from your point of view.

4. Give one or two examples from your own life in which your knowledge of physics was inadequate to the requirements (exclude school experiences).

5. Compare the growth of a city to the growth of science. Does the growth of a city have many similarities to the growth of science? Perhaps new homes correspond to new facts. Perhaps new roads correspond to new theories. Point out similarities and differences. Is the city a good analogue model for science in this respect?

6. Compare the growth of science to various other growth processes. Point out similarities and differences. Are these other examples more or less helpful than the one discussed in Problem 5?

7. Use a dictionary to trace the definition of the word matter. Look up the definition of each major word used to define matter, and so on, until you discover where this process leads. Discuss your discovery and compare it with the approach of this text, which is to leave "matter" as an undefined term (see note in margin on p. 6).

8. Express your preferences with regard to the corpuscular and wave theories of light.

9. Compare scientific "truth" with truth in another domain.

10. Tell which of your senses are most effective in detecting events at the lower limit of the macro domain in space and time. Estimate the magnitude of the smallest length and shortest time interval your senses can detect directly.

11. Tell which of your senses are most effective in detecting events at the upper limit of the macro domain in space and time. Estimate the magnitude of the largest length and longest time interval your senses can detect directly.

12. List examples of indirect evidence (not directly perceived by your sense organs) of phenomena in the macro domain.

13. List examples of direct sensory evidence of phenomena in the micro and cosmic domains. What are some tools used to extend the senses to enable them to cope with phenomena in these domains? Describe the use of these tools and explain whether it leads to direct or indirect evidence.

14. Explain the similarities and differences between a scientific "working model" (such as considering the earth as a uniform, smooth

sphere) and each of the following examples of a "model":

(a) A scale model, such as a model airplane.

(b) A small-scale architectural model of a proposed building.

(c) A model home.

(d) An individual who poses for photographs or paintings, a fashion or artists' model.

15. Carefully examine the system illustrated in Fig. 1.4a.

(a) Propose two (or more) working models that are compatible with all the information given in Fig. 1.4.

(b) Describe one (or more) thought experiments in which your two models exhibit different outcomes. (Such experiments can be used in real experimental tests to eliminate models that lead to a wrong prediction.)

16. Describe two or more working models that apply in an academic field of your choice or in everyday life. For each model, describe some of its properties, how it functions, what observations it explains successfully, and where it fails.

> EXAMPLE. Protein-carbohydrate-and-fat model for food. All foods consist of these three materials, in various proportions. The energy (Calorie) value of any food can be found from its content of the three materials by a mathematical model. The planning of a balanced diet takes into account the human body's need for the three materials. Gain or loss of weight can be planned on the basis of the Calorie value.
>
> Limitation: it is possible to have a well-balanced diet in terms of proteins, carbohydrates, and fats, yet suffer nutritional deficiencies. The model does not include all the contributions that food makes. Vitamins and minerals are also important, even though they do not contribute to the energy (Calorie) value of food.

Suggested models: computer model for the human brain, gene model for inheritance, "free/efficient market" model for world economy, "economic" model for human beings, demon model for the source of disease.

17. Five blind men investigated an elephant by feeling it with their hands. One felt its tail, one a leg, one a tusk, one an ear, one its side. Describe the analogue models for an elephant they might create individually and by pooling their observations. Describe the implications of this fable for science.

18. Interview three or more children (between ages 7 and 10) to ascertain their ideas as to the source of knowledge and the creation of new knowledge. Ask questions such as, How do we know that 3 + 3 = 6? How do we know that the sun will rise tomorrow? How do we know the earth is round? How do we know how to make a watch (car, rocket, cake....)? Ask questions to probe beyond the first responses. (If possible, undertake this project jointly with several other students so as to obtain a larger collection of responses.) Comment on the responses.

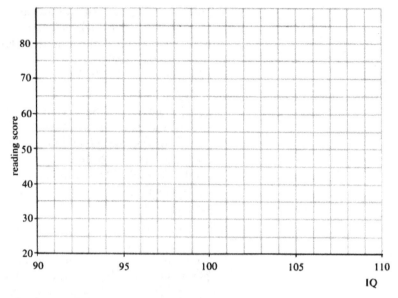

Figure 1.12 Coordinate grid for graph from Problem 19.

TABLE 1-3 READING AND INTELLIGENCE TEST SCORES
(PROBLEM 19)

School	Reading	IQ		School	Reading	IQ
A	33	93		J	51	99
B	59	103		K	31	92
C	57	104		L	51	98
D	46	99		M	69	107
E	48	99		N	73	108
F	54	100		O	48	98
G	52	100		P	75	108
H	52	101		Q	64	105
I	61	103		R	79	111

19. Reading tests and intelligence tests were given sixth graders in a large state. Table 1.3 (above) lists the average scores for schools in eighteen different communities, in order from the largest to the smallest enrollment. Display the data on a graph (Fig 1.12, above), and, if there is a relationship between the two scores, make a mathematical model (in either graphical or algebraic form) for this relationship. Interpret this model. Be careful about making interpretations not actually supported by the given data; explain and criticize whatever assumptions you make, as well as the assumptions that are "hidden" in the data (the test scores).

20. State a formal definition and describe an operational definition for each of the following.
    (a) chair   (d) life
    (b) gift   (e) person
    (c) teacher (f) scientific literacy

Comment on the advantages and limitations of the definitions you have constructed.

21. To be constitutional, laws must be applicable to real cases with a minimum of ambiguity. Therefore, they often include operational definitions of the terms that are used in them. Find and report three operational definitions that are part of laws. Discuss the extent to which the inclusion of these operational definitions promotes or restricts the achievement of justice.

22. State three or more operational definitions that you use in your everyday life. The definitions should not deal with profound ideas but may be as simple as: ironing temperature (of a flatiron) is measurable by the "sizzling rate" of a water drop that touches the iron.

23. Write a critique of the hypothesis (beginning of Section 1.2) that the foundations of a person's sense of space and time are laid before birth.

24. Identify one or more explanations or discussions in this chapter that you find inadequate. Describe the general reasons for your judgment (conclusions contradict your ideas, steps in the reasoning have been omitted, words or phrases are meaningless, equations are hard to follow, . . .), and make your criticism as specific as you can.

Bibliography

Introductory physics or physical science textbooks and surveys. These references cover the same material as this text but from a different point of view. Many of them present a more mathematical treatment that uses algebra and trigonometry, but none employs calculus.

K. R. Atkins, *Physics*, Wiley, New York, 1966. Emphasizes interaction point of view, more mathematical.

D. Cassidy, G. Holton, J. Rutherford, *Understanding Physics*, Springer Verlag, 2002. This text, an update of Holton's outstanding 1952 book, is particularly recommended for its historical perspective.

L. N. Cooper, *An Introduction to the Meaning and Structure of Physics*, Harper and Row, New York, 1968. Includes much on quantum physics and relativity, many quotations.

E. R. Huggins, *Physics 1*, W. A. Benjamin, New York, 1968. Presents the interaction point of view. Strong on quantum physics and relativity.

R. B. Lindsay and H. Margenau, *Foundations of Physics*, Wiley, New York, 1936.

V. L. Parsegian, A. S. Meltzer, A. S. Luchins, and K. S. Kinerson, *Introduction to Natural Science, Part One: The Physical Sciences*, Academic Press, New York, 1968. Relates biological, historical, philosophical, sociological, and humanistic material to physical science.

U. Haber-Schaim *et al, PSSC Physics*, Kendall/Hunt, Dubuque, Iowa, 1991. Especially good on wave physics.

E. Rogers, *Physics for the Inquiring Mind*, Princeton University Press, Princeton, New Jersey, 1960. Contains informal explanations, quaint diagrams, and many suggestions for simple experiments.

L. W. Taylor, *Physics, the Pioneer Science*, Dover Publications, New York, 1941. Includes a great deal of historical material.

V. F. Weisskopf, *Knowledge and Wonder: The Natural World as Man Knows It*, 2nd Edition, MIT Press, Cambridge, Mass. 1979. An excellent qualitative lecture series on many facets of physical science. Authored by one of the central players in the discovery of quantum mechanics and quantum electrodynamics, a theorist noted both for his brilliance as well as for his infectious, vibrant personality.

Books and essays on the history and philosophy of science, especially physics:

M. J. Aitkin, *Physics and Archeology*, Wiley (Interscience), New York, 1961.

Francis Bacon. *Novum Organum*, Collier, New York, 1902.

W. I. B. Beveridge, *The Art of Scientific Investigation*, W. W. Norton. New York, 1957.

P. W. Bridgman, *The Logic of Modern Physics*, Macmillan, New York, 1946.

J. Bronowski, *Science and Human Values*, Harper and Row, New York, 1965.

J. Bronowski, *The Common Sense of Science*, Harvard University Press, Cambridge, Massachusetts, 1953.

G. B. Brown, *Science, Its Method and Philosophy*, W. W. Norton, New York, 1950.

H. Butterfield, *The Origins of Modern Science*, Macmillan, New York, 1965.

N. Campbell, *What is Science?*, Dover Publications, New York, 1952.

B. K. Cline, *The Questioners*, Crowell, Collier and Macmillan, New York, 1965.

J. B. Conant, *Science and Common Sense*, Yale University Press, New Haven, Connecticut, 1951.

G. de Santillana, *Origins of Scientific Thought*, University of Chicago Press, Chicago, Illinois, 1961.

A. Einstein and L. Infeld, *Evolution of Physics*, Simon and Schuster, New York, 1938.

A. R. Hall, *The Scientific Revolution 1500-1800*, Beacon Press, Boston, Massachusetts, 1954.

W. Heisenberg, *Physics and Philosophy*, Harper and Row, New York, 1958.

E. H. Hutten, *The Ideas of Physics*, Oliver and Boyd, Edinburgh and London, 1967.

S. L. Jaki, *The Relevance of Physics*, University of Chicago Press, Chicago, Illinois, 1966.

T. S. Kuhn, *Structure of Scientific Revolutions*, University of Chicago Press, Chicago, Illinois, 1962.

S. K. Langer, *Philosophy in a New Key*, Harvard University Press, Cambridge, Massachusetts, 1942.

L. Leprince-Ringuet, *Atoms and Men*, University of Chicago Press, Chicago, Illinois, 1961.

H. Margenau, *The Nature of Physical Reality*, McGraw-Hill, New York, 1950.

E. Nagel, *The Structure of Science*, Harcourt, Brace and World, New York, 1961.

H. Poincaré (S. Gould, Editor), *The Value of Science: Essential Writings of Henri Poincaré* (includes three books: Science and Hypothesis, 1903; The Value of Science, 1905; and Science and Method, 1908), Modern Library, 2001.

J. W. N. Sullivan, *The Limitations of Science*, The New American Library, New York, 1952.

Historical source materials and collections of biographies. These references contain short articles appropriate to many of the later chapters, but these articles are not listed again in the chapter bibliographies. If a topic arouses your interest, you should examine one or more of these references for information supplementary to the text.

A. Beiser, Ed., *The World of Physics*, McGraw-Hill, New York, 1960.

B. Brody and N. Capaldi, Ed., *Science: Men, Methods, Goals*, W. A. Benjamin, New York, 1968.

T. W. Chalmers, *Historic Researches*, Scribner's, New York, 1952.

J. B. Conant and L. K. Nash, *Harvard Case Histories in Experimental Science*, Harvard University Press, Cambridge, Massachusetts, 1957.

L. Hamalian and E. L. Volpe, Ed., *Great Essays by Nobel Prize Winners*, Noonday Press, New York, 1960.

W. S. Knickerbocker, *Classics of Modern Science*, Beacon Press, Boston, Massachusetts, 1962.

H. Lipson, *The Great Experiments in Physics*, Oliver and Boyd, Edinburgh and London, 1968.

O. Lodge, *Pioneers of Science*, Dover Publications, New York, 1926.

W. F. Magie, *A Source Book in Physics*, Harvard University Press, Cambridge, Massachusetts, 1963.

G. Schwartz and P. W. Bishop, *Moments of Discovery*, Basic Books, New York, 1958.

M. H. Shamos, *Great Experiments in Physics*, Holt, Rinehart, and Winston, New York, 1959.

Articles from Scientific American. Some or all of these, plus many others, can be obtained on the Internet at http://www.sciamarchive.org/.

A. V. Astin, "Standards of Measurement" (June 1968).

T. G. R. Brower, "The Visual World of Infants" (December 1966).

D. D. Kosambi, "Scientific Numismatics" (February 1966). A study of the fascinating results yielded by the application of modern science to ancient coins.

H. Zuckerman, "The Sociology of the Nobel Prizes" (November 1967).

So thou mayst say,
the king lies by a beggar, if
a beggar dwell near him . . .

WILLIAM SHAKESPEARE
Twelfth Night,
1602

Reference frames

The word position has several meanings, two of which, "posture or attitude" and "site or location," are easily confused. We will use the word position always with the latter meaning.

In physics, the terms "relative position" and "relative motion" refer to the fact that position and motion must be defined "relative to" or "in relation to" something other than the object itself.

In common speech, we often simply use ourselves or the earth as the reference object without saying so. For example, we say, "The moon is far away," or "The train is moving." It usually seems unnecessary to say "The moon is far away from me." or "The train is moving with respect to the station." However, in physics we often must describe motion from various reference frames, including especially those in which we are not at the origin and/or in which we are not at rest. Therefore, whenever there is any possibility of confusion, we will explicitly name the reference frame, and you should do the same.

You may associate the word "relativity" with mathematical mystery and scientific complexity, yet the basic concept, which we will try to explain in these pages, is simple. The matters of concern in relativity are the position (location) and motion of objects. The basic concept is that position and motion of an object can only be perceived, described, and recognized with reference to (that is, "relative" to) other objects. When you say, "The physics books are at the left rear of the book store," you refer the position of the books to the entrance and outline of the store. Objects such as the store entrance, to which position or motion are related, are called *reference objects*. Several reference objects used in combination to describe position are said to form a *reference frame (or frame of reference)*, and we speak of the position or motion of the original object relative to the reference frame.

If we know the position and motion of an object relative to one reference frame, we might ask about the position and motion relative to a second reference frame. This is the root of the theory of relativity: development of specific mathematical models for relating position and motion as observed relative to one reference frame to position and motion as observed relative to another reference frame. Einstein's theory of relativity is the most complete theory of these relationships. We will describe some aspects of Einstein's work in Section 7.3, but we will not go into the mathematical details in this text.

2.1 Relative position

Look at the girl in the field of daisies (Fig. 2.1). How would you tell someone where she is? Most directly, you could go to the edge of the field and point at her, saying, "The girl is there." By this action, you indicate the position of the girl relative to your outstretched arm and finger.

If you had to describe the girl's position to someone who was not watching, you could say, "She is a little way in from the south edge of the field, near the southeast corner." This statement indicates her position relative to the edges and corners of the field. In other words, it is impossible to describe the position of the girl (or of anything else) without referring to one or more other objects. Even if you were to draw a map of the girl's position, you would have to include on it some objects that could be used to align it with the actual field.

Reference objects and reference frames. For practical purposes, the reference objects must be easy to locate and identify, or they cannot be used as guides in finding the object whose position is being described. It would be hopeless, for example, to try to find the girl in the daisy field if her position were described by saying, "The girl is between two daisy blossoms." Something more distinctive is needed: the edges and corners of the field, as used earlier, or possibly a scarecrow at the center of the field.

The use of reference objects in everyday life is highly varied and adapted to many special circumstances. A piece of furniture, corners of

Figure 2.1 Can you find 3 children hiding among the daises? Describe their relative positions.

a room, street intersections, or a tall building can be used as a reference frame for the complete description of the location of a residence, restaurant, or mailbox.

Examples. An imaginary conversation is recounted in Fig. 2.2. What happened in this conversation? What finally allowed Percy to communicate the location of the hawk without confusion, ambiguity, or absurdity? First, he established a reference frame by selecting a large, easily identifiable branch on the tree and pointing in the direction of the tree. Clyde could grasp this reference frame. Next, Percy specified the direction ("above it") and distance ("the second branch") from the branch to the hawk.

The reference frame first used by Percy consisted of a reference *point* (Percy's body) and a reference *direction* (along Percy's pointing finger). These two components are necessary parts of a reference frame and are defined through more or less easily identified reference objects or earth-based directions, such as north and up. Percy's initial attempt to use the tree as reference point failed because there were several trees.

One of the most difficult communications problems is to give instructions for locating a book to a person who is not acquainted with the room in which the book is kept. In such a case it is most helpful to use the person's body as the reference frame by telling him to stand in the door to the room, look for the bookcase on his left, and then scan the middle of the second shelf of that bookcase. This example illustrates how you might use large elements of the environment (the room) to locate smaller ones (the person, and directions defined by the body), and then still smaller ones (the bookcase, ultimately the book) by a narrowing down process.

Describing position is more difficult when you do not have any reference objects to use for a narrowing down process. For example, a passenger on a ship who observes something in the ocean faces this problem. In these circumstances you would have to start with the ship you

An overdose of relativity.

Mr. Jones was going to a doctor's office and had never been there before. He called the doctor's office to ask for directions. After the receptionist told him how to get there, he asked whether it was on the north or south side of the street.

The response: "It depends which way you're walking."

are on and work outward. You may sight a flying fish "500 yards off the starboard bow" (ahead and to the right), using the ship as reference frame. You could say instead that the fish is "500 yards northwest," using the ship as reference point and compass directions to complete the reference frame.

One-particle model. So far we have been content with describing the position of a very small object that is located at a certain point in space. Real objects, of course, actually occupy an entire region of space, which may be small or large, round or thin, upright or slanted. For a complete description of an object, you therefore should take into account its shape and orientation as well as its location. A useful approach that avoids much unnecessary detail is to make a *one-particle model* for each object of interest. A particle is a very small object

Figure 2.2 Percy and Clyde took a long walk through the county of McDougall.
   Percy: Clyde, do you see the falcon sitting in that tree over there?
   Clyde: What falcon? In which tree? Where?
   Percy: In that big, broken tree over there (pointing his finger).
   Clyde: Oh, that tree in front of us! I see it, but I don't see the falcon.
   Percy: It's on the branch.
   Clyde: There are too many branches. I give up. Let's forget it.
   Percy: No, let's start over again. Do you see that broken branch about halfway
      up the trunk on the right side?
   Clyde: Yes, I do.
   Percy: Fine. Now look at the second branch above it, on the same side of the
      tree. Now move to the right and you just have to see the falcon.
   Clyde: Oh sure, but that's a hawk.

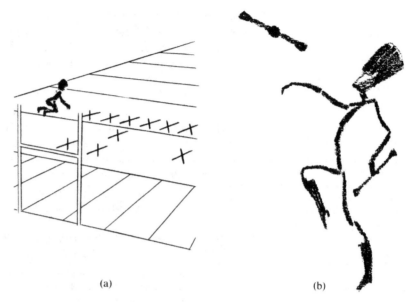

Figure 2.3 Two applications of a one-particle model.
(a) An X represents each football player. Eleven "particles" represent the team.
(b) A working model for the baton is the particle at its center.

that is located at the center or midpoint of the region occupied by the real object (Fig. 2.3). This working model greatly oversimplifies most objects, but is nevertheless accurate enough for most purposes in this text.

Coordinate frames. The laboratory scientist, who tries to describe natural phenomena in a very general way, avoids using incidental objects such as the laboratory walls or table surfaces as reference objects. Instead, a scientist frequently uses a completely artificial reference frame consisting of an arbitrarily chosen reference point and reference direction. The only requirement is to be able to describe the position of objects by using numbers. The numerical measures are called *coordinates;* the reference frame is called a *coordinate frame.* Two coordinate frames in common use are degrees of latitude and longitude, to define position on the earth relative to the equator and the Greenwich meridian, respectively, and distances measured in yards from the end zones and the sidelines on a football gridiron.

Polar coordinates. The procedure of giving the distance from the reference point and the direction relative to the reference direction gives rise to two numbers called *polar coordinates.* The distance may be measured in any unit, most commonly in meters, centimeters, or millimeters. The relative direction is usually measured in angular degrees. How this works is shown in Fig. 2.4. The necessary tools are a ruler to measure distance and a protractor to measure angles. Polar coordinates provide an operational description of the relative position of a point.

A polar coordinate grid, from which you may read the polar

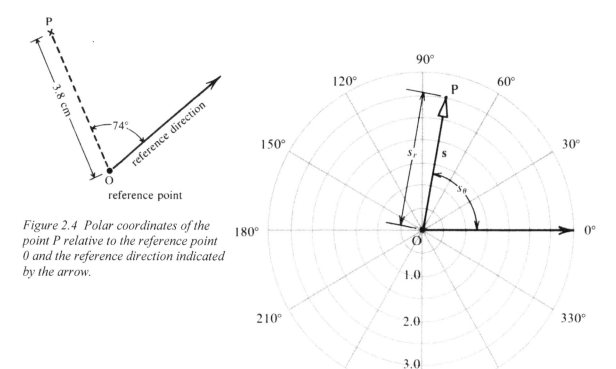

Figure 2.4 *Polar coordinates of the point P relative to the reference point O and the reference direction indicated by the arrow.*

Figure 2.5 Polar coordinates of the point P which has relative position $\mathbf{s} = (s_r, s_\theta) = (3 \text{ cm}, 80°)$.

coordinates of a point directly, is shown in Fig. 2.5. We shall indicate the relative position of the point P by an arrow in the diagram and by the boldface symbol **s** in the text. The polar coordinates will be indicated by the length of the arrow s_r *(r* for radius) and the direction of the arrow s_θ (Greek θ, theta, for angle), as indicated in the figure. Angles are customarily measured counterclockwise from the reference direction.

Examples. An example of the application of polar coordinates is shown in Fig. 2.6. The surveyor is using the direction of the road as the reference direction and the location of his tripod as the reference point. Some of the measurements he has made are given in the figure. These measurements may be used to make a map by locating the objects on a polar coordinate grid, as in Fig. 2.7. Polar coordinates have the intuitive advantage that they represent relative position in the way a person perceives them, namely, with the objects at various distances in various directions from the observer at the center.

Polar coordinates are used to direct aircraft to airports, with the control tower as reference point and north as the reference direction, as well as in other situations where a unique central point exists (e.g., in radar surveillance with the transmitter acting as reference point).

Figure 2.6
Polar coordinates.

	direction	distance
tree A	80°	10 m
tree B	270°	6 m
tree C	10°	31 m
windmill	40°	38 m

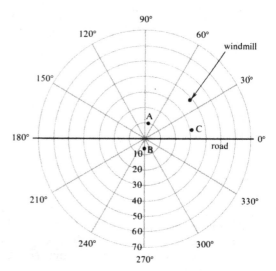

Figure 2.7 The position of the objects in Fig. 2.6
is mapped on a polar coordinate grid.

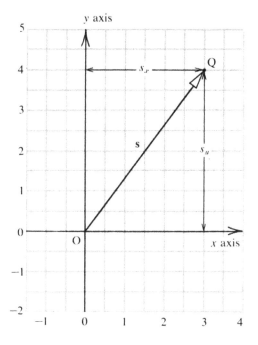

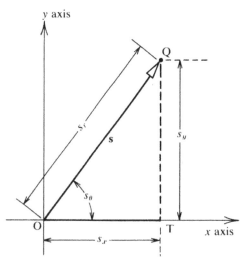

Figure 2.8 Rectangular coordinates s_x and s_y of the point Q, whose relative position is $\mathbf{s} = [s_x, s_y] = [3.0, 4.0]$. The two rectangular axes are indicated by arrows, and the origin of the coordinates by O.

Figure 2.9 Here Fig. 2.8 is redrawn to show the rectangular and polar coordinates of the point Q. These are related by means of the right triangle QTO.

Rectangular coordinates are sometimes called Cartesian coordinates, after René Descartes. Descartes, a French philosopher, first described this coordinate frame in the Discourse on Method, published in 1637.

Rectangular coordinates. A particularly useful technique for describing relative position that we will employ extensively later in the course makes use of two lines at right angles to each other called *rectangular coordinate axes* (Fig. 2.8). They are usually labeled the x-axis and y-axis. Their point of intersection is called the *origin of coordinates*. Distances are measured to the desired point Q along lines perpendicular to each axis, and the two measurements obtained are called the *rectangular coordinates* of the point Q relative to the two axes. As before, we introduce the boldface symbol **s** for the relative position of a point and use ordinary letter symbols with subscripts for the rectangular coordinates, this time s_x and s_y. The relative position of point Q is indicated by an arrow from O to Q in Fig. 2.8, just as it was in Fig. 2.5. Sometimes, for the sake of brevity, we will write the rectangular position coordinates in square brackets, with s_x first and s_y second: $[s_x, s_y]$ (Fig. 2.8).

You can see that the rectangular coordinates, unlike polar coordinates, do not give directly the distance s_r of a point from the origin. You can find the distance, however, by applying the Pythagorean theorem (Appendix, Eq. A.5) to right triangle QTO in Fig. 2.9, where the distance s_r *is* the length of the hypotenuse OQ:

$$s_r = \sqrt{s_x^2 + s_y^2}, \text{ see Ex. 2.1, below.}$$

EXAMPLE 2.1. Relate the polar coordinates of the point Q to its rectangular coordinates (Fig. 2.9) $s_x = 3.0$, $s_y = 4.0$.

Solution:

(a) To find s_r use the Pythagorean theorem (Appendix, Eq. A.5).

$$s_r = OQ = \sqrt{s_x^2 + s_y^2} = \sqrt{(3.0)^2 + (4.0)^2} = \sqrt{25.0} = 5.0$$

(b) To find s_θ, use the definition of the trigonometric ratios (Appendix, Eq. A.6):

$$tangent(s_\theta) = \frac{s_y}{s_x} = \frac{4.0}{3.0} = 1.33$$

$$s_\theta \approx 53° \text{ (from Table A.7)}$$

Rectangular coordinate frames, unlike polar coordinate frames, do not have a single center, as you may observe easily when you compare the polar and rectangular coordinate grids in Fig. 2.5 and 2.8. Whereas there is a unique reference point, the "pole," in the polar grid, it is possible to use any point in a rectangular grid as reference point by selecting the horizontal and vertical lines passing through this point as x axis and y axis. With R as reference point in Fig. 2.10, for instance, the point Q has the rectangular coordinates [*5, 1*], as you may verify in the figure. Rectangular coordinates, therefore, are advantageous when you are interested in the position of two points or objects relative to one another, rather than only relative to the origin of the coordinate frame. We will use this feature when we calculate changes in the position of a moving object.

2.2 Relative motion

So far we have considered ways of describing the relative position of objects that are stationary. Now, consider objects that change position. When an object changes position, you commonly say that it "moves," or that it is "in motion." But what do you mean by "motion?" Since position is defined relative to a reference frame, it is plausible to expect that motion would also be defined relative to a reference frame. It is therefore customary to use the phrase "relative motion."

Examples of relative motion. Imagine that a truck moving on a roadway is being described relative to two different reference frames. Reference frame A is attached to the roadway, reference frame B to the truck. To make the description simpler and more concrete, we will introduce two observers, one representing each reference frame (Fig. 2.11).

As the truck moves down the road, Observer A reports its position first on his left, then in front of him, then on his right. The position of

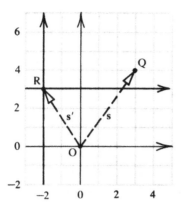

Figure 2.10 (above) The point R is chosen as reference point for describing the relative position of Q. The coordinates of Q relative to R are [5, 1]. (Note the reduced scale of the diagram compared to Figs. 2.8 and 2.9.)

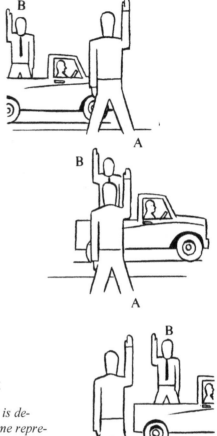

Figure 2.11 The motion of the truck is described relative to the reference frame represented by Observer A and relative to the reference frame represented by Observer B.

the truck relative to Observer A has changed. But Observer B always reports the truck as being in the same position, with the platform under his feet and the driver's cab on his left. The position of the truck relative to Observer B, therefore, has not changed from the beginning to the end of the experiment. Thus, you find that you have two different sets of data, one from each reference frame. In one reference frame you would conclude that the position had changed, and in the other that it had not. If we define relative motion as the change of position relative to a reference frame, then the two observers disagree (but each is correct) not only with regard to the relative position of the truck at various times in the experiment, but also about the truck's relative motion.

We can extend the discussion to motion of other objects. For instance, does the earth move? This depends on the reference frame used to define the earth's position. Relative to a reference frame attached to the earth, to which we are all accustomed, the earth is stationary. Relative to a sun-fixed reference frame, which was introduced by Copernicus and about which you probably studied in school, the earth moves in its orbit. In Chapter 15 we shall describe the resistance which Copernicus and Galileo encountered when they took the sun-fixed reference frame seriously.

Look at some of the consequences of this concept of relative motion. (You may find these consequences fascinating or merely strange, depending on how willing you are to break out of habitual modes of thinking.) In order to determine the position and motion of objects in an experiment, it is important to decide upon a reference frame. You have already noted that observers may disagree about the relative motion of an object. You must also recognize that an observer will always report an object to be stationary in his own reference frame so long as he is attached to that object. Such an observer's report about the motion of the remainder of the world will seem unusual indeed, if the observer's reference frame is attached to a merry-go-round, a satellite in orbit, or even a sewing machine needle.

Speed and relative motion. Another consequence of the relative motion concept is that different observers might disagree about the direction and the speed of a moving object they both observe. Think about the following example, in which an object is reported to travel at different speeds relative to different reference frames.

The speed of a riverboat going upstream as reported by its passengers looking at the shoreline is a snail-like 1 mile per hour, but the speed as reported by the captain is a respectable 10 miles per hour. Who is right? All steering and propulsion take place in the reference frame of the water. The motion of a riverboat relative to the shore is different from its motion relative to the water unless the water is still. Since the water is flowing downstream at a speed of 9 miles per hour (relative to the shore), and the riverboat is traveling upstream at a speed of 10 miles per hour (relative to the water), the speed of the riverboat (relative to the shore) is 1 mile per hour. Thus, the conflicting reports of the two observers are understandable and correct, for they are observing from different reference frames. The question, "Who is right?" can only be answered, "Each is right from his own point of view."

Here is a second example, in which we would like you to imagine you are each of the observers in turn:

Three cars, A, B, and C, are traveling north on a highway at speeds of 55, 65, and 75 miles per hour, respectively. Observers attached to each car make the following reports.

Observer in car A: Relative to me, car B is traveling north at 10 miles per hour, car C is traveling north at 20 miles per hour, car A (my car) is stationary, and the roadside is traveling south at 55 miles per hour.

Observer in car B: Relative to me, car A is traveling south at 10 miles per hour and car C is traveling north at 10 miles per hour.

Observer in car C: Relative to me, car B is traveling south at 10 miles per hour, and car C is stationary.

What does the observer in car B report about the speed of the roadside? What does the observer in car C report about the speed of car A and the roadside?

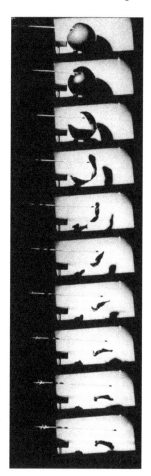

Recording and reproducing relative motion. Since relative motion is a transitory phenomenon, it cannot be recorded on a diagram or map with the ease with which relative position can be recorded. The motion-picture film is the most familiar way of recording and recreating relative motion.

Motion pictures. Motion pictures consist of a strip of photographs (Fig. 2.12) that show a scene at very short intervals (approximately 1/24 second). Of course, the scene changes, but does not change much in this short time. When the pictures are rapidly projected in the correct order, the viewer's eye and mind perceive smooth motion. If the pictures are taken with too great a time interval, so that the scene changes significantly between, then the smooth motion becomes jerky.

Flip books. Another way to represent relative motion is through flip books. Flip books create the illusion of motion through the same device as motion pictures, a series of pictures that show the same scene with slight changes in appearance. The pictures are bound in a book and are viewed when the pages of the book are flipped. With a flip book, you can examine each individual scene more easily than with a motion picture, you can control the speed, and you can view the sequence both "forward" and "backward" by starting at the front or the back of the book.

Multiple photographs. Still another technique for representing but not recreating motion is the multiple photograph. This is produced by taking many pictures at equal short time intervals on the same piece of film as in the example shown in Fig. 2.13. The multiple photograph gives a record of the path of the racquet and of the ball during a serve.

Blurred photographs. Even a single photograph can give evidence of motion when the camera shutter remains open long enough for the image projected onto the photographic film to change appreciably.

Figure 2.12 A section of motion-picture film taken at high speed, showing an arrow bursting a balloon. Harold Edgerton, the master of high speed photography, took the photograph by means of a rotating prism synchronized with a strobe [regularly flashing] light. In 1940, Edgerton won an Academy Award for movies made with this type of camera. For other photos by Edgerton see Stopping Time *by G. Kayafas, listed in the bibliography at the end of this chapter.*

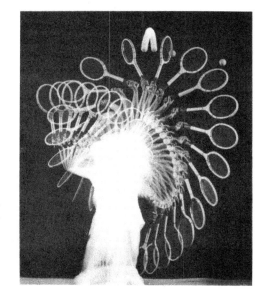

Figure 2.13 Multiple exposure photograph of a tennis serve, taken by Harold Edgerton. Can you estimate the time interval between exposures?

Figure 2.14 How many people are shown in this picture?

This can happen either because the photographic subject moves (Fig. 2.14), or, as every photographer knows, because the camera moves. In other words, the significant fact is motion of the subject relative to the camera. Indeed, an experienced photographer can "stop" the motion of a racing automobile by purposely sweeping his camera along with the automobile (Fig. 2.15). The automobile appears sharply in this picture, but the background, which was moving relative to the automobile and therefore relative to the camera, is blurred.

Example. An illustration of how significant relative motion can be was discovered by Berkeley physicist Luis W. Alvarez in a magazine reproduction of part of a motion-picture film showing the assassination of President John F. Kennedy. The President was riding in a motorcade, and Alvarez noticed something exciting in photograph 227: the motorcade in the photo was blurred, but the background and foreground were sharp. This was in contrast to most other photos, where the background was blurred and the motorcade was sharp. Apparently, Alvarez reasoned, the photographer had been sweeping his camera sideways to keep it lined up with the moving car but had suddenly stopped the

Figure 2.15 The racing car is stationary relative to the camera. Were its wheels stationary?

motion for a fraction of a second. The film, which showed the relative motion of camera and photographed objects by a steady change in picture from frame to frame, showed a change in this relative motion, a change that Alvarez ascribed to the photographer's neuromuscular reaction to the sound of a rifle shot. Further investigation of the original film in the National Archives, and of human flinching reactions to sudden sounds, confirmed Alvarez's interpretation of the blurred motorcade as evidence that a rifle was fired at that instant.

Definition of speed. Up to this point we have used the word "speed" without defining it. Automobile speed in miles per hour is usually read directly off the dial of a speedometer, the speed of a runner is expressed in his or her times for a particular distance, wind speed is indicated by a device called an anemometer, and so on. To be comparable with one another, these speeds must all be derived from the same definition. As the unit of speed (miles per hour) suggests, the generally accepted definition (at left) is the rate at which distance is traversed (Eq. 2.1).

This definition can be applied whenever a distance and a time measurement have been made. You are probably familiar with the highway "speedometer checks," roadside markers that identify a one-mile stretch; by driving at a steady speed and observing the time required to traverse the distance, you can calculate the speed (Example 2.2).

FORMAL DEFINITION
The average speed is equal to the ratio of the distance traversed divided by the time interval required to traverse the distance.

Equation 2.1

distance traversed $= \Delta s$
(pronounced "delta ess")
time interval $\quad = \Delta t$
(prounced "delta tee")
average speed $\quad = v_{av}$

$$v_{av} = \frac{\Delta s}{\Delta t}$$

Note: The "Δs" and "Δt" symbols stand for single quantities and <u>do not</u> indicate multiplication of Δ by "s" or by "t". The meaning of the Δ symbol will be explained further below (Section 2.3).

Units of speed:
meters per second (m/sec)
miles per hour $\quad$ *(mph)*
feet per second $\quad$ *(ft/sec)*

1 m/sec ≈ 2.2 mph ≈ 3.3 ft/sec

1 mph ≈ 0.45 m/sec ≈ 1.5 ft/sec

($\approx$ indicates approximately equal)

EXAMPLE 2.2. Find the speed if the distance traversed and the time interval are given.

(a) $\Delta s = 2.5\,m$ $\qquad\qquad \Delta t = 4.5\,\sec$

$$v_{av} = \frac{\Delta s}{\Delta t} = \frac{2.5\ m}{4.5\ sec} \approx 0.56\ m/sec$$

(b) $\Delta s = 140\,m$ $\qquad\qquad \Delta t = 0.4\,\sec$

$$v_{av} = \frac{\Delta s}{\Delta t} = \frac{140\ m}{0.4\ sec} \approx 350\ m/sec$$

(C) $\Delta s = 1\,mile$ $\qquad\qquad \Delta t = 4\,\min$

$$v_{av} = \frac{\Delta s}{\Delta t} = \frac{1\ mile}{4\ min} \approx 0.25\ mile/min$$

$$= 15\ mph \approx (15 \times 0.45)\ m/sec = 6.8\ m/sec$$

The definition can be applied in many other circumstances, too, where the distance and time interval may be measured in any convenient unit. The speed of a taxicab in a large city, for instance, may be described as six blocks per minute, the speed of a bus may be only three blocks per minute, the speed of an elevator may be one floor in three seconds (one-third floor per second or 20 floors per minute), and so on.

FORMAL DEFINITION
*The instantaneous speed is
equal to the average speed
measured during an "in-
stant." An instant is a time
interval short enough that
the speed does not change
to a significant extent.*

OPERATIONAL DEFINI-
TION
*The instantaneous (or ac-
tual) speed is equal to the
number shown by a speed-
ometer.*

Equation 2.2

instantaneous speed = v

$$v = \frac{\Delta s}{\Delta t}$$

*(Δt is an instant, an
interval of time chosen
short enough so that the
speed does not change
to a significant extent.)*

You can convert from one to another of these units of speed if you know
how the units of distance and time compare (number of feet per floor,
number of seconds per minute).

Curiously, the work of Galileo, who was the first to investigate moving
bodies systematically and quantitatively, contains no reference to this
idea of speed as a numerical quantity (*v*) equal to the ratio of distance
divided by time. Instead, he always compared two or more speeds with
one another (often by comparing the times to go equal distances, or by
comparing the distances traveled in equal times), and he was able to derive and state his results by using ratios of distances (or times) to one
another. One of Galileo's major contributions was a clear understanding
of what we now call "average speed" and "instantaneous speed." We now
explain these two key concepts.

Average speed. When you think of a bus making its way in city traffic,
you immediately realize that the speedometer reading has little direct
connection with a measured speed of, say, three blocks per minute. After
all, the bus is stopped a good fraction of the available time. The speedometer needle may swing from 0 miles per hour (the bus is stopped) up
to 20 or even 30 miles per hour while the bus is moving, and then back to
0 miles per hour again at the next stop. If you count how many blocks the
bus travels in a minute, you include the stops and the motion. The speed
determined in this way is called the *average speed*, because it is an average value intermediate between the maximum and minimum values. The
average speed is always referred to a certain distance or time interval,
such as the average speed over a mile of highway (speedometer check) or
in a minute of city driving (bus and taxi examples). A car that required a
minute to drive 1 mile on the highway was traveling at the average speed
of 1 mile per minute, 60 miles per hour, or about 90 feet per second.

Instantaneous speed. After this explanation, you may wonder what the
car's speedometer indicates. The speedometer indicates the "actual
speed" of the car. The actual speed is equal to the average speed if the car
is driven steadily without speeding up or slowing down. In this way the
speedometer check can be used as intended by the highway builders. In
other words, the average speed, which can be measured in the standard
units of distance and time, is used to calibrate the speedometer dial.

There is a second relation between average speed and actual speed – a
relation that has led to the term *instantaneous speed* for the latter. Imagine the average speed measured during a very short time interval, such as
1 second or less. During such a short time interval, the car has barely any
possibility of speeding up or slowing down. Hence the average speed in
this short time interval is practically equal to the actual speed. Since a
very short time interval is called an instant, the name instantaneous speed
is generally used (Eq. 2.2).

How short is an "instant"? The instant is defined to be so short that the
speed of the moving object does not change appreciably. Just how short
it must be depends on the motion that is being studied. For a car that accelerates from a standing start to 60 miles per hour in 10 seconds, the
instant must be considerably shorter than 1 second. For a bullet being
fired, an instant must be very much shorter yet, for the entire time interval during which the bullet accelerates inside the gun barrel is much,

much shorter than 1 second. At the other extreme, consider the ice in a glacier slowly gliding down a mountain valley. For this motion, even a day may be a brief instant because years elapse before the speed changes.

Applications. Since the average speed is defined by means of a mathematical formula ($\Delta s/\Delta t$), you can use mathematical reasoning (Section A.2) to solve a variety of problems. For instance, you can compute the distance traversed by a moving object if you know its average speed and the travel time (Section 1.3, Eq. 1.2, and Fig. 1.6). Or you can compute the time required for a trip if you know the average speed and the distance to be covered. These ideas are illustrated in Example 2.3.

EXAMPLE 2.3

(a) Find the distance if the average speed and time interval are given. How far does a pedestrian walk in 1.6 hours?

Solution: We wish to use Equation 2.1 to find Δs; thus we multiply both sides of Equation 2.1 by Δt to get: $\Delta s = v_{av}\,\Delta t$

For a pedestrian we can estimate $v_{av} \approx 3$ mph and $\Delta t = 1.6$ hours. Thus $\Delta s = v_{av}\,\Delta t = 3$ mph x 1.6 hours ≈ 5 miles.

(b) Find the time interval required if the average speed and distance are given. If a bullet's average speed is 700 m/sec, how long does a bullet take to travel 2000 meters?

Solution: We wish to find Δt; thus we multiply both sides of Equation 2.1 by Δt and divide by v_{av} to get: $\Delta t = \dfrac{\Delta s}{v_{av}}$

For the bullet,

$$v_{av} = 700\,m/sec = 7\times10^2\ m/sec$$
$$\Delta s = 2000\,m = 2\times10^3\ m$$
$$\Delta t = \frac{\Delta s}{v_{av}} = \frac{2\times10^3\ m}{7\times10^2\ m/sec} \approx 0.28\times10 = 2.8\,sec.$$

2.3 Displacement

The concept connecting relative position with relative motion is the change of relative position, which enables you to apply coordinate techniques to motion. For example, when a particle moves from one point R to another point Q (Fig. 2.16), then its position relative to any reference point fixed on the page changes. The change in position of a moving object is called the *displacement* because you can think of the moving object being displaced from one point to the other, from R to Q. By marking the successive displacements of a moving object, you can trace its path in space (Fig. 2.17).

The symbol for displacement is the boldface Δs with the Greek Δ

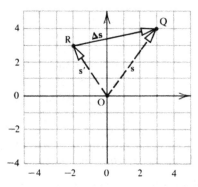

Figure 2.16 *The dashed arrows represent the positions of points Q and R relative to the coordinate axes. The solid arrow represents the displacement* **Δs** *from R to Q.*

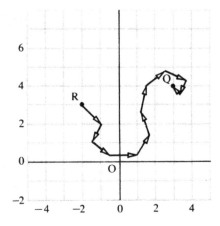

Figure 2.17 *Arrows represent the successive small displacements of a particle whose overall (or net) displacement is from R to Q.*

Equation 2.3

Relative position of Q

$$s = [3, 4] \begin{cases} s_x = 3 \\ s_y = 4 \end{cases}$$

Relative position of R

$$s' = [-2, 3] \begin{cases} s_x' = -2 \\ s_y' = 3 \end{cases}$$

Equation 2.4

Change of relative position from R to Q (displacement **Δs***)*

$$\Delta s_x = s_x - s_x'$$
$$= 3 - (-2) = 5$$
$$\Delta s_y = s_y - s_y'$$
$$= 4 - 3 = 1$$
$$\Delta s = s - s'$$
$$= [s_x - s_x', s_y - s_y']$$
$$= [5, 1]$$

Note that the x- and y- components of **Δs** *are independent of one another and are calculated separately.*

(delta) standing for "difference" and **s** standing for relative position. The boldface symbol **Δs** is a single entity and does not signify multiplication of two factors. It should be distinguished from the symbol Δs, the distance traveled, which was used in the definition of speed (Eq. 2.1). The displacement (**Δs**) is a more complex quantity than simply the distance traveled (Δs). The displacement **Δs** includes the distance traveled (Δs) **and** the direction of that movement in space.

With the help of the coordinate grid in Fig. 2.16, you can find the co-ordinates of Q and R relative to the origin O (Eq. 2.3). The coordinates of the change in relative position, which are called the components of the displacement from R to Q, are written Δs_x and Δs_y. They are equal to the differences of the coordinates of Q and of R relative to O (Eq. 2.4).

In a diagram, a position or displacement will be indicated by an arrow (with an open arrowhead, Figs. 2.16 and 2.17) from the reference or starting point at the tail to the actual or final point at the head. The length of the arrow represents the magnitude of the displacement, and the direction of the arrow represents the direction of the displacement. If several arrows have to be drawn, then their tails, magnitudes, and directions must be properly related. As you will discover in later chapters, the arrow description of a magnitude and a direction in space will be used for force, velocity, and other physical quantities, as well as for displacements

Placement of arrows. Consider, for instance, the surveyor (Fig. 2.6), whose measurements were represented by relative position arrows in Fig. 2.7. In this diagram the tails of all the arrows would be placed at the same point, which represents the surveyor's benchmark. If, however, you want to track a sailboat moving on a zigzag course

(Fig. 2-18), then you must represent the displacement on each straight part by an arrow whose tail is placed at the head of the arrow representing the preceding displacement. Thus, you obtain a map of the boat's path, and you can find the overall displacement from the starting point to the finish point by interpreting the map.

Addition of displacements. The process of combining the displacements to find the overall displacement by placing the tail of one arrow at the head of the previous one is called *addition of displacements,* and the overall displacement is called the *sum.* This is analogous to the

Figure 2.18 The sailboat proceeds from the starting point to the finishing point along a six-part zigzag course. The individual displacements Δs_1, Δs_2, Δs_3, Δs_4, Δs_5, and Δs_6 combine (add together) to give the sum, or overall displacement Δs. (Scale: length of side of small square in graph below = 1/2 mile.)

First, we find the various displacements by counting squares on the figure below:

$\Delta s_1 = [+2.0, +3.0]$ $\Delta s_3 = [+5.0, +3.0]$ $\Delta s_5 = [+2.5, -1.5]$

$\Delta s_2 = [+1.0, -3.0]$ $\Delta s_4 = [-0.5, -2.0]$ $\Delta s_6 = [-0.5, +5.5]$

$\Delta s = [\Delta s_x, \Delta s_y] = [+9.5, +5.0]$ (counted directly from figure below)

We can check the above result by adding the individual displacements to find the sum: $\Delta s_1 + \Delta s_2 + \Delta s_3 + \Delta s_4 + \Delta s_5 + \Delta s_6 = \Delta s$

Now, to actually find the x- and y-components of Δs, we must add the individual x- and y-components separately, so they do not get mixed up with one another!

x components : $\Delta s_x = 2.0 + 1.0 + 5.0 - 0.5 + 2.5 - 0.5 = 9.5$ (agrees with above)

y components: $\Delta s_y = 3.0 - 3.0 + 3.0 - 2.0 - 1.5 + 5.5 = 5.0$ (agrees with above)

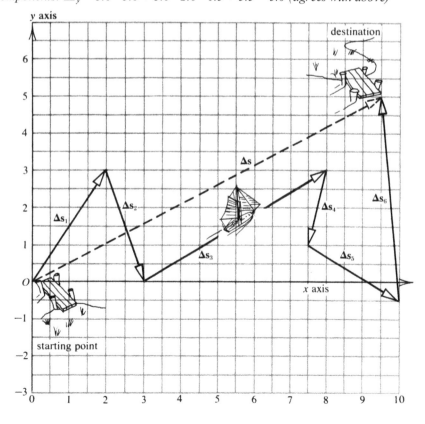

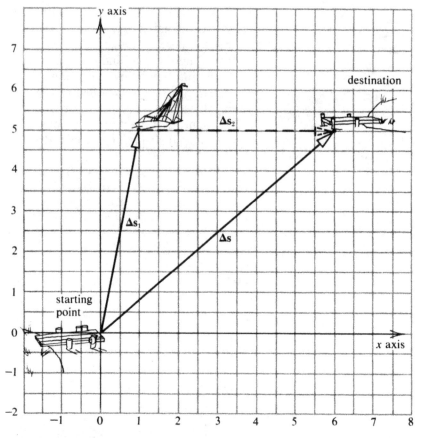

Figure 2.19 The sailboat has accomplished displacement Δs_1 and still needs to make displacement Δs_2 to reach its destination. The displacement Δs_2 is the difference between Δs and Δs_1. (Scale: 1 square = 1/2 mile.)
$\Delta s_1 = [1, 5]$, $\Delta s_2 = [5, 0]$, $\Delta s = [6, 5]$.
$\Delta s_2 = \Delta s - \Delta s_1 = [6, 5] - [1, 5] = [6 - 1, 5 - 5] = [5, 0]$. (**Note that the x- and y-components are subtracted separately so they do not get mixed together!**) As you can see in the figure, Δs_2 indeed has an x-component of 5 and a y-component of 0.

Finding the sum (or difference) of displacements: the x- and y-components are independent and must be kept track of separately; thus x-components are only added to (or subtracted from) other x-components, and y-components are only added to (or subtracted from) other y-components.

addition of numbers, where $100 combined with $60 gives the sum of $160. The graphical process of adding displacements is illustrated in Fig. 2.18. You can also use normal arithmetic to do this if you know the rectangular components of each displacement. The rectangular components may be read off Fig. 2.18 and are listed in the legend to that figure. It is clear that the x component of the overall displacement is the sum of the x components of the individual components, and the same is true of the y components. It is essential to keep the x- and y-components separate.

Subtraction of displacements. The course of the sailboat in Fig. 2.18 gave a natural illustration of the sum of displacements. To find illustrations of the difference of displacements, consider first two ways of interpreting the difference of two numbers: what is left over after part is removed and what is needed to obtain a larger quantity. The first way applies when you have $100 and spend $60; you are left with the difference, which is $40. The second way applies when you want $100 and have $60; you still need the difference, which is $40. The second interpretation can be applied to displacements. If you are in the sailboat, are aiming for a destination at a displacement **Δs** from the starting point, but have only made the progress described by the displacement **Δs₁**, the displacement **Δs₂** must still be traversed (Fig. 2.19). The displacement **Δs₂** is the difference between the goal **Δs** and the partial achievement **Δs₁**, that is, **Δs₂** = **Δs** - **Δs₁**. The difference may be found either graphically or arithmetically from the rectangular displacement components by subtraction as shown in Fig. 2.19.

Equation 2.5

$$\frac{\Delta \mathbf{s}}{b} = \left[\frac{\Delta s_x}{b}, \frac{\Delta s_y}{b} \right]$$

EXAMPLE
Δs = *[14, 5] and b = 4*

Then

$$\frac{\Delta \mathbf{s}}{b} = \frac{[14,5]}{4} = [3.5, 1.25]$$

Multiplication and division. Certain other arithmetic operations can be carried out with displacements by performing these operations on all the rectangular components of the displacements, just as you have calculated sums and differences by applying the appropriate arithmetic operation to the rectangular components. By adding a displacement to itself repeatedly (Fig. 2.20a), you obtain a multiple of the displacement. You can also divide a displacement into equal parts, such that each part is a fraction of the original displacement (Fig. 2.20b). Finally, you can find the negative of a displacement, which is just a displacement of equal magnitude and in the opposite direction from the original displacement (Fig. 2.20c).

An important algebraic concept is the division of a displacement by a number (Eq. 2.5), which will be used in the definition of velocity and

Figure 2.20 Arithmetic operations with displacements.
(a) Multiple displacements.
(b) Fractional displacement.
(c) Negative displacement.

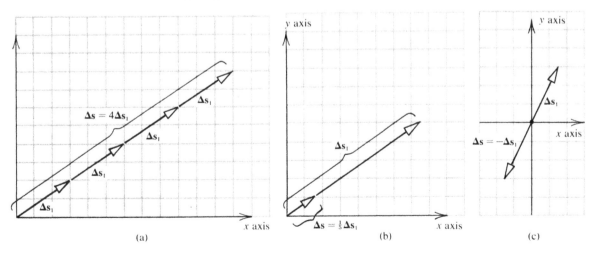

acceleration in Chapter 13. Specifically, a displacement divided by a number is just another, smaller displacement in the same direction; to calculate this, simply divide each rectangular component by the number, as illustrated in Eq. 2.5.

Summary

Position and motion of objects can only be observed and described relative to reference frames. Position or motion of an object may differ when described relative to different reference frames. A reference frame may be centered on you or on any other point in space that is stationary or in motion relative to you. Quantitative ways of describing relative position make use of polar coordinates (s_r, s_θ) and rectangular coordinates [s_x, s_y]. A quantitative description of relative motion makes use of the average speed, which is defined as the distance traveled divided by the time interval required (Eq. 2.1).

The change of an object's relative position is called the displacement. The displacement has a magnitude and a direction in space. It is represented in diagrams by an arrow and is described quantitatively by polar coordinates or rectangular components, [Δs_x, Δs_y]. The processes of arithmetic (addition, subtraction, multiplication and division) can be carried out with displacements in rectangular coordinates by calculating the x- and y-components separately.

List of new terms

reference object	coordinate frame	displacement
reference frame	polar coordinates	component
reference point	rectangular coordinates	average speed
reference direction	coordinate axis	instantaneous speed
particle	origin of coordinates	

List of Symbols

s (bold) relative position in space

[s_x, s_y] rectangular position coordinates

[s_r, s_θ] polar position coordinates

Δ**s (bold)** displacement

[Δs_x, Δs_y] displacement components (or displacement rectangular coordinates)

Δs distance traversed

Δt time interval

v speed

Problems

1. Lie on your bed on your back. Use your head as reference point and the directions front, back, right, left, above, below (or combinations of these) to describe the position of objects relative to this reference point. Give the approximate direction and distance of several of the following: pillow, lamp, radio, door, your feet, and so on.

2. Describe the position of your bedroom by using the building as reference frame. (Do *not* give detailed instructions as to how a person could walk to your bedroom.)

3. Describe the position of the children in Fig. 2.1 by using the picture edge as reference frame.

4. Find an alternate way of locating the hawk for Clyde (Fig. 2.2).

5. (a) Estimate the polar coordinates of the house and trees D and E in Fig. 2.6.
(b) Mark the location of the house and trees D and E on the map in Fig. 2.7.

6. Find the distance from R to Q in Fig. 2.10.

7. Find an arithmetic relationship among the coordinates of Q relative to O, R relative to O, and Q relative to R in Fig. 2.10.

8. A man on the earth travels 10 miles south, then 10 miles east, then 10 miles north. After the 30-mile trip is finished, he is back at his starting point. Identify his starting point.

9. Give two examples from everyday experience where you intuitively identify motion relative to a reference frame that is moving relative to the earth.

10. Explain how a motion-picture strip can be used to determine the speed of a moving object. Refer to average and instantaneous speeds. Apply your method to Fig. 2.12. You should estimate the approximate distances in the figure.

11. Explain how a multi-flash photograph like Fig. 2.13 can be used to determine the speed of the moving object at various points along the path. Apply your method to an object in Fig. 2.13. You should estimate the approximate distances in the figure.

12. Measure the average speeds of two or three objects in everyday life. You may use any convenient units, but you should convert to standard units (meters per second). Explain for each example how you chose to define "average."

Problems 13-16 ask you to compare or describe motion. You should describe the path(s) of the motion in words, name the reference frame(s) and make a drawing(s). You should also describe the speed of the motion.

13. A record is being played on a phonograph. Compare the motion of the needle relative to the phonograph base with the needle's motion relative to the record on the turntable.

14. Describe the motion of the moon relative to a reference frame fixed on the sun. Use a one-particle model for the moon.

15. Describe the motion of the earth relative to a reference frame fixed on the surface of the moon. Use a one-particle model for the earth.

16. Describe the motion of the sun relative to a reference frame fixed on the moon. Use a one-particle model for the sun.

17. A girl is pedaling a bicycle in a straight line.
(a) Describe the motion of the tire valve relative to the axle of the wheel.
(b) Describe the motion of the tire valve relative to the road.
(c) Describe the motion of one pedal relative to the road.
(d) Describe the motion of one pedal relative to the other pedal.

18. Using the two different methods outlined below, find the displacement from tree A to the windmill in Fig. 2.7.
(a) Use ruler and protractor to solve the problem geometrically. State the result in polar coordinates. You answer should have both a distance (s_r) and an angle (s_θ), expressed as [s_r , s_θ].
(b) Impose a rectangular coordinate frame (graph paper) on Fig. 2.7 and solve the problem arithmetically. You answer should have both an x-component (s_x) and a y-component (s_y), expressed as [s_x , s_y].

19. These questions deal with Fig. 2.18.
(a) Find the combined displacement of the second and third legs of the sailboat's course.
(b) Find the combined displacement of the fourth, fifth, and sixth legs of the sailboat's course.
(c) Find the combined displacement of the first, third, and fifth legs of the course.
(d) Find the displacement still required for the sailboat to reach its destination after the first leg of the course.
(e) Find the displacement still required for the sailboat to reach its destination after the fourth leg of the course.

20. Identify one or more explanations or discussions in this chapter that you find inadequate. Describe the general reasons for your dissatisfaction (conclusions contradict your ideas, or steps in the reasoning have been omitted; words or phrases are meaningless; equations are hard to follow; etc.) and pinpoint your criticism as well as you can.

Bibliography

G. Galilei, *Dialogues Concerning Two New Sciences,* Dover Publications, New York, 1952.

D. A. Greenberg, *Mathematics for Introductory Science Courses: Calculus and Vectors,* W. A. Benjamin, New York, 1965.

E. R. Huggins, *Physics I,* W. A. Benjamin, New York, 1968. Chapters 3 and 4 present a clear discussion of vectors.

M. Jammer, *Concepts of Space,* Harper and Row, New York, 1960.

G. Kayafas, *Stopping Time: The Photographs of Harold Edgerton,* published by Harry N. Abrams, Inc. (paperback), 2000, ISBN: 0810927179. A bullet seen the instant it explodes through an apple...a milk-drop splash transformed into a perfect coronet...a golf swing shown in a procession of exposures taken milliseconds apart—some of the world's most famous photographs were taken by Harold Edgerton, the late MIT scientist who invented the electronic flash. This great classic, first published in 1987 and now back in print in paperback, celebrates Edgerton's genius and his lasting influence on photography.

J. Piaget, *The Child's Conception of Geometry,* Basic Books, New York, 1960.

J. Piaget, *The Child's Conception of Space,* Humanities Press, New York, 1948.

Great has ever been the fame
of the loadstone and of amber in
the writings of the learned:
many philosophers cite the loadstone
and also amber whenever,
in explaining mysteries,
their minds become obfuscated
and reason can no farther go . . . .
Medical men also . . . in proving
that purgative medicines
exercise attraction through likeness of
substance and kinship of juices
(a silly error and gratuitous!),
bring in as a witness the
loadstone, a substance of
great authority. . . .

WILLIAM GILBERT
De Magnete,
1600

The interaction concept

The interaction concept is being used more and more widely to explain social and scientific phenomena. At conferences, strong interaction may be evident among some participants, weak interaction among others. At the ocean shore, erosion is caused by the interaction of wind and water with rock. In the laboratory, magnets interact even when they are not touching.

A dictionary provides the following definitions:

interact (verb): to act upon each other ...;

interaction (noun): action upon or influence on each other.

To say that objects interact, therefore, is to say that they have a relationship wherein they jointly produce an effect, which is the result of their action upon each other. In the examples cited above, anger may be the effect caused by strong (and irritating) interaction among the conference participants; crumbling and wearing away is the effect of the interaction of wind and water with rocks; and movement toward one another followed by sticking together is the effect of the interaction of the magnets.

3.1 Evidence of interaction

We take the point of view that influence and interaction are abstractions that we cannot observe directly. What we *can* observe are the effects or results of interaction. Congressional passage of unpopular legislation requested by the President would be an observable effect of the President's influence and therefore would be called evidence of his influence. The change in direction of motion of a struck baseball is an observable effect of its interaction with the bat and therefore can be called evidence of interaction.

You may believe that you can sometimes observe the interaction itself, as when a bat hits a baseball or a typewriter prints a letter on a piece of paper. These examples, which include physical contact and easily recognized effects, seem different from those where magnets interact without contact or where erosion is so slow that the effects are imperceptible. This apparent difference, however, is an illusion. You observe only the close proximity of bat and ball, a sound, and the change of the ball's direction of motion, all of which are so closely correlated and so familiar that you instantly interpret them as evidence of interaction between bat and ball. The interaction of bat and ball is, however, merely the relationship whereby the observable effects are brought about, and relationships are abstractions that cannot be observed directly.

Indirect evidence of interaction. An example where the evidence is very indirect was described in Section 2.2. In his analysis of the films of President Kennedy's assassination, Alvarez interpreted a blurred photograph as evidence of interaction between the photographer and a rifle being fired. No one can question the blurring in the photograph, which is directly observable. But it is clear that not everyone may agree with Alvarez's interpretation that there is a relationship between the blur and a rifle shot.

Another example of interpretation of indirect evidence for interaction is in the relationship between cigarette smoking and lung cancer. Lung cancer and cigarette smoking are separately observable, and the statistical evidence from the 1950s showed a very strong correlation, sufficient to warrant the conclusion of a pathological interaction between cigarette smoke and lung tissue (publicized in the Surgeon General's Report for 1964). There is now, in 2003, a much larger body of evidence for this interaction. Yet many smokers do not take this interpretation of the evidence seriously enough to believe that they are slowly committing suicide.

Alternate interpretations. The critical problem in interpreting evidence of interaction is that any one of several different interactions might be responsible for the same observed effect. The typed letter in the example of the typewriter and the paper does not furnish conclusive evidence as to which typewriter made the letter, a question that sometimes arises in detective stories. A direct way to overcome this weakness is to find evidence that supports one hypothesis. For instance, the paper might be beside a typewriter, the ribbon on one typewriter might match the shade of the typed letter, or a defect in the machine's type might match one that appears in the typed letter. If so, the original identification of the typewriter is supported.

An indirect way to support one hypothesis is to eliminate alternatives. By checking many typewriters and finding how poorly they match the ribbon color and type impression, the detective may be able to eliminate them from further consideration. Supporting evidence for one alternative and/or evidence against other alternatives will enable you to establish one hypothesis conclusively or may only lead you to decide that one of them is more likely than the others. A procedure for finding such evidence by means of control experiments is described in Section 3.4.

When you suspect that there might be interaction, you should make a comparison between what you observe and what you would expect to observe in the absence of interaction. If there is a difference, you can interpret your observation as evidence of interaction and seek to identify the interacting objects; if there is no difference, you conclude that there was no interaction or that you have not observed carefully enough.

3.2 Historic background

Mankind has not always interpreted observed changes or discrepancies as evidence of interaction. In ancient times, some philosophers took the view that changes were brought about by a fate or destiny that was inherent in every object. In our own day, many people ascribe specific events

to supernatural or occult forces. These forms of explanation and the interaction concept we are advocating, however, *do* share a common feature: they are both attempts to explain regular patterns in nature so as to anticipate the future and possibly to influence and control future events.

Cause and effect. When you observe two happenings closely correlated in space and time, you tend to associate them as cause and effect. Such a conclusion is reinforced if the correlation of the happenings persists in a regular pattern. The person who strikes a match and observes it bursting into flame infers that the striking caused the fire. The primitive man who performs a rain dance infers that the dance causes the ensuing rain. Even the laboratory pigeon that receives a pellet of grain when it pecks a yellow card becomes conditioned to peck that card when hungry. These individuals will repeat their actions - striking the match, dancing, or pecking the card - if they wish to bring about the same consequences again. After a sufficient number of successful experiences, all three will persist in their established behavior, even though some failures accompany their future efforts.

The interaction viewpoint. We may state the distinction between the modern scientific approach and other types of explanations for events in the following way: the scientist ascribes happenings to interactions among two or more objects rather than to something internal to any one object. Thus, the falling of an apple is ascribed to its gravitational interaction with the earth and not to the heaviness inherent in the apple. The slowing down of a block sliding on a table is ascribed to friction between the block and the table and not to the power or "desire" of the block to come to rest by itself. Fire is the manifestation of combustion, that is, the interaction of fuel and oxygen, and is not itself an element. The rain dance and the rain, however, cannot be put into this framework; therefore, this association is nowadays considered a superstition.

At any one time, however, science cannot provide explanations for all possible happenings. When a new phenomenon is discovered, the interacting objects responsible for it must be identified, and this may be difficult. The origin of some of the recently discovered radiation reaching the earth, for example, is yet to be found. On the other hand, we now have identified specific chemical substances in cigarette smoke that cause cancer. We are beginning to understand the specific biochemical mechanisms by which such substances cause lung cells to start the explosive multiplication that manifests itself as cancer. Research is continuing to further elucidate the details of this dangerous interaction.

3.3 Systems

The word "system" has entered our daily lives. Communication systems, computer systems, and systems analysis are discussed in newspapers and magazines and on television. In all these discussions, and in this text as well, the word "system" refers to a whole made of parts.

The systems concept is applied whenever a whole, its parts, and their inter-relationships must all be kept clearly in mind, as illustrated in the following two examples. Traffic safety studies take into account an entire driver-car system and do not confine themselves merely to the engineering of the car or the health of the driver. A physician realizes that the human heart, though a single organ, is really a complex system composed of muscles, chambers, valves, blood vessels, and so on. The system is physically or mentally separated from everything else so that the relations among the parts may be studied closely.

To simplify our terminology, we will often refer to the whole as "system" and to the parts as "objects." Thus, the car and the driver are the objects in the driver-car system, and the muscles, chambers, and so on are objects in the system called "the heart." By using the word "object" to refer to any piece of matter (animate or inanimate, solid, liquid, or gaseous), we are giving it a broader meaning than it has in everyday usage.

Sometimes one of the parts of a system is itself a system made of parts, such as the car (in the driver-car system), which has an engine, body, wheels, and so forth. In this case, we should call the part a subsystem, which is a system entirely included in another system.

In a way, everyone uses the systems concept informally, without giving it a name. At times, everyone focuses attention temporarily on parts of the environment and ignores or neglects other parts because the totality of incoming impressions at any one moment is too complex and confusing to be grasped at once. The system may have a common name, such as "atmosphere" or "solar system," or it may not, as in the example of the jet fuel and liquid oxygen that propel a rocket. The systems concept is particularly useful when the system does not have a common name, because then the group of objects under consideration acquires an individual identity and can be referred to as "the system including car and driver" or more briefly as "the driver-car system" once the parts have been designated.

Conservation of systems. Once we have identified a system, changes may occur in the system. We must have a way to identify the system at later times in spite of the changes. A chemist uses the conservation of matter to identify systems over time. This means that no matter can be added to or removed from the matter originally included in the system. For example, when jet fuel burns, the fuel and oxygen become carbon dioxide and water. Therefore, the chemist thinks of the carbon dioxide and water as being the same system as the jet fuel and oxygen, even though the chemical composition and temperature have changed.

The psychotherapist and the economist do not use the same criteria as the chemist for following the identity of a system over time. The psychotherapist focuses his attention on a particular individual with a personality, intellectual aptitudes, and emotions. A therapist, therefore, selects this individual as a system that is influenced by its interaction with other individuals and by its internal development. The person as a system retains its identity even though it exchanges matter with its environment (breathing, food consumption, waste elimination). For the

"... in all the operations of art and nature, nothing is created; an equal quantity of matter exists both before and after the experiment, ... and nothing takes place beyond changes and modifications in the combinations of these elements. Upon this principle, the whole art of performing chemical experiments depends."

Antoine Lavoisier
Traite Elementaire de Chimie, *1789*

economist, all the production, marketing, and consuming units in a certain region constitute an economic system that retains its identity even though persons may immigrate or emigrate and new materials and products may be shipped in or out.

The physicist studying macro-domain phenomena finds the matter-conserving system most useful. This is, therefore, the sense in which we will use the systems concept throughout this text. In the micro domain, however, the concepts of matter and energy have acquired new meanings during the last few decades, and, if you study physics further, you will learn how to expand and modify these criteria for defining systems.

You can apply conservation of matter to the selection of systems in two ways. By watching closely, you can determine whether you see the same system before and after an event. For instance, when a bottle of ginger ale is opened, some of the carbon dioxide gas escapes rapidly. The contents of the sealed bottle (we may call it System A), therefore, are not the same system as the contents of the opened bottle, which may be called System B. The escaping bubbles are evidence of the loss of material from the bottle.

In the second kind of application, you seek to keep track of the system even though its parts move from one location to another. Thus, after the bottle is opened, System A consists of System B plus the escaped gas; the latter, however, is now mixed with the room air and can be conveniently separated from the room air only in your mind. For this reason we stated at the beginning of this section that a system of objects need only be separated mentally from everything else; sometimes the physical separation is difficult or impossible to achieve, but that is immaterial for purposes of considering a system.

State of a system. To encompass the continuity of the matter in the system as well as the changes in form, it is valuable to distinguish the identity of the system from the state of the system. The identity refers to the material ingredients, while the state refers to the form or condition of all the material ingredients (Fig. 3.l). . Variable factors, such as the distance between objects in the system, its volume, its temperature, and the speeds of moving objects, are used to describe the state. In

Place two identical pieces of clean writing paper in front of you. Pick up one piece and call it System P.

(1) Wrinkle the paper in your hand into a ball. Is what you now hold in your hand System P?

(2) Is the paper lying on the table System P?

(3) Tear the wrinkled paper in half, and hold both pieces. Is what you now hold in your hand System P?

(4) Put down one of the two torn pieces. Is what you now hold in your hand System P?

Figure 3 1 Change in the state of a system.

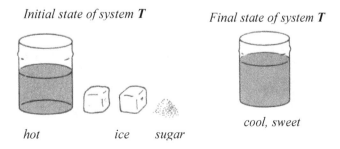

*Initial state of system **T***

*Final state of system **T***

hot   *ice*   *sugar*

cool, sweet

Chapter 4 we will relate matter and energy, which are of central concern to the physical scientist, to changes in the state of a system. There we will describe the ways in which a system may store energy and how energy may be transferred as changes occur in the state of a system. From an understanding of energy storage and transfer has come the extensive utilization of energy that is at the base of modern technology and current civilization.

Investigations of interacting objects. In their research work, physicists study systems of interacting objects in order to classify or measure as many properties of the interactions as they can. They try to determine which objects are capable of interacting in certain ways, and which are not (e.g., magnetic versus nonmagnetic materials). They try to determine the conditions under which interaction is possible (a very hot wire emits visible light but a cold wire does not). They try to determine the strength of interaction and how it is related to the condition and spatial arrangement of the objects (a spaceship close to the earth interacts more strongly with the earth than does one that is far away from the earth). Physicists try to explain all physical phenomena in terms of systems of interacting objects or interacting subsystems.

Working models for systems and the structure of matter. There are some happenings, however, such as the contraction of a stretched rubber band that involve only a single object and appear to have no external causes. In such cases, the scientist makes a working model in which the object is made of discrete parts. A working model for the rubber band is made of parts called "rubber molecules." The properties of the entire system are then ascribed to the motion and the interaction of the parts. Some models are very successful in accounting for the observed behavior of the system and even suggest new possibilities that had not been known but that are eventually confirmed. Such a model may become generally accepted as reality: for instance, everyone now agrees that rubber bands are systems made of rubber molecules. Also, further model building may represent the rubber molecules as subsystems composed of parts called "atoms" and explain the behavior of the molecules in terms of the motion and interaction of the atoms.

This kind of model building is called the search for the structure of matter - how ordinary matter in the macro domain is composed of interacting parts, and these parts in turn are composed of interacting parts, and so on into the micro domain. One of the frontiers of science is the search for ultimate constituents, if such exist. Since we will always find more questions to ask, it is unlikely that we will ever accept the concept of an "ultimate constituent."

3.4 Collecting evidence of interaction

Interactions are recognized by their effects, that is, by the difference between what is actually observed and what would have been observed in the absence of interaction. Such a difference is evidence of interaction. The systems concept is of great value here because it enables you to designate and set apart (at least mentally) the objects that are being compared as you look for a difference. One approach is to compare a system before an event (in its so-called initial state) with the same system after an event (in its so-called final state). For example, you compare a section of bare skin on the morning and the evening of a day at the beach (Fig. 3.2). The section of skin is the system. In this experiment you assume based on your experience that the skin color would not have changed in the absence of interaction. The observed change in skin color is therefore evidence of interaction with the sun.

As another example, take some sugar and let it dissolve in water in a glass beaker to form a solution (Fig 3.1). At the beginning of the experiment, the water-sugar system consists of dry crystals and colorless, tasteless water. At the end, there are no crystals and the liquid tastes sweet. The change in the state of this system is evidence of interaction between sugar and water.

Control experiment. Consider now an experiment in which you put yeast into a sugar solution in a glass and let this system stand in a warm place for several days. You will observe bubbles, an odor, and a new taste - that of ethyl alcohol. These changes can be interpreted as evidence of interaction within the water-sugar-yeast system. Can you narrow down the interacting objects more precisely or are all three parts necessary?

For comparison, suppose you can conduct experiments in which one ingredient is omitted. You dissolve sugar in water without yeast, you dissolve yeast in water without sugar, and you mix sugar and yeast. Each of these is called a control experiment; from their outcomes, you can answer the question above. By designing other control

Figure 3.2 The skin shows evidence of interaction with the sun only where it was exposed to sunlight. The exposed skin can be compared to the unexposed areas.

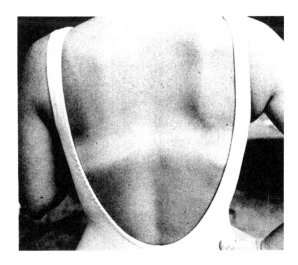

Figure 3.3 When you try to determine which electric circuit breaker supplies power to a particular light fixture, you turn on the switch of the fixture and then turn off the circuit breakers one at a time. In one of these "experiments" the bulb darkens, in the others it does not. Each turning off serves as a control experiment to be compared to the situation in which all circuits are turned on.

"It frequently happens, that in the ordinary affairs and occupations of life, opportunities present themselves of contemplating some of the most curious operations of Nature . . . I have frequently had occasion to make this observation; and am persuaded, that a habit of keeping the eyes open to everything that is going on in the ordinary . . . business of life has oftener led, as it were by accident . . . to useful doubts and sensible schemes for investigation and improvement, than all the most intense meditations of philosophers in the houses expressly set apart for study."

Benjamin Thompson,
Count Rumford
Philosophical Transactions, 1798

experiments, you can try to determine whether the glass container was necessary, and whether the temperature of the environment made any difference.

By carrying out control experiments, you try to identify those objects in the system that interact and those whose presence is only incidental (Fig. 3.3).

Inertia. One other important concept in the gathering of evidence of interaction is the concept of inertia. Inertia is the property of objects or systems to continue as they are in the absence of interaction, and to show a gradually increasing change with the elapse of time in the presence of interaction. For example, you expected the pale skin on the girl's back (Fig. 3.2) to remain pale as long as it was not exposed to the sun. You expect a rocket to remain on the launching pad unless it is fired. You expect sugar crystals to retain their appearance if they are not heated, brought into contact with water, or subjected to other interactions. You expect an ice cube to take some time to melt even when it is put into a hot oven.

Your everyday experience has taught you a great deal about inertia of the objects and systems in your environment. When you compare the final state with the initial state of a system and interpret a difference as evidence of interaction, you are really using your commonsense background regarding the inertia of the system. You must be careful, however, because occasionally your commonsense background can be misleading.

"... we may remark that any velocity once imparted to a moving body will be rigidly maintained as long as the external causes of acceleration or retardation are removed..."

Galileo Galilei
Dialogues Concerning
Two New Sciences, 1638

OPERATIONAL DEFINITION

Inertial mass is measured by the number of standard units of mass required to give the same rate of oscillation of the inertial balance.

Figure 3.4 An air track (below).
(a, below left) Small holes in the track emit tiny jets of air. When a close fitting metal piece passes over an opening, the air is trapped and forms a thin film over which the metal piece can slide with very little friction.
(b, below right) The closeness of fit can be seen in this end view.

Inertia of motion. The motion of bodies also exhibits inertia. Curiously, motion is one of the most difficult subjects to treat scientifically because of commonsense experience. When you see a block gliding slowly on an air track (Fig. 3.4), you almost think it must contain a motor because you expect such slowly moving objects to come to rest after a very short time. In fact, the block is only exhibiting its inertia of motion because the frictional interaction with the supporting surface is very small. You must, therefore, extend your concept of inertia to cover objects in motion (such as the block), which tend to remain in motion and only gradually slow down if subject to a frictional interaction. You must also extend it to objects at rest (such as the rocket), which tend to remain at rest and only gradually acquire speed if subject to an interaction. Change in speed from one value to another - where the state of rest is considered to have "zero" speed - is therefore evidence of interaction. Galileo already identified inertia of motion even though he did not give it a name. Isaac Newton framed a theory for moving bodies in which he related their changes in speed and direction of motion to their interactions. The "laws of motion," as Newton's theory is called, will be described in Chapter 14.

A key concept in the laws of motion is that of the *inertial mass*. This is an extension of Galileo's idea that, in the absence of external influences, objects maintain their state of motion, whether at rest or moving; it is useful to have a numerical quantity which measures the extent to which an object does this: "inertial mass" is the name for this quantity. Speaking roughly, inertial mass is the degree to which a body tends to maintain its state of motion. More specifically, an object with a large inertial mass takes longer to speed up (or slow down) than an object with small inertial mass. It is important to keep in mind the *difference* between *inertial* mass and *gravitational* mass. The latter (Section 1.5) is connected with the downward pull of gravity (the weight) and can be measured with an equal-arm balance. In contrast, inertial mass can be defined and measured with a device called the *inertial balance* (Fig. 3.5) to compare two objects or to compare an object of unknown inertial mass with standard units of inertial mass.

The inertial balance operates *horizontally*, thus eliminating the effects of gravity. The body attached to the end of the steel strip is repeatedly speeded up and slowed down by the oscillation of the strip. The inertia of the body, therefore, strongly influences the rate

(a)

(b)

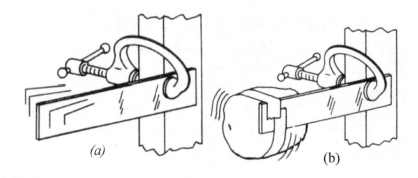

(a)

(b)

Figure 3.5 The inertial balance.
(a) The inertial balance consists of an elastic steel strip, which oscillates back and forth after the free end is pulled to the side and released.
(b) When objects are attached to the end of the strip, the oscillations take place more slowly. The inertial mass of a stone is equal to the number of standard objects required to give the same count of oscillations per minute. To measure the inertial mass of the stone, it is attached to the end of the steel strip and set into oscillation. The number of oscillations in 1 minute is counted. Then the stone is taken off, a number of standard objects are attached, and their number adjusted until the count of oscillations is equal to the count obtained with the stone.

FORMAL DEFINITION
Momentum is the product of inertial mass multiplied by instantaneous speed.

The unit of momentum does not have a special name; it is a composite unit, kilogram-meters per second (kg m/sec) that combines mass and speed.

Equation 3.1
momentum $= \mathcal{M}$
speed $= v$
inertial mass $= M_I$

$$\mathcal{M} = M_I\, v$$

of oscillation of the strip: large inertia (resistance to change of speed) means slow oscillations, small inertia means rapid oscillations.

The generally accepted standard unit of inertial mass is the kilogram, represented by the same platinum-iridium cylinder as the unit of gravitational mass. Even though inertial and gravitational masses are measured in the same units, they are different concepts and have different operational definitions. Inertial and gravitational mass are both important for understanding motion, particularly bodies falling under the influence of gravity. We will focus on this in Chapter 14.

A second important concept in the laws of motion is the *momentum* of a moving body. The word is commonly applied to a moving object that is difficult to stop. A heavy trailer truck rolling down a long hill may, for instance, acquire so much momentum that it cannot be brought to a stop at an intersection at the bottom. By contrast, a bicycle coasting down the same hill at the same speed has much less momentum because it is less massive than the truck.

The physical concept of *momentum* is defined formally as the product of the inertial mass multiplied by the speed of the moving object (Eq. 3.1). This concept was used by Newton to formulate the laws of motion (Chapter 14), and it plays an important role in the modern models for atoms (Sections 8.3, 8.4, and 8.5). We will elaborate on the momentum concept in Chapter 13, where we will describe how it depends upon the direction of motion as well as on the speed.

If we want to use changes of motion as evidence of interaction, we must be careful because, as we have pointed out in Chapter 2, motion must be defined relative to a reference frame. An object moving relative to one reference frame may be at rest relative to another. Evidence of interaction obtained from observation of moving objects, therefore, will depend on the reference frame. We will ordinarily use a reference

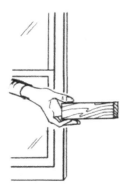

frame attached to a massive body such as the earth (for terrestrial phenomena) or the sun (for the solar system).

Combined interaction. A block held in your hand does not show evidence of interaction (i.e., it remains at rest), yet it is clearly subject to interaction with the hand and with the earth. This is an example of what we must describe as two interactions combining in such a way that they compensate for one another and give the net effect of no interaction. Situations such as this raise the question of the strength of interaction; how can you compare two interactions to determine whether they can compensate exactly or not, other than to observe their combined effect on the body? We will take up this question in Chapter 11.

Figure 3.6 Four steps in the investigation of the interaction of a match flame with a detector show the effects of a shield placed in various locations.

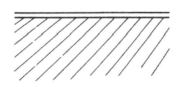

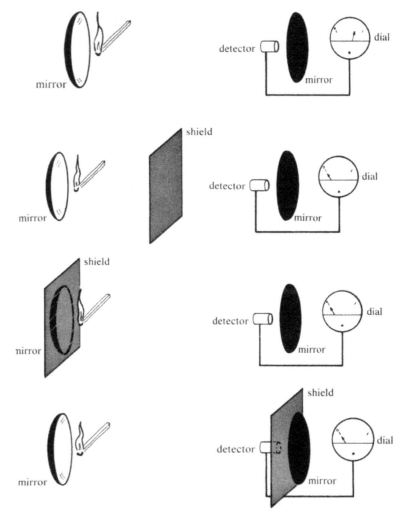

Radiation. A situation that contains a different element of mystery is illustrated in Fig. 3.6. Two mirrors are facing each other at a separation of several meters. There is no mechanical connection between them. At a central point near one mirror is a device called a detector, which is connected to a dial. If a lighted match is placed at a central point in front of the opposite mirror, you see a deflection on the dial (Fig. 3.6a). After a little experimentation, you recognize that the placement of the match and the dial deflection are definitely correlated. This is the evidence of interaction between the match and the detector. If, now, the match is held in position and a cardboard shield is placed in various positions in the apparatus, the deflection falls to zero [Figs. 3.6 (b), (c), and (d)]. Without anyone touching any of the visible objects used for the experiment, an effect was produced. The inference is that something

Figure 3.7 Four steps in an investigation of material X show evidence of interaction between the material and the detector.

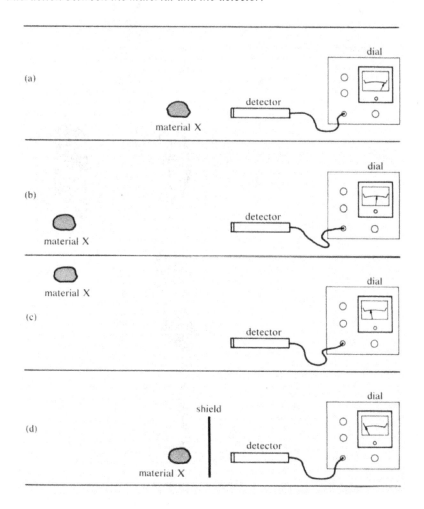

was passing from the match to the detector by way of the mirrors, and that the shield somehow blocked or interrupted this passage.

We, therefore, construct a working model that is just like the experimental system but includes in addition an "object" that passes from the match to the first mirror, the second mirror, and the detector. The scientist calls this "model object" radiation. In terms of this model, he can describe the effect of the shield on the dial reading as evidence of interaction between the shield and the radiation, he can describe the path of the radiation, he can describe the match as a radiation source, and he can describe the detector as a radiation detector.

Another experiment, with a rocklike material X and a detector with a dial, is illustrated in Fig. 3.7. From the evidence you may conclude that material X is not an ordinary inert rock but is a source of radiation, and you make a working model that includes an "object," again called radiation, that passes between material X and the detector. After this discovery, you can study the spatial distribution of the radiation by holding the detector in various directions and at various distances from the rock, you can study the interaction of the radiation with various shields (cardboard, glass, iron, and aluminum) placed to intercept it, and so on. From this kind of investigation you become more familiar with the radiation from material X and may, eventually, think of it as a real object and not only as part of a model.

The discovery of evidence of interaction is a challenge to identify the interacting objects and to learn more about the interaction: the conditions under which it occurs, the kind of objects that participate, the strength and speed with which the evidence appears, and so on. It can be the beginning of a scientific investigation.

3.5 Interaction-at-a-distance

Consider now a common feature of the two experiments with radiation. In both cases, you observed evidence of interaction between objects that were not in physical contact. We speak of this condition as *interaction-at-a-distance* because of the distance separating the interacting objects. The idea that objects interact without touching seems to contradict our intuition based on physical experience and the sensations of our bodies; therefore, we construct working models that include radiation to make interaction possible between the two objects. The shields intercept the radiation and show the effect of its presence and absence; this confirms the usefulness of our working models.

An experiment that significantly resembles the radiation experiments can be carried out with the system shown in Fig. 3.8. A spring is supported at the ends by rigid rods. If a ruler strikes the spring at point A, you see a disturbance in the spring, which is evidence of interaction, and then movement of the flag at B, another piece of evidence of interaction. The first movement is evidence of interaction between the ruler and the spring. The second is evidence of interaction of the spring with the flag.

The experiment with the spring and the flag becomes another example of interaction-at-a-distance, however, if you choose to focus

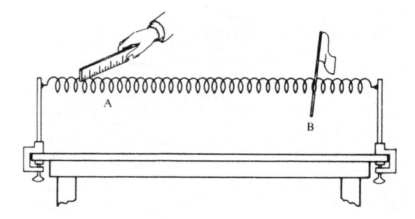

Figure 3.8 The ruler interacts with the flag by way of the long spring. Is this an example of interaction-at-a-distance?

on the system including only the ruler and the small flag. The motion of the flag correlated with the motion of the ruler is evidence of interaction-at-a-distance between these two objects. Of course, in this experiment you can see the spring and a disturbance traveling from the ruler along the spring to the flag. You do not need to construct a working model with a "model object" to make the interaction possible. You can, therefore, use the disturbance along the spring as an analogue model to help you visualize the radiation traveling from the match or material X to the detectors in the two other experiments.

The field model. Familiar examples of interaction-at-a-distance are furnished by a block falling toward the earth when it is not supported, by a compass needle that orients itself toward a nearby magnet, and by hair that, after brushing on a dry day, extends toward the brush. The intermediaries of interaction-at-a-distance in all these examples are called *fields*, with special names, such as *gravitational field* for the block-earth interaction, *magnetic field* for the compass needle-magnet interaction, and *electric field* for the brush-hair interaction. We may call this approach the *field model* for interaction-at-a-distance.

Radiation and fields. Do radiation and fields really exist, or are they merely "theoretical objects" in a working model? As we explained in Section 1.3, the answer to this question depends on how familiar you are with radiation and fields. Since radiation carries energy from a source to a detector, while the field does not accomplish anything so concrete, radiation may seem more real to you than fields. Sunlight, the radiation from the sun to green plants or to the unwary bather, is so well known and accepted that it has had a name for much longer than has interaction-at-a-distance. Nevertheless, as you become more familiar with the gravitational, magnetic, and electric fields, they also may become more real to you.

"The physicist ... accumulates experiences and fits and strings them together by artificial experiments ... but we must meet the bold claim that this is nature with ... a good-humored smile and some measure of doubt."

Goethe
Contemplations of Nature

For the scientist, both radiation and fields are quite real. In fact, the two have become closely related through the field theory of radiation, in which the fields we have mentioned are used to explain the production, propagation, and absorption of radiation. More on this subject is included in Chapter 7.

Gravitational field. Two fields, the gravitational and the magnetic, are particularly familiar parts of our environment. At the surface of the earth the gravitational field is responsible for the falling of objects and for our own sense of up and down. The plumb line (Section 1.4) and the equal-arm balance (Section 1.5) function because of the gravitational interaction between the plumb bob or the weights and the earth. We, therefore, use a plumb line to define the direction of the gravitational field at any location. Because the earth is a sphere, the direction of the gravitational field varies from place to place as seen by an observer at some distance from the earth (Fig. 3.9). More about the gravitational field will be described in Chapter 11.

Magnetic field. The magnetic field is explored conveniently with the aid of a magnetic compass, which consists of a small, magnetized needle or pointer that is free to rotate on a pivot (Fig. 3.10). When the compass is placed near a magnet, the needle swings back and forth,

OPERATIONAL DEFINITION
The direction of the gravitational field is the direction of a plumb line hanging freely and at rest.

Figure 3.9 (to right) The gravitational field near the earth is directed as indicated by plumb lines. The field appears to converge on the center of the earth

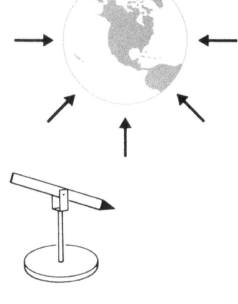

Figure 3.10 Examples of magnetic compasses. (below)
(a) The compass needle is often enclosed in a case for better protection.
(b) The pivot may permit the needle to rotate in a horizontal plane.
(c) The pivot may permit the needle to rotate in a vertical plane.

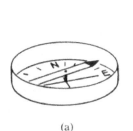

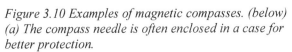

(a)   (b)   (c)

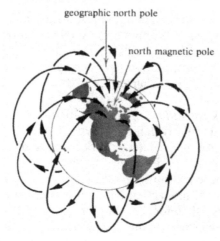

geographic north pole

north magnetic pole

Figure 3.11 The magnetic field of the earth, represented by the arrows, lies close to the geographic north-south direction, but does not coincide with it. In the magnetic dipole model for the earth, the two magnetic poles lie near the center of the earth on a line through northern Canada and the part of Antarctica nearest Australia.

and finally comes to rest in a certain direction. Because of its interaction with other magnets, the compass needle functions as detector of a magnetic field at the point in space where the compass is located. It is most commonly used to identify the direction of the magnetic field at the surface of the earth, which lies close to the geographic north-south direction (Fig. 3.11). Since the compass needle has two ends, we must decide which end indicates the direction of the magnetic field. The accepted direction of the magnetic field is that of the geographic north-seeking end of the needle (henceforth called the "direction of the needle"), as shown by the arrows in Fig. 3.11.

Figure 3.12 The arrows represent the compass needles that indicate the magnetic field near the bar magnet.

Figure 3.13 A bar magnet is cut in half in an effort to separate the north pole from the south pole. Arrows represent compass needles. Each broken part still exhibits two poles, the original pole and a new one of the opposite kind.

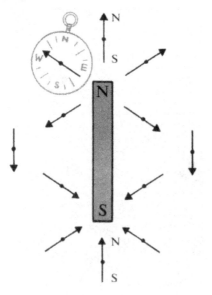

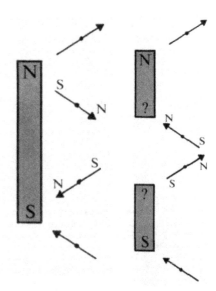

William Gilbert (1544-1603), an Elizabethan physician and scientist, wrote the first modern treatise on magnetism, De Magnete. *Gilbert worked with natural magnets (lodestones). In one chapter of this work, Gilbert introduced the term* electric *(from the Greek elektron for amber).*

Strength of the magnetic field. When you place a compass near a magnet, you notice that the needle swings back and forth rather slowly if it is far from the magnet and quite rapidly if it is close to the magnet. You can use this observation as a rough measure of interaction strength or magnetic field strength: rapid oscillations are associated with a strong field, slow oscillations with a weak field. You thereby discover that the magnetic field surrounding a magnet has a strength that differs from point to point; the field strength at any one point depends on the position of that point relative to the magnet.

Magnetic pole model. When you explore the magnetic field near a bar magnet, you find that there are two regions or places near the ends of the magnet where the magnetic field appears to originate. This common observation has led to the magnetic pole model for magnets as described in 1600 by William Gilbert (quoted to left). In this model, a magnetic pole is a region where the magnetic field appears to originate. The magnetic field is directed away from north poles and toward south poles according to the accepted convention (Fig. 3.12). All magnets have at least one north pole and one south pole. Opposite poles of two magnets attract one another, and like poles repel. If you apply these findings to the compass needle itself, you conclude that the north-seeking end of the needle contains a north pole (it is attracted to a magnetic south pole, Fig. 3.12).

"...thus do we find two natural poles of excelling importance even in our terrestrial globe . . . In like manner the lodestone has from nature its two poles, a northern and a southern . . . whether its shape is due to design or to chance . . . whether it be rough, broken-off, or unpolished: the lodestone ever has and ever shows its poles."
William Gilbert
De Magnete, 1600

An obvious question now suggests itself: can a magnetic pole be isolated? So far, physicists have failed in all their attempts to isolate magnetic poles (Fig. 3.13), in that they have not been able to narrow down the regions inside magnets where the magnetic field originates. They have found instead that the magnetic field appears to continue along lines that have no beginning or end but loop back upon themselves (Fig. 3.14). Thus, magnetic poles appear to be useful in a working model for magnets when the magnetic field outside magnets is described, but they fail to account for the field inside magnets.

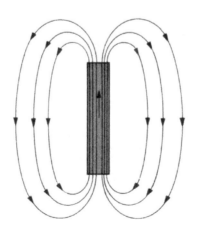

Figure 3.14 So-called magnetic field lines indicate the direction of the magnetic field. Lines inside the magnet close the loop made by the lines outside the magnet.

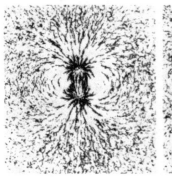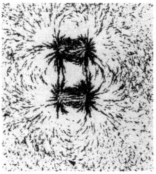

Figure 3.15 Iron filings were sprinkled over a piece of paper that concealed one or two bar magnets.

Hans Christian Oersted (1777-1851) inherited his experimental acumen from his father, an apothecary. His famous discovery in 1819 of the magnetic field accompanying an electric current took place as he was preparing for a lecture demonstration for his students. Until that time, electricity and magnetism were considered unrelated. Thus Oersted's discovery initiated intensive study of the relationships between electricity and magnetism, and these two separate disciplines gradually merged into the branch of physics known as electromagnetism.

Display of the magnetic field. Another technique for exploring a magnetic field is to sprinkle iron filings in it (Fig. 3.15). The filings become small magnets and, like the compass needle, tend to arrange themselves along the direction of the magnetic field. They produce a more visual picture of the magnetic field. This method is less sensitive than the compass, because the filings are not so free to pivot.

Electromagnetism. Not quite 150 years ago, Hans Christian Oersted, while preparing for a lecture to his students, accidentally found evidence of interaction between a compass needle and a metal wire connected to a battery. Such a wire carries an electric current (see Chapter 12). One of the most startling properties of the interaction was the tendency of the compass to orient itself at right angles to the wire carrying the electric current. Oersted's discovery is the basis of the electromagnet, a magnet consisting of a current-carrying coil of wire that creates a magnetic field. The distributions of iron filings near current-carrying wires are shown in Fig. 3.16.

Electric field. Somewhat less familiar than gravitational or magnetic fields is the electric field, which is the intermediary in the interaction of a hairbrush and the brushed hair. Electric fields also are intermediaries

Figure 3.16 Iron filings near current-carrying wires clearly show the closed loops of the magnetic field lines.

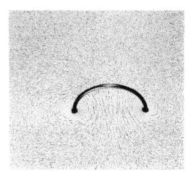

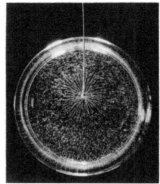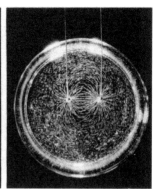

Figure 3.17 Grass seeds suspended in a viscous liquid indicate the direction of the electric field near charged objects.

"... for men still continue in ignorance, and deem that inclination of bodies to amber to be an attraction, and comparable to the magnetic coition ... Nor is this a rare property possessed by one object or two, but evidently belongs to a multitude of objects..."

William Gilbert
De Magnete, 1600

Benjamin Franklin (1706-1790) was born in Boston, Massachusetts. His father was an impoverished candle maker. Ben was apprenticed to an older brother in the printing trade. When his apprenticeship was terminated, he left for Philadelphia where he supported himself as a printer and eventually earned the fortune that freed him for public service. His political career as one of the "founding fathers' of democracy is celebrated, but it is less widely known that he was also one of the foremost scientists of his time. Franklin proposed the one-fluid theory of electricity, and he introduced the terms "positive electricity" and "negative electricity."

in the interaction of thunderclouds that lead to lightning, the interaction of phonograph records with dust, and the interaction of wool skirts with nylon stockings. Objects that are capable of this kind of interaction are called electrically charged.

Electrically charged objects. Objects may be charged electrically by rubbing. Many modern plastic materials, especially vinyl (phonograph records), acetate sheets, and spun plastics (nylon and other man-made fabrics) can be charged very easily. Electric fields originate in electrically charged objects and are intermediaries in their interaction with one another.

Display of the electric field. Individual grass seeds, which are long and slender in shape, like iron filings, orient themselves when they are placed near charged objects (Fig. 3.17). Their ends point toward the charged objects. The patterns formed by the seeds are very similar to the iron filing patterns in a magnetic field. Using the more familiar magnetic field as an analogue model for the electric field, we define the direction of the electric field to be the direction of the grass seeds.

Early experiments with electrically charged objects. Benjamin Franklin and earlier workers conducted many experiments with electrically charged objects. Gilbert had already found that almost all materials would interact with charged objects. One important puzzle was the ability of charged objects to attract some charged objects (light seeds, dry leaves, etc.) but to repel certain others. It was found, for instance, that two rods of ebonite (a form of black hard rubber used for combs and buttons) rubbed with fur repelled one another. The same was true of two glass rods rubbed with silk. But the glass and ebonite rods attracted one another (Fig. 3.18). Since a glass rod interacted differently with a second glass rod from the way it interacted with an ebonite rod, it followed that glass and ebonite must have been charged differently.

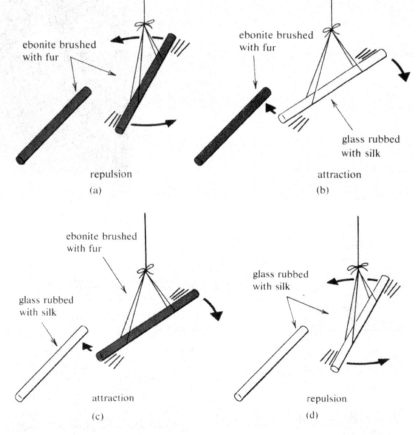

Figure 3.18 Hard rubber rods brushed with fur and glass rods rubbed with silk are permitted to interact. One rod is suspended by a silk thread and is free to rotate. The other rod is brought near until movement gives evidence of interaction.

Two-fluid model for electric charge. Two working models for electrically charged objects were proposed. One model assumed the existence of two different kinds of electric fluids or "charges" (a word for electrical matter) that could be combined with ordinary matter. Charges of one kind repelled charges of the same kind and attracted charges of the other kind.

Franklin's experiment. In a highly original experiment, Franklin found that the two kinds of charges could not be produced separately, but were formed in association with one another. Thus, when an uncharged glass rod and silk cloth are rubbed together, both objects become charged, but with different charges (Fig. 3.19). When the silk is wrapped around the rod, however, the rod-silk system does not have an observable electric field, even though the two objects separately do. Therefore, Franklin concluded that the two kinds of charges were opposites, in that they could neutralize each other. Accordingly, he called them "positive" and "negative," the former on the glass rod, the latter on the silk (and on the ebonite). Combined in equal amounts in one object, positive and negative charges add to zero charge. An object with zero charge is uncharged or electrically neutral.

One-fluid model for electric charge. Franklin's model, to explain this observation, provided for only a single "electric fluid." Uncharged objects have a certain amount of this fluid. Positively charged objects have an excess of the fluid, whereas negatively charged objects have a deficiency. When two uncharged objects are charged by being rubbed together, fluid passes from the one (which becomes negative) to the other (which becomes positive). The fluid is conserved (neither

Figure 3.19 Franklin's experiment.
(a) A glass rod is charged by rubbing a piece of silk.
(b) The silk is tested for electric charge by interaction.
(c) The glass rod is tested for electric charge.
(d) The rod-silk system is tested for electric charge.

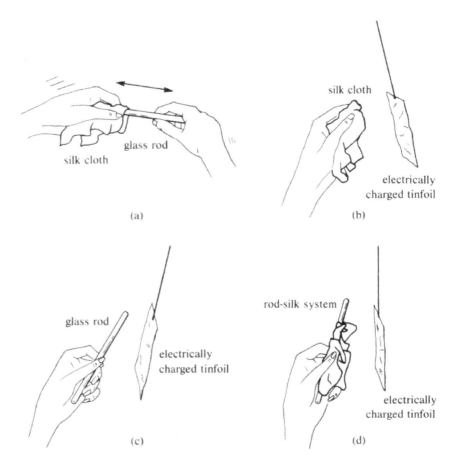

created nor destroyed), so the two objects together have just as much fluid as at the beginning of the experiment; hence they form an electrically neutral system. Clearly the isolation of one or two electric fluids would be an exciting success of these models. We will   pursue this subject further in Chapters 8 and 12.

Summary

Pieces of matter (objects) that influence or act upon one another are said to interact. The changes that occur in their form, temperature, arrangement, and so on, as a result of the influence or action are evidence of interaction. For the study of interaction, pieces of matter are mentally grouped into systems to help the investigator focus his attention on their identity. As he gathers evidence of interaction, the investigator compares the changes he observes with what would have happened in the absence of interaction. Sometimes he may carry out control experiments to discover this; at other times he may draw on his experience or he may make assumptions.

Pieces of matter that interact without physical contact are interacting-at-a-distance. Radiation and fields have been introduced as working model intermediaries for interaction-at-a-distance. The gravitational field, the magnetic field, and the electric field are the fields important in the macro domain. All three fields have associated with them a direction in space.

List of new terms

interaction	variable factor	field
system	inertia	gravitational field
conservation	inertial mass	magnetic field
state	inertial balance	electric field
evidence of interaction	radiation	magnetic pole
control experiment	interaction-at-a-distance	electric charge

Problems

1. What evidence of interaction might you observe in the following situations? Identify the interacting objects, identify systems (or subsystems) that show evidence of interaction, and describe what would have happened in the absence of interaction.

   (a) A man steps on a banana peel while walking.

   (b) A young man and a young woman pass each other on the sidewalk.

   (c) Two liquids are poured together in a glass.

   (d) A professor lectures to his class.

   (e) A comet passes near the sun.

   (f) Clothes are ironed.

2. Give two examples of interaction where the evidence is very indirect (and possibly unconvincing).

3. Do library research to find what role the interaction concept or other concepts of causation played (a) in Greek philosophy, (b) during the Middle Ages, (c) during the Renaissance, (d) in an Asiatic culture, (e) in a contemporary culture of your choice, and (f) in biblical literature.

4. Interview four or more children (ages 6 to 10 years) to determine their concepts of causation. Raise questions (accompanied by demonstrations, if possible) such as "What makes the piece of wood float?" "What makes the penny sink?" "What makes the clouds move?" "What makes an earthquake?" "What makes rain?" "Can rainfall be brought about or prevented?" (If possible, undertake this project cooperatively with several other students so as to obtain a larger collection of responses.)

5. Give two examples from everyday life of each of the following, and describe why they are appropriate. (Do not repeat the same example for two or more parts of this problem.)

    (a) Systems of interacting objects

    (b) Systems of objects that interact-at-a-distance

    (c) "Control experiments" you have carried out as part of an informal investigation

    (d) Systems that have social inertia

    (e) Systems that have thermal inertia

    (f) Systems that have economic inertia

    (g) Systems that have inertia of motion

6. Analyze three or four common "magic" tricks from the viewpoint of conservation of matter (rabbit in a hat, liquid from an empty glass, etc.).

7. Answer the questions in the margin on p. 59 and explain your answers.

8. Interview four or more children (ages 5 to 8 years) to determine their concept of the conservation of matter. (For suggestions, refer to B. Inhelder and J. Piaget, *The Growth of Logical Thinking from Childhood to Adolescence*, Basic Books, New York, 1958.)

9. Explain how various professions might define systems that are "conserved." Do not use the particular examples in Section 3.3.

10. Compare the use of the word "state" in the phrases "state of a system" and "state of the nation."

11. Propose a systematic series of experiments, other than the one described in Fig. 3.3, to identify the "guilty" circuit breaker.

12. Radiation and fields are introduced as models for intermediaries in interaction-at-a-distance. Describe your present preference for treating interaction-at-a-distance with or without such a model.

13. Four compasses are placed on a piece of cardboard that conceals some small bar magnets. How many magnets are under the cardboard? Locate their poles. Justify your answer. The compass needle directions are shown in Fig. 3.20, below.

bar magnet size:

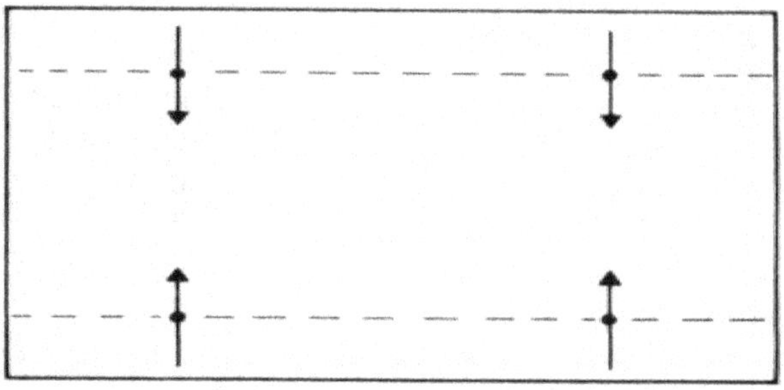

Figure 3.20 Compass needles near concealed magnets (Problem 13).

14. Comment on your present preference for one or the other of the two models described for electrical interaction: the two-fluid model and Franklin's one-fluid model.

15. Describe necessary features of a one- or two-fluid model for magnetized objects. Point out its advantages and disadvantages compared to the pole model.

16. Construct an operational definition for the direction of the electric field.

17. In Section 1.1, "matter" was left as an undefined term. It has been suggested that matter is "anything capable of interaction." Compare this definition with your intuitive concept of matter in the light of Chapter 3, especially Section 3.5. Comment on the logic of this definition, keeping in mind the definition of "interaction." Comment also on the effect of this definition on the conservation of matter principle.

18. Identify one or more explanations or discussions in this chapter that you find inadequate. Describe the general reasons for your dissatisfaction (conclusions contradict your ideas, or steps in the reasoning have been omitted, words or phrases are meaningless, equations are hard to follow, etc.) and pinpoint your criticism as well as you can.

Bibliography

F. Bitter, *Magnets: The Education of a Physicist*, Doubleday (Anchor), Garden City, New York, 1959.

I. B. Cohen, (Ed.), *Benjamin Franklin's Experiments*, Harvard University Press, Cambridge, Massachusetts, 1941.

B. Dibner, *Oersted and the Discovery of Electromagnetism*, Blaisdell, Waltham, Massachusetts, 1963.

W. Gilbert, *De Magnete*, Dover Publications, New York, 1958.

M. B. Hesse, *Forces and Fields*, Littlefield, Totowa, New Jersey, 1961.

E. R. Huggins, *Physics 1*, W. A. Benjamin, New York, 1968. The four basic interactions are discussed in Chapter 11.

E. H. Hutten, *The Ideas of Physics*, Oliver and Boyd, Edinburgh and London, 1967. Chapter 3 is devoted to the field concept.

J. Piaget, *The Child's Conception of Physical Causality*, Littlefield, Totowa, New Jersey, 1960.

J. Piaget, *The Child's Conception of the World*, Littlefield, Totowa, New Jersey, 1960.

J. Piaget and B. Inhelder, *The Growth of Logical Thinking from Childhood to Adolescence*, Basic Books, New York, 1958.

Articles from Scientific American. Some or all of these, plus many others, can be obtained on the Internet at http://www.sciamarchive.org/.

G. Burbridge, "Origin of Cosmic Rays" (August 1966).

A. Cox, G. B. Dalrymple, and R. Doell, "Reversals of the Earth's Magnetic Field" (February 1967).

R. Dulbecco, "The Induction of Cancer by Viruses" (April 1967).

W. P. Lowry, "The Climate of Cities" (August 1967). A discussion of the interactions between human activities and the variables of climate. [Editor's Note: There are many more recent publications on this general topic, with increasing emphasis on "global warming" – the growing evidence that the mean temperature of the atmosphere is increasing and that this is correlated with the use of fossil fuels as well as with an increase in the concentration of carbon dioxide in the atmosphere. For additional information visit the web site of the National Academy of Science (www.nationalacademies.org) and search on "global warming."]

P. J. E. Peebles and D. T. Wilkenson, "The Primeval Fireball" (June 1967).

*In numberless cases we see
motion cease without having caused
another motion or the lifting of a weight;
but [energy] once in existence
cannot be annihilated,
it can only change its form;
and the question therefore arises,
what other forms is [energy],
which we have become
acquainted with as [separation
energy] and motion, capable
of assuming? Experience
alone can lead us to a conclusion
on this point.*

JULIUS ROBERT MAYER
*Philosophical Magazine,
1842*

Matter and energy

Matter and energy are of central concern to the physicist. From our ability to make successful theories has come understanding of the ways in which a system may store energy and how energy may be transferred by interaction of objects or systems with one another. From this understanding has come the extensive and effective utilization of energy that is at the base of modern technology and our civilization.

Everyone forms qualitative concepts of matter and energy as a result of everyday experience. *Matter* is represented by the solid objects, liquids, and gases in the environment. Matter is tangible; it is capable of interacting with the human sense organs, and various pieces of matter are capable of interacting with one another. Matter appears to be conserved: if an object is once observed in a certain place and later is not there, you are convinced that it has been removed to another location or that it has been made unrecognizable by changes in its appearance. No one believes that it could be annihilated without a trace remaining.

By *energy* is usually meant the inherent power of a material system, such as a person, a flashlight battery, or rocket fuel, to bring about changes in the state of its surroundings or in itself. Some sources of energy are the fuel that is used to heat water, the wound-up spring (or charged battery) that operates a watch, the storage battery in an electric toothbrush, the spinning yo-yo that can climb up its string, the dammed water that drives a hydroelectric plant, and the food that results in the growth of the human body. Your experience with energy is that it appears to be consumed in the operation of the energy sources. Thus, the fuel turns to ashes and becomes useless, the spring unwinds and must be rewound, the battery needs to be recharged, the yo-yo gradually slows down and stops unless the child playing with it pulls properly on the string, and so on. At first glance, therefore, you might conclude that energy, unlike matter, is not conserved.

We will describe two operational definitions of energy in Chapter 9. In the meantime, we will use this term and refer to energy sources, energy receivers, and energy transfer from source to receiver in the expectation that you have an intuitive understanding of these concepts.

4.1 Conservation of energy

Energy transfer. When Sir Edmund Hillary and Tenzing Norgay climbed to the top of Mt. Everest, they expended a great deal of energy. Everyone knows that the two consumed food and breathed air containing oxygen to make this possible. The food, which served them as energy source, came from plants or animals that in turn depended on an energy source (plants or other animals) in a chain of interdependence that is called a food chain. Ultimately, the energy being transferred from organism to organism in the *food chain* can be traced to the sun, which produces energy in the form of light and other radiation.

Energy transfer along the food chain occurs as a result of one

James Prescott Joule (1818-1889) was born near Manchester, the son of a well-to-do brewer whose business he inherited. He devoted himself to science early in life. At 17, he was a pupil of John Dalton, and at 22 he had begun the series of investigations that was to occupy the greater part of his life—the proof that when mechanical energy gives rise to heat, the ratio of energy consumed to heat evolved has a constant and measurable value. Joule's work had supreme significance because it solidly established the principle of the conservation of energy. It was, in Joule's words, "manifestly absurd to suppose that the powers with which God has endowed matter can be destroyed."

organism's eating another, but this is not the only mechanism of energy transfer. You are familiar with other chains of energy transfer. For instance, water escaping from a dam rushes down gigantic pipes to operate turbines, the turbines drive electric generators, and the energy is then distributed by means of transmission lines to factories and residences where some of it may be used to charge the storage battery in an electric toothbrush. The battery finally operates the toothbrush. Depending on the selection of the systems that make up the chain, energy transfer may occur from one system to another (from rushing water to the turbine) or it may occur from one form to another form in the same system (dammed water to rushing water).

Historical background. During the seventeenth century, many natural philosophers studied rigid-body collisions, such as occur between bowling balls and pins. The bowler transfers energy to the bowling ball, which rolls to the pins and hopefully knocks many of them over, perhaps with so much force that they knock over other pins. The recognition of energy transfer during collisions led Huygens (1629-1695) to a quantitative study from which he concluded that the energy of motion (at that time called *vis viva* or living force, now called *kinetic energy*) was conserved under some conditions. It took many years, however, before scientists realized that the *vis viva* could be transformed into other types of energy and back again with very little loss.

James Watt developed many of the foundations of our modern concept of energy from 1763 on as he invented the steam engine and transformed it from a huge, dangerous monster into the efficient, reliable marvel powering the Industrial Revolution. Basically, the concept of energy provided a way to measure (and helped maximize) the amount of work one could get out of the coal that fueled the early railroad locomotives, textile mills and other factories. Scientists had also simultaneously been discovering the connection between "animal heat" and chemical reactions (metabolism). The third important piece of the puzzle was the recognition that the shaping and drilling of metals by machine tools resulted in a temperature rise.

These three factors inspired the physician Julius Robert Mayer (1814-1878) to speculate on the inter-convertibility of all forms of energy. James Prescott Joule then discovered the quantitative relation between thermal energy and various other forms of energy, establishing a solid experimental foundation for Mayer's theory that is still accepted today.

Since then the *law of conservation of energy* has become generally accepted as one of the most fundamental laws of nature. According to this law, energy may be changed in form but it cannot be created or destroyed. Whenever existing theories have failed to account for all the energy, it has been possible to modify the theory by including new forms of energy that filled the gap. Making these modifications is analogous to your applying the law of conservation of matter to a missing book; you include more locations in your search, and you try to remember who may have borrowed it. You do not believe the book has disappeared from the face of the earth.

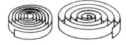

Energy storage. Because energy is conserved, it acquires the same permanence as matter. Just as material objects are kept or stored in certain containers, so we may say that energy is stored in systems (batteries, wound-up springs) that can act as energy sources. During interaction, energy is transferred from the source, which then has less energy, to one or more receivers, which then increase in energy. A system that acted as energy receiver in one process may, during later interaction, act as source and transfer to another system the energy that was temporarily stored in it (Fig. 4.1).

How much energy a system has stored depends on the state of the system. The spring in the state of being tightly coiled has more energy than in the uncoiled state. The storage battery in the charged state has more energy than in the discharged state. We will show in later chapters how the energy stored in a system is related by mathematical models to the variable factors (distance, temperature, speed, and so on) that describe the state of the system.

Energy degradation. These observations suggest that the apparent energy consumption of your everyday experience may actually be only a transfer of energy to a form less easily recognized and less easily transferred further to other systems. During the successive interactions in an energy transfer chain, some energy is transferred to receivers that are difficult to use as energy sources. Examples of these are the warm breath exhaled by Hillary and Tenzing in their climb and the very slightly heated bowling pins that become warm when their motion is stopped by friction with the bowling alley floor or walls. For practical purposes, therefore, their energy is no longer available and appears to have been consumed. In the framework of energy conservation, it is customary to refer to the apparent energy consumption of everyday experience as *energy degradation*.

Figure 4.1 Temporary storage of energy in the bow.
(a) The bow acts as energy receiver.
(b) The bow acts as energy source.

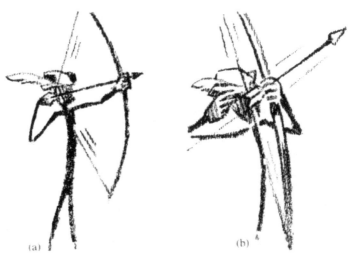

(a)   (b)

Equation 4.1

energy (initial state)
$$= E_i$$
energy (final state)
$$= E_f$$
energy transfer
$$= \Delta E$$

$$\Delta E = E_f - E_i$$

Identification of systems as energy sources. When you consider the examples mentioned in the introduction to this chapter, you find that we were rather careless in ascribing energy to some of the items mentioned. The food consumed by human beings is not really an energy source capable of sustaining human activity, since it cannot by itself undergo the transformation needed for the release of energy. Instead, the system of food and oxygen is the energy source, whose state can change until the materials have been converted to carbon dioxide and water. This system has more stored energy in its initial state, before digestion and metabolism, than in the final state (Fig. 4.2 and Eq. 4.1).

You see, therefore, that a system must be properly chosen if it is to function as energy source. Just how the system is to be chosen depends on the changes that lead to a decrease or increase of the stored energy. An automobile storage battery functions as the energy source when it is used to operate electrical devices. When the battery is used in a very unusual way—when, for example, it is dropped on walnuts to break them—then the appropriate energy source includes the storage battery, the earth, and their gravitational field. The physicist's definition of systems is geared to ensure conservation of a system (Section 3.3) that functions as energy source or energy receiver. The description of the state of a system includes all the variable factors whose numerical values determine the quantity of energy stored in the system.

4.2 Systems and subsystems

When a system is thought of as energy source or receiver, it becomes worthwhile to choose as simple a system as possible so as to localize the energy. Thus, the spring is only a small part of the watch; the storage battery is only a small part of the electric toothbrush, and so on. This choice of system then will not encompass the entire phenomenon or process being investigated. A system for the entire process will include the energy sources, energy receivers, and other objects participating in the interaction of the source with the receivers. The smaller systems of objects within larger systems are called *subsystems*.

Applications. A simple application of this idea can be made to a slingshot whose rubber band hurls a stone. Both the rubber band and the stone are subsystems of the larger system including slingshot and

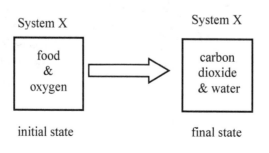

Figure 4.2 Change in the state of System X, which serves as an energy source.

stone. The rubber band acts as energy source and the stone as energy receiver in this instance.

Another application can be made to the South Sea Islander, who starts a fire by twirling a sharp stick against a piece of coco palm bark (Fig. 4.3). A system for the entire process would include the man, the stick, the bow and string, and the piece of bark. The man is the subsystem that acts as energy source. The piece of bark, which gets hot from rubbing, is the subsystem that receives the energy. The stick, bow, and string form another subsystem of interacting objects that facilitate energy transfer.

Selection of subsystems. Before concluding this section, we should add that there are often reasons other than energy transfer for selecting subsystems in a system. For instance, it may be that evidence of interaction is revealed particularly by one subsystem, as when a meat thermometer is used in a large piece of roast beef in the oven. Or, a sample of ocean water may be divided into a water subsystem and a salt subsystem. The subsystems concept is used whenever it is convenient to identify systems that are completely contained within other systems.

4.3 Passive coupling elements

In the slingshot example of the previous section, energy was transferred from the rubber band to the stone by direct interaction. In the making of a fire, however, energy was transferred from one subsystem (man) to the other (bark) through an intermediate subsystem, the bow, string, and stick (Fig. 4.3). This intermediate subsystem never acquires an appreciable amount of energy of its own. A subsystem like the bow, string, and stick, which facilitates energy transfer in a passive way, is called a *passive coupling element*.

Example. There are many situations in which there is a chain of interactions, with energy being transferred from one system to a second,

Figure 4.3 Do the important energy source and receiver in the process interact directly with one another?

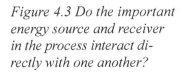

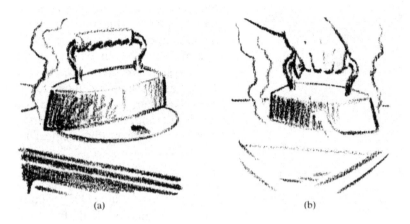

(a)   (b)

Figure 4.4 Temporary storage of energy in the old-fashioned flatiron.
(a) The iron is placed on stove and acts as an energy receiver.
(b) The iron acts as an energy source. What is the receiver?

then to a third, and so on. Some of the intermediate systems do not fluctuate in energy but, as it were, pass on the energy they receive as fast as they receive it. A modern electric iron is a good example. Once it has been plugged in for a time and has reached its operating temperature, the energy is supplied from the power line at the same rate as that at which it is passed on to the clothes and the room air surrounding the iron. The energy and temperature of this iron do not fluctuate appreciably. An intermediate system that acts in this way is a passive coupling element. After the iron is unplugged, however, it becomes an energy source and cools off at the same time as it heats the air; then it is no longer only a passive coupling element.

Counter-example. Some of the intermediate systems in an energy transfer chain act as energy receivers and energy sources and are not passive coupling elements; that is, their stored energy fluctuates up and down. An old-fashioned flatiron (which has no electrical heating element) is an example (Fig. 4.4). While it is standing over the fire, it acts as energy receiver and becomes hot. While it is being used to iron clothes, it acts as an energy source and gradually cools off. In other words, its energy content fluctuates along with its temperature.

Passive coupling elements. The concept of a passive coupling element is an idealization that is useful when long chains of energy transferring interactions are to be described and analyzed. The concept can be applied when the energy fluctuation of the intermediate systems are small compared to the energy transmitted. Because the passive coupling elements neither gain nor lose energy, they can be ignored insofar as measurements of energy conservation are concerned. Examples of additional systems that can be treated as passive coupling elements are illustrated in Fig. 4.5.

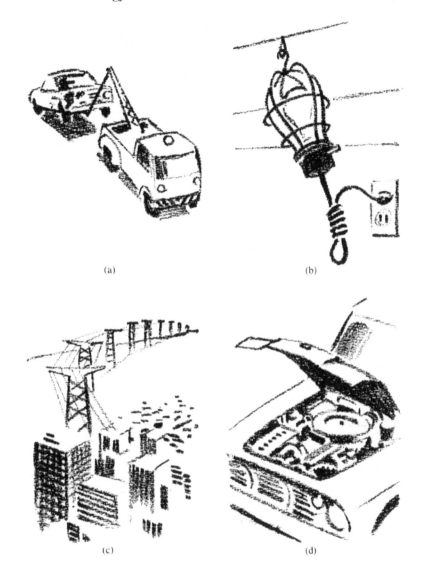

(a)

(b)

(c)

(d)

Figure 4.5 Passive coupling elements.
(a) The chain is a passive coupling element between the truck and the smashed car.
(b) The glowing light bulb is a passive coupling element between the power line and the radiated light.
(c) The high-tension line is a passive coupling element between the power station and the city.
(d) An automobile engine is approximately a passive coupling element between the exploding gasoline-oxygen system and the car.

4.4 Forms of energy storage

Consider a rubber band that is stretched and then let fly across a room. This action illustrates an energy transfer, but does not fit into the scheme of subsystems and coupling elements described in the preceding two sections. The energy is always stored in the rubber band, at first because the rubber band is stretched, and after its release because the rubber band is moving relative to the room. To differentiate the different ways in which one system or subsystem may store energy, we introduce the *forms of energy storage*. For the rubber band, we speak of two forms: elastic energy (stretched vs. unstretched rubber band) and kinetic energy (moving vs. stationary rubber band).

The concept of a form of energy storage is useful whether energy is transferred from one form to another within one system (rubber band example) or from one subsystem to another (Section 4.2). The forms of energy stored in a system are associated with the changes that can lead to an increase or decrease of the energy stored in the system. Thus, a pot of water whose temperature may drop has *thermal energy*. The food and oxygen system whose chemical composition may change has *chemical energy*. The steam that may turn to liquid water has *phase energy*. A stretched spring or rubber band that may snap together has *elastic energy*. A moving bullet whose speed can change has *kinetic energy*.

Field energy. In the forms of energy storage just enumerated, the energy was directly associated with one or more concrete objects. This is not the case when you examine systems of objects that interact-at-a-distance. Consider, for instance, the energy that is stored when a stone is raised off the ground. Because of its gravitational interaction with the earth, the stone can fall down and acquire kinetic energy. But from where does the energy come? Neither the stone nor the earth separately is the energy source here; the entire interacting earth-stone system is the energy source.

In Section 3.5 we introduced the concept of the gravitational field as intermediary in the interaction-at-a-distance between the earth and the stone. As the stone is being raised and while it falls, the gravitational field between the two objects changes. Thus, the changes of energy in the earth-stone system are correlated with changes of the gravitational field and not with changes in the stone or the earth separately. We will therefore attribute the energy of the earth-stone system to the gravitational field. In other words, we will take the view that the energy of the raised stone is actually stored in the gravitational field of the earth-stone system. This is an example of a new form of energy storage, *gravitational field energy*.

Electric and magnetic interaction-at-a-distance may be handled in exactly the same way as the gravitational interaction. Thus, a pair of interacting magnets has *magnetic field energy*, and two electrically charged clouds during a thunderstorm have *electric field energy*. The magnetic example is easiest to explore in the macro domain, since you

The word "phase" is used in physics with two meanings. The first refers to stages in a repeated motion, as in the "phases of the moon." The second, which we use in this text, refers to "solid," "liquid" and "gaseous phases of matter." That is, phase will refer to the three distinct "states" that matter can take, depending on the temperature and pressure.

Scientists have also discovered a fourth phase, called the "plasma" phase. A plasma occurs at extremely high temperatures, such as at the center of a star, where the particles of a gas break apart into electrically charged particles. Thus a plasma is, essentially, a gas in which each individual particle carries an excess of positive or negative charge. We will not discuss the plasma phase further in this text.

can take two small magnets and manipulate them near each other. Your sensations suggest that the magnets are linked by a spring that pulls them together or pushes them apart (depending on their relative orientation). When you exert yourself, more energy is stored in the magnetic field; when you relax and the magnets spring back, energy is transferred from the magnetic field.

Radiant energy. In Sections 3.4 and 3.5 we described radiation as an intermediary in interaction-at-a-distance. In the example of the candle and the detector, the candle acted as an energy source. When the sun shines on green plants, the sun functions as energy source and the plants as energy receivers. In both cases, radiation carries the energy from the source to the receiver. It is therefore customary to include radiation as a form of energy, called radiant energy.

Radiant energy is similar to field energy in the sense that it is not associated with a material object or system. Nevertheless, it is necessary that you recognize radiant energy if you wish to maintain energy conservation, for there is a time interval after the sun radiates the energy and before the plant receives it. Where is the energy during this time interval? It is not stored in the sun or in the plant; if energy is conserved, it must be stored temporarily as *radiant energy*.

As we pointed out in Section 3.5, the field theory of radiation represents radiation in terms of fields that vary in space and time. In this theory, radiant energy may be classified as field energy. However, since radiant energy manifests itself quite differently from the energy stored in the fields described in the previous subsection, we will refer to radiant energy as a separate form of energy.

Examples. Several examples of how the energy transfer in some common phenomena can be described are illustrated in Fig. 4.6. In these descriptions of energy transfer we have combined the ideas of systems and of forms of energy storage by identifying the forms of energy storage that are important in each system. The idea of a form of energy storage is especially necessary, however, in those examples in which the energy is transferred from one form to another form within the same system or subsystem, as in the rubber band.

4.5 The many-interacting-particles (MIP) model for matter

The atomic model for matter. The question of what happens when matter is subdivided into smaller and smaller pieces has fascinated philosophers for thousands of years. They have also speculated about the existence of a few elementary substances, from which all others were built up. In the eighteenth and nineteenth centuries, the modern science of chemistry was established when the concepts of *element* and *compound* were given operational definitions. According to Lavoisier (1743-1794), a substance was considered to be a chemical element if it could not be decomposed into other substances by any

Figure 4.6 Examples of energy storage and transfer (* indicates passive coupling elements).

(a) A long pass in a football game.

System	Type of energy storage
passer's arm	chemical (muscle)
football (just after throw)	kinetic
football & earth system (at top of flight)	gravitational field (& kinetic)
football (just before catch)	kinetic
receiver's hands & football	thermal (& kinetic)

(b) Automobile coasts downhill at a steady speed.

System	Type of energy storage
earth & car system	gravitational field
car*	kinetic
brake lining	thermal

(c) An archer shoots an arrow.

System	Type of energy storage
Robin Hood	chemical (muscle)
bow	elastic
arrow	kinetic

means. On the other hand, the substances that could be decomposed were considered to be chemical compounds composed of several elements.

According to this definition, a substance believed to be an element might later be decomposed by some new procedures. It would then be reclassified as a compound. Lavoisier's definition is no longer satisfactory because the development of modern techniques has made it possible to decompose even chemical elements into more primitive components (see Chapter 8). The presently accepted definition of a chemical element is based on properties that are most useful to the chemist.

John Dalton (1766-1844) was largely self-taught. He was a retiring person, son of a humble handloom weaver, and from the age of 12 he barely supported himself as a teacher and general tutor in Manchester, England. Dalton possessed strong drive and a rich imagination and was particularly adept at developing mechanical models and forming clear mental images. His astonishing physical intuition permitted him to reach important conclusions despite being only "a coarse experimenter," as his contemporary Humphry Davy called him. Dalton's atomic theory was set forth in A New System of Chemical Philosophy, *published in 1808 and 1810.*

Dmitri Ivanovich Mendeleev (1834-1907) was a professor of chemistry at the University of St. Petersburg in the 1860's when he first noticed that the known elements could by systematized according to chemical properties. His method of classifying and arranging the elements gave us the periodic table, one of chemistry's fundamental conceptual tools. Mendeleev was a compassionate man, deeply involved in the great issues of his time. In 1890, he courageously resigned his chair in protest against the Czarist government's oppression of students and the lack of academic freedom.

Dalton's model. In the nineteenth century it became possible to make a quantitative study of the proportions in which elements combine to form compounds. It was found that elements combine in fixed proportion by weight, and that gaseous elements combine in fixed and small-number ratios by volume. These observations can be explained on the basis of the following working model, based on the ideas of John Dalton and Amadeo Avogadro (1776-1856). Elements are composed of small particles of definite weight called *atoms*; atoms combine in simple numerical ratios to form particles called *molecules*; and every volume of gas under the same conditions of pressure and temperature contains the same number of molecules (Fig. 4.7). This atomic model for matter has been remarkably successful in stimulating chemical research and in accounting, with additional refinement, for the observations made by scientists since Dalton's time.

Existence of atoms and molecules. The atoms and molecules, their interaction, and their motion make up the phenomena in the micro domain described in Section 1.2. Even though atoms and molecules were introduced originally as parts of a working model, they have been so useful in interpreting phenomena in the macro domain that almost everyone now believes that they really exist. Theories based on atoms and molecules have greatly furthered the scientist's understanding of the macro domain. Atoms and molecules are described by many specific and detailed properties, such as mass, size, shape, magnetism, ability to emit light, and so on. Mendeleev found it possible to arrange the elements in a sequence (the Periodic Table) that highlighted similarities in their chemical activity. This sequence was later expanded and slightly revised, and each element was given an atomic number according to its place in the sequence from hydrogen (atomic number 1) to

Figure 4.7 Two chemical reactions according to Dalton and Avogadro's model. Each volume of gas at the same pressure and temperature contains 14 atoms. Atoms combine in simple ratios to form new particles called molecules.

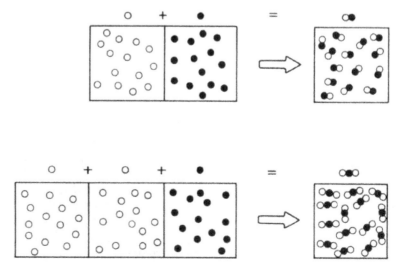

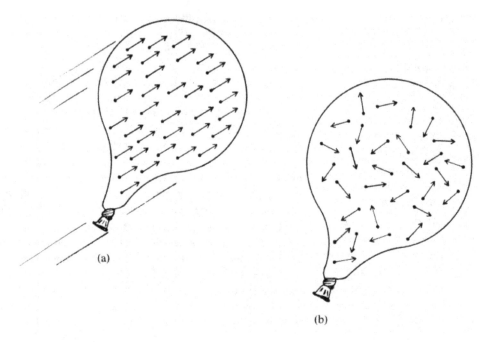

Figure 4.8 Concerted and random action of particles in a balloon.
(a) Macro-domain motion of the balloon is associated with concerted motion of the particles.
(b) Macro-domain failure to move is associated with random motion of the particles.

uranium (atomic number 92). To explain these properties, working models for atoms themselves have been invented; we will describe several of these in Chapter 8.

The MIP model. Many physical phenomena in the macro domain can be explained with the help of a working model in which matter is composed of micro-domain particles in interaction-at-a-distance with one another. The nature of the interactions, the intermediary fields, the sizes of the particles, and other details will be described in Chapter 8 but are not as important as the particles' great number and their ability to interact. We will therefore speak of the many-interacting-particles model (abbreviated MIP model) for matter. We will not identify the particles with atoms or molecules except for associating a different kind of particle with each substance. In the remainder of this section we will describe a few simple examples that illustrate the usefulness of the MIP model. Some of these will be elaborated and made quantitative in later chapters.

Random and concerted action. At the heart of the various applications of the MIP model is the essential idea that the particles are so numerous as to make the action of any single one of little consequence. Instead, only the average action or the most likely combination of actions of many particles is significant in the macro domain. The situation

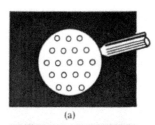

(a)

Figure 4.9 MIP model for solids, liquids, and gases.

(a) In a solid material, the particles can only oscillate about their equilibrium positions, which are arranged in a regular pattern.

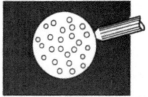

(b)

(b) In a liquid, the particles can move about, but they also interact and are strongly attracted to one another.

(c) In a gas, the particles are so far apart that they rarely interact.

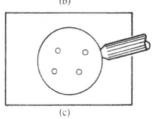

(c)

is analogous to that of a human mob, whose members may stampede and act in concert or may be confused and act at cross-purposes. The former is an example of concerted action, the latter of random action.

If, for instance, a blown-up balloon is thrown in a certain direction, then the particles inside the balloon move, on the average, in the same direction (Fig. 4.8a). If, however, the balloon is stationary, the particles move completely at random and their impacts with the rubber skin keep the balloon inflated but do not result in macro-domain motion (Fig. 4.8b).

The MIP model and the phases of matter. The solid, liquid, and gaseous phases of matter in the macro domain can be contrasted particularly easily by use of the MIP model. In a solid material the interacting particles are locked in a rigid pattern, and they can only vibrate a slight distance. In a liquid material the particles are in contact but are not locked into a rigid pattern; in a gas the particles are so widely separated from one another that they hardly interact at all (Fig. 4.9). This explains the rigidity of the solid, the fluidity of the liquid, and the ability of a gas to permeate the entire region of space accessible to it. It also explains the relatively low density of gases compared to solids and liquids (there are large empty spaces between the particles in a gas) and the relatively high compressibility of gases (the particles in a gas can relatively easily be forced into a smaller volume because of the spaces).

The MIP model and mixing. The MIP model also furnishes a ready description of mixing processes where several separate phases in the macro domain combine to form one phase, as when alcohol mixes with water, sugar dissolves in water, or water evaporates. According to the

Figure 4.10 Separate sugar and water phases mix to form a single phase.
(a) Separate phases.
(b) Solution phase.

model, the particles of the separate phases mix with one another to produce the uniform solution phase, which includes particles of several kinds. Thus, sugar solution contains sugar and water particles (Fig. 4.10), humid air contains water and "air" particles. (Air is actually a solution containing nitrogen, oxygen, and several other gases, but this fact is unimportant for many physical purposes.) Thus, the interaction of separate phases that leads to solution formation in the macro domain is explained by a mixing of particles in the micro domain.

The MIP model and energy storage. A third use of the MIP model is related to energy storage. In Section 4.4 we identified seven forms of energy storage: kinetic energy, thermal energy, chemical energy, phase energy, elastic energy, field energy (gravitational, electric, magnetic), and radiant energy. These forms of energy storage are appropriate to the macro domain. With the help of the MIP model, thermal energy, chemical energy, phase energy, and elastic energy of systems in the macro domain can be explained as kinetic energy and field energy of the interacting particles (Table 4. l). In the micro domain, therefore, thermal energy, chemical energy, phase energy, and elastic energy do not have separate meanings. The elimination of these four forms of energy storage is an important unifying feature of the MIP model. We will explain how this unification is achieved in later chapters, focusing on the various forms of energy storage.

TABLE 4.1 ENERGY STORAGE ACCORDING TO THE MIP MODEL

Macro domain	Micro domain
kinetic energy	kinetic energy
thermal energy	kinetic energy and field energy
chemical energy	field energy
phase energy	field energy
elastic energy	field energy
field energy	field energy
radiant energy	radiant energy

The MIP model and interaction. A fourth use of the MIP model is to explain macro-domain interaction through properties of the particles. When two solid objects touch one another (macro domain), then the particles at the surface of one interact with the particles at the surface of the other (micro domain). The interaction that leads to the formation of solutions (macro domain) has already been described as a mixing of the particles (micro domain). When two objects interact-at-a-distance (macro domain), then some or all of the particles in one object interact-at-a-distance with some or all of the particles in the other object (micro domain). A magnet, for instance, is made up of many particles, some of which are magnets; an electrically charged body is made up of many particles, some of which carry an electric charge.

Limitations of the MIP model. Other applications of the MIP model will also be valuable. In all of them, however, it is important for you to remember that the particles of the model are not little metal or plastic balls that roll, bounce, spin, rub, and scrape the way real metal or plastic balls do. You may picture the particles as little balls, but you should be aware that such particles do not have properties beyond the ones assigned to them in the model.

4.6 Equilibrium and steady states

The equilibrium state. It is common experience that systems show changes in their state but that these changes do not continue forever if the system is kept in a uniform environment. A tray of water placed in the freezer, for example, freezes and eventually comes to the temperature of the freezer, but then its state does not change further. A swinging pendulum continues to swing back and forth for some time, but the length of arc of the swing decreases until it comes to rest. A flashlight operates on its battery, but after several hours the light gets dim and finally goes out.

The state of a system that no longer changes in the absence of new environmental interactions is called an *equilibrium state*. A system may come to an equilibrium state in interaction with its environment, such as the water in the freezer, or it may come to equilibrium in isolation from the environment, as did the flashlight. The equilibrium concept is applied to both cases. Since no further change of any kind takes place, no energy transfer occurs either. The equilibrium state of a system, therefore, is a state of maximum energy degradation for that system.

Partial equilibrium. For practical purposes, the equilibrium idea is often applied to changes with respect to only one form of energy storage at a time. In the freezer example, for instance, the equilibrium concept is applied to the temperature of the water, while motion of the water (or ice) is disregarded. In the flashlight example, the charge state of the battery is of interest, and its kinetic or gravitational field energies are not. Therefore, we speak of partial equilibrium, such as mechanical equilibrium (position and motion), thermal equilibrium, chemical equilibrium, and phase equilibrium, whenever equilibrium is reached with respect to the corresponding form of energy storage. Another example is a hot swinging pendulum bob, which may come to mechanical equilibrium (its motion stops) before it comes to thermal equilibrium with the room. Or an ice cube may melt and come to thermal equilibrium with the room air long before its water has evaporated and come to phase equilibrium as a gas mixed with the air. In all these situations, the observer's interest determines which aspect of the system he particularly notes and which details he chooses to overlook.

Phase equilibria. An especially important example of equilibrium states is phase equilibrium, in which two or more phases (solid, liquid, gas) of one substance can coexist indefinitely. At ordinary atmospheric pressure, a pure substance, such as pure water, changes from solid to liquid (melting) or from liquid to solid (freezing) at a certain fixed temperature called the melting temperature. For the substance "water," the melting temperature at atmospheric pressure is 0° Celsius on the internationally accepted scale (formerly called Centigrade) or 32°

Figure 4.11 Phase equilibria.
(a) The equilibrium state of solid and liquid water at atmospheric pressure (reference temperature for Celsius thermometer, 0° Celsius) defines the melting temperature of ice, which is equal to the freezing temperature of water.
(b) The equilibrium state of liquid and gaseous water at atmospheric pressure (reference temperature for Celsius thermometer, 100° Celsius) defines the boiling temperature of water (gaseous water is inside the bubbles).
(c) Equilibrium state of liquid and gaseous water. At room temperature (68°F, 20°C) the gas contains about 2% gaseous water and 98% air.
(d) Equilibrium state of liquid water and salt. At room temperature the liquid contains about 35% salt and 65% water.

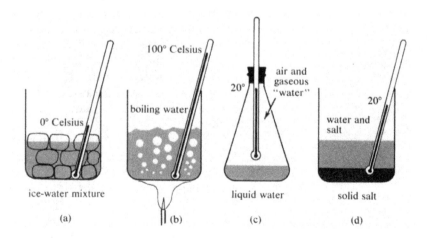

Fahrenheit. At this temperature, liquid water and solid water (ice) are at equilibrium with one another and can coexist indefinitely. If you attempt to raise the temperature of such a system by supplying heat, the solid water (ice) will all melt before the system begins to get warmer; if you attempt to lower the temperature by removing heat, the liquid water will all solidify before the system gets colder. The melting temperature of pure solid water (ice) has been used to define a reference temperature for thermometer scales, on which it is indicated as 0° Celsius. This and other examples of phase equilibria are illustrated in Fig. 4.11.

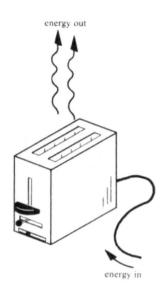

energy out

energy in

The steady state. There are many systems that do not seem to change but that are *not* in equilibrium, for example, the heating coil of an electric toaster, which starts to glow when the toaster is turned on and reaches a steady glow after a few seconds. The state of the coil does not change any more, but the coil is continuously receiving electric energy from the power line and, at the same time, transferring energy at the same rate to the room air. The toaster coil is now acting as a *passive coupling element* between the power line and the room air. Its own state does not fluctuate, and it experiences no net change of its own energy. Such a state, in which a system acts as a passive coupling element transferring energy between other systems, is called a *steady state*. It is to be contrasted with a genuine equilibrium state, in which neither the state of the system nor the state of the system's environment is changing.

An exact steady state is rarely achieved in practice, but it is a useful idealization that is approximated in many practical situations. Even in the toaster example, the room gradually gets warmer and warmer. As a result, the rate of energy transfer from toaster to room is altered, with a consequent gradual change in the "steady" state of the toaster coil. The significance of the steady state is that it represents a balance between energy input and output. In the next section we will describe ways in which a steady state may be maintained, if that is a desirable objective.

How do you tell whether a system is in equilibrium or a steady state?
1) Equilibrium systems do not gain energy from nor lose energy to their environment. Steady state systems steadily gain energy from or lose energy to their environment.
2) An equilibrium system does not tend to change the state of its environment. A steady state system generally does change the state of its environment (possibly gradually).

Analogue models for equilibrium and steady states. Analogue models for equilibrium and steady states can be created with tanks of water. The water level in a tank represents the state of the system. The approach to equilibrium is modeled by two connected tanks of water (Fig. 4.12), with the connecting valve between them closed. Initially, all the water is stored in the tank on the left. As soon as the valve is opened, water rushes into the second tank and fills the latter up to a level that changes no further. This level represents the equilibrium state.

The steady state is modeled by a different arrangement. Only one tank is supplied with water from a faucet and the water is permitted to escape through a hole near the bottom (Fig. 4.13). Then the water in the tank reaches a level at which the water inflow and outflow are equal. This water level represents the steady state.

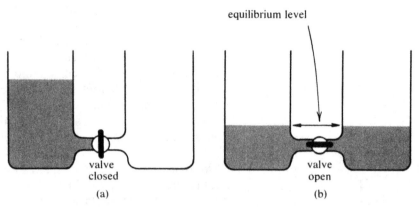

Figure 4.12 Approach to equilibrium is represented by the water level in two connected tanks.
 (a) Before interaction.
 (b) In equilibrium state with interaction.

Working models for systems in equilibrium and steady states. In studying equilibrium and steady-state conditions, it is frequently useful to construct working models that idealize the actual physical conditions. We took this approach in the example of the toaster, when we assumed that the slowly changing room temperature did not affect the energy output of the toaster wires. Even though real physical systems may never reach equilibrium or steady states, they often come sufficiently close so that a working model, in which certain small interactions or changes in environmental conditions are ignored, does lead to valuable predictions.

The total energy of an isolated system is conserved because the system does not interact and, therefore, does not transfer energy to any other system.

One idealized working model is that of the *isolated system*, which does not interact with its environment at all. An example to which this model may be applied is a system of ice and hot tea interacting with one another in a glass. Treating the ice and tea as an isolated system, which does not interact with the room air, enables you to predict an

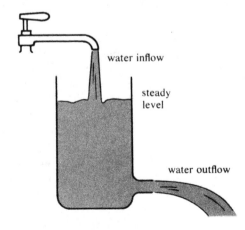

Figure 4.13 Steady state is represented by the water level in one tank with inflow and outflow

equilibrium state for the system (the temperature of the iced tea), even though interaction with the glass and the room air means that this state is approached but never reached by the actual iced tea.

A second idealized working model is that of a system interacting with an unchanging environment. Under these conditions the system may come to equilibrium or to a steady state. For example, a real sailboat on a real lake usually encounters rapidly fluctuating wind and wave conditions. Yet a boat designer will first evaluate the boat's performance in mechanical equilibrium—under a steady wind and in the absence of waves—as he determines the size of the sail, length of keel, and so on.

The temperature at any point on the earth shows great day-night and seasonal variations. By the earth's mean temperature we mean the average temperature calculated from all points on the earth's surface at one instant of time.

Applications. An important application of the steady-state concept is to the energy balance of the earth. The earth receives energy from the sun and radiates energy into space. The warmer the earth, the more it radiates. In a working model where the sun is a steady source, the earth's mean temperature will reach a steady value such that the earth's rate of energy loss through radiation is equal to the rate at which it receives energy from the sun.

The steady-state concept has many applications in other sciences as well as in physics. Food supplies remain steady as long as agricultural production is equal to consumption. Excess consumption leads to scarcity and higher prices, excess production to surpluses, lower prices, and loss of farm income. A baby's weight (and yours) increases only as long as the food it eats, water and milk it drinks, and air it inhales exceed losses through elimination, evaporation, and exhaling. When the baby's weight is steady or decreases, there is serious cause for alarm. Deer populations remain steady only as long as the birthrate is equal to the death rate. When deer are protected and their predators are killed, the population is likely to grow until other effects, such as competition for food, result in a new balance. College enrollment increases only as long as more students enter college than are graduating or dropping out. When these two rates come into balance, college enrollments will stabilize.

4.7 The feedback loop model

Stabilization of steady states. In this section we will discuss natural and man-made ways for stabilizing a steady state so that conditions are maintained despite fluctuating external influences. Among the most important natural mechanisms for stabilizing a steady state are the biological systems by means of which warm-blooded animals, including man, regulate their body temperature. A simpler but similar non-biological example is the regulation of a room's temperature by means of a room thermostat, which controls the operation of a furnace or air conditioner (or both). The thermostat ensures that the room remains at a comfortable temperature level no matter what extremes of temperature may occur outdoors.

Another example of stabilization is furnished by the trained sea lion that balances a ball on its nose by carefully timed movements of its head. Still another illustration of stabilization can be found when you drive a car in heavy traffic and try to maintain a safe distance between your car and the car in front of you in spite of changes in traffic speed. When the space in front of your car widens, your foot goes on the accelerator; when the space narrows, your foot goes on the brake.

These examples illustrate how a condition may be maintained at a steady value even though there are influences that would tend to disturb it. The intermittent operation of the furnace, the head movements of the sea lion, and your braking or acceleration have a stabilizing influence. The *feedback loop model* provides a way to analyze and understand such stabilized situations in which several interactions operate in a loop or circular pattern (Fig. 4.14).

Feedback loop systems. To apply the feedback loop model to a stabilized phenomenon that you observe, you have to take three steps. First, you have to identify the important state or condition that is being stabilized, such as the position of the ball on the sea lion's nose. Second, you have to identify the interaction whereby deviations from the steady state are detected by a system called a *detector* (the sea lion's eyes or the tactile nerves in its nose). Third, you have to identify the interaction that brings about a correction, called *feedback*, to counteract the deviation and restore the desired condition (achieved by the sea lion's head movements). Because the detection and feedback interactions are separate and distinct, the entire regulating process can be diagrammed in loop form (Fig. 4.15), whence its name is derived.

Negative and positive feedback. The stabilizing influence we described in the specific examples came about because the feedback counteracted the deviation from steady state that triggered the detector. You can conceive of a malfunction of the coupling elements so that the feedback actually makes the deviation worse rather than counteracting it. This situation may lead to catastrophe when, for example, the car in front of you slows down and you step on the gas instead of the brakes.

There are also natural situations where the feedback enhances the deviation rather than counteracting it. When water flows off a mountain, for instance, it erodes the land to form a channel for itself. The channel is the deviation from the previously uniform mountain slope. After further rainfall, more water is gathered in the channel and it flows more rapidly down the steep walls, causing more and more erosion. The process feeds itself until the mountain is deeply eroded. In this example the water combines the roles of detector and "corrector" (by gathering in the previously formed valleys and there concentrating its erosive interaction with the surface soil). Of course, as we said above, the feedback in this example is not corrective but rather increases the deviation.

It is customary to use the term *positive feedback* in situations where the feedback enhances the deviation (erosion example). By contrast, *negative feedback* is used to denote situations where the feedback counteracts the deviation (room thermostat example). Positive feedback, which enhances the deviation, results in catastrophe or brings into play new factors that were omitted from the original feedback loop model.

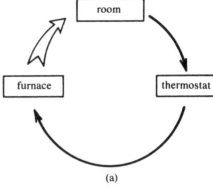

Figure 4.14 Loop or circular pattern of interactions that lead to stabilization of a steady state.
(a) Maintaining room temperature.
(b) Balancing a ball.
(c) Maintaining car separation in traffic.

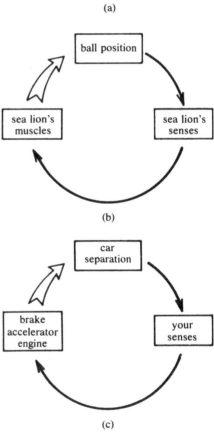

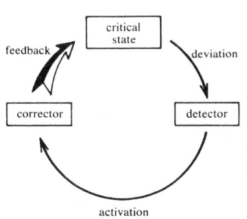

Figure 4.15 The detector identifies a deviation from the critical state. It then activates the corrector, which supplies feedback.

Feedback loop models in the social sciences. Positive and negative feedback are used in the training of animals. Psychologists use the term reinforcement rather than feedback in this context. Nevertheless, the concept is the same. For instance, consider how a pigeon is conditioned to peck at a round shape and not at a square one. Before training, the pigeon does not peck often at any shape. When it deviates from this behavior by pecking at the square shape, a "peck detector" behind the square activates an electric shock, which tends to reestablish the non-pecking behavior (negative feedback). When the pigeon deviates by pecking the round shape, however, it receives food, which tends to enhance this deviation (positive feedback) until the pigeon pecks at the round shape constantly.

Many social phenomena can be analyzed fruitfully with the aid of the feedback loop model. It is important for you to identify the steady condition of the subsystem that is at the focus of your analysis and to describe the deviations that actuate the feedback loop. If the feedback enhances the deviation, it is positive; if the feedback counteracts the deviation, it is negative. Whether positive or negative feedback is more desirable depends on your social values. Negative feedback maintains the status quo, whereas positive feedback leads to evolutionary or even revolutionary change.

Distribution of wealth. In complex phenomena, you must often consider several feedback loops, some with positive and some with negative feedback. Their net effect then depends on the relative effectiveness of the opposing feedback loops. In a social system with a fairly broad distribution of wealth, for example, laissez-faire capitalism seems to provide positive feedback for changes toward a polarized

Figure 4.16 Stabilization of a broad distribution of wealth by a negative and a positive feedback loop.

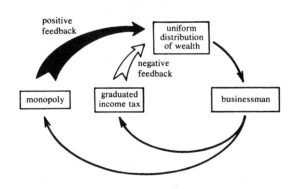

Figure 4.17 Schools can provide positive socioeconomic feedback to the population distribution in a city. When middle-class families leave the city, the schools change (decreased funding, fewer middle-class students) in a way that may repel additional middle-class residents and therefore further reduce the breadth of the population spectrum.

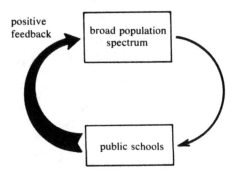

class structure of very rich and very poor individuals. In contrast, a graduated income tax (with higher tax rates for higher incomes) provides negative feedback for such changes and partially stabilizes the original distribution of wealth (Fig. 4.16).

Urban populations. Another currently important social feedback loop leads to the polarization of big city populations into slum dwellers and the very rich. The public schools are one of the important detecting and "correcting" subsystems here (Fig. 4.17). If the middle-class population decreases, the public schools tend to become less attractive to the remaining middle-class residents, who then tend to migrate to the suburbs in still greater numbers. On the other hand, urban schools, if they are funded and organized well enough to offer a superior educational environment, can provide negative feedback and thus contribute to maintaining, or restoring, a broad population spectrum.

4.8 Efficiency of Energy Transfer

When we use energy, we want as much of the energy as possible to go towards achieving our objective. This involves the idea of the *efficiency* of energy transfer. Specifically, *efficiency* of energy transfer refers to the percentage of the energy from the source that is delivered to the system that we define as the intended receiver, rather than being degraded and/or transferred to other objects. For example, in a toaster, the intended receiver is the bread; the heating of the air is an unintended consequence, and the energy that ends up heating the air can be considered as degraded or "lost." Thus we expect the efficiency of a toaster to be much less than 100%. The toaster itself is a passive coupling element, which simply transfers energy and does not retain any. The total amount of energy is still conserved.

The fact that energy is degraded or considered to be "lost" does *not* mean that the law of conservation of energy is being violated; energy is *always* conserved, and the law of conservation of energy has always turned out to be true. (This is in spite of the stream of people claiming to have invented perpetual motion machines or other ways to generate energy from nothing; in fact, no such claim has ever been substantiated.). In the case of the toaster, a passive coupling element, the total amount of electrical field energy drawn from the source (the electrical utility service) can be measured to be exactly equal to the energy output, that is, the sum of the energy delivered to the toast, air, and other receivers. However, from the point of view of the *user*, the efficiency of a toaster is low because only a small part of the original energy drawn from the power source ended up in the intended receiver (the toast). We will explain efficiency of energy transfer in more detail in Chapter 16.

Summary

The conservation of energy is a powerful physical principle that leads to the concept of energy transfer. Since energy cannot be created or destroyed, all changes in the energy stored by one system must be accompanied by the transfer of energy to or from other systems. The systems that supply the energy are called energy sources, those that receive the energy are called energy receivers, and those that transmit the energy without appreciably increasing or decreasing it are called passive coupling elements.

In addition to being transferred from one system to another, energy can be transformed from one form to another. The major forms of energy storage are kinetic, thermal, chemical, phase, elastic, field, and radiant energy. The many-interacting-particles model for matter helps to explain a large number of properties of matter, including certain forms of energy storage. This model introduces a micro-domain description of physical phenomena to supplement and unify the macro-domain descriptions based on observations made by our sense organs.

The energy stored by a system in equilibrium or steady state does not change. In the former, there is actually no energy transfer between the system and its environment; in the latter there is energy transfer, but no net change because the energy input and output are equal. A system in a steady state, therefore, is a passive coupling element for energy transfer with other systems or the environment. The mechanisms by which a system maintains its steady state are analyzed most effectively by means of the feedback loop model. Negative feedback results in a stable steady state, whereas positive feedback enhances small deviations from a steady state and thereby destroys it. The efficiency of energy transfer is of interest to a user of energy; the efficiency of energy transfer is the percentage of the energy from a source has actually reached its intended receiver or been devoted to a desired end.

List of new terms

matter
energy
energy source
energy receiver
energy transfer
energy conservation
energy storage
energy degradation
subsystem
passive coupling element
kinetic energy
thermal energy
phase energy
chemical energy
elastic energy
field energy
efficiency

radiant energy
atomic model for matter
element
compound
many-interacting-particles
    model (MIP model)
equilibrium state
steady state
partial equilibrium
phase equilibrium
melting temperature
boiling temperature
isolated system
feedback loop model
positive feedback
negative feedback

Problems

1. Give three examples from everyday life of the apparent degradation of energy.

2. A stage magician performs many amazing tricks. Discuss two or three magic tricks as observations that appear to violate conservation of matter or energy (or both) but really do not.

3. Interview four or more children (ages 8 to 12 years) to explore the meaning they attach to the word "energy" and to the phrase "energy source." Compare their understanding with the modern scientific view.

4. Compare the meanings of the word "law" in the phrases "law of conservation of energy" and "law-abiding citizen."

5. Explain how the law of conservation of energy applies to the examples you gave in response to Problem 1.

6. Give three examples from everyday life of systems that are used to store energy temporarily. Be careful to identify the complete system and compare the state of each system when it stores energy with the state when it has no energy.

7. Consider the following statement about the bow and string in Fig. 4.l: "This intermediate subsystem never acquires an appreciable amount of energy of its own." Explain and criticize this statement. Describe one or two thought experiments that can be used to argue its validity.

8. Discuss (qualitatively) two or more of the examples of coupling elements described in the text with respect to how passive they are (i.e., how much or little of the energy they transfer is stored in them temporarily) and how much the stored energy fluctuates.

9. Give three examples (distinct from those mentioned in the text) from everyday life of systems that usually serve as passive coupling elements. Compare the three examples qualitatively with respect to how passive they are (i.e., how much or little of the energy they transfer is actually stored in them temporarily).

10. Describe how energy is stored and how it is transferred from one form to another in the following examples (see Fig. 4.6). Refer to the forms of energy storage in the macro domain. Begin each description with an energy source.
    (a) The ball is kicked off during a football game.
    (b) A child hops around on a pogo stick.
    (c) A plugged-in electric toaster is used to toast bread.
    (d) A pole vaulter vaults over a bar 5 meters high.
    (e) An automobile drives up a hill at a steady speed.
    (f) A tiger leaps on a mouse.
    (g) A photograph is made of a snowy landscape.

11. Interview four or more children (ages 10 to 14 years) to investigate their concept of the structure and phases of matter. Evaluate their responses in relation with the modern scientific view presented in Section 4.5. (Hint: Prepare a few demonstrations where new phases are formed, such as dissolving fruit punch powder and pouring vinegar over baking soda; ask children to explain how the new phases appear from the old ones.)

12. Formulate an acceptable modern definition for "chemical element." You may consult any references you like. Classify the definition as operational or formal and give your reasons.

13. Compare the atomic model for matter with the MIP model described in Section 4.5.

14. Apply the MIP model to illuminate two phenomena from everyday life.

15. Compare the five general applications of the MIP model described in the text and rank them in order from the one that is most meaningful to you to the one that is least meaningful. Explain briefly.

16. Give two examples from everyday life of systems in (approximate) equilibrium with their environment.

17. Give two examples from everyday life of systems in (approximate) steady states.

18. Use the MIP model to construct a micro-domain description of phase equilibrium (e.g., the liquid water, air, and gaseous water in the flask in Fig. 411c).

19. Find an analogue model that clarifies the distinction between equilibrium and steady states. (See the water tank analogue, Figs. 4.12 and 4.13, as an example.)

20. Describe applications of the steady-state concept to two or three phenomena in the social sciences.

21. Describe applications of the equilibrium concept to two or three phenomena in the social sciences.

22. Construct a feedback loop model for:
    (a) the temperature of a refrigerator interior;
    (b) the light intensity reaching the retina of your eyes;
    (c) an example of your choice.
    In each case, identify the system and steady state, describe the feedback mechanism, and explain whether the feedback is positive or negative.

23. Write a critique of one of the feedback loop examples described in the text. Point out the limitations and incorrect conclusions that are implied.

24. The point is made in the text (Fig. 4.17) that public schools could provide positive feedback for changes in the socioeconomic population spectrum of a community.
(a) Apply this model to the conversion of a village into a suburb.
(b) Apply this model to an example that reveals its limitations (i.e. one for which it does not make the correct prediction).

25. Suppose you are an owner of a small store that has a certain annual income from sales as well as expenses such as rent, employees' salaries, purchase of merchandise, and taxes. Whatever is left from the income after expenses is your profit. As the owner, you might think of the percentage of the income that ends up as profit as the efficiency (or profit ratio) of your business. The income is all accounted for; none of the money disappears or is lost, but you consider part of it (the profit) as especially important, and you may want to consider how to increase the efficiency so as to maximize your profit. Compare this situation with the process of energy transfer; explain the similarities with the law of conservation of energy and with the efficiency of energy transfer.

26. Consider the process of energy transfer in an automobile: the original energy is represented by the chemical energy of the fuel; the engine transforms this energy into heat and kinetic energy and thus enables the vehicle to move from place to place. Explain how the law of conservation of energy and the concept of efficiency apply to this process. How would you expect the efficiency to be related to the number of miles per gallon of fuel? How would you expect the efficiency of an SUV and of a racecar to compare with that of a conventional automobile? What might a car designer do to increase the efficiency of a vehicle? How would you expect the efficiency to be related to the expense of running a vehicle? How would efficiency be related to the distance you could travel on a gallon of fuel? How would efficiency be related to the amount of pollution generated by the car? If you wish, find out the efficiency and miles per gallon of various types of vehicles.

25 Identify one or more explanations or discussions in this chapter that you find inadequate. Describe the general reasons for your judgment (conclusions contradict your ideas, steps in the reasoning have been omitted, words or phrases are meaningless, equations are hard to follow, . . .), and make your criticism as specific as you can.

Bibliography

S. W. Angrist and L. G. Hepler, *Order and Chaos, Laws of Energy and Entropy*, Basic Books, New York, 1967.

A. Eddington, *The Nature of the Physical World*, University of Michigan, Ann Arbor, Michigan, 1958.

A. Holden and P. Singer, *Crystals and Crystal Growing*, Doubleday (Anchor), Garden City, New York, 1960.

H. Roscoe, *John Dalton and the Rise of Modern Chemistry*, Macmillan, New York, 1895.

J. Rose, *Automation, Its Anatomy and Physiology*, Oliver and Boyd, Edinburgh and London, 1968. Chapter 2 explores the applications of the feedback concept in automation.

W. Weaver, *Lady Luck: The Theory of Probability*, Doubleday, Garden City, New York, 1963.

N. Wiener, *The Human Use of Human Beings: Cybernetics and Society*, Doubleday, Garden City, New York, 1954.

Articles from Scientific American*.* Some or all of these, plus many others, can be obtained on the Internet at http://www.sciamarchive.org/.
S. W. Angrist, "Perpetual Motion Machines" (January 1968).

A. J. Ayer, "Chance" (October 1965).

A. H. Cottrell, "The Nature of Metals" (September 1967).

K. Davis, "Population" (September 1963).

E. S. Deevey, Jr., "The Human Population" (September 1960).

G. Feinberg, "Ordinary Matter" (May 1967).

D. Fender, "Control Mechanisms of the Eye" (July 1964).

M. Gardner, "Can Time Go Backward?" (January 1967).

N. Mott, "The Solid State" (September 1967).

E. J. Opik, "Climate and the Changing Sun" (June 1958).

C. S. Smith, "Materials" (September 1967).

R. Stone, "Mathematics in the Social Sciences" (September 1964).

Part two: waves and atoms

Models for light and sound

The nature of light and sound has intrigued mankind over the centuries. Light and sound are connected with sight and hearing and are therefore vital sources of information about our environment as well as essential for survival. The control of sound has led to spoken language and music. The use of light has led to written language and the visual arts. Hearing and seeing, sound and light, enable us to communicate with one another and to derive pleasure from the natural world as well as from music and art.

5.1 Properties of light and sound

Primary sources. We tend to associate light and sound because both are important for sense perception through interaction-at-a-distance. In this way sight and hearing are different from touch, taste, and smell: these three latter senses depend on physical contact of the sense organ and the material (solid, liquid, or gaseous) to be sensed. Both light and sound originate in so-called primary sources: a candle flame, a lightning flash, and the sun emit light; a bowed or plucked violin string, a thunderclap, and a croaking bullfrog produce sound. Both light and sound interact with detector systems: light with the human eye, photographic film, or a video camera; sound with the human ear or a microphone. Both light and sound, therefore, function as intermediaries in interaction-at-a-distance.

Information and energy. Sound and light transmit not only information but also significant amounts of energy. The energy the earth receives from the sun maintains the earth's temperate climate and makes possible the photosynthesis of food material by green plants. The sound blast from dynamite explosions (or Joshua's trumpets at Jericho) breaks windows and may even topple buildings. Even though sound and light always transmit some energy, the amounts involved in seeing and hearing are very small. Therefore it is worthwhile to distinguish between situations in which the transmitted information is more important (such as in sense perception) and those in which the energy transferred is of greater interest (such as in photosynthesis, sunlamps, and dynamite blasts).

Reflection. In describing the similarity of sound and light, we mentioned the existence of primary sources for each. Even though you recognize these primary sources, you are also aware of the reflection of sound and light by all the objects in your environment. As a result of this reflection process, you receive light and sound from all directions and thus appear to be bathed in sound and illumination. Reflecting objects may be called "secondary sources" because they do serve as sources, but their action depends on that of the primary sources.

In other words, the primary sources of light and sound lose energy, but the reflecting objects do not. Rather, the reflecting objects often absorb some of the energy that strikes them and reflect only a part of

111

it. The difference in level of illumination in a room with light-colored walls and in one with dark-colored walls is due to the poor reflection, and greater absorption, of light by the dark walls. A similar effect with respect to sound is achieved by adding rugs and drapes to a bare room. The rugs and drapes are poor sound reflectors (and better absorbers) compared to the bare walls, which reflect most of the sound energy that strikes them.

Sound sources. Sound is generated by vibrating or suddenly moving objects, such as a plucked violin string, a bursting balloon, or a jet engine in operation. Sound travels in air and is blown aside by the wind, but it also travels along a stretched string, through water, and along steel rails. Sound does not propagate in the absence of matter. You can recreate a speaking tube, like the one used by a ship's captain to communicate with the engine room, with the help of an air-filled garden hose. You can also put your ear against a table and listen to the sound produced when you scrape the table surface with your fingernail. Sound is transmitted well by almost all materials, but it is not transmitted well from one material to a very dissimilar material, as from a tuning fork to air.

Loudness and pitch. A sound signal has certain recognizable properties, such as its loudness and, in the case of musical notes, its pitch. The loudness of a sound is related to the energy being transferred from the source to the air and then to the ear. You can reduce the loudness of a sound by weakening the primary source, increasing the distance from the source, or placing obstacles near the source. The pitch of a note is not affected by any of these procedures, but it can be altered by a process called "tuning" of the primary source. A violinist, for instance, tunes his instrument by tightening or loosening the strings. A complex sound, which is a mixture of notes, can be changed by nearby reflecting objects because such objects are usually more efficient reflectors of low notes than of high ones. The sound therefore becomes muffled. You are probably familiar with the effect of a long pipe on the sound of the human voice as well as the fact that the sound of a speaker system depends as much on the room surroundings as on the speakers themselves.

Light sources. Light is usually generated by extremely hot objects, such as a candle flame, an incandescent light bulb filament, or the sun, but it may also come from a fluorescent tube or a firefly. Light travels through air, transparent liquids (like water), and transparent solid materials (like glass), but it also travels through interstellar space, where there is no appreciable amount of material present (vacuum). Most solid objects are not transparent and do not transmit light. They may be white or light-colored and thereby act as efficient reflectors. Photographers have to be especially alert to the color of objects surrounding their photographic subject, because these objects act as "secondary light sources" and influence the exposure conditions. Snow or beach sand may result in overexposure, and dark foliage in underexposure, unless precautions are taken.

Can you find evidence that most of the sound is transmitted by the air column inside the hose and not by the walls of the hose?

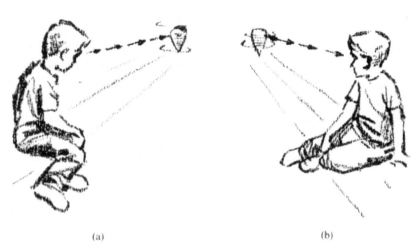

Figure 5.1 Two models for vision.
(a) Child's model that eyes reach out.
(b) Scientific model that light reflected by the object reaches the eye.

(a)   (b)

Properties of a light signal are intensity and color. As with sound, the intensity is related to the energy being transmitted. You can decrease the perceived intensity of light by weakening the source, going to a greater distance from it, or interposing obstacles. The color depends on the nature of the light of the primary light source, and it can be modified by interposing color filters (green, red, or yellow transparent plastic sheets), or by the presence of strongly colored reflectors whose hue influences the general illumination.

Images. For purposes of sense perception, the most important difference between light and sound is that your eyes give you an incredibly detailed picture (image) of the position, shape, color, and so on, of the objects that reflect light (secondary light sources) in your surroundings, whereas your ears mainly give you information about the primary sound sources. Related to this circumstance is the fact that you possess a primary sound source (your vocal cords) to enable you to communicate with others, but your body is only a reflector and not a source of light. Therefore, your body can be seen only if there is an external source of light and is invisible in the dark. Even when you do control a primary light source, such as automobile headlights, you use it to illuminate and enable you to see *other* objects that reflect light. In other words, you rarely look directly at a primary light source, and you rarely pay attention to secondary sound sources.

Misconceptions about vision. This difference between sound and light, which results in your attending mainly to primary sound sources and secondary light sources, gives rise to a curious misconception among many children. They believe that their eyes (or something from them) "reaches out" to the objects they see, much as their hands reach out to touch objects they wish to feel (Fig. 5.1). In fact, infants must both see and touch objects to develop their visual perception and hand-eye coordination. All of us at one time or other have probably had the feeling that our eyes "reach out." The control we have over selecting the objects we look at and the great detail we can perceive when we look closely, plus the apparent similarity with the way we use our hands, seem to create a strong, but false, impression that the eye is an "active" organ rather than a passive receiving system like the ear. Can you cite any specific evidence that shows that the eye simply receives and senses light, rather than "reaching out"? How might you go about convincing a child who claims his eyes "reach out"?

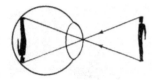

Image on the retina. The sharp visual images are created by the composite effect of many light signals striking the retina of the eye (as shown to the left). Light signals can create an image of a reflecting object because adjacent parts of the object register on adjacent portions of the retina. The retinal image is therefore usually a reliable indicator of the shape of the object. An exception to this rule occurs when the light on its way to your eye passes through a rain-covered automobile windshield or the hot exhaust of a jet engine; then the object appears blurry and you do not register a sharp image.

Ray model for light. Our ability to have visual images, the obstruction of images by objects intervening between object and eye, the silhouette-like shadows cast by objects in bright sunlight, and the projection of film pictures on a screen, all suggest that light usually travels in straight lines. In the ray model, light signals are made up of light rays that travel in straight lines and obey simple geometrical rules as they propagate (move) from source to detector. In the next two sections we will describe details of the ray model and how it may help in the analysis or construction of optical instruments using lenses (eyeglasses, projectors, cameras, microscopes, and telescopes). The ray model, which explains the formation of images and shadows very directly, does not directly explain the existence of a background level of illumination on a cloudy day, when the sun is concealed and there are no sharp shadows. This problem, as well as the problem of the color of light, can nevertheless be solved within the framework of the ray model.

Wave model for sound. It is quite clear that a ray model is not adequate for describing sound because obstacles placed between the source and receiver do not block a sound signal in the same way as they block a light signal. Sound does not appear to travel in straight lines. Instead, it appears to diffuse through space around obstacles much like water waves, which radiate out from a dropped pebble, pass on both sides of an obstacle, and then close again behind it. Sound (and water waves) rarely give rise to a distinct "shadow," that is, a region behind an obstacle from which the effect of the source is completely blocked. This analogy has given rise to the wave model for sound. We will introduce the concept of wave motion in Chapter 6 and apply it to sound in Chapter 7. There you will see how the pitch of a note and the construction of musical instruments are explained by a wave model.

Combination of two sound or light signals. So far we have been describing generally well-known properties of sound and light. We will now describe what happens when light or sound from two sources is combined. The resultant effects of such combinations are surprising

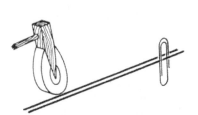

Figure 5.2 Low notes on the piano have two strings. Change the tuning of one string by clipping a paper clip on it. Then hit the key and listen for beats. To compare the sound of two strings with the sound of one, stop the vibration of one string with your finger.

and the implications derived from them, though not widely known, are decisive in formulating working models to explain the phenomena.

Beats. The first phenomenon is the beats that may be heard from a poorly tuned piano. When a single note with two or three strings is struck, there is a rhythmic pulsation of the sound level (so-called beats) even though the tone does not change (Fig. 5.2). To produce beats, it is necessary to have two primary sound sources with pitches that are almost the same. The closer the pitches of the two notes, the slower the beats. The detection of beats is of great importance to the piano tuner, who adjusts the tension of the strings until no beats are heard.

The scientific significance of beats is that there are instants when the combined sound produced by the two sources together (strange as it may seem) is *weaker* than the sound produced by either one alone. This is a very surprising observation, since you would ordinarily expect two similar sources to produce about twice the sound intensity of one alone. You are forced to conclude that the two sources somehow act "in opposition" to one another at the time when the combined sound intensity is very low. To be acceptable, the wave model for sound must therefore make possible this "opposition" as well as the expected reinforcing of two sound sources.

Interference. The second phenomenon involves light. To observe it, look through a handkerchief, a fabric umbrella, or window curtains at a bright, small source of light such as a distant street lamp at night (Fig. 5.3). You will see the light source broken into a regular array of

Figure 5.3 Look through a hand-kerchief at a bright but distant light. Identify the interference pattern

bright spots. The array of spots will be much larger than the source seen directly, and cannot be explained as arising because the light source is seen through the holes in the fabric. The array of spots may even have touches of color. It is called an *interference pattern*.

An interference pattern is visible whenever one small primary light source is viewed through an obstacle, such as a piece of fabric that has many holes or slits in a regular pattern. Each of the slits functions as a separate "source" and the light passing through each slit "interferes" with the light from all the other slits to produce the observed effect. If you compare the interference patterns caused by coarse and fine fabrics, you will find that the fine fabric creates a larger-scale pattern than the coarse, exactly the opposite from what would happen if you saw the source through the holes in the fabric. With very coarse fabrics (1-millimeter spacing or more) the pattern is too closely spaced to be seen. You can also see rainbow-colored interference effects when you look along a compact disc (CD) or record and let the grooves catch the light, or when you look at thin layers of material in a bright light, such as soap bubbles, mother-of-pearl, or an oil film on water, all of which appear iridescent in sunlight.

5.2 The ray model for light

Many common conditions create the appearance that light rays are visible objects. A powerful searchlight at night is the source of a beam that stabs the sky. A crack in a shutter or curtain admits a shaft of light into a darkened room. Gaps between clouds on a hazy day or trees in mist allow the sun to create streaks of illumination (Fig. 5.4). Actually, light is registered by your eyes only if it strikes the retina. What you see in these examples, therefore, are the massed dust particles or water particles, which reflect the light striking them to your eyes. The form of a beam, shaft, or streak is created by the pattern of the incident light, which is restricted to a slender region of space. The most impressive property of this region of space is that it is straight. Hence the concept of a light ray that travels in straight lines.

Figure 5.4
Streaks of sunlight at dawn.

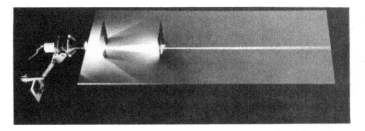

Figure 5.5 A beam and a pencil of light are formed by the small holes in the two screens illuminated from the left.

Figure 5.6 A laser producing a pencil of light.

The phrases "light beam," "pencil of light," and "shaft of light" refer to light traveling in a particular direction, usually directly from a primary source. The distinction among these phrases is a subjective one, with "light beam" referring to the widest lit area and "pencil of light" referring to the narrowest. Their meaning is to be compared with that of "illumination," which refers to light traversing a region of space from all directions, usually coming from reflecting surfaces, so that there are no dark shadows and all objects in the space are visible.

In the ray model, a light beam is represented as a bundle of infinitesimally thin light rays. The width and shape of the bundle determines the area and shape of the beam. The propagation of each ray determines the behavior of the entire beam. In other words, a ray of light is to a beam just as a particle in the MIP model is to a piece of matter (Section 4.5). The function of the ray model is to explain the observations made on light in terms of the assumed properties of light rays. The ray model does not attempt to explain the assumed properties of rays, which are justified (or undermined) by the successes (or failures) of the model.

Properties of light beams. As is true of all models, the observations to be explained by the ray model are built into its assumptions in a simplified and/or generalized form. We therefore begin with the observation of light beams, which the model will have to explain.

Light beams are usually produced by letting light from a powerful source impinge on a screen in which there is a small hole (Fig. 5.5). The light passing through the hole forms the beam. A modern device, the laser (Fig. 5.6), is a primary source that produces a pencil of light.

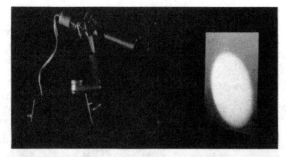

Figure 5.7 The light source and the light spot on the screen can be seen, but not the light beam between them.

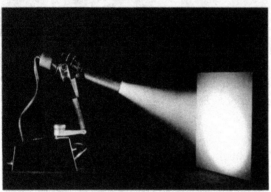

Figure 5.8 Fine smoke particles in the path of the light beam reflect the light and make the beam visible.

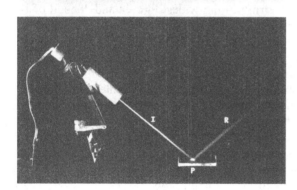

Figure 5.9 A beam of light is reflected from a polished metal mirror (P). Note the incident beam (I) and the reflected beam (R). Why is the mirror mostly dark? Why is the reflected beam dimmer than the incident beam?

If you try this experiment with a flashlight, you wouldn't expect to be able to see the paths of the two light beams as in this photograph. Why not? How might you make the beams visible?

A beam of light cannot be seen from the side, because our eye detects light only when the light strikes our retina (Fig. 5.7). To make the beam "visible," chalk, smoke, or dust particles must be introduced into the region of space (in air or in water) traversed by the beam. The illuminated particles act as secondary light sources and can be seen by reflection, while the un-illuminated particles outside the beam remain dark and unseen (Fig. 5.8). Another technique for tracing a light beam's path is to let it strike a white sheet at a glancing angle (Fig. 5.5).

Three properties of light can be easily observed with beams rendered visible by one of these procedures. 1) A light beam travels in straight lines. 2) A light beam is *reflected* by a polished surface or mirror (Fig. 5.9), and 3) A light beam is partially reflected and partially *refracted*, when it

TABLE 5.1 INDEX OF REFRACTION FOR TRANSPARENT MATERIALS

Material	Index of refraction
air	1.00
water	1.33
glass	1.5
diamond	2.42
ethyl alcohol	1.36

***Equation 5.1
(Law of reflection)***

angle of
  reflection $= \theta_R$
angle of
  incidence $= \theta_i$
*(Figure 5.11, below,
shows θ_R and θ_i.)*

$\theta_R = \theta_i$

***Equation 5.2
(Snell's law of refraction)***

angle of
  refraction $= \theta_r$
angle of
  incidence $= \theta_i$
*(Figure 5.11, below,
shows θ_r and θ_i.)*
index of
  refraction in
  material of
  refraction $= n_r$
index of
  refraction in
  material of
  incidence $= n_i$

n_r sine $\theta_r = n_i$ sine θ_i

*See Appendix for defini-
tion (Eq. A.6) and values
(Table A.7) of sine θ.*

crosses the boundary between two transparent materials such as air and glass or air and water (Fig. 5.10). Simple mathematical models have been found to fit the observations on the relationships of the angles of reflection, refraction, and incidence (Fig. 5.11, Eqs. 5.1 and 5.2). The symbol "n" in Eq. 5.2 stands for the index of refraction, a number that is determined experimentally for each transparent material (Table 5.1 and Example 5.1).

Figure 5.10 A beam of light enters a semicircular slab of glass at various angles. Note the incident beam, reflected beam, and refracted beam.

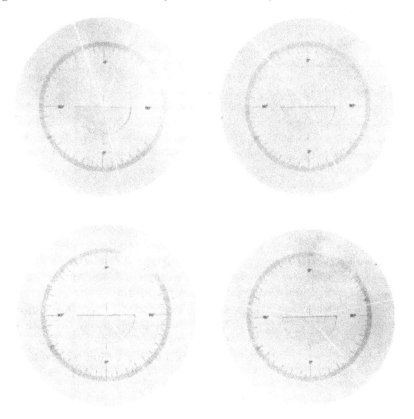

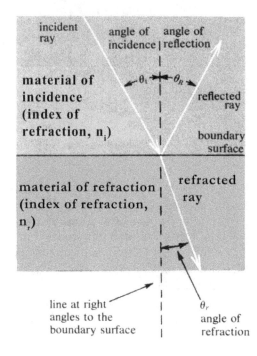

incident ray

angle of incidence | angle of reflection

θ_i | θ_R

material of incidence (index of refraction, n_i)

reflected ray

boundary surface

material of refraction (index of refraction, n_r)

refracted ray

line at right angles to the boundary surface

θ_r
angle of refraction

Figure 5.11 The angles of incidence (θ_i), reflection (θ_R), and refraction (θ_r) are defined in relation to the line at right angles to the boundary surface between two media.

Willebrord Snell (1591-1626). Snell's law is an example of an empirical law that summarizes data but does not rest on a theoretical foundation.

Claudius Ptolemy (approx. 140 A. D.) is much better known for his work in astronomy (see Chapter 15).

Equation 5-3 (Ptolemy's Law of Refraction)
$n_r\theta_r = n_i\theta_i$
(less adequate than Snell's Law of Refraction)

EXAMPLE 5-1. Applications of Snell's law.

(a) Light incident on glass: $\theta_i = 40°$, $n_i = 1.00$, $n_r = 1.50$

$$\text{sine } \theta_r = \frac{n_i}{n_r} \text{ sine } \theta_i = \frac{1.00}{1.50} \text{ sine } 40° = 0.67 \times 0.64 = 0.43$$
$$\theta_r = 25°$$

(b) Light incident on diamond: $\theta_i = 60°$, $n_i = 1.00$, $n_r = 2.42$

$$\text{sine } \theta_r = \frac{n_i}{n_r} \text{ sine } \theta_i = \frac{1.00}{2.42} \text{ sine } 60° = 0.41 \times 0.87 = 0.36$$
$$\theta_r = 21°$$

Refraction. Even though the index of refraction of each material must be found, Eq. *5.2* is a useful mathematical model because it uses only one empirical datum (the index of refraction) and yet predicts an angle of refraction for each angle of incidence. This mathematical model is called *Snell's law,* and was formulated by Snell early in the seventeenth century on the basis of experiments conducted with air, water, and glass.

Long before Snell, Ptolemy had tabulated and proposed a mathematical model (Eq. 5.3) for refraction, but the Arabian investigator Alhazen (965-1038) pointed out the inadequacy of Ptolemy's model. The two models are represented graphically in Fig. 5-12, where their similarity for small angles can be recognized. Johannes Kepler (1571-1630), better known for

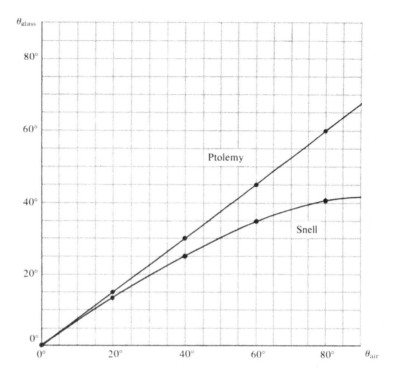

Figure 5.12 *Refraction of light passing between air and glass (n_{glass} = 1.5). Ptolemy's model (Eq. 5.3) is compared with Snell's model (Eq. 5.2). For experimental results, see Fig. 5.10. The graph applies for light passing from air into glass or from glass into air. Note that θ_{glass} is always less than θ_{air}.*

"I procured me a triangular glass prism, to try therewith the celebrated phenomena of colours... having darkened my chamber, and made a small hole in my window-shuts, to let in a convenient quantity of the sun's light, I placed my prism at its entrance... It was ... a very pleasing divertissement, to view the vivid and intense colours ... I [thereafter] with admiration beheld that all the colours of the prism being made to converge ... reproduced light, entirely and perfectly white, and not at all sensibly differing from the direct light of the sun...."

Isaac Newton
Philosophical Transactions,
1672

his planetary models for the solar system (see Section 15.2), made measurements of refraction, but was not able to construct a mathematical model better than Ptolemy's (Eq. 5.3). Long after Snell's time, refraction was a key to the acceptance of new models that replaced the ray model for light (Section 7.2).

White and colored light. Isaac Newton achieved a breakthrough in the understanding of light. He found (as others had before) that a glass prism refracted a pencil of light in such a way that a rainbow-colored streak appeared (Fig. 5.13). Newton theorised that the white light was composed of a mixture of various colors. He tested this idea by using a second prism to refract the colored streak back toward its original direction of propagation (Fig. 5.14). The original white light was restored, confirming his theory! When Newton used a screen to isolate only one color in his streak to impinge on the second prism, he found that further refraction of this one color did not alter the color of the light (Fig. 5.15). These findings led Newton to elaborate on the ray model generally accepted in his time to include an explanation of color.

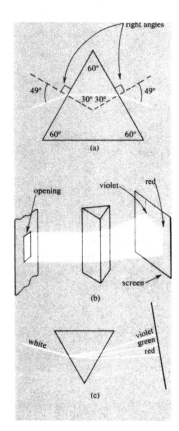

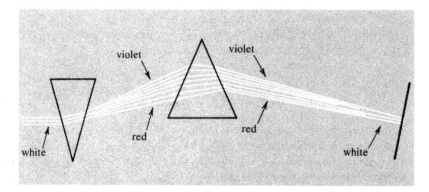

Figure 5.13 (to left) Refraction of light by a glass prism (index of refraction, n = 1.5).
(a) Refraction of a ray of light according to Snell's Law.
(b) and (c) Refraction of a beam of white light. Violet light is bent (or refracted) more than red light.

Figure 5.14 Refraction of a pencil of white light in opposite directions, by two prisms. The colors recombine to make white light. This experiment was first carried out and reported by Isaac Newton.

Figure 5.15 Refraction of light of a single color does not change the color. The color is selected from the spectrum by an opening at the appropriate place in the opaque shield.
(a) Red light selected for passage through second prism.
(b) Yellow light selected for passage through two further prisms.

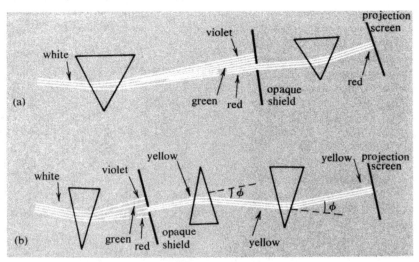

Assumptions of the ray model. Isaac Newton gathered the assumptions of the ray model in his *Opticks, or a Treatise on Reflections, Refractions, Inflections , and Colours of Light*. In his version of the ray model, the elemental light rays were colored (monochromatic = single color), and their color was an intrinsic, unchangeable property. Visual impressions of color were produced by monochromatic rays or by combinations of monochromatic rays, as we will explain below.

The assumptions of the ray model as adapted for this text are summarized in Fig. 5.16 to the right. It is clear that the assumptions were selected so as to be consistent with the observation on light beams described in the previous paragraphs. The task of the model is to explain, in addition, more complicated phenomena, such as the formation of images, the concept of illumination, why not all surfaces act as mirrors, and how colors are produced in mixtures of paint pigments. In all these situations, light is not a single bundle of rays, but a combination of diverging, converging, and crossing rays of various colors and directions of propagation. Assumptions 6 and 7 enable us to apply the model in situations where many rays, following different paths, combine to form an *image* (Section 5.1 and Figure 5.17). Their essence is that only rays that begin by diverging from one point on an object can later converge to form an image of the object point where they started.

Color. The ray model provides three techniques for the production of colored light. One of these is by selective transmission through a color filter. The color filter absorbs rays of some colors and transmits rays of other colors. The colored beam obtained in this way may happen to be monochromatic in Newton's sense, or it may be composed of several monochromatic rays. The second technique is selective reflection by an opaque object. The surface of the object absorbs rays of some color and reflects rays of other colors. The third technique is refraction, through a glass prism or other suitably

Assumption 1. Each point in a primary light source emits rays which diverge in all directions from that point.

Assumption 2. Light rays in a uniform medium travel in straight lines.

Assumption 3. Each point in an object may transmit, absorb, or reflect the rays incident upon it.

Assumption 4. Light rays are reflected, with the angle of incidence equal to the angle of reflection, $\theta_i = \theta_R$.

Assumption 5. Light rays are refracted, with the angles of incidence and refraction of monochromatic rays related by the formula $n_r \sin \theta_r = n_i \sin \theta_i$. The index of refraction may differ for different colored rays (dispersion).

Assumption 6. Whenever rays that diverge from one point in an object meet again at another point on a white screen as the result of reflection and refraction, they make an image of the point in the object.

Assumption 7. An object seen directly, by reflection, or by refraction, appears to be in that place from which the rays diverge as they fall into the observer's eye.

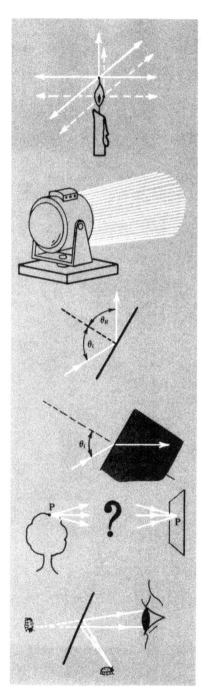

Figure 5.16 Assumptions of the ray model for light.

*Figure 5.17 (to left) Specular reflection from a mirror. At first glance this photo may seem unremarkable. However, notice that the mirror is dark and the screen behind it is bright. In addition, both the candle and its image are bright. Finally, notice the light source at the right side: why don't you see it in the mirror? If you can make a ray diagram that shows clearly what is going on here, you understand the ray model very well. (This is **not** in any way a trick photograph; no hidden lenses, mirrors or other apparatus were used.)*

shaped object, such as described earlier (Fig. 5.13). For example, the colors of the rainbow are produced by refraction of light in rainwater droplets. Only refraction is sure to give rise to monochromatic light. Selective transmission and reflection may or may not, depending on the materials in the filter or reflector.

Addition of colors. The ray model can explain the modification of colored light by addition or subtraction of colors. Color addition occurs when several colored beams illuminate the same screen. Where they overlap, the screen acts as a reflecting surface of all the incident colors. This is the process by which Newton obtained white light from the colored streak (Fig. 5.14); after refraction through the second prism, the various colored rays were brought to overlap on the screen.

Subtraction of colors. Color subtraction occurs when filters are inserted into the path of a beam of light, or pigments are mixed in paint.

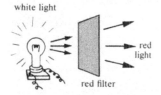

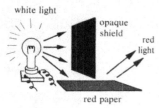

selective reflection

Figure 5.18 Diffuse reflection.
(a) A pencil of light strikes a dull white surface. Note the diffuse reflected light near surface.
(b) Working model for a dull surface that shows diffuse reflection of light.

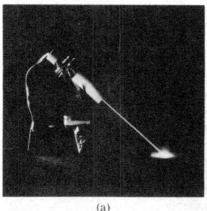

(a)

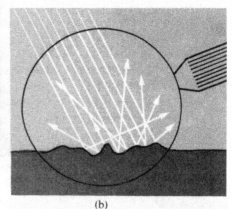

(b)

The light absorption in the filter removes some colors while others are transmitted. The insertion of successive filters results in the subtraction of more and more colors, until none (that is, no light at all) may be left. Inferences about the absorption by color filters are indirect, because the observer sees only the transmitted light and not what is absorbed (or reflected). To find out what a filter absorbs, you have to illuminate it with monochromatic light and determine whether or not any light at all is transmitted. All the monochromatic light that is not transmitted is absorbed or reflected.

Interference colors. There is a fourth technique that produces colored light: interference. This was mentioned briefly at the end of Section 5.1. It is an observation that cannot be explained by the ray model, since the experiment uses no refraction or selective absorption. Curiously enough, Newton carried out investigations of the color of thin films (air between glass plates, soap bubbles) and tried to adapt the ray model by including multiple reflections and refractions back and forth between the two surfaces of the film. This attempt, however, while suggestive, was not successful and the observations remained unexplained for more than 100 years (Section 7.2).

Models for reflecting surfaces. We have already explained how the color of surfaces arises from selective reflection and absorption of monochromatic light rays. Still to be discussed is the difference between mirrors on the one hand and dull surfaces that reflect light but do not reflect images on the other hand.

Mirror surfaces. A mirror is a surface that maintains the divergence or convergence of rays incident upon it. When you look into a mirror, therefore, you do not see the mirror itself. Rather, you see the light rays reflected from the mirror to your eye, or, more precisely, you see the apparent *sources* of those reflected light rays. (Fig. 5.17). A mirror has a smoothly polished surface to ensure the orderly reflection of all incident rays. This is an example of *specular reflection.* It is well known that fingerprints on a mirror interfere with the specular reflection.

Dull surfaces. A model for reflecting surfaces that are not mirrors must be designed to destroy the divergence or convergence of rays incident upon it. Accordingly, the model surface is highly irregular, with minute irregularities. As light rays strike various points of the surface, they are absorbed or reflected according to Assumption 6. Since the surface is irregular, however, the reflected rays diverge in all directions from a small area on the surface and do not propagate in a direction simply related to the placement of the primary source (Fig. 5.18). This phenomenon is called *diffuse reflection.* According to Assumption 7, an observer will see the surface from which the rays diverge after reflection.

Diffuse reflection, selective reflection and the straight-line propagation of light from primary source to reflecting surface and to the eye account for most of the everyday properties of light and vision. The phenomenon

---

Experiment A. Observations of light absorption by single-color filters

red filter absorbs blue, green, and yellow

yellow filter absorbs blue

green filter absorbs red

blue filter absorbs red and yellow

Experiment B. Observations of light transmission by combinations of single-color filters

red and yellow filters: red light transmitted

red and green filters: no light transmitted

yellow and green filters: yellow and green light transmitted

yellow and blue filters: green light transmitted

On the basis of the ray model and the information in Experiment A above, can you predict the results summarized in Experiment B?

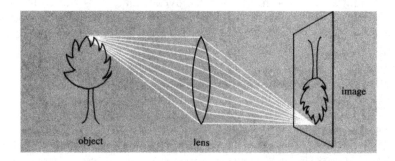

Figure 5.19.
The convex glass lens in air pro-
duces an image.

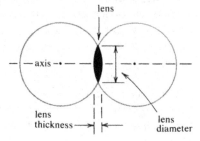

Figure 5.20 A thin spherical lens
has the shape of the overlapping
region of two slightly interpene-
trating spheres.

of refraction enters mainly through the widespread use of lenses in eyeglasses and optical instruments. The next section contains an introduction to the theory of lenses, as derived from the ray model for light.

5.3 Application of the ray model to lenses

Binoculars, telescopes, cameras, and projectors are so common today that it is difficult to believe that none of them existed 400 years ago (and most of them not even 150 years ago). Even eyeglasses were invented only in the fourteenth century, and their use spread very slowly. Until about the time of Galileo the world must have been a blur to

Figure 5.21 Two rays that diverge from an object are refracted by the lens to converge at one point of the image. One ray passes along the axis.

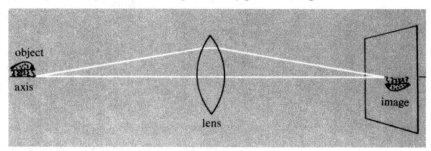

In contrast to a lens, a sheet of glass with parallel faces (see diagram below) refracts light rays in such a way that their convergence or divergence is not altered. Snell's law applied at the two points labeled B leads to refraction which exactly compensates the refraction at the two points labeled A. Each ray emerges in a direction parallel to its incident direction (see Problem 17).

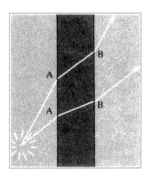

*The concept of focal **point** applies to all lenses, not only thin ones. The focal **length**, however, is definable only for __thin__ lenses, which do not extend far from the lens center on either side.*

quite a few people, considering how many are now wearing glasses. The principles of the ray model occupied scientists in a great flurry of discovery during the sixteenth and seventeenth centuries and led to the development of optical instruments.

The converging lens. A convex glass lens in air has the property of taking rays that diverge from one point on a source object and refracting them so they converge to a point on the other side of the lens (Fig. 5.19). This point is called an *image* of the source point for the following reason. Once the rays have passed through the image point, they diverge again, just as though they had been emitted by a source at the image point. To an observer, therefore, the rays diverging from the image point appear to come from an object located at the image point.

The vast majority of lenses are spherical; that is, their two surfaces are sections of two spheres (Fig. 5.20). The axis of the lens is the line that joins the centers of those spheres. We will restrict our discussion to lenses that are thin compared to their diameter.

Image formation. Consider two rays emitted from a small object on the lens axis, one traveling along the axis of the lens, the other at an angle to the axis, striking the lens at an off axis point (Fig. 5.21). The ray on the axis goes through the lens undeviated (Snell's law, Fig. 5.12, Eq. 5.2) because it strikes the glass at right angles to its surface. As the oblique ray strikes the first surface and enters the glass, it is refracted toward the right angle direction (Fig. 5.11). Upon emerging from the second surface into air, it is refracted again. Both times refraction bends the ray toward the axis, similar to refraction by a prism (Fig. 5.13). Because the two rays in Fig. 5.21 are *converging* to one point after passing through the lens even though they were diverging before, the lens is called a *converging lens*. The lens has focused the two rays to produce an image. Other rays from the same point of the object are also refracted to the same image point.

The basic problem to be solved in the operation of a converging lens is to find the position and size of the image, given the position and size of the source object and some properties of the lens. The problem is solved in the ray model by taking the rays diverging from each point on the source object and finding where they converge after passing through the lens.

Focal length. It turns out that the shape and index of refraction of the lens, which clearly must influence the image position, can be used to derive a property of a thin spherical lens called its *focal length*. The focal length is the distance from the center of the lens to the point called the *focal point* where all rays parallel to the axis converge (Fig. 5.22(a)). In addition, as shown in fig. 5.22 (b), rays diverging from the focal point are refracted by the lens so that they emerge parallel to the axis.

The lens formula. The lens problem is solved by the geometrical construction illustrated in Fig. 5.23. This construction is based on a thought experiment in which a source object to the left of the lens produces an image to the right of the lens. In the thought experiment, you consider

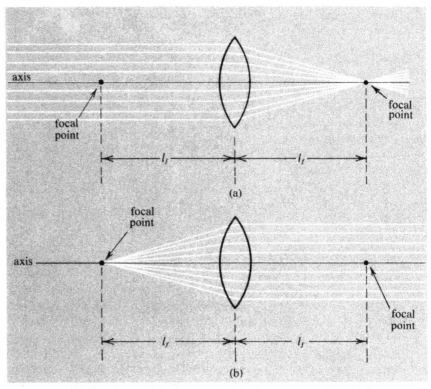

Figure 5·22 Focal length (l_f) of a convergent lens.
(a) Parallel rays entering a convergent lens converge at the focal point.
(b) Rays leaving the focal point are refracted so they emerge parallel to the axis.

Equation 5.4

object distance = l_o
image distance = l_i
focal length = l_f

$$\frac{1}{l_o} + \frac{1}{l_i} = \frac{1}{l_f}$$

Equation 5.5

image size = s_i
object size = s_o

$$\frac{s_i}{s_o} = \frac{l_i}{l_o}$$

three rays that diverge from the top of the source object and whose refraction by the lens is easy to predict. You trace those rays to find where they converge after passing through the lens. The point where they converge forms the "top" of the image. The result is a mathematical model called the lens formula relating the object distance, image distance, and focal length (Eq. 5.4). In words, the reciprocal of the object distance plus the reciprocal of the image distance equals the reciprocal of the focal length. The formula applies only to thin, spherical lenses because the concepts of focal length, image distance, and object distance have no consistent meaning otherwise.

Since Eq. 5.4 does not make reference to the distance between the axis and the source point at the top of the object, the lens formula predicts that light diverging from all other points of the object will converge to form the image of the object as indicated in Fig. 5.24. The size of the image can therefore be found from the same thought experiment, which is illustrated in a simplified form in Fig. 5.25. The result is a mathematical model relating object and image sizes to object and image distances (Eq. 5.5). The ratio of image size to object size is equal to the ratio of image distance to object distance. The properties of converging lenses are illustrated in Examples 5.2 and 5.3.

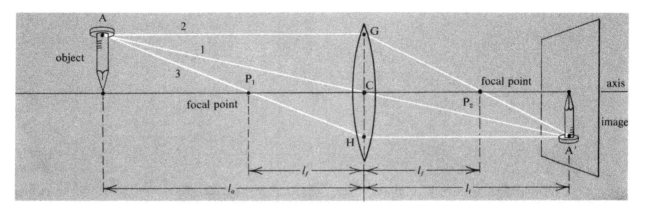

(a) A nail at a distance l_o from the converging lens produces an image at the distance l_i from the lens. To find the position of the image, three rays diverging from the top of the nail A are traced to their point of convergence A'. Ray 1 passes through the center of the lens C. It is not deflected because it traverses two glass-air surfaces that are parallel. Ray 2 propagates parallel to the lens axis and is refracted by the lens G to pass through the focal point P_2. Ray 3 propagates through focal point P_1 and is refracted by the lens at H to emerge parallel to the axis.

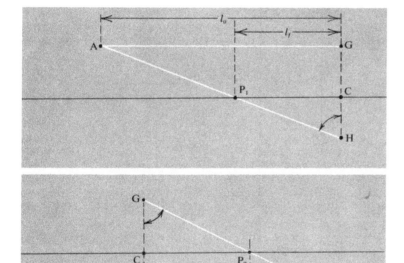

(b) The two right triangles P_1CH and AGH are similar, because they have a common acute angle at H. Corresponding sides P_1C (length l_f) and AG (length l_o) have the same ratio as sides CH and GH. Therefore,

$$\frac{l_f}{l_o} = \frac{\text{length of line } CH}{\text{length of line } GH} \quad (1)$$

(c) The two right triangles P_2CG and A'HG are similar, because they have a common acute angle at G. Corresponding sides P_2C (length l_f) and A'H (length l_i) have the same ratio as sides CG and GH.

$$\frac{l_f}{l_o} = \frac{\text{length of line } CG}{\text{length of line } GH} \quad (2)$$

Figure 5.23 Solution of the thin lens problem and derivation of Equation 5.4. The construction in parts (a), (b), and (c) leads to Eqs. (1) and (2). By adding Eqs. (1) and (2) and noting that segments CH and CG add up to GH, you obtain Eq. (3) below:

$$\frac{l_f}{l_o} + \frac{l_f}{l_i} = \frac{\overline{CH}}{\overline{GH}} + \frac{\overline{CG}}{\overline{GH}} = \frac{\overline{CH} + \overline{CG}}{\overline{GH}} = \frac{\overline{GH}}{\overline{GH}} = 1 \quad (3)$$

When both sides of Eq. (3) are divided by l_f, the result is Eq. 5.4.

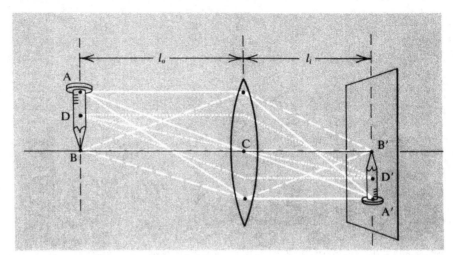

Figure 5·24   Rays diverging from A, B, and D converge at A', B', and D', respectively, to form the image.

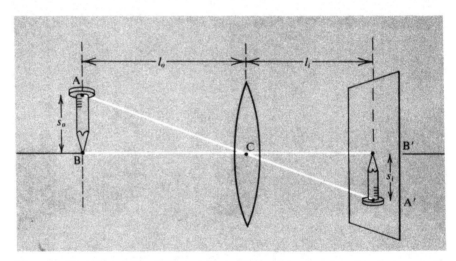

Figure 5·25   The ratio of image size to object size is related to image and object distances because triangles A'B'C and ABC are similar.

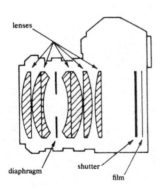

Typical 35 mm camera.

Applications. Based on this abbreviated discussion of a lens, we can describe several optical instruments.

Cameras. Let us consider a camera. A camera consists of a lens system of fixed focal length (usually about 50 millimeters) with some movement possible for focusing the image on the film. Since most objects are thousands of millimeters distant from the lens, Eq. 5.4 predicts that the image will be very close to the focal point, and will vary only slightly in position even though the object distance varies from 5 meters to 100 meters (see also Example 5.2). Movement of the lens of only a few millimeters in or out is needed to focus the image on the film unless the subject of the photograph is very close to the lens. The very fact that a camera lens has to move at all (as opposed to the more

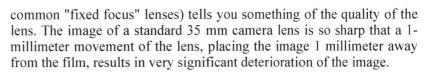

common "fixed focus" lenses) tells you something of the quality of the lens. The image of a standard 35 mm camera lens is so sharp that a 1-millimeter movement of the lens, placing the image 1 millimeter away from the film, results in very significant deterioration of the image.

EXAMPLE 5.2. A lens with focal length of 0.10 meter is used to focus the image of a light bulb on a piece of paper. The lens is 0.15 meter from the light bulb. Where do you have to hold the paper? How big is the image compared to the bulb itself?

$Solution: l_f = 0.10\,m,\ l_o = 0.15\,m, l_i = ?,\ \dfrac{s_i}{s_o} = ?$

$Equation\,5.4: \dfrac{1}{l_o} + \dfrac{1}{l_i} = \dfrac{1}{l_f},$

$or\ \dfrac{1}{l_i} = \dfrac{1}{l_f} - \dfrac{1}{l_o} = \dfrac{1}{0.10\,m} - \dfrac{1}{0.15\,m} = 10.0 - 6.7 = 3.3$

$thus: l_i = 0.30\,m$

$Equation\,5.5: \dfrac{s_i}{s_o} = \dfrac{l_i}{l_o} = \dfrac{0.30\,m}{0.15\,m} = 2.0$

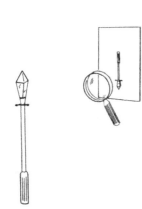

The paper must be held 0.30 meter from the lens. The image is two times as large as the bulb.

EXAMPLE 5.3. The same lens is used to focus the image of a street lamp 30 meters away on a piece of paper. Where do you have to hold the paper? How big is the image compared to the street light?

$Solution: l_f = 0.10\,m,\ l_o = 30\,m, l_i = ?,\ \dfrac{s_i}{s_o} = ?$

$Equation\,5.4: \dfrac{1}{l_o} + \dfrac{1}{l_i} = \dfrac{1}{l_f},$

$or\ \dfrac{1}{l_i} = \dfrac{1}{l_f} - \dfrac{1}{l_o} = \dfrac{1}{0.10\,m} - \dfrac{1}{30\,m} = 10.0 - 0.033 \approx 10$

$thus: l_i = 0.10\,m$

$Equation\,5.5: \dfrac{s_i}{s_o} = \dfrac{l_i}{l_o} = \dfrac{0.10\,m}{30\,m} = \dfrac{1}{300}$

The paper must be held at the focal point 0.10 meter from the lens. The image is 1/300 as large as the street lamp itself.

If the diameter of the camera lens is large, more rays from every source point on the object pass through the lens. That is, a large lens collects and focuses on the film more of the light energy emitted by the object than does a small one. For poorly lit objects, the size of the lens can therefore be of great significance to the photographer. He may want

A lens has a focal length of 50 mm and a diameter of 25 mm. Its f-number is f = 50 mm/25 mm = 2. Opening or closing the camera iris increases or decreases the effective f-number of the lens, since only the non-covered part is used.

to gather all the light from the object he possibly can to interact with the film. Lens size is given by the f-number, which is equal to the focal length divided by the diameter of the lens. Thus high f-numbers describe small lenses and low f-numbers describe large lenses. To make a lens of large diameter with good focusing properties is very difficult, thus accounting for the high cost of lenses with low f-number.

A telephoto lens enlarges the image of distant objects. It does so by having a longer focal length than an ordinary camera lens. The reason the focal length is so significant is that the image distance of a very distant object is almost equal to the focal length (Fig. 5.22). A longer focal length therefore leads to a larger image distance, and this in turn leads to a larger image size according to Eq. 5.5. Telephoto lenses are physically large and expensive. Powerful ones have a focal length of 200 millimeters or more, and therefore enlarge the image fourfold or more over the size obtainable with a regular lens (focal length about 50 mm.). Wide-angle lenses function in the opposite way. They are of short focal length, say 20 millimeters, and are used to make the image of all objects smaller so as to include a broader panorama on the film.

Magnifying instruments. The magnifying glass, in its usual application, does not project an image by converging the rays to a focus, but transmits divergent rays, which appear to come from a larger object (Fig. 5.26). Refracting telescopes and binoculars are a little more complicated, consisting of two lenses. Since the object is enormously distant with respect to the focal length, the image produced by the first

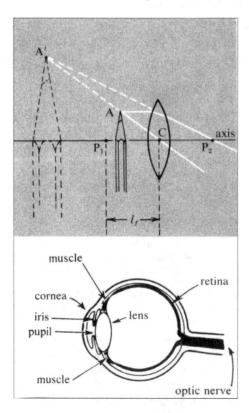

Figure 5.26 A magnifying glass (lens) is placed so near the object (A) to be examined that the latter is closer than the focal point P_1. The two rays traced from the point A are refracted by the lens and appear to diverge from the point A' behind the object. The mathematical model in Eqs. 5.4 and 5.5 can still be used. However, these equations will yield negative numbers for the image distance l_i and for the image size s_i. This means that the image is on the opposite side of the lens compared to the way they were defined in Fig. 5.23a. The distance from the eye to point A' has to be large enough so the eye can focus the incoming divergent rays on the retina.

Figure 5.27 Cross section of the eye.

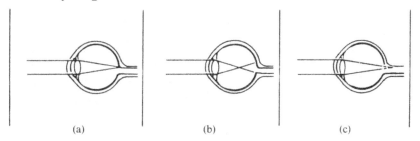

*Figure 5·28   Focusing of images on the eye's retina. (a) normal eye.
(b) nearsighted eye. (c) farsighted eye.*

lens is just slightly beyond its focal point. This image then acts as "object" just inside the focal point of the second lens, which functions like a magnifying glass.

The diameter of the outer (or objective) lens determines the light-gathering ability of a telescope or a pair of binoculars. This is very importance for astronomers who look at distant, faint stars. Binoculars are usually characterized by two dimensions, the magnifying power and the diameter of the outer (objective) lenses. Thus, a 7 x 35 rating means the image is magnified 7 times and the lens has a diameter of 35-millimeters. A larger magnification (such as 10) makes it easier to view small, distant objects; however, a larger magnification also means that the image jumps around more. For hand-held binoculars, most people find that a magnification of 7 or 8 is most practical. For greater magnifications, a tripod must be used.

The human eye. Human eyes function somewhat like a camera with the retina taking the place of the film (Fig. 5.27). The eye's lens focuses the image on the retina. To accommodate objects at various distances, muscles in the eye modify the shape of the lens so as to change its focal length. This process differs from that in a camera, where the distance from lens to film can be adjusted but the focal length cannot.

Eyeglasses have become an indispensable optical instrument. For various reasons, the rays from an object are often brought to focus in front of or behind the retina, in spite of the changes of shape of which the eye's own lens is capable (Fig. 5.28). If the rays are focused in front of the retina, the rays are too strongly converged by the eye's lens and the convergence must be reduced by what is called a negative correcting lens in front of the eye. For the other case, when the image falls behind the retina, the rays are not converged enough and a converging lens must be put in front of the eye.

An ophthalmologist expresses the strength of the necessary correction in the unit *diopters* by taking the reciprocal of the focal length (in meters) of the necessary lens. For example, a 0.33-meter focal length lens gives a correction of 3 diopters. The thickest glasses made give a correction of about 15 diopters. Eyeglasses can also correct for other defects in your vision. These defects are due to the asymmetric shape of your cornea, lens, or retina. To correct for such defects, the lens is not ground to a simple spherical shape. Often the inside is ground to a cylindrical contour while the outside surface may be spherical.

Summary

Light and sound are intermediaries in the interaction-at-a-distance of light or sound sources (candle, loudspeaker) and human sense organs or other detectors (photographic film, microphone). Light and sound transmit information and energy. The very detailed images transmitted by light have led to the development of the ray model. The lack of images and the analogy with water waves have led to the wave model for sound, which will be described further in Chapter 7. Assumptions of the ray model for light provide for emission, propagation, absorption, reflection, and refraction of monochromatic light rays, and the formation of images by converging and diverging rays. These assumptions may be used to explain color, the eye's ability to see objects that reflect light, and the function of mirrors, lenses, cameras, magnifiers, and telescopes.

List of new terms

primary source	wave model	diffuse reflection
reflecting surface	beats	lens
loudness	refraction	converging lens
pitch (of sound)	interference pattern	focal point
light intensity	Snell's law	focal length
color	monochromatic	f-number
image	selective transmission	negative lens
retina	selective reflection	diopter
ray model	specular reflection	

List of symbols

θ_i	angle of incidence	l_f	focal length of a lens
θ_R	angle of reflection	l_o	length (distance) of object from lens
θ_r	angle of refraction	l_i	length (distance) of image from lens
n	index of refraction	s_o	size of object
		s_i	size of image

Problems

1. List similarities and differences of light and sound.

2. Obtain an empty bottle and make a sound by blowing across the top. Do you have control over the intensity and/or the pitch of the sound? Explain.

3. Whistle through your lips. Describe how you control the intensity and the pitch of the whistle.

4. Experiment at talking to yourself through a tube, placing one end at your mouth and the other end over one ear. Use as many different materials and lengths of tubing as you can find (garden hose, rubber tubing, plastic tubing, and so on). Describe the effect of the material and length of the tube on the transmitted sound.

5. Collaborate with a colleague to find whether you can produce sound "shadows."

6. Interview four or more children (ages 6-10) to investigate their concept of vision and the role of light in vision. (A useful aid is an opaque piece of flexible tubing. Bend the tube into various shapes and orientations and ask whether a flashlight at the far end would be seen through the tube in each case.)

7. Use the ray model for light to explain the concept of illumination.

8. Investigate beats produced by an untuned piano. If you can, use paper clips to change the pitch (as in Fig. 5.2), try various notes and various positions of the paper clip (end of string, center of string, two paper clips on one string, and so on). Report your findings, including the rate of beats. (Count the number of beats in 5 or 10 seconds whenever they are not too fast.)

9. Investigate the interference pattern produced when a small, strong light source (a very distant street light, for example) is viewed through a piece of woven fabric (fig 5.3). Draw and describe your observations. (Variations to try: hold the fabric at various distances from your eye; stretch the fabric; distort the fabric so the threads do not cross at right angles; compare dense and loose fabrics. Give evidence that the pattern you see is not simply a "shadow" cast by the threads and holes of the fabric. Include measurements as well as you can.

10. Look at the reflection of a bright light source at medium distance (10-50 meters) from the grooved area of a phonograph record or compact disc (CD). Describe your observations.

11. Describe conditions under which you have "seen" light rays (Fig. 5.4).

12. Inside the tube at the left of Fig. 5.8 is the primary light source from which the beam originates. Use the shape of the visible beam to make inferences about the position and size of the primary source.

13. Measure the angles of incidence and reflection of the pencil of light in Fig. 5.9. (See Fig. 5.11 for the definitions.)

14. Tabulate the angles of incidence, reflection, and refraction in the four parts of Fig. 5.10. Explain why the refracted beam is not refracted again when it emerges from the glass. Find the index of refraction of the glass piece used.

15. (a) A beam of light is generated in an underwater lamp and hits the smooth water-air surface at an angle of incidence of 30°. Make a diagram to show the incident, reflected, and refracted beams. Calculate the angles of reflection and refraction and indicate

them in the diagram. Make the diagram true to the real angles.
(b) Do the same for an underwater angle of incidence of 40°.
(c) Do the same for an underwater angle of incidence of 50°.
(Hint: You will find that part (c) of the problem does *not* have the same kind of solution as parts (a) and (b). You should use a ray diagram to interpret your calculations. Discuss your conclusions.)

16. Compare the magnitude of refraction effects you expect to observe at an air-water, air-glass, air-diamond, water-glass, water-diamond, and water-alcohol (separated by a thin plastic sheet) boundary surface. Investigate one of more of these situations experimentally.

17. A pencil of light enters a pane of glass with parallel faces. Calculate the angle at which it emerges for various angles of incidence. Relate your result to your observations of noticeable refraction (or lack thereof) by the glass in a window. (Hint: the index of refraction can be found in Table 5.1.)

18. Look at the fish or other objects in a rectangular aquarium from all side directions and the top. Describe your observations, especially when you are looking into the aquarium along a diagonal line.

19. A pencil of light is incident on a glass prism (Fig. 5.13) at an angle of 60°. Find the angle at which the pencil of light emerges from the prism. The index of refraction is in Table 5.1.

20. Stick a pencil into a glass of water. Look at the submerged part of the pencil through the water-air boundary and through the water-glass-air boundary. Report the refraction effects you observe and explain them by the ray model.

21. Experiment with pieces of colored cellophane combined with one another and held over colored pictures to explore the subtraction of colors. Report your findings and relate them to Newton's model.

22. The transparent film in a 35-millimeter slide is about 1 inch high. A student wishes to buy a projector that will give a 20-inch-high picture in his room, which is 12 feet long. What focal length projector lens should he buy?

23. Use a magnifying lens for the following experiments.
(a) Find the focal length by focusing a distant object outside your window on a piece of paper at right angles to the lens axis.
(b) Hold the lens between a lit light bulb and a piece of paper so that the image of the bulb is focused on the paper. Measure the object and image distances and check whether they satisfy Eq. 5.4 with the focal length measure in (a). Comment whether discrepancies are due to experimental error or to limits of the mathematical model. If you get very poor results, repeat the measurement with a different distance between bulb and paper.
(c) Hold the lens very close to your eye and bring a small object (pencil point, fingertip, print) into focus (Fig. 5.26). Have a friend measure the distance from lens to object. Relate this distance to the

focal length.

(d) Look at an object (about 2 feet away) through the lens held close to the object. Slowly move the lens away from the object and toward your eye. Observe and describe the changes in the image seen through the lens and explain them in terms of the lens theory described in Section 5.3.

(e) Hold the lens close to your eye and look toward a very distant object. Gradually move the lens away from your eye until it is at arm's length. Observe and describe the image you see through the lens. Explain in terms of the lens theory of Section 5.3.

(f) When you use the lens as magnifier, the image of the object is projected on your retina. No measurable image is projected on a screen. It is difficult to measure the "magnification," since object size and image size cannot be compared directly. Formulate an operational definition of the magnification of a lens used as a magnifier and apply the definition to your lens.

24. Assumption 2, Fig. 5.16, refers to a "straight line." Describe how you would define "straight line" operationally and/or formally. Discuss the application of your definition to the assumption.

25. Identify one or more explanations or discussions in this chapter that you find inadequate. Describe the general reasons for your judgment (conclusions contradict your ideas, steps in the reasoning have been omitted, words or phrases are meaningless, equations are hard to follow, . . .), and make your criticism as specific as you can.

Bibliography

E. N. daC. Andrade, *Sir Isaac Newton - His Life and Work*, Doubleday (Anchor), Garden City, New York, 1958.

Sir William Bragg, *The Universe of Light*, Dover Publications, New York, 1959.

M. Minnaert, *Light and Color in the Open Air*, Dover Publications, New York, 1954.

Isaac Newton, *Opticks*, Dover Publications, New York, 1952.

U. Haber-Schaim *et. al.,* PSSC *Physics, Seventh Edition,* Kendall/Hunt, Dubuque, Iowa, 1991.

Michael Rodda, *Noise and Society*, Oliver and Boyd, Edinburgh and London, 1968.

R. A. Weale, *From Sight to Light*, Oliver and Boyd, Edinburgh, London, 1968.

Article from Scientific American (see archives on the Internet at http://www.sciamarchive.org/.)

L. L. Beranek, "Noise" (December 1966).

*We have still to consider,
in studying the spreading out
of these waves, that each particle
of the matter in which a wave
proceeds not only communicates
its motion to the next
particle to it, which is on
the straight line drawn from
the luminous point, but that
it also necessarily gives a motion.
The result is that around
each particle there arises
a wave of which this particle
is the center.*

CHRISTIAN HUYGENS
Traité de la Lumierè,
1690

The wave theory

Waves on a water surface are such a familiar and expected occurrence that a completely still, glassy pool excites surprise and admiration (Fig. 6.1). You can also observe waves on flags being blown by a strong wind. In this chapter you will be concerned with how waves propagate, what properties are used to describe them, and how waves combine with one another when several pass through the same point in space at the same time. In the wave theory, which was formulated by Christian Huygens during the seventeenth century, the space and time distribution of waves is derived from two assumptions, the superposition principle and Huygens' Principle. The wave theory is very "economical" in the sense that far-reaching consequences follow from only these two assumptions.

Waves are important in physics because they have been used in the construction of very successful working models for radiation of all kinds. You can easily imagine that dropping a pebble into a pond and watching the ripples spread out to the bank suggests interaction-at-a-distance between the pebble and the bank. The waves are the intermediary in this interaction, just as radiation was the intermediary in some of the experiments described in Sections 3.4 and 3.5. In Chapter 7, we will describe wave models for sound and light and how these models can explain the phenomena surveyed in Chapter 5. The success of these models confirms Huygens' insight into the value of wave theory. However, Huygen's contributions and wave theory were not fully appreciated and exploited until the nineteenth century.

Waves were originally introduced as oscillatory disturbances of a material (called the *medium*) from its equilibrium state. Water waves and waves on a stretched string, the end of which is moved rapidly up and down, are examples of such disturbances. The waves are emitted by a source (the pebble thrown into the pond), they propagate through the medium, and they are absorbed by a receiver (the bank). Even though waves are visualized as disturbances in a medium, their use in certain theories nowadays has done away with the material medium. The waves in these applications are fluctuations of electric, magnetic, or

Figure 6.1 The reflected image gives information about the smoothness of the water surface. Why are the reflections of the sails dark and not white?

gravitational fields, rather than oscillations of a medium. The use of such waves to represent radiation has unified the radiation model and the field model for interaction-at-a-distance (Section 3.5). Our discussion here, however, will be of waves in a medium and not of waves in a field.

6.1 The description of wave trains and pulses

Oscillator model. We will analyze the motion of the medium through which a wave travels by making a working model in which the medium is composed of many interacting systems in a row. Each system is capable of moving back and forth like an oscillator, such as the inertial balance shown below and described in Section 3.4. You may think of the oscillators in a solid material as being the particles in an MIP model for the material.

Amplitude and frequency. Each oscillator making up the medium has an equilibrium position, which it occupies in the absence of a wave. When an oscillator is set into motion, it swings back and forth about the equilibrium position. The motion is described by an *amplitude* and a *frequency* (Fig. 6.2). The amplitude is the maximum distance of the oscillator from its equilibrium position. The frequency is the number of complete oscillations carried out by the oscillator in 1 second.

Interaction among oscillators. When waves propagate through the medium, oscillators are displaced from the equilibrium positions and are set in motion. The wave propagates because the oscillators interact with one another, so that the displacement of one influences the motion of the neighboring ones, and so on. Each oscillator moves with a frequency and an amplitude. It is therefore customary in this model to identify the frequency and amplitude of the oscillators with the frequency and amplitude of the wave. In addition, as you will see, there are properties of the wave that are not possessed by a single oscillator but that are associated with the whole pattern of displacements of the oscillators.

Conditions for wave motion. The oscillator model described above has two general properties that enable waves to propagate. One is that the individual oscillator systems interact with one another, so that a displacement of one influences the motion of its neighbors. The second is that each individual oscillator has inertia. That is, once it has been set in motion it continues to move until interaction with a neighbor slows it down and reverses its motion. These two conditions, interaction and inertia, are necessary for wave motion.

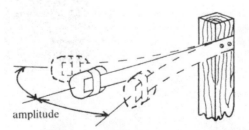

Figure 6·2 *An oscillator in motion.*

amplitude

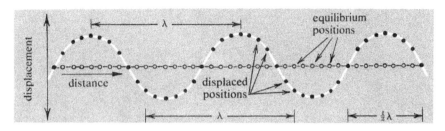

Figure 6.3 Row of oscillators in a medium, showing equilibrium positions and displaced positions in a wave. The wavelength is the distance after which the wave pattern repeats itself.

Equation 6.1

wavelength (meters) = λ

wave number (per meter)

$= k$

$\lambda \times k = 1, \ k = \dfrac{1}{\lambda}, \ \lambda = \dfrac{1}{k}$

EXAMPLES

$\lambda = 0.25$ m

$k = \dfrac{1}{\lambda} = \dfrac{1}{0.25\,m} = 4\,/\,m$

This is 4 wavelengths/m.

$\lambda = 5.0$ m

$k = \dfrac{1}{\lambda} = \dfrac{1}{5.0\,m} = 0.2\,/\,m$

This is 0.2 wavelengths/m.

$\lambda = 0.0001$ m $= 10^{-4}$ m

$k = \dfrac{1}{\lambda} = \dfrac{1}{10^{-4}\,m} = 10^{4}\,/\,m$

This is 10^{4} or $10,000$ wavelengths/m.

Wave trains. Look more closely now at the pattern of the oscillators in the medium shown in Fig. 6.3. As a wave travels through the medium, the various oscillators have different displacements at any one instant of time. The wave is represented graphically by drawing a curved line through the displaced positions of all the oscillators (shown above in Fig. 6.3). This curved line, of course, changes as time goes on because the oscillators move. Note, however, that the individual oscillators in the model move only up and down.

Wavelength and wave number. You can see from Fig. 6.3 that the wave repeats itself in the medium. This pattern of oscillators is called a wave train, because it consists of a long train of waves in succession. A complete repetition of the pattern occupies a certain distance, after which the pattern repeats. This distance is called the *wavelength*; it is measured in units of length and is denoted by the Greek letter lambda, λ. Sometimes it is more convenient to refer to the number of waves in one unit of length; this quantity is called the *wave number* and it is denoted by the letter **k**. Wavelength and wave number are reciprocals of one another (Eq. 6.1).

Period and frequency. We have just described the appearance of the medium at a particular instant of time. What happens to one oscillator as time passes? It moves back and forth through the equilibrium position as described by a graph of displacement vs. time (Fig. 6.4) that is very similar to Fig. 6.3. The motion is repeated; each complete cycle requires a time interval called the *period* of the motion, denoted by a script "tee," $\mathcal{T}$. The number of repetitions per second is the

Figure 6.4 Graph of the motion (displacement) of one oscillator over time. The period $(\mathcal{T})$ is the time internal after which the motion repeats itself.

one period

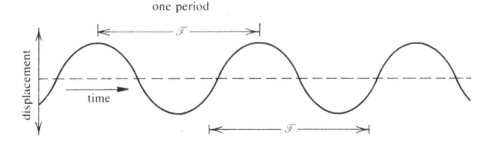

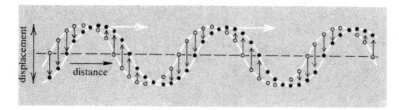

Figure 6·5   The wave moves to the right as the oscillators move up and down. The black circles and black dots represent the displacements of the oscillators at two different times.

Figure 6·6   In one period, oscillators A and B carry out a full cycle of motion from crest to trough and to crest again. The crest initially at A moves to B in this time interval.

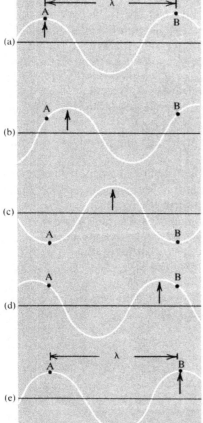

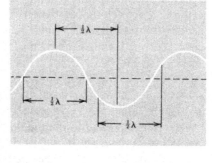

Figure 6·7   Oscillators in a wave train have opposite displacements if their separation is $\frac{1}{2}$ wavelength.

Figure 6·8   Pulse patterns of disturbance in a medium. The approximate length of the pulse is denoted by Δs.

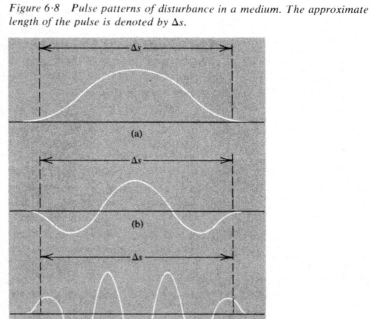

Equation 6.2 (period and frequency of a wave)

period (time for one complete repetition, in seconds) $= \mathcal{T}$
frequency (number of complete repetitions in one second, per second) $= f$

$$\mathcal{T} \times f = 1, \; f = \frac{1}{\mathcal{T}}, \; \mathcal{T} = \frac{1}{f}$$

EXAMPLES
If $\mathcal{T} = 0.05$ sec,

$$f = \frac{1}{\mathcal{T}} = \frac{1}{0.05} = 20/\text{sec}.$$

If $\mathcal{T} = 3.0$ sec,

$$f = \frac{1}{\mathcal{T}} = \frac{1}{3.0} = 0.33/\text{sec}.$$

If $\mathcal{T} = 10^{-6}$ sec,

$$f = \frac{1}{\mathcal{T}} = \frac{1}{10^{-6}} = 10^6/\text{sec}.$$

Equation 6.3 (wave speed)

wave speed $= v$

$$v = \frac{\Delta s}{\Delta t} = \frac{\lambda}{\mathcal{T}} \qquad (a)$$

$$v = \lambda f \qquad (b)$$

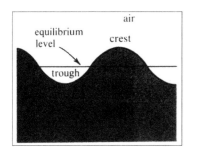

frequency (symbol f). The period and frequency are reciprocals of one another (Eq. 6.2), just as are the wavelength and wave number. The period and frequency describe the time variation of the oscillator displacements, while the wavelength and wave number describe the spatial variation.

Wave speed. One of the most striking properties of waves is that they give the appearance of motion along the medium. If you look at the pattern of displacements at two successive instants of time (Fig. 6.5), you see that the wave pattern appears to have moved to the right (along the medium), although the individual oscillators have only moved up and down. Since the pattern actually moves, you can measure its speed of propagation through the medium. The wave speed is usually represented by the symbol v (Section 2.2).

You can conduct a thought experiment with the oscillator model for the medium to find a relationship among period, wavelength, and wave speed. Imagine the oscillator at a wave crest carrying out a full cycle of its motion (Fig. 6.6). While this goes on, all the other oscillators also carry out a full cycle, and the wave pattern returns to its original shape. The wave crest that was identified with oscillator A in Fig. 6.6, however, is now identified with oscillator B. Hence the wave pattern has been displaced to the right by 1 wavelength. The wave speed is the ratio of the displacement divided by the time interval (Eq. 2.2), in this instance the ratio of the wavelength divided by the period ($\lambda/\mathcal{T}$, Eq. 6.3a). By using Eq. 6.2, $f = 1/\mathcal{T}$, you can obtain the most useful form of the relationship: $v = \lambda f$, or wave speed is equal to wavelength times frequency (Eq. 6.3b).

Positive and negative displacement. Waves are patterns of disturbances of oscillators from their equilibrium positions. The displacement is sometimes positive and sometimes negative. In Fig. 6.3, the open circles and the horizontal line drawn along the middle of the wave show the equilibrium state of the medium. Displacement upward may be considered positive, displacement downward negative. In water waves, for example, the crests are somewhat above the average or equilibrium level of the water and the troughs are somewhat below the average or equilibrium level of the water. In fact, the water that forms the crests has been displaced from the positions where troughs appear.

By definition, the pattern in a wave train repeats itself after a distance of 1 wavelength. It therefore also repeats after 2, 3, ... wavelengths. Consequently, the oscillator displacements at pairs of points separated by a whole number of wavelengths are equal. If you only look at a distance of 1/2 wavelength from an oscillator, however, you find an oscillator with a displacement equal in magnitude but opposite in direction (Fig. 6.7).

Wave pulses. In the *wave trains* we have been discussing, a long series of waves follow one another, and each one looks just like the preceding one. On the other hand, a *wave pulse* is also a disturbance in the medium but it is restricted to only a part of the medium at any one time (Fig. 6.8). It is not possible to define frequency or wavelength for a pulse since it does not repeat itself. The concept of wave speed, however, is applicable to pulses since the pulse takes a certain amount of time to travel from one place to another. In Section 6.2 we will describe how wave trains and wave pulses can be related to one another.

Examples of wave phenomena. The oscillator model for a medium can be applied to systems in which small deviations from a uniform equilibrium arrangement can occur. One such system is a normally motionless water surface that has been disturbed so that water waves have been produced. Another example is air at atmospheric pressure in which deviations from equilibrium occur in the form of pressure variations: alternating higher or lower pressure. Such pressure variations are called sound waves. A third example is an elastic solid such as Jell-O, which can jiggle all over when tapped with a fork. In the oscillator model, movement results from oscillating displacements within the Jell-O after the fork displaced the oscillators at the surface.

Oscillator model for sound waves. Since sound in air is of special interest, we will describe an oscillator model for air in more detail. Visualize air as being made up of little cubes of gas (perhaps each one in an imaginary plastic bag). When acted upon by a sound source, the first cube is squeezed a little and the air inside attains a higher pressure (Fig. 6.9). The first cube then interacts with the next cube by pushing against it. After a while the second cube becomes compressed and the first one has expanded back to and beyond its original volume. The second cube then pushes on the third, and so on. In this way the sound propagates through the air.

The initial pressure increase above the equilibrium pressure may be

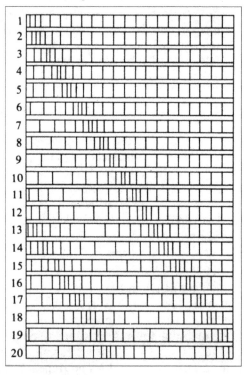

Figure 6.9 A gas bag model for air is used to represent the propagation of a sound wave. An individual bag of gas is alternately compressed and expanded. Its interaction with adjacent bags of gas leads to propagation of the compression and expansion waves.

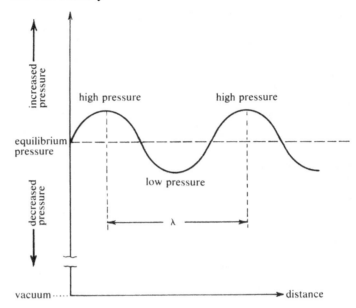

Figure 6.10 Pressure profile in a sound wave. The graph shows deviations from the equilibrium pressure.

created by a vibrating piano string or a vibrating drumhead. In addition to regions of increased pressure, the sound wave also has regions of deficient pressure where the air has expanded relative to its equilibrium state.

Thus the sound wave consists of alternating high-pressure (above equilibrium) and low-pressure (below equilibrium) regions. A pressure profile (pressure versus distance) for a pure tone has the typical wave pattern shown in Fig. 6.10.

6.2 Superposition and interference of waves

The superposition principle. Can you visualize what happens when two waves overlap? In the oscillator model, it is easy to describe the medium at a place where there are two or more waves at the same time. Each oscillator is displaced from its equilibrium position by an amount equal to the sum of the displacements associated with the waves separately (Fig. 6.11). In other words, you visualize the oscillator displacements associated with each of the wave patterns and add them together. This procedure takes for granted that the waves do not interact with one another, but that each propagates as though the others were not present.

The property of non-interaction we have just described is called the *superposition principle*. It makes the combination of waves simple to carry out in thought experiments, and it has been exceedingly valuable for this reason. Fortunately, a wave model that incorporates the superposition principle describes quite accurately many wave phenomena in nature.

Figure 6.11 Superposition of two waves leads to interference. One wave is represented by black dashes, the other by dots. The combination wave is the sum of both waves and is represented by the solid line.

(a) Constructive interference occurs when dotted and dashed waves reinforce each other.

(b) Destructive interference occurs when dotted and dashed waves cancel each other.

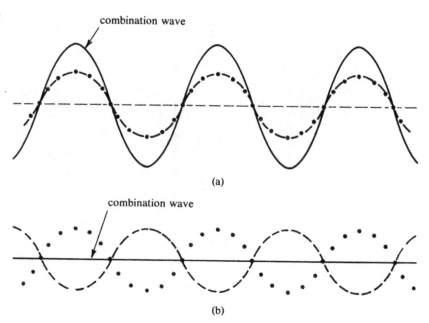

combination wave

(a)

combination wave

(b)

Interference of waves. Consider now what may happen to the oscillator motion as a result of the superposition of two waves. The two waves may combine in various ways. Perhaps each of two wave patterns has an upward displacement of an oscillator at a certain time and at a certain place. In such a case, the upward displacement in the presence of the combined wave will be twice as big as that from one wave alone (Fig. 6.11(a)). If there are simultaneous downward displacements in the two waves separately, the combined displacement will be twice as far down. Suppose you consider a point in space where one wave has an upward displacement and the other wave has an equal downward displacement at the same time. Now, the upward (positive) displacement and the downward (negative) displacement add to give zero combined displacement (zero amplitude of oscillation). In fact, it is possible for

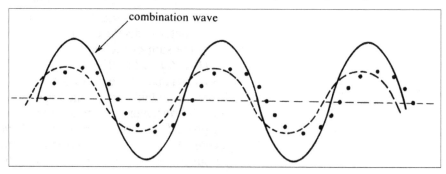

combination wave

Figure 6.12 Superposition of two waves leading to partially destructive interference. The displacements of the dashed and dotted waves are added together at each point to yield the displacement of the combination wave (represented by the solid line). Note that displacement below the line is negative.

The "one-particle model" for a real object is a "very small object that is located at the center... of the region occupied by the real object." (Section 2.1) This is a way to think about an object so as to focus on the object's position, motion, and inertia without considering its shape and orientation. Complex objects can be thought of as two or more particles that interact in defined ways, or with the many-interacting-particles (MIP) model. In such models, each particle is thought of as a single, tiny bit of matter. The matter itself is thought of as indestructible, or "conserved." Such particles cannot "cancel" one another to cause destructive interference.

A wave, on the other hand, is quite different. A wave, as explained in this chapter, is thought of as a disturbance or oscillation that passes <u>through</u> matter. The displacement of the particles can be positive or negative and, as with (+1) + (−1) = 0, two waves <u>can</u> cancel one another.

In the 20th century, physicists found that matter in the micro-domain behaves in ways that conform with <u>neither</u> the particle <u>nor</u> the wave model. This led to the "wave-particle duality" and quantum mechanics. (Chapter 8).

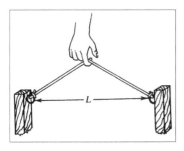

two waves to combine in such a way that they completely cancel one another, as in Fig. 6.11b.

This characteristic of waves makes their behavior different from what we expect of material objects, particularly when we think of them as single particles (Section 2.1) or as made up of particles. If one particle and another particle are combined, you have two particles, and you cannot end up with zero particles. Two or more waves, however, may combine to form a wave with larger amplitude, a wave with zero amplitude, or a wave with an intermediate amplitude (Fig. 6.12).

This result of the superposition of waves is a phenomenon called *interference*. If waves combine to give a larger wave than either one alone, you have *constructive interference*. If waves tend to cancel each other, you have *destructive interference*. There is a continuum of possibilities between the extremes of complete constructive interference shown in Fig. 6.11(a) and complete destructive interference shown in Fig. 6.11(b). With particles, the concept of destructive interference is meaningless in that the presence of one particle can never "cancel" the presence of another.

Standing waves. When two equal-amplitude wave trains of the same frequency and wavelength travel through a medium in opposite directions, their interference creates an oscillating pattern that does not move through the medium (Fig. 6.13). Such an oscillating pattern is called a *standing wave*. The points in a standing wave pattern where there are no oscillations at all are called *nodes*. At a node, there is always complete destructive interference of the two wave trains; the displacements associated with the two waves at the nodes are always equal and opposite. Because the waves move in opposite directions at the same speed, each node remains at one point in space and does not move; this is the reason behind the choice of name: a *standing* wave does not move.

You can see in Fig. 6.13 that the distance between two nodes must be exactly ½ wavelength. This holds true not only for the illustration but also for *all* standing wave patterns. The reasoning is as follows. At any node, the two wave displacements must always be equal and opposite to produce complete destructive interference. At a distance of ½ wavelength, the displacement associated with each wave has exactly reversed (as illustrated in Fig. 6.7). Thus, the two displacements must again be equal and opposite and again produce a node.

An easy way to set up standing waves is to place a reflecting barrier in the path of a wave. The reflected wave interferes with the incident wave to produce standing waves. The nodes are easy to find because the oscillators remain stationary at a node. This offers a convenient way to determine the wavelength: measure the distance between nodes and multiply by 2.

Tuned systems. It is very fruitful to pursue the standing wave idea one step further. Suppose an elastic rope is tied to a fixed support at each end and the middle is set into motion by being pulled to the side and released (see drawing to left). How will the rope oscillate? To solve this problem, think of the pattern as being made up of wave trains in

Equation 6.4 (Possible number of half wavelengths that fit within L)

$$L=\frac{1}{2}\lambda$$

or

$$L=2\times\left(\frac{1}{2}\lambda\right)=\frac{2}{2}\lambda$$

or

$$L=3\times\left(\frac{1}{2}\lambda\right)=\frac{3}{2}\lambda$$

or

$$L=4\times\left(\frac{1}{2}\lambda\right)=\frac{4}{2}\lambda$$

or

$$L=5\times\left(\frac{1}{2}\lambda\right)=\frac{5}{2}\lambda$$

. . . and so on

Equation 6.5 (wavelengths permitted on a tuned system, from above)

$$\lambda=\frac{2}{1}L, \text{ or}$$

$$\lambda=\frac{2}{2}L = L, \text{ or}$$

$$\lambda=\frac{2}{3}L, \text{ or}$$

$$\lambda=\frac{2}{4}L = \frac{1}{2}L, \text{ or}$$

$$\lambda=\frac{2}{5}L$$

and so on . . .

Equation 6.6 (finding frequency for a given speed and wavelength)

$$f=\frac{v}{\lambda}$$

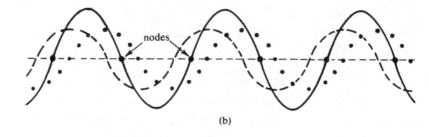

(a)

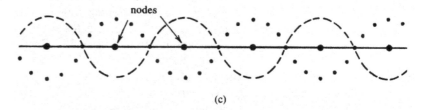

(b)

(c)

Figure 6.13 The formation of standing waves by the superposition of two wave trains propagating in opposite directions (dotted wave towards right and dashed wave toward left). The combination wave is the solid line. Note the stationary position of the nodes, marked by the large dots.
(a) Constructive interference of the two wave trains.
(b) Partially destructive interference after 1/8th of a period.
(c) Destructive interference after 2/8ths (1/4th) of a period.
Can you draw the pattern after 3/8ths of a period? After 4/8ths (1/2) of a period?

combinations, some moving to the right, others to the left. Because the ends are fixed, the wave pattern must be such that the ends of the rope are its nodes. The length of the rope is the distance between the nodes, which must be an integral multiple of ½ wavelength (Eq. 6.4). It follows that the wavelengths of the waves that can exist on this rope are related to the length of the rope by Eq. 6.5 to satisfy the conditions of nodes at the ends.

A system such as the rope with fixed ends is called a *tuned system*, because it can support only waves of certain wavelengths (Eq. 6.5) and the frequencies related to them by Eq. 6.6 (derived from Eq. 6.3b). The wave speed is a property of the medium from which the tuned system is constructed.

Musical instruments. Musical instruments employ one or more tuned systems whose frequencies are in a suitable relation to one another. For stringed instruments, such as the violin and guitar, the tuned system is a wire or elastic cord; for wind instruments, it is an air column in a pipe closed at one end; for drums, it is an elastic membrane whose edge is fixed; and so on.

The tone of the instrument is determined by the oscillation frequency of the tuned system. It is possible to change the frequency either through changing the length of the tuned system (and therefore changing the wavelength of the allowed standing waves) or through changing the wave velocity by modifying the medium in the tuned system.

Sound waves of a single frequency can be produced in closed pipes of a certain length. Longer pipes produce lower tones. A pressure wave starts at one end of the pipe and travels down the pipe, confined by the walls. When the wave reaches the other end of the pipe, it is reflected back and interferes with waves coming down the pipe. The interference forms a standing wave. This standing wave is of the characteristic wavelength determined by the length of the pipe and has the frequency that we hear.

Beats. Standing waves are created by the interference of waves with the same frequency. What will be the combined effect of two waves of differing frequencies? To answer this question, apply the superposition principle in a thought experiment in which two such waves are combined. Suppose the two waves are in constructive interference at one instant of time. Since one wave has shorter cycles than the other before repeating, they will soon get out of step. After a while, the two waves will be in destructive interference, and a little later in constructive interference again. So the net effect is an alternation from constructive interference (loud) to destructive interference (soft) and back again. These alternations in volume are called beats.

It is easily possible to calculate the time interval between two beats from the difference in frequency of the two interfering wave trains. During this time interval the two waves must go from constructive interference to destructive interference and back to constructive interference. Therefore, the higher-frequency wave must vibrate exactly once more than the lower frequency wave. The additional oscillation restores the constructive interference of the two waves, since waves repeat exactly after a whole oscillation. Hence the wave amplitude after the interval is equal to its value before, meaning that the next beat is ready to begin.

The number of oscillations made by either of the two waves is equal to its frequency (oscillations per second) times the time interval ($N_1 = f_1\Delta t$ and $N_2 = f_2\Delta t$, Eq. 6.7). The two numbers, according to the condition, must differ by one ($N_1 - N_2 = (f_1 - f_2)\Delta t$, Eq. 6.8). The conclusion is that the frequency difference times the time interval is equal to one ($\Delta f\, \Delta t = 1$, Eq. 6.9). The frequency of the individual waves determines the overall pitch of the sound, not the beat frequency; in fact, the beat frequency is $f_1 - f_2$.

Wave packets. Standing waves and beats are wave phenomena that are observable when two wave trains are combined. You may, of

---

Equation 6·7

frequencies of the two wave trains (per second) f_1, f_2
time interval (seconds) Δt
number of oscillations N_1, N_2

$$N_1 = f_1\Delta t. \qquad N_2 = f_2\Delta t$$

Equation 6·8

$$1 = N_1 - N_2$$
$$= f_1\Delta t - f_2\Delta t$$
$$= (f_1 - f_2)\Delta t$$

Equation 6·9

frequency difference Δf

$$\Delta f = f_1 - f_2 \quad \text{(a)}$$
$$\Delta f\Delta t = 1 \qquad \text{(b)}$$

EXAMPLE
Frequencies of 255/sec and 257/sec

$$\Delta f = 2/sec$$

$$\Delta t = \frac{1}{\Delta f} = \frac{1}{2/sec} = 0.5 \; sec$$

course, use the superposition principle and the rules for constructive and destructive interference to combine as many different wave patterns as you wish. In the early nineteenth century, it was discovered by Joseph Fourier (1768-1830) that any wave pattern could be formed by a superposition of one or more wave trains, as illustrated below. All wave phenomena can thereby be related to the frequencies, amplitudes, wavelengths, and velocities of the component wave trains in a wave pattern.

To illustrate Fourier's discovery, we will construct a wave pulse close to the one shown in Fig. 6.8a by combining the four wave trains drawn in Fig. 6.14. You are invited to read off the wave amplitudes from the graphs, to add the wave amplitudes of the four waves, and to verify that the combined wave drawn in Fig. 6.14 really is obtained by superposition of the four wave trains. By combining more and more wave trains of other wavelengths and successively smaller and smaller amplitudes, you can achieve further constructive and destructive interference at various locations in the pulse. In this way you could obtain a closer and closer approximation to the wave pulse shown in Fig. 6.8a and Fig. 6.14 (see Fig. 6.15).

The representation of wave pulses by a superposition of wave trains has led to the introduction of the suggestive phrase *wave packet* (instead of wave pulse), which we will also adopt. The superposition procedure can be quite tedious to work out in detail if many wave trains must be combined to achieve success. The essence of the procedure, however, is to select wave trains that interfere destructively in one wing of the wave packet, constructively at the center, and destructively again in the other wing. This can be achieved if one wave train has one more

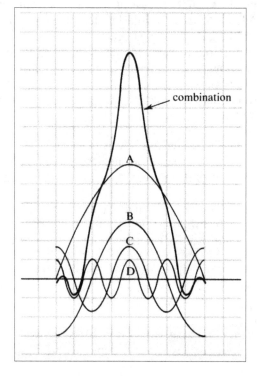

Figure 6.14 (left) The superposition of four wave trains to produce a wave pulse.

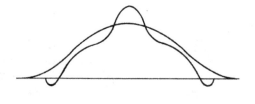

Figure 6.15 (below) The wave packet in Fig. 6.14 and the pulse in Fig. 6.8a have been drawn to the same scale for easier comparison.

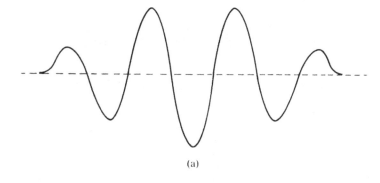

(a)

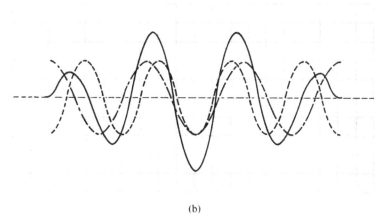

(b)

Figure 6.16 The superposition of wave trains to produce a wave packet.
(a) The wave packet pictured in Fig. 68c, enlarged.
(b) A very similar wave packet constructed by the superposition of two wave trains.

full wave over the length of the packet than does the other one. Look, for example, at the wave packet with about four ripples shown in Fig. 6.8c, and reproduced here (Fig. 6.16a). We can combine two wave trains (one with four full waves over the length of the packet and one with three) to find a first approximation to the desired wave packet (Fig. 6.l6b).

Uncertainty principle. We will now formulate a general principle governing the superposition of wave trains to form wave packets. It is called the *uncertainty principle*, and it has played a very important role in the application of the wave model to atomic phenomena, which we will describe in Chapter 8.

Physical significance. The content of the uncertainty principle is that a wave packet that extends over a large distance in space (large Δs) is obtainable by superposition of wave trains covering a narrow range in wave numbers, but that a wave packet that extends over only a short distance in space (small Δs) must be represented by the superposition of wave trains covering a wide range in wave number. It is consequently impossible to construct a wave packet localized in space (small Δs) out of wave trains covering a narrow range in wave number. This idea is known as the "uncertainty principle" because it means that there is an inherent uncertainty in our ability to measure the exact position of a wave packet; Δs represents the "length" of the wave packet and thus the range of uncertainty in our measurement of the packet's position. The size of Δs is closely related to the range of wave numbers included in the packet. We cannot specify the range of wave numbers (Δk) precisely, but we can relate it to the size (Δs) of the wave packet. We will now derive a mathematical model that expresses this relationship.

Mathematical model. The calculation proceeds in the same way as the calculation for the time interval between beats in Eqs. 6.7, 6.8, and 6.9. First, we select two wave trains with different wave numbers k_1 and k_2, one a little larger and one a little smaller than the average wave number of all the waves needed. Each wave train has a certain number of waves ($N_1 = k_1 \Delta s$ and $N_2 = k_2 \Delta s$) within the length (Δs) of the wave packet (Eq. 6.10). By how much do these two numbers have to differ? They have to differ sufficiently so that the two wave trains are in destructive interference in the regions to the left and to the right of the wave packet's center, where they are in constructive interference. The distance between the two regions is approximately the spatial length Δs of the wave packet. Now, to achieve the desired destructive interference in both regions, the wave train with the shorter wavelength has to contain at least one more whole wave than the other in the distance Δs, that is: $1 + N_2 = N_1$, or $1 = N_1 - N_2$. This condition is applied in Eq. 6.11 to yield an important result: the range of wave numbers (Δk) times the width of the wave packet (Δs) is equal to one (Eq. 6.12b).

Comparison of beats and wave packets. It is clear that Eqs. 6.9 and 6.12b are closely similar. You may consider both of them as statements of an uncertainty principle for wave packets if you are willing to think of one beat pulsation as a wave packet. Equation 6.12 refers to the size of the wave packet in space. Equation 6.9 refers to the duration of the wave packet in time. The wave trains included in a wave packet have a certain average wave number or frequency, and extend above and below these average values by an amount equal to about one half of the wave number difference Δk or frequency difference Δf. The wave packet includes wave trains of substantial amplitude within this range of wave number or frequency, and wave trains of progressively smaller and smaller amplitude outside this range. The exact amplitude distribution of the included wave trains is determined by the shape of the wave packet and can be calculated by more complicated mathematical procedures developed by Fourier and later workers. We apply the uncertainty principle to wave packets below in Examples 6.1 and 6.2

Equation 6·10

wave number of the two wave
  trains (per meter) k_1, k_2
wave packet length (meters) Δs
number of waves N_1, N_2

$$N_1 = k_2 \Delta s, \qquad N_2 = k_2 \Delta s$$

Equation 6·11

$$1 = N_1 - N_2$$
$$= k_1 \Delta s - k_2 \Delta s$$
$$= (k_1 - k_2)\Delta s$$

Equation 6·12

wave number difference Δk

$$\Delta k = k_1 - k_2 \quad \text{(a)}$$
$$\Delta k \Delta s = 1 \qquad \text{(b)}$$

EXAMPLE 6.1. A telegraph buzzer operates at a pitch of 400 vibrations per second. A sound wave packet is formed by depressing the key for 0.1 second. What is the frequency range in the wave packet?

Solution:

$\Delta f \, \Delta t = 1$, $\Delta t = 0.1$ sec., hence $\Delta f = (1/\Delta t) = (1/0.1 \text{ sec}) = 10/\text{sec}$.

The frequency range is about 395 per second to 405 per second.

EXAMPLE 6.2. The wave packet pictured here is 0.08 meter long and contains approximately 16 ripples. What is the wave number range in this wave packet?

Solution : Average wave number $k = \dfrac{16}{0.008} = 200/m$

$\Delta k\,\Delta s = 1,\; \Delta s = 0.08\ m,\; \Delta k = \dfrac{1}{\Delta s} = \dfrac{1}{0.08\,m} = 12/m$

The wave number range is 194/m to 206 /m

6.3 Huygens' Principle

Ripple tank. Let us now return to study the propagation of waves by experimenting with water waves. A ripple tank is a useful device for observing water waves. It is a shallow tank with a glass bottom through which a strong light shines onto a screen (Fig. 6.17). Dipping a wire or paddle into the water, creates waves on the water surface; the crests of the waves create bright areas on the screen and troughs create shadows. The patterns of disturbance of the water surface may be observed (Fig. 6.18). A wide paddle generates straight waves (Fig. 6.18a), while the point of a wire generates expanding circular waves (Fig. 6.18b).

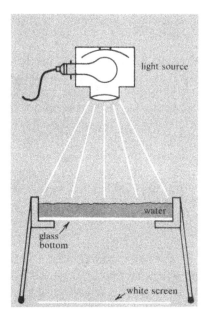

Figure 6.17 (left) Diagram of a ripple tank used for the production and observation of water waves. The wave crests and troughs create bright areas and shadows on the screen.

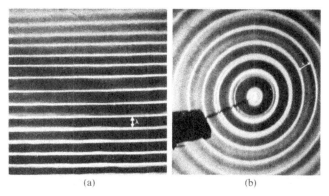

(a)   (b)

Figure 6.18 (above) Water waves in a ripple tank. (a) Waves generated by a wide paddle. (b) Waves generated by the point of a wire.

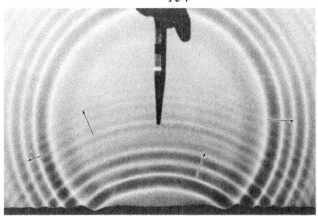

Figure 6.19 *The bright lines of the wave crests indicate the wave fronts. The arrows at right angles to the wave fronts indicate the direction of propagation. The waves were originally produced by the tip of the pointer at the center of the photo. The wave fronts form circles centered on the point where they were created until they reflect from the barrier at the bottom of the photo. Where does the wave appear to be diverging from* after *it is reflected? Can you relate this to what you see in a plane (flat) mirror, as in Fig. 5.17?*

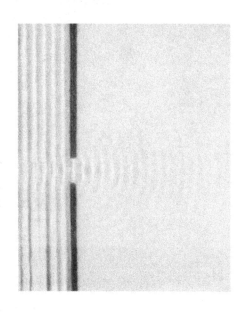

Figure 6.20 *Straight waves from the left impinge on a barrier with a hole. Note the curved, circular shape of the wave front to the right of the barrier*

The point of a wire may be considered as a point source of waves. The reflection of a circular wave pulse created by a pencil point touched to the water surface is shown in Fig. 6.19.

Wave patterns. To describe the pattern, we identify the *wave front*, which is the line made by each wave crest or trough, and the *propagation direction* in which the wave is traveling. The wave always travels in the direction at right angles to the wave front. Therefore, the wave travels in different directions at different parts of a curved wave front such as the one shown in Fig. 6.19.

You can make an interesting discovery if you use a barrier to block off all but one small section of the water. The waves passing through the hole from one side of the barrier to the other spread out in ever increasing circles (Fig. 6.20). This shows a very important result: the small section of the wave front acts as if it were itself a point source of waves.

Huygens' wavelets. In the oscillator model, the oscillator in the small section moves in rhythm with the waves impinging from the source side of the barrier; it also interacts with the oscillators on the other side and

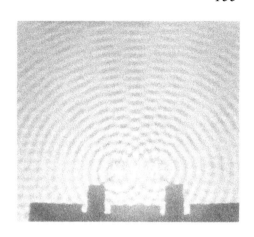

Figure 6.21 Two point sources produce an interference pattern. Note the lines of "nodes" fanning out from the sources.

sets them in motion as though it were a point source. In fact, you can think of every point of a wave front as the source of *wavelets* (numerous mini-waves generated by another wave) that radiate out in circles. That is, each oscillator interacts equally with the other oscillators in all directions from it. This principle is called *Huygens' Principle*. The wavelets have the same frequency of oscillation as their source points in the old wave front. When a wave front encounters a barrier, then most parts of the wave front are prevented from acting as wave sources. What remains is the circular wavelet originating from that part of the wave front that passes through the hole in the barrier.

Two-hole interference. When the barrier has two holes, the waves not only pass through both holes and spread out, but there also is interference between the waves coming from these two "sources." The observable result is very similar to the interference produced by waves from two adjacent point sources (Fig. 6.21). Note the lines of "nodes" fanning out at various angles from the sources, forming what is known as a "two-hole (or double-slit) interference pattern." This pattern demonstrates the existence of interference and can be observed in all waves (including light and sound), not just those in a ripple tank.

Construction of wave fronts. The position of the wave front at successive times may be found by seeking the region of constructive interference of the wavelets emanating from all the source points in a wave front. When there is no barrier, the complete circular wavelets originating from each point in the wave front are not seen because of destructive interference among them.

Schematic diagrams for the procedure of locating the constructive interference are drawn in Fig. 6.22. These diagrams show a wave crest at three successive instants. Huygens' Principle is applied to source points a in the initial wave crest AB to obtain the circles b, c, d. The destructive and constructive interference of all these wavelets results in a new wave crest at the position of the common tangent line CD of all the circles. After a second equal time interval, all the circles

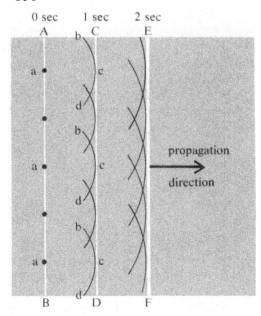

Figure 6.22 The wave front AB (thick white line) advances to CD and then EF, which are the common tangent lines of all the circular wavelets (thin black lines) from Huygens' sources (black dots) in the wave front at AB.

are twice as large, but again the interference effects result in a wave crest EF at the position of the common tangent line of all the larger circles. In this way the straight wave crest advances.

6.4 Diffraction of waves

It is clear from Fig. 6.20 that waves do not necessarily travel in straight lines. Even though the incident wave is headed to the right, the wave transmitted through the hole has parts that travel radially outward

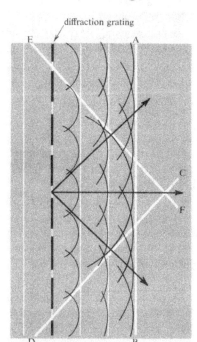

Figure 6.23 Huygens' Principle is used to find the waves transmitted by a diffraction grating. Note the wavelets (thin, curved black lines) centered on the slits. The white lines indicate the undiffracted wave crests (along common tangent line AB) and the diffracted wave crests (common tangent lines CD and EF). The black arrows show the directions of propagation of the observable waves.

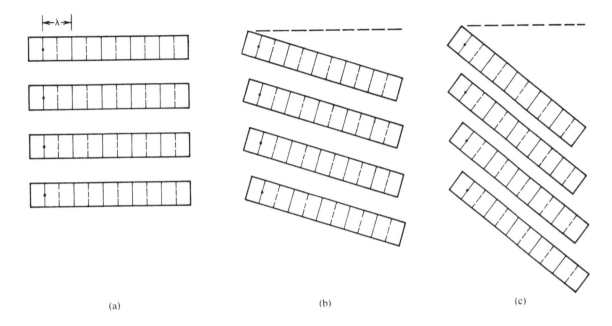

(a)   (b)   (c)

Figure 6.24 Paper strip analogue model for diffraction of waves by a grating. Four paper strips are marked at equal intervals to represent wave troughs and crests. The four strips are then pinned in a row to represent four wave trains passing through equidistant slits in a grating. The strips may be rotated, but are always kept parallel so that the strip direction represents the propagation direction. Interference is determined by the superposition of crests and troughs on the strips.
(a) Constructive interference in the un-diffracted direction is indicated by the alignment of crests with crests, troughs with troughs.
(b) Destructive interference is indicated by the alignment of the crests of one "wave" and the troughs of the adjacent "wave."
(c) Constructive interference in the diffracted direction is indicated by the alignment of crests with crests, troughs with troughs.

from the hole. In other words, the wave was deflected (or bent) by the barrier. This process of deflection of waves passing beside barriers is called *diffraction*. Diffraction makes it possible for waves to bend around a barrier.

Diffraction grating. Let us now apply Huygens' Principle to a device called a *diffraction grating*. A diffraction grating has many evenly spaced slits through which waves can travel. Between the slits, waves are absorbed or reflected. A wave coming through the slits radiates out from each slit in circular wavelets according to Huygens' Principle (Fig. 6.23). You do not observe simple circular waves, however, because the many waves interfere, sometimes constructively and sometimes destructively. A convenient analogue model for diffraction that can be constructed from four strips of paper is described in Fig. 6.24.

158

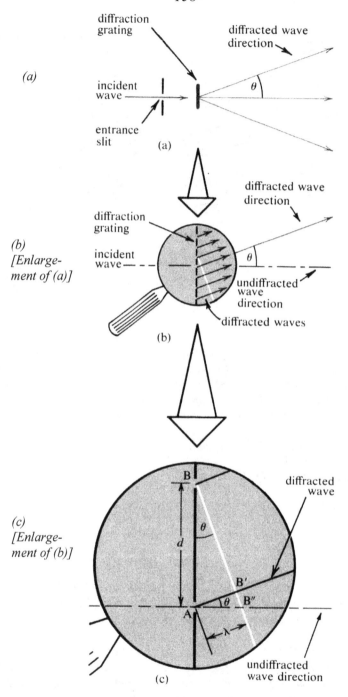

Figure 6·25   Construction of a mathematical model for diffraction by a large diffraction grating.
(a) Waves impinge on the grating from the left. Part of the wave pattern is diffracted at an angle θ, part continues in the undiffracted direction.
(b) Enlarged view of the grating shows that waves passing through adjacent slits travel different distances to contribute to the same wave front (white line).
(c) Constructive interference of diffracted waves occurs if the wave trains from adjacent slits are exactly 1 wavelength out of step as they contribute to one wave front (white line).

distance between slits	d
wavelength	λ
diffraction angle for constructive interference	θ
additional path length for waves passing slit A compared to waves passing slit B	$\overline{AB}'$

Diffraction condition: $\overline{AB}' = \lambda$

Step I: By definition, sine $\angle ABB' = \overline{AB}'/\overline{AB} = \lambda/d$.
Step II: Prove $\angle ABB' = \theta$.
(i) Extend line BB' to the undiffracted wave direction at B".
(ii) θ·is complementary to $\angle AB"B'$ in right triangle AB"B'.
(iii) $\angle ABB'$ is complementary to $\angle AB"B$ in right triangle AB"B.
(iv) Hence $\angle ABB' = \theta$.
Step III: It follows from I and II that sine $\theta = \lambda/d$.

With a large diffraction grating of many slits (perhaps 10,000 slits or more), constructive interference of the waves from all the slits occurs only when the adjacent strips are exactly one, two, or three waves out of step. For all other directions, you can find pairs of close or distant slits that give complete destructive interference and thereby cancel one another's wavelets. Waves are therefore diffracted by the grating only

Equation 6.13
(diffraction grating)

distance between slits
(meters) $= d$
diffraction angle $= \theta$

$$sine\ \theta = \frac{\lambda}{d}$$

into certain special directions. The diffraction angle can be calculated from the condition for constructive interference (Fig. 6.25).

The diffraction grating formula states that the sine of the angle of diffraction is equal to the ratio of the wavelength to the distance between slits. (See Eq. 6.13 and Example 6.3.) The most important practical application of the diffraction grating has been to the study of light, which will be described in the next chapter.

EXAMPLE 6·3. Use of the diffraction formula.

(a) $\lambda = 0.2$ m, $d = 0.3$ m, $\theta = ?$

$$sine\ \theta = \frac{\lambda}{d} = \frac{0.2m}{0.3m} = 0.67$$
$$\theta = 42°$$

(b) $d = 10^{-6}$ m, $\theta = 25°$, $\lambda = ?$

$$sine\ 25° = 0.42$$
$$\lambda = d\ sine\ \theta = 10^{-6}\ m \times 0.42 = 0.42 \times 10^{-6}\ m$$

(c) $\lambda = 10^3$ m, $\theta = 15°$, $d = ?$

$$sine\ \theta = 0.26$$
$$d = \frac{\lambda}{sine\ \theta} = \frac{10^3 m}{0.26} = 3.9 \times 10^2\ m$$

(d) $\lambda = 10^{-4}$ m, $d = 10^{-2}$ m, $\theta = ?$

$$sine\ \theta = \frac{\lambda}{d} = \frac{10^{-4}}{10^{-2}} = 10^{-2}$$
$$\theta = 0.6°$$

(e) $\lambda = 0.3$ m, $d = 0.2$ m, $\theta = ?$

$$sine\ \theta = \frac{\lambda}{d} = \frac{0.3}{0.2} = 1.5$$

θ does not exist.

Diffraction by single slits and small obstacles. Huygens' Principle can also be applied to diffraction by a single slit opening (Fig. 6.20) and to diffraction by a short barrier. The result of the theory suggests that the ratio of the wavelength to a geometrical dimension of the diffracting barrier is of decisive importance for diffraction. In fact, if this ratio is very small (short wavelength, large slit, or large obstacle), the angles of diffraction are very small, so that diffraction is hardly noticeable. If the ratio is large (long wavelength, small slit, or small obstacle),

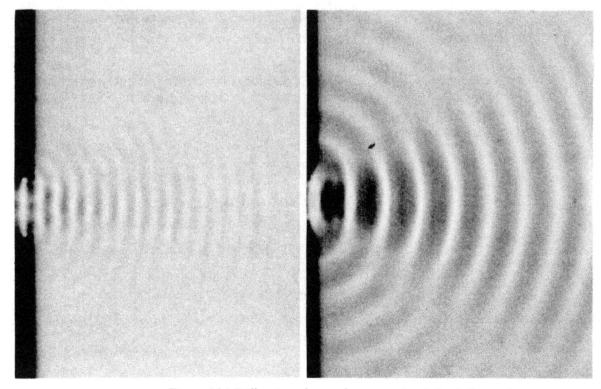

Figure 6.26 Diffraction of waves by an opening. In both photos, the waves are moving from left to right. In the left photo, the wavelength is relatively short (1/3 the width of the opening), so there is little diffraction, and only very weak waves are diffracted away from the original direction of propagation. In the right photo, the wavelength is longer (2/3 the width of the opening), and the waves experience substantial diffraction, spreading out in all directions. After passing through the opening, the wave fronts are essentially semicircles, showing that the waves are now moving in various directions away from the opening. This is a practical demonstration of Huygen's Principle: the waves passing through the opening act as point sources of new waves, which then travel in all directions away from the sources.

then diffraction covers all angles, but the amplitudes of the diffracted waves are very small because the slit or obstacles are small. For intermediate values of the ratio (wavelength comparable to the slit or obstacle in size), diffraction is an important and easily noticeable phenomenon. Two photographs of waves in a ripple tank (Fig. 6.26) show long and short wavelength waves passing through an opening and being diffracted when they pass through an opening. The greater diffraction of the longer wavelength waves is obvious.

6.5 Reflection of waves

The ripple tank photograph to the left (from Fig. 6.19) shows reflection of an expanding circular wave packet. We picked one point on the barrier and drew arrows showing the approximate direction of propagation before and after reflection from that point. The angles of incidence and reflection (as defined in Fig. 5.11) are shown. You can measure the angles to test whether they are equal; we measured one to be 43.5° and the other to be 45°; this is satisfactory agreement given the accuracy of our measurements.

We can also use Huygens' Principle to investigate the relation of these angles in a more general way. According to this principle, each point in a wave front acts like a source of wavelets propagating outward. The wavelets have the same frequency and wavelength as the original waves. The common tangent line of the wavelets is the wave front they produce by constructive interference.

The reflection process is illustrated in Fig. 6.27. A straight wave is incident on the reflecting barrier obliquely from the left. Between the

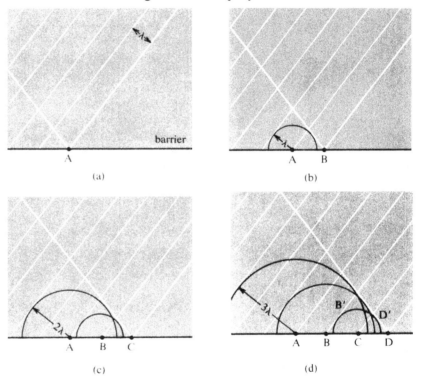

Figure 6.27 Reflection of waves by a barrier. The incident wave crests (white lines) are advancing toward the lower right. Only one reflected wave, moving toward the upper right, is shown. To avoid clutter in the diagram, we have not drawn the other reflected waves.

(a) Three wave crests are striking the barrier, which has reflected a section of the first crest. Point A on the first crest acts as a source of Huygens' wavelets.

(b) The wave crests advance by a distance of one wavelength (λ), and the wavelet from Point A has expanded into a semicircle of radius λ. Point B becomes a source of wavelets.

(c) The wave crests advance by another wavelength; the wavelet from Point A now has a radius of 2λ; the wavelet from B has a radius λ and the Point C becomes a source of wavelets.

(d) The wavelet from A has radius 3λ; the wavelet from B has radius 2λ, the wavelet from C has radius λ, and Point D becomes a source of wavelets.

The wavelets constructively interfere all along the common tangent line DD', which defines the location and direction of the reflected wave fronts.

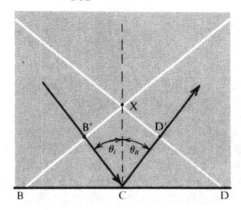

Figure 6·28 Construction of a mathematical model for wave reflection based on Fig. 6·27. Arrows B'C and CD' represent, respectively, the incident and reflected propagation directions. They are at right angles to the corresponding wave fronts BB' and DD' (white lines). Consider the two right triangles XCB' and XCD'. They share the common hypotenuse XC and have two sides equal, CB' = CD' = λ. Hence the two triangles are congruent. It follows that corresponding angles are equal, $\theta_i = \theta_R$.

four successive instants shown in Fig. 6.27, the wave advances between each drawing by 1 wavelength. The Huygens wavelets formed by the first wave crest passing through points A, B, C, and intermediate points on the barrier have a common tangent, which is the reflected wave front. The crests of the Huygens' wavelets all fall on the common tangent, where they interfere constructively; at all other points, the wavelets interfere destructively and cancel one other.

To relate the angles of incidence and reflection, the directions of propagation have to be taken into account. This is done in Fig. 6.28, where only one incident and one reflected wave crest from Fig. 6.27d are included. The application of Huygens' Principle in Fig. 6.28 results in a familiar conclusion: the angle of incidence is equal to the angle of reflection (Eq. 6.14). This statement may be called the *law of wave reflection*.

Equation 6.14 (Law of Reflection)

angle of incidence $= \theta_i$

angle of refraction $= \theta_R$

$$\theta_i = \theta_R$$

6.6 Refraction of waves

When a wave propagates from one medium into another, its direction of propagation may be changed. An example of this happening with water waves is shown in Fig. 6.29. The boundary here is between deep water above and shallow water below. Even though water is the

Fig 6.29 Water waves passing from a deeper region to a shallower region are refracted and travel in a different direction at the boundary. Huygens' Principle does not reveal which direction the waves are traveling. Can you figure this out? (Hint: Look carefully for reflected waves!)

Equation 6.3b

$$v = f\lambda$$

material on both sides of the boundary, it acts as a different medium for wave propagation when it has different depths. You can see that the wavelength is shorter in the shallow water and can infer from this that the wave speed is slower there (Eq. 6.3b).

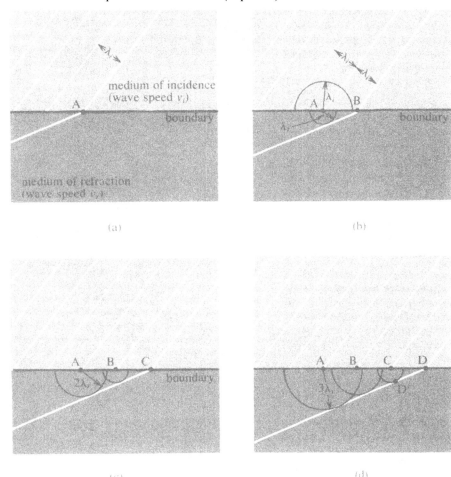

*Figure 6.30 Refraction at a boundary between two media, in which the wave speeds are v_i (above boundary) and v_r (below boundary). The wave speed is assumed to be **less** below the boundary than above it. (v_r is **less** than v_i). The waves are moving downward to the right.*

(a) Three wave crests (white lines) are incident on the boundary, which has refracted a section of the waves. Point A on the first crest acts as a source of Huygens' wavelets.

(b) The wavelet from point A has expanded into a semicircle of radius λ_r below the boundary and a semicircle of radius λ_i above the boundary. The latter gives rise to a reflected wave (see Fig. 6.27) and will not be described further. The wave crests advance by the distance λ_i. Point B becomes a source of wavelets.

(c) The wavelet from A has reached radius $2\lambda_r$, the wavelet from B has radius λ_r. Point C becomes a source of wavelets.

*(d) The wavelet from A has radius $3\lambda_r$, the wavelet from B has radius $2\lambda_r$, and the wavelet from C has radius λ_r. The common tangent line DD' coincides with the refracted wave front. Note the change in the direction of propagation (which is perpendicular to the wave front): we can see from the diagram that the refracted wave (below the boundary) is traveling **slower** and in a direction **farther** from the boundary surface than the incident wave (that is, **closer** to the perpendicular to the boundary).*

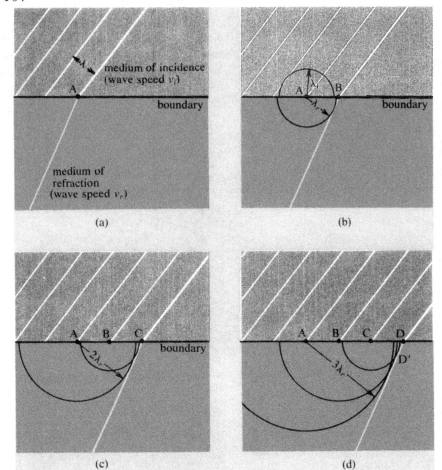

*Figure 6.31 Refraction at a boundary between two media, in which the wave speeds are v_i (above boundary) and v_r (below boundary). The wave speed is assumed to be **greater** below the boundary (v_r is **greater** than v_i). (This is similar to Figure 6.30 but the speeds are reversed.) The waves are assumed to be moving downward to the right.*

(a) Four wave crests (white lines) are incident on the boundary, which has refracted a section of the waves. Point A on the first crest acts as a source of Huygens' wavelets.

(b) The wavelet from point A has expanded into a semicircle of radius λ_r below the boundary and a semicircle of radius λ_i above the boundary. The latter gives rise to a reflected wave (see Fig. 6.27) and will not be described further. The wave crests advance by the distance λ_i. Point B becomes a source of wavelets.

(c) The wavelet from A has radius $2\lambda_r$; the wavelet from B has radius λ_r. Point C becomes a source of wavelets.

*(d) The wavelet from A has radius $3\lambda_r$, the wavelet from B has radius $2\lambda_r$, and the wavelet from C has radius λ_r. The common tangent line DD' coincides with the refracted wave front. Note the change in the direction of propagation (which is perpendicular to the wave front): we can see from the diagram that the refracted wave (below the boundary) is traveling **faster** and in a direction that is **closer** to the boundary surface than the incident wave (that is, **further** from the perpendicular to the boundary).*

Note: Although we have assumed above that the waves are traveling toward the right, this demonstration can also be carried out using the same diagram with the waves traveling in the opposite direction. Thus wave theory based on Huygens' Principle predicts that refracted waves will follow the same path in either direction. Does this seem reasonable to you? Can you suggest any observations or experiments that would confirm or refute this?

The refracting boundary. The change in the direction of propagation is called refraction, the same term that was introduced in Section 5.2. We will now find the law of refraction of waves by applying Huygens' Principle to the propagation of the wave across the boundary between two media with different wave velocities. Each point in the wave front that touches the medium of refraction acts like a source of wavelets that propagate into that medium. These waves have the same

Equation 6·15

*wave speed in medium of
   incidence v_i
wave speed in medium of
   refraction v_r
wavelength in medium of
   incidence λ_i
wavelength in medium of
   refraction λ_r*

$$v_i = \lambda_i f$$

$$v_r = \lambda_r f$$

Equation 6.17

$$\frac{\text{sine } \theta_i}{\text{sine } \theta_r} = \frac{\lambda_i}{\lambda_r} \quad (a)$$

$$\frac{\text{sine } \theta_i}{\text{sine } \theta_r} = \frac{v_i}{v_r} \quad (b)$$

frequency as their source, and therefore the same frequency as the wave in the medium of incidence. The wave in the medium of refraction, however, where the speed is different, has an altered wavelength, because wavelength, frequency, and speed are related by $v = \lambda f$ (Eq. 6.15). The ratio of the wavelengths in the two media is equal to the ratio of the wave speeds, since these two properties of the wave are directly proportional as long as the frequency remains the same (Eq. 6.16). Thus, the change in medium results in a changed wavelength.

Construction of the refracted wave front. The procedure for finding the law of refraction is very similar to that used in the preceding section to find the law of reflection. A straight wave is incident on the refracting boundary obliquely from the left. Between each of the four successive instants shown in Figs. 6.30 and 6.31, the wave advances by 1 wavelength. The Huygens' sources on the boundary generate wavelets that propagate into the second medium with the wave speed and therefore the wavelength appropriate to that medium. The case of reduced wave speed and wavelength is illustrated in Fig. 6.30, while the case of increased wave speed and wavelength is illustrated in Fig. 6.31. In both cases the wavelets originating in points A, B, and C (and intermediate points on the boundary) have a common tangent that is the refracted wave front.

Law of refraction. To relate the angles of incidence and refraction, the directions of propagation have to be taken into account. This is done for both cases above in Fig. 6.32, where only one incident and one refracted wave crest from the previous figures are included. The conclusion from the application of Huygens' Principle is that the sines of the angles of incidence and refraction have the same ratio as the wavelengths (Eq. 6.17a) and, therefore, the same ratio as the wave speeds in the two media (Eq. 6.17b). This result is similar in form to Snell's Law of Refraction: (n_i sine $\theta_i = n_r$ sine θ_r, Eq 5.2, Section 5.2), a key assumption in Newton's ray model of light. We shall study this further below in Section 7.2, where we will compare and evaluate the ray and wave models in some detail.

If you look at the propagation direction of the refracted waves in Figs. 6.30 and 6.31, you will recognize that the effect of crossing the boundary can be described as follows. In the medium with the slower wave, the propagation direction is farther away from the boundary surface; in the medium with the faster wave, the propagation direction is closer to the boundary surface. You may use the tables of the sine functions (Appendix, Table A.7) to solve problems on the refraction of waves.

Reflection at the boundary. The application of Huygens' Principle to the boundary between the two media leads to reflected wavelets as well as refracted ones. One reflected wavelet is indicated in Fig. 6.30b and one is indicated in Fig. 6.31b. Since these wavelets are in the medium of incidence, their speed and wavelength are appropriate to that medium. By pursuing their formation further, we could have obtained the same sequence of diagrams as are shown in Fig. 6.27. The wavelets would interfere constructively to form a reflected wave according to the law of reflection (Eq. 6.14). Thus wave theory suggests that we should also look for reflected waves, and, in fact, by looking carefully, you can indeed identify reflected waves in the deeper water of Fig. 6.29! In other

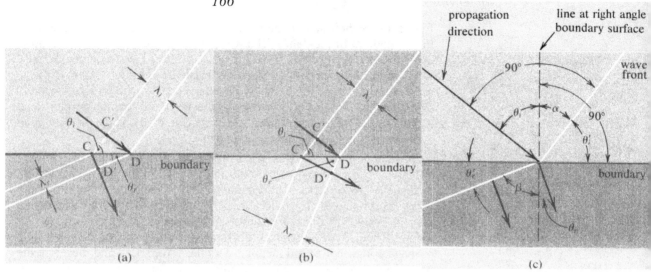

Figure 6.32 Construction of a mathematical model for wave refraction, based on Figs. 6.30 and 6.31. Note that in (a) the medium with the longer wavelength (faster speed) is at top in lighter shading. However, in (b) the medium with the longer wavelength (faster speed) is at bottom, also in lighter shading.

In both (a) and (b), Arrows C'D and CD' represent, respectively, the incident and refracted propagation directions. They are at right angles to the corresponding wave fronts (white lines) CC' and DD'. ASSUMPTION: Angle C'CD is equal to the angle of incidence θ_i and D'DC is equal to the angle of refraction θ_r ; we will prove this assumption in (c) below. The definition of the sine functions can be applied to right triangles CDC' and CDD' with the following results:

$$\text{sine } \theta_i = \frac{C'D}{CD} = \frac{\lambda_i}{CD} \quad (1)$$

$$\text{sine } \theta_r = \frac{D'C}{CD} = \frac{\lambda_r}{CD} \quad (2)$$

Divide Eq. (1) by Eq. (2)

$$\frac{\text{sine } \theta_i}{\text{sine } \theta_r} = \frac{\lambda_i}{\lambda_r} \quad (3)$$

(c) Proof of ASSUMPTION asserted above: the angle of incidence θ_i equals the angle between the boundary and the wave front θ_i'. Two overlapping right angles in the medium of incidence are indicated in the figure (c) above. Both angles θ_i and θ_i' are complementary to the angle α. Consequently the two angles are equal, $\theta_i = \theta_i'$. The same construction with respect to the angle β in the second medium leads to the conclusion that $\theta_r = \theta_r'$.

words, the incident wave appears to be split by the boundary into a reflected wave and a refracted wave. Huygens' Principle has indeed served us well; however, it does not reveal how the energy carried by a wave is divided between reflection and refraction.

The occurrence of partial reflection gives an important clue about the sense of the direction of propagation of waves. By examining only the incident and refracted waves, such as in Fig. 6.29, you would not be able to determine whether the waves were incident as described at the beginning of this section (from upper left), or whether the waves were incident from the lower right and passed from the shallow water to the deeper water. The observation of reflected waves in the upper part of the photograph is evidence that the waves were incident from above.

Summary

Equation 6.3b
(wave speed)

$$v = \lambda f$$

The concept of waves has its roots in water waves. More generally, waves are oscillatory displacements of a medium from its equilibrium state. Two important forms that such disturbances can take are the wave train, in which the displacement pattern repeats over and over, and the wave pulse, in which the displacements are localized in space and time. The wavelength, wave number, period, frequency, amplitude, and speed of the waves can be defined for a wave train, but only the last two of these can be defined for a pulse. The frequency, wavelength, and speed of a wave train are related by $v = \lambda f$ (Eq. 6.3b).

The wave theory is built upon the above ideas and applies to a wide variety of types of waves. The goal of wave theory is the construction of mathematical models to describe the behavior and propagation of waves. Wave theory explains and clarifies a large variety of phenomena. Such phenomena include sound, music, water waves, radio, light, constitution of the atom, traffic flow, and earthquakes.

The wave theory rests on two key assumptions about waves: 1) the superposition principle and 2) Huygen's Principle. The theoretical deductions from these assumptions can be compared with observation to identify the successes and the limitations of the wave theory.

According to the superposition principle, the displacement of the combination of two or more waves passing through the same point in space at the same time is the sum of the displacements of the separate waves. The result is constructive or destructive interference, depending on whether the separate waves reinforce or oppose one another.

Huygens' principle is used to investigate the propagation of waves. Each point in a wave front is considered as a source of circular outgoing wavelets. The amplitude and frequency of the wavelets are determined by the amplitude and frequency of the wave at the source point. The wavelets interfere constructively along their common tangent line, which is therefore the front of the propagating wave. Elsewhere, the wavelets interfere destructively and are not separately observable.

Huygens' Principle allows us to conduct thought experiments on the propagation of waves and furnishes a procedure for determining the results. We have used Huygens' Principle to understand diffraction, reflection, and refraction of waves.

The wave theory does not attempt to relate the wave speed, amplitude, and energy to properties of the medium, the wave source, and the wave absorber. These matters require more detailed working models for the three systems; their treatment is beyond the scope of this text.

List of new terms

medium	superposition	Huygens' Principle
(for wave propagation)	interference	wave front
wave train	constructive	propagation direction
wave pulse	interference	Huygens' wavelets
amplitude	destructive	diffraction
frequency (f)	interference	diffraction grating
wavelength (λ)	node	reflection of waves
wave number (k)	tuned system	refraction of waves
period ($\mathcal{T}$)	beats	standing waves
wave speed (v)	wave packet	
	uncertainty	
	principle	

List of symbols

k	wave number	Δf	frequency range *(f₁ - f₂)*
λ	wavelength	Δk	wave number range *(k₁ - k₂)*
f	frequency	v	wave speed
$\mathcal{T}$	period	θ	diffraction angle
N	number of waves	θ_i	angle of incidence
Δs	pulse width	θ_R	angle of reflection
Δt	time for one beat	θ_r	angle of refraction

Problems

Here are some suggestions for problems that have to do with water waves. Observations on a natural body of water are made most effectively from a bridge or pier overhanging the water. You may observe wind-generated wave trains or pulses generated by a stone. By dipping your toe rhythmically into the water, you may be able to generate a circular wave train.

Experiments can be conducted in a bathtub or sink if natural bodies of water are not available. A pencil or comb dipped horizontally into the tub near one end can generate straight wave pulses. Dipping your finger, a pencil or a comb vertically will generate circular waves. To observe the waves, place a lamp with one shaded bulb over the bathtub so as to direct the light at the water surface and not into your eyes. You should also avoid looking at the reflected image of the bulb. Under these conditions, waves cast easily visible shadows on the bottom of the tub or on the ceiling. **Caution: You must be careful when using electricity near the bath or sink; an electrical shock from household**

current can be dangerous. Keep water away from the lamp and do not under any circumstances touch the lamp with wet hands nor while any other part of your body is wet or touching something wet.

1. Measure the speed of water waves by measuring how long they take to traverse a given distance. Describe the conditions of your observations, especially the depth of the water and the amplitude of the waves. If you observe wave trains, determine their frequency and wavelength and test Eq. 6.3b.

2. Identify the interaction(s) that are involved in the propagation of waves on a water surface.

3. Observe waves at the seashore and report qualitatively about as many of the following as you can observe.
   (a) Differences in speed of various waves.
   (b) Differences in direction of propagation.
   (c) Applicability of the superposition principle.
   (d) Effect of the depth of the water on the wave motion.
   (e) Reflection of wave fronts.
   (f) Refraction of wave fronts.
   (g) Diffraction of waves.
   (h) Transfer of energy from the waves to other systems.

4. Sand ripples are frequently observed on the ocean or lake bottom in shallow water. They are formed by the interaction of sand and water just as water waves are formed by the interaction of water and wind. Measure the wavelength of sand ripples that you observe. Comment on their propagation speed.

5. Observe reflection of water waves in your sink or bathtub. Estimate the angles of incidence and reflection as well as you can and compare your results with the law of reflection for waves.

6. Observe single-slit diffraction of water waves in your sink or bathtub. Report the slit width you found most suitable and other conditions that helped you to make the observations.

7. Several different diffraction gratings diffract water waves with a wavelength of 0.03 meter. Find the diffraction angle for a diffraction grating with a slit spacing of (a) 0.30 meter; (b) 0.10 meter; (c) 0.05 meter; (d) 0.025 meter.

8. Water waves are diffracted by a grating with a slit spacing of 0.30 meter. Find the wavelengths for the waves when the diffraction angle is (a) 10°; (b) 25°; (c) 60°.

9. Find the result of superposing the following three waves:
   wave A -- wavelength (λ) = 6 centimeters (cm), amplitude = 3 cm;
   wave B -- λ = 3 cm, amplitude = 2 cm;
   wave C -- λ = 2 cm, amplitude = 1 cm.
   Start from a point where all three waves interfere constructively; keep plotting until all three waves again interfere constructively.

10. Sound waves in air have a wave speed of 340 meters per second. Find the wavelength and wave number of the following sound waves: (a) middle C, frequency (f) = 256 per second; (b) middle A, f = 440 per second (c) high C, f = 1024 per second.

11. Use the paper strip analogue (Fig. 6.24) to study diffraction of waves. Report the wavelength, "slit" separation, and diffraction angle(s) for three different "gratings." Choose λ/d small (0.5), medium (2.0), and close to one for the three cases. (Note: one grating may give several diffraction angles, according to whether the waves from adjacent slits are 1, 2, 3, ... wavelengths out of step.) Make as many paper strips as you feel necessary to help you.

12. Use the paper strip analogue (Fig. 6.24) to study diffraction from only two slits. Measure and/or use geometrical reasoning to find the angles of diffraction amplitude maxima (constructive interference) and diffraction amplitude minima (destructive interference). Compare your results with those obtained for a diffraction grating and describe qualitatively the reasons for similarities and differences.

13. Identify one or more explanations or discussions in this chapter that you find inadequate. Describe the general reasons for your judgment (conclusions contradict your ideas, steps in the reasoning have been omitted, words or phrases are meaningless, equations are hard to follow, . . .), and make your criticism as specific as you can.

Bibliography

W. Bascom, *Waves and Beaches*, Doubleday, Garden City, New York, 1964.

C. Huygens, *Treatise on Light*, Dover Publications, New York, 1912.

U. Haber-Schaim, *PSSC Physics, Seventh Edition,* Kendall/Hunt, Dubuque, Iowa, 1991.

R. A. Waldron, *Waves and Oscillations*, Van Nostrand, Princeton, New Jersey, 1964.

Articles from Scientific American. Some or all of these, plus many others, can be obtained on the Internet at http://www.sciamarchive.org/.

E. D. Blackham, "The Physics of the Piano" (December 1965).

E. E. Helm, "The Vibrating String of the Pythagoreans" (December 1967).

From a comparison of various
experiments, it appears
that the breadth of
the undulations constituting the
extreme red light must be
supposed to be, in air,
about one 36 thousandth of an inch,
and those of the extreme
violet about one 60 thousandth; . . .
almost 500 million of millions of
such undulations must enter
the eye in a single second.

THOMAS YOUNG
Course of Lectures on
Natural Philosophy and
the Mechanical Art,
1807

Wave models for sound and light

In this chapter we will return to the discussion of sound and light that we began in Chapter 5. This time we will describe both phenomena from the point of view of the wave model. As we already explained in Section 1.1, the history of models for light was full of controversy and the currently accepted models are still undergoing change. By contrast, the understanding of sound as wave motion dates to the seventeenth century and has advanced steadily with significant contributions by many physicists and mathematicians.

7.1 Applications of wave theory to sound

Early history. Sound and motion had been associated with one another since ancient days, but Galileo was the first person to note clearly the connection between the frequency of vibration of a sound source and the pitch of the note that was produced. This, in spite of the fact that musical instruments had existed for millennia!

Early experiments on the speed of sound. Many of the early experiments reveal a simplicity, an ingenuity, and occasionally a misconception that are charming. Thus, Galileo's idea originated in his scraping a knife at various speeds over the serrated edge of a coin. The first measurements of the speed of sound were made by timing the interval between the flash and sound of a distant gun being fired. A value of the speed very close to the present 344 meters per second was found. It was also noticed that sound travels faster in water (experiments were conducted in Lake Geneva, Switzerland) and in steel wires than in air.

Frequency of sound. Marin Mersenne, sometimes called the "father of acoustics," was the first individual to measure the frequency associated with the musical note emitted by a particular organ pipe. He tuned a short brass wire to the same musical note (and therefore the same frequency) as the organ pipe by hanging on weights to adjust the tension. Then he repeated the experiment with wire of the same material and thickness, and under the same tension, but 20 times as long. This wire vibrated so slowly that he could count ten vibrations per second. Because the material, thickness and tension of the two wires were identical, the wave speed along the two wires was the same. The relationship between frequency and wavelength (Eq. 6.3, $f = v/\lambda$) shows that, if v is constant, the frequency is *inversely proportional* to wavelength. (To review inverse proportions see Appendix A.2.) Thus Mersenne concluded that the original short wire (and the organ pipe) had executed 200 vibrations per second, 20 times as many as the long wire.

Medium for sound propagation. One of the big questions that had to be resolved was whether sound needed a medium. Vacuum pumps had been invented about the middle of the seventeenth century, and it was a simple matter to suspend an alarm clock or a bell in a jar to be evacuated. Unfortunately, a decade was required before Robert Boyle (1627-1691) successfully observed that the alarm in the jar could not be heard outside when the jar did not contain air. Other investigators had arrived at the contrary conclusion, perhaps because they had failed to remove

Marin Mersenne (1588-1648). The scientific movement in France was characterized by a tradition of informal gatherings of interested scientists. One such informal group was held together by the personality of Marin Mersenne, a friar in the Palais Royale. Mersenne and his students disseminated the discoveries of Galileo, popularized the Cartesian coordinate frame, and publicized the work of such men as Pascal. Moreover, Mersenne succeeded (where Galileo had failed) in identifying the path of a falling body as having the shape of a parabola.

the air completely enough, or perhaps because the bell's support conducted the sound to the observers.

By the time of Isaac Newton, the wave model for sound as vibration of an elastic medium was generally accepted. In fact, it was so well accepted that the differences between sound and light, which we described in Section 5.1, convinced Newton that light could not also be a wave phenomenon (see Section 7.2)!

Speed of sound. In Section 6.1 we described the conditions for wave motion in terms of the oscillator model for the wave medium. The two key factors were the inertia of the oscillators making up the medium and the interaction among them. The gasbag model for air and the MIP model for solids and liquids represent these materials as composed of subsystems that have inertia and interact with one another. Sound waves, therefore, propagate in all materials, but with a predicted speed that is high if the oscillators have low inertia and/or strong interaction and low if the conditions are opposite. Values for the speed of sound in various materials are listed in Table 7.1.

According to the MIP model, hard materials, in which there is strong interaction between oscillators, should exhibit a higher sound velocity than "soft" materials. You can see a trend compatible with your expectation; rubber, lead, paraffin, and water have a relatively low sound speed, while glass, iron, and aluminum have a high sound speed.

Speed of sound in gases. Gases are more difficult to include in the comparison, because both the interaction and the inertia in these low-density materials are much smaller than in liquids and solids, and these two differences may compensate for one another. From the fact that the sound speed in gases is lower than that in solids or liquids, you can conclude that the reduced interaction strength is more significant than the reduced inertia. This analysis of sound speed in gases is an example where the model does not lead to an unambiguous prediction, but where the model and experimental data may be combined to yield more insight into the properties of matter.

TABLE 7.1 SPEED OF SOUND IN SOLIDS, LIQUIDS, AND GASES

Material	Speed (m/sec)
metals:	
aluminum	5100
brass	3500
copper	3560
gold (soft)	1740
iron	5000
lead	1230
brick	3650
glass	5000
marble	3800
paraffin	1300
rubber	54
liquids:	
alcohol	1240
water	1460
gases (room temperature):	
air	344
carbon dioxide	277
helium	960
hydrogen	1360

Frequency of musical notes. The wave model explains musical notes of different pitch as vibrations of different frequency. This relation was first investigated quantitatively by Mersenne. The presently accepted standard frequency is 440 vibrations per second for the "middle A" note. The entire musical scale is divided into octaves, which are two notes with a frequency ratio of two to one. Thus, various A notes have vibration frequencies of 110 per second, 220 per second, 440 per second, 880 per second, and so on. In Western music since about 1800 the octave interval is generally divided into twelve notes ("semitones"). The frequency ratio of adjacent semitones is slightly less than 1.06. In other words, each semitone has a frequency almost 6% larger than the next lower semitone. The notes in one octave and their frequencies are listed in Table 7.2. All frequencies in the table are multiples of the standard A-440 frequency. You can calculate the frequencies of the corresponding notes in higher or lower octaves by successively doubling or halving the frequencies in the table.

TABLE 7.2 PROPERTIES OF SOUND WAVES FOR MUSICAL NOTES

Note	Frequency (f, /sec) (approximate)	Frequency ratio to C note (approximate)	Wavelength in air (λ, m) (= 344/f, approx.)
C (middle C)	262	1/1	1.31
C# = D^b	277	-	-
D	294	-	-
D$^{\#}$ = E^b	311	-	-
E	330	5/4	1.04
F	349	4/3	0.99
F$^{\#}$ = G^b	370	-	-
G	392	3/2	0.88
G$^{\#}$ = A^b	415	8/5	0.83
A (standard)	440	5/3	0.78
A$^{\#}$ = B^b	466	-	-
B	494	-	-
C	523	2/1	0.66

Equation 7.1

$$v = f \lambda,$$

$$or, \ \lambda = \frac{v}{f}$$

$$or, \ f = \frac{v}{\lambda}$$

EXAMPLE:
$v = 344 \, m/sec,$

$f = 262 /sec$

$$\lambda = \frac{344}{262} m = 1.31 \, m$$

Wavelengths of sound waves. Air is, of course, the most important medium for the transmission of sound on earth. You can calculate the wavelengths, λ, in air of musical notes from their frequency (f, Table 7.2, second column) and the known sound speed, v, in air (344 m/sec from Table 7.1), by using the equation v = f λ in the form λ = v/f (Eq. 7.1, from Eq. 6.3b). The results are included in Table 7.2, fourth column. It is clear that audible sound waves, especially the ones used in speech, have a wavelength comparable to the size of the human body and to objects in our environment. This result (wavelengths of a few feet) is not surprising—as explained above (Section 6.2), the lengths of organ pipes range from a few inches to many feet, which is also the approximate size of the wavelengths of the notes they produce.

The magnitude of wavelengths of audible sound in air explains, in the context of the wave theory, why it is impossible to form a sharp acoustic image of the placement and shape of primary sound sources or reflectors. We pointed out in Section 6.4 that obstacles whose size is comparable to the wavelength diffract waves most strongly. Diffraction by persons, furniture, doors, and buildings, therefore, bends the sound waves so much that their direction and intensity is related only remotely to the placement of the primary sound sources and the reflecting surfaces. Sound transmits certain information about the primary source, for instance, intensity, pitch, and duration, but no sharp image of the location of the sound sources. Only in a clear space, and with the help of both ears, which receive somewhat different information (stereophonic), can we determine the position of sound sources in even an approximate way.

Musical instruments. In Section 6.2 we explained standing waves in an organ pipe and on a violin string as standing waves on tuned systems. The musical octave is simply related to the musical intervals between notes that can exist in a tuned system. For illustration, the various notes generated by standing waves in an organ pipe of 0.657-meter

TABLE 7.3 STANDING WAVES IN AN ORGAN PIPE WITH LENGTH (L) = 0.657 METER

Wave-length (meters) (λ)	Fre-quency (/sec) (f = v/λ = 344/λ)	Note on mus-ical scale
1.31=2 L	262	C
0.66=2/2 L	524	C'*
0.44=2/3 L	785	G'*
0.33=2/4 L	1047	C"*
0.26=2/5 L	1309	~E"†
0.22=2/6 L	1571	G"*

* C' indicates a note exactly one octave above middle C with frequency = 2 x 262 = 524; C" indicates a note exactly two octaves above middle C with frequency = 4 x 262 = 1048. G', G" and E" are defined similarly with respect to G and E in Table 7.2.

† The note with frequency of 1309 corresponds only approximately with E".

length are listed in Table 7.3. The lowest frequency wave (longest wavelength) has nodes at both ends of the pipe with half the wavelength equal to the length of the pipe, or a full wavelength equal to 2L (first line of Table 7.3). The frequency is determined from Eq. 7.1, $f = v/\lambda = 344/1.32 = 262$ /sec. This wave is known as the fundamental; it has the lowest frequency possible on this system. Other waves (overtones) must also have nodes at the ends of the pipe, but additional half wavelengths can be fitted within the length of the pipe; this determines the remaining wavelengths (and higher frequencies) in Table 7.3, as explained in Sect. 6.2 (Eqs. 6.4 and 6.5).

Stringed instruments. We will explain stringed instruments (such as the guitar and violin) using the wave model. A single vibrating string does not transfer energy and sound effectively to the air; therefore, all stringed instruments require amplification and/or a well-designed sound chamber (the instrument's hollow body), which acts as a coupling element (Section 4.3) to the air. The sound chamber is passive in the sense that it does not affect the transfer of energy, but it is very important in determining the "quality" of the sound we hear. The best instruments are made from wood; the specific characteristics of the wood and its finish (the surface of which actually transfers the sound to the air) are critical.

In a guitar or violin, the lengths of all the strings (and thus the wavelengths of the sounds traveling along the strings) are determined by the length of the instrument and, therefore, are all the same. For the instrument to be able to produce a sufficiently wide range of pitches, the various strings, played at full length, must, therefore, vibrate at different frequencies. Waves of equal wavelengths but different frequencies must travel at different wave speeds along the various strings (Equation 7.1, $v = f\lambda$). Differing speeds means the various strings must have different inertia (weight or density) and/or a different strength of interaction (tension) along the string.

The first column of Table 7.4 lists the notes sounded by the six strings of a guitar when vibrating at full length; the second column lists the frequencies of these notes. To "tune" the strings so they vibrate at

TABLE 7.4 GUITAR WITH STRINGS 0.65 METER LONG (= 1/2 λ)

Note sounded by string (at full length)	Frequency (f, /sec) (from Table 7.2)	Wave speed on string (v, m/sec) (= fλ) (= f x 1.3m)	Wavelength in air (λ, m) (= v_{air}/f)
E	82.5	107	4.17
A	110	-	-
D	147	-	2.34
G	196	-	-
B	247	-	-
E	330	429	1.04

exactly the correct frequency, the player turns the pegs, slightly adjusting the tension and thus the wave speed.

We can use the information in Table 7.4 to find the actual speed of the waves on the strings. The strings are all 0.65m long, so there must be nodes at both ends; thus half the wavelength must be 0.65m, and $\lambda = 1.30$ m. Putting this and the frequency into Equation 7.1 yields the wave speed on the string (Table 7.4, third column). Alternatively, using the frequency plus the speed of sound *in air* in Eq. 7.1 gives the wavelength in air (fourth column). Note that it is easy to forget that the wave speed (v) in Eq. 7.1 depends on the medium (sound or air). You should be careful to use the appropriate value, depending upon the medium (the substance that is actually vibrating and carrying the wave).

Other sound phenomena. We began the discussion of sound in Chapter 5 with the unexpressed operational definition of sound, "sound is what people can hear." It is now appropriate to redefine sound with a formal definition, as displacement waves in a medium, and thereby to extend the concept of sound beyond the limitations of the human ear. As a matter of fact, the human ear is capable of detecting sound waves only between frequencies of approximately 20 and 20,000 vibrations per second, with a great deal of variation among individuals. Lower-frequency vibrations are sensed as rapid knocking, while higher-frequency vibrations are not detected at all, except possibly as pain if they are very intense. Nevertheless, these other waves do occur naturally and/or have been exploited technologically.

Ultrasonics. Ultrasonics refers to sounds with frequencies too high for the human ear. Therefore, ultrasonic sound waves have a much shorter wavelength than audible sound, only about 0.01 meter or less. Ordinary size obstacles therefore diffract ultrasonic sound waves much less than audible sound; hence ultrasonic sound waves can be directed into narrow beams that are reflected by environmental objects. The reflected beam furnishes information about the position of the reflecting object.

Sonar is a method for locating objects under water using ultrasonic sound waves. A high frequency sound source emits wave pulses at a frequency of 20,000 vibrations per second or more; a detector records the reflected pulses (echoes). The relative position (direction and distance) of the reflecting object is determined from the direction of the reflected pulse and the time delay of its arrival after the original pulse was emitted. Sonar depth gauges, which measure the distance to the ocean floor by the time delay of reflected pulses, are now standard equipment on many pleasure boats and commercial craft.

In an industrial application of the sonar principle, sound with several million vibrations per second is used to locate flaws in steel pieces, rubber tires, and so on. The flaw is an irregularity that reflects sound waves and can thereby be detected.

The sonar principle has also been applied in many beneficial ways to health care, most notably to form images of a developing human fetus within the mother's womb. The baby's tissues and bones reflect the sound waves, and computer-assisted detectors can then, amazingly enough, produce an image from the reflected waves and display it on a conventional TV monitor. Naturally the potential effects of the sound waves on the human body must be investigated carefully and, so far, such effects have been found to be negligible.

The bat is an unusual mammal that can use the sonar principle to locate objects and avoid obstacles in dark spaces (such as caves and belfries). A bat can make sounds with frequencies close to 100,000 vibrations per second. Bats use these sounds to perform amazing feats locating tiny insects (their food) while flying in pitch darkness.

Sub-audible waves. At the other end of the sound spectrum from ultrasonic waves are waves with sub-audible frequency. Most interesting to the scientist are seismic waves, which are generated by the movement of large bodies of rock during earthquakes. The frequencies of seismic waves are in the range of a few vibrations per minute (0.1 per second). There are two kinds of seismic waves, which differ in the direction of the oscillator displacement relative to the direction of propagation. One kind, called the primary wave, travels about 6500 meters per second in the earth's crust, twice as fast as the other kind, called the secondary wave.

Seismic waves are the best source of information about the interior of the earth. They are refracted inside the earth because their speed in various layers is greater or less than it is at the surface. The earth therefore acts like a huge, complicated lens whose properties are inferred from the geographic distribution of seismic waves emitted in earthquakes. One inference is that the material changes abruptly at a depth of about 50 kilometers. This change, which defines the boundary of the earth's crust, is called the Mohorovicic discontinuity ("Moho" for short).

Shock waves. The final item we will take up in this section is shock waves, which are a form of sound with extremely large amplitude and very sudden onset. Whenever an object moves with supersonic speed (faster than the speed of sound in the surrounding air or other medium), the air is displaced very abruptly. What happens then is analogous to what happens at the bow of a speedboat that pushes the water aside

Figure 7.1 A shock wave created by a plastic sphere traveling through air at ten times the speed of sound.

suddenly. The sudden displacement of the air by the moving object cannot communicate itself to other parts of the air in the form of sound waves, because sound travels too slowly. For example, at the front of the moving object, the sound can't get "ahead" of the object because the object itself is moving faster than sound.

Consequently, there is a very large change of air pressure, the ordinary wave model breaks down, and the frequently destructive shock wave is formed (Fig. 7.1). Supersonic airplanes generate shock waves (sonic boom) in air very much in the way the speedboat generates shock waves on the water surface. The boom is caused by the sudden increase in air pressure. Explosions also generate shock waves. The very hot material near the site of the explosion expands into the surrounding material with a speed faster than the speed of sound in that material.

7.2 Application of the wave model to light

The wavelength of light. Your observation of an interference pattern when you looked through a piece of cloth at a distant light source (Section 5.1) seems to be understandable only with a wave model for the light (Figure 7.2). When you apply the wave model to your everyday experience with light, you conclude, from the absence of noticeable diffraction under ordinary circumstances, that the wavelength of light must be much smaller than the size of the objects around you. Only when you looked through finely woven fabric were the effects of diffraction noticeable, and even then they were quite small.

You can understand the appearance of the interference pattern by thinking of the threads in the fabric as forming two diffraction gratings, one with its slits and barriers at right angles to those of the other one.

Figure 7.2 A handkerchief serves as diffraction grating (Fig. 5.3).
(a) Thread pattern of a handkerchief.
(b) Diagram of the interference pattern from a distant lamp observed through a handkerchief.

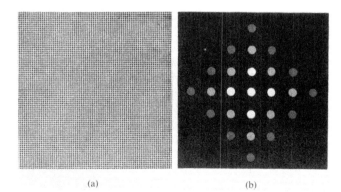

(a)   (b)

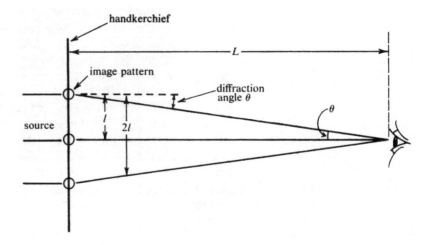

Figure 7·3 *Measuring the wavelength of light with a handkerchief held at arm's length, a distant light source, and a ruler (example included for illustration).*

(1) Measure arm length: L ≈ 70 cm = 700 mm.

(2) Measure the spacing of the three bright image spots at a distance of one arm's length: 2 l ≈ 3 mm.

(3) Use Eq. A·6 to calculate:

$$sine\ \theta \approx \frac{l}{L} \approx \frac{1.5}{700} \approx 2 \times 10^{-3}$$

(4) Measure the "slit distance" of the handkerchief, which has three threads per millimeter (use a magnifier):

d ≈ 1/3 mm ≈ 0.3 mm = 3 × 10⁻⁴ m

(5) Calculate the wavelength (Eq. 6·13):

λ = d sine θ ≈ 3 × 10⁻⁴ × 2 × 10⁻³ m = 6 × 10⁻⁷ m

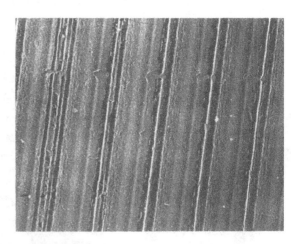

Figure 7.4 Electron microscope photograph of a diffraction grating.

*The first machines for mak-
ing the slits in a grating
were designed and built by
Henry Rowland of Johns
Hopkins University at the
end of the nineteenth century; his gratings, made by
scribing many precise
scratches on metal or
glass, were expensive and
prized scientific tools.
Nowadays, very inexpen-
sive gratings are manufac-
tured by impressing the
rulings on a sheet of plastic, in much the same way
that CDs or auto parts are
stamped out from a master
mould.*

The interference of the light diffracted by the vertical threads produces
images of the source displaced in the horizontal direction. The horizontal threads diffract the light to produce images that are displaced in the
vertical direction. The combination of both then results in the checker-
board array of images that is observed (Fig. 7.2). How a simple measurement can be used to calculate the wavelength of visible light is explained in Fig. 7.3. The wavelength is indeed very short, only about 6 x
10^{-7} meters.

Diffraction gratings. Diffraction gratings for the study of light have to
be made with a spacing between slits that is comparable to the wavelength. Then the light is diffracted at angles that can be observed easily.
A commercial diffraction grating is a transparent sheet with many narrow scratches on its surface (Fig. 7.4). The scratches, which are too
small to be seen, are slight obstacles to the propagation of light. The
narrow regions between the scratches therefore act as narrow slits.
Most of the light incident on the grating passes through unaffected. A
small portion of the light, however, is diffracted by the many slits and
emerges traveling in a direction at an angle to the incident light. As was
shown in Section 6.4, Eq. 6.13 relates the diffraction angle to the distance between slits and the wavelength of the light. Light of

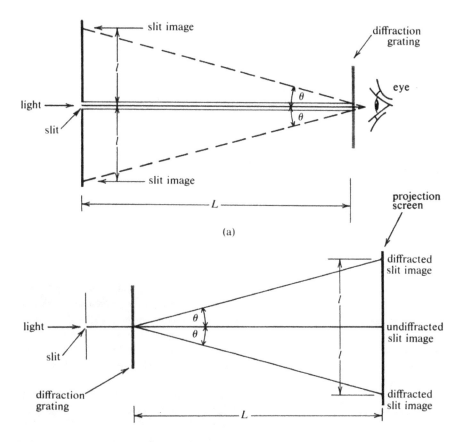

*Figure 7.5 Observation
of the light that comes
from a slit and passes
through a diffraction
grating.*

*(a) Viewing the dif-
fracted light directly.*

*(b) Projecting the dif-
fracted light on a screen.*

Equation 6·13

wavelength (meters) λ
distance between slits
$\quad\quad\quad$ *(meters)* d
diffraction angle θ

$$sine\ \theta = \frac{\lambda}{d}$$

Equation 7·2

(Symbols defined in Fig. 7·5.)

$$sine\ \theta = \frac{l}{\sqrt{l^2 + L^2}} \approx \frac{l}{L}$$

Equation 7·3

$$\frac{\lambda}{d} = \frac{1}{\sqrt{l^2 + L^2}} \approx \frac{l}{L}$$

short wavelength, therefore, is diffracted through a smaller angle than light of long wavelength.

To observe the diffracted light, you may either look through the grating at the light source or let the diffracted light impinge on a screen at a substantial distance behind the grating (Fig. 7.5). A shield with a slit is usually placed in front of the grating to act as a narrow rectangular light source whose diffracted images can be recognized easily by their shape. You can use the measurements to calculate the diffraction angle (Eq. 7.2) or you can calculate the wavelength directly (Eq. 7.3).

If the wavelength is very much smaller than the grating spacing, then the diffraction angle is very small and you cannot observe the diffracted light separately from the undiffracted light because the two images of the slit overlap. For good observations, the grating spacing must be almost as small as the wavelength of light. The manufacture of such gratings clearly requires precision apparatus.

Color and wavelength. If a beam of white light, which includes light of many different wavelengths strikes a diffraction grating, the various wavelengths emerge at different angles and thus strike a screen (as shown in Fig. 7.5) at different locations. What your eye observes on the screen, however, is a display of all the colors of the rainbow side by side (as shown in Fig. 7.6). Such a display of light, similar to the one obtained by Newton with a prism (Section 5.2), is called a *spectrum*. Each portion of the light with a single wavelength has a particular color and is called monochromatic light. In the wave model, therefore, *color of light is associated with its wavelength.*

The wavelength of visible light ranges around the value we reported above from the crude experiment with the handkerchief. The association of color and wavelength is given in Table 7.5. As shown in the Table, the wavelength range of visible light is actually quite narrow (from 4 to 7 x 10^{-7} m). This seems somewhat paradoxical: the colors detectable by the human eye seem, intuitively, to span a huge range (think about the number of colors available in a paint store, or, even more impressive, the complex shades and hues discovered by the Impressionist painters), yet this extraordinary complexity is confined within such a narrow range of wavelengths! However, on reflection,

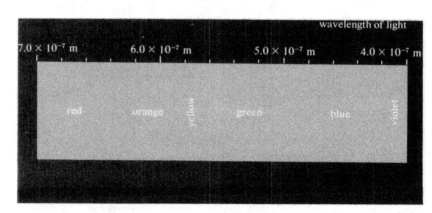

Figure 7.6 Diagram of the spectrum of light, with an indication of the wavelength ranges of the various colors. The transitions are gradual and vary depending on the eye of the observer

TABLE 7.5 WAVELENGTH AND COLOR OF LIGHT (APPROXIMATE)

Color	Wavelength range (m)
violet	4.0 to 4.2 x 10^{-7}
blue	4.2 to 4.9 x 10^{-7}
green	4.9 to 5.7 x 10^{-7}
yellow	5.7 to 5.8 x 10^{-7}
orange	5.8 to 6.4 x 10^{-7}
red	6.4 to 7.0 x 10^{-7}

Ole Roemer (1644-1710) was a prominent Danish scientist who served as a member of the French Academy and as the tutor of the son of King Louis XIV. With the revocation of the Edict of Nantes in 1681, Roemer, like Huygens and others prominent in the French Academy, fled France for the safety of Protestant Northern Europe.

Equation 7.4 (Roemer's measurement of the speed of light in 1676)

$$v_{light} = \frac{\Delta s}{\Delta t}$$

$\Delta s = diam., earth orbit$
$= 3 \times 10^{11} m$

$\Delta t = 22 \min.$
$= 1.3 \times 10^3 \sec$

$$v_{light} = \frac{3 \times 10^{11} m}{1.3 \times 10^3 \sec}$$

$= 2.3 \times 10^8 m/\sec$

Modern value:
$v_{light} = 3.0 \times 10^8 m/sec$

there is a refreshing lesson here; our eyes (and brains) indeed are capable of quite extraordinary feats as detectors of light, able to easily distinguish colors that have almost identical wavelengths (or slightly different mixtures of wavelengths). Modern spectroscopes and other instruments are extraordinarily good at spreading light out and detecting even the faintest components, but our eyes are also extremely capable. The human visual system (the eye and brain) is also capable of making extraordinarily fast judgments in real time; while the best optical instruments probably could match or exceed the human eye in the narrowly-defined task of distinguishing between slightly different wavelengths; there would be no contest with regard to speed.

We can also think of the differences among the various colors of light in terms of their frequencies. The frequency of visible light (calculated from $f = v/\lambda$) ranges from about 1.3 to 2.3 x 10^{15}! This is an extraordinarily high frequency. The complex information about colors that can be conveyed by light is a example of the huge information-carrying ability of a wave with such a high frequency. It is indeed generally true that higher frequency waves can carry more information. Another example of this is in the capacity of fiber optic cables (which use light) to convey information, which far exceeds the capacity of coaxial cable or ordinary telephone wires (both of which use much lower-frequency waves in the radio range).

By using energy detectors other than the human eye it is possible to identify diffracted radiation of shorter and of longer wavelength than visible light. This radiation is called *ultraviolet* light and *infrared* light, respectively. In other words, the implicit operational definition for light, "radiation detected by the human eye," should be extended to forms of radiation that are not detected by the eye, but are functionally very similar to visible light.

The speed of light. Ancient philosophers speculated about the speed of light and variously held the opinion that light propagated with a finite speed and that it propagated instantaneously. Galileo made the first attempt to measure the speed of light by having two distant observers flash lanterns back and forth. The experiment failed because the time required by the light was much less than the reaction time of the participants.

Roemer's measurement. Ole Roemer made the first successful measurement of the speed of light in 1676. He studied the revolution of Jupiter's satellites (the four "Galilean moons," discovered by Galileo in 1610). Roemer noticed something strange: the moons' orbital periods all became gradually shorter while the earth was approaching Jupiter and longer while the earth was moving away from Jupiter. Roemer figured out that these changes must be due to the fact that light did not travel at infinite speed. In fact, light took some time to travel from Jupiter to the earth, and this delay would gradually become shorter (or longer) when the earth was approaching (or moving away from) Jupiter. Roemer measured the maximum time difference in the revolution periods to be 22 minutes (1.3 x 10^3 sec), during which time the light would have to cross the earth's orbit, a distance that was then thought to be 3.0 x 10^{11} meters. Even though Roemer's result (Eq. 7.4) is considerably lower than the presently accepted value, it is a truly remarkable achievement because it occurred only a century after the planetary model for the solar system was introduced by Copernicus, only 66 years after Jupiter's moons were discovered by Galileo, and at a time when the diameter of the earth's orbit around the sun was not known accurately.

Foucault's measurement. Modern methods for measuring the speed of light basically make use of Galileo's concept but substitute a mirror rotating at a known high speed (Foucault's contribution) for the man flashing the lantern. The rotating mirror flashes a beam of light to a distant mirror, which returns the light with a delay equal to the time required for the light to travel to and from the distant mirror. Depending on the time delay, the rotating mirror has assumed a new position, which reflects the returning light to a detector. Because the mirror rotates at high speed, even the short travel time of the light flash finds the mirror in a measurably changed position. This change in the mirror's position is compared with the known speed of rotation to yield the travel time. The speed of light is then found by dividing the distance the light traveled by the time. (Table 7.6).

It is ironic that modern methods for measuring the dimensions of the solar system represent a reversal of Roemer's procedure. Now that the speed of light is known accurately from terrestrial measurements, the travel time of light to other planets is observed (as it was by Roemer) and the distance to them is calculated.

Speed of light in water. Foucault also measured light's speed in water and found that it travels considerably *slower* than in air. This behavior is contrary to the behavior of sound, which travels *faster* in dense materials than in air (Table 7.1). For this reason, as we explain below, Foucault's measurements of the speed of light were very significant in the history of the theory of light; Foucault's measurements provided a critical test for models of light and dramatized the inadequacy of the wave model. Later measurements confirmed Foucault's results for water and found that light also travels more slowly in glass than in air (Table 7.6).

Jean Bernard Lion Foucault (1819-1868) studied medicine before changing to physics. In 1851 his celebrated Foucault pendulum experiment demonstrated the earth's rotation relative to the fixed stars. In later years, he invented the gyroscope and made a determination of the velocity of light by using a revolving mirror.

The ray model and the wave model for light. The ray model described in Section 5.2 was based on a set of assumptions that were not further justified. It was sufficient that they were successful in explaining the observed properties of light, such as formation of shadows, operation of lenses, combination of colors, and so on. The model described but did not explain refraction (Fig. 5.16, Assumption 5) or the difference among monochromatic rays of various colors. Furthermore, it did not specify how a single ray might be separated from a light beam; that is, the ray was a formal concept in the model and did not have an operational definition.

Isolation of a light ray. Since the ray model says that light is composed of rays, you may well be curious to see a single such ray. Suppose we attempt to isolate a single ray as follows: We place an opaque shield in front of a light source and puncture it. Through the tiny hole, a

TABLE 7.6 SPEED OF LIGHT

Material	Speed (m/sec)
vacuum	3.0×10^8
air	3.0×10^8
glass	1.9×10^8
water	2.3×10^8

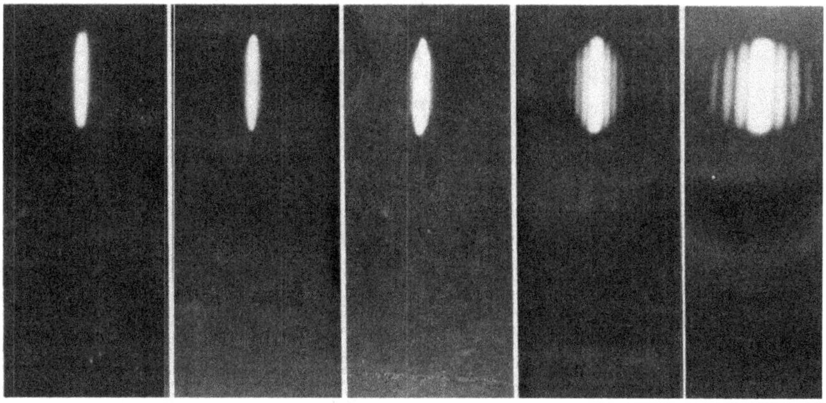

Figure 7.7 The attempts to isolate a single light ray by passing light through a narrow slit fail. The slit widths (in order from left to right) are 1.5, 0.7, 0.4, 0.2, and 0.1 millimeters, respectively.

Equation 6·14

angle of incidence θ_i
angle of reflection θ_R

$\theta_i = \theta_R$

Equation 6·17b

angle of refraction θ_r
speeds of light v_i, v_r

$$\frac{sine\ \theta_i}{sine\ \theta_r} = \frac{v_i}{v_r}$$

Equation 5·2

indices of refraction n_i, n_r

$$n_r\ sine\ \theta_r = n_i\ sine\ \theta_i$$

slim beam of light passes. Now we make successively smaller and smaller punctures in the shield. The ray model predicts that we should get thinner and thinner shafts of light. However, the physical world doesn't always act the way we expect: Figure 7.7 shows what actually happens.

The light actually *spreads out* as the hole is reduced below a fraction of 1 millimeter in width (Fig. 7.7)! This behavior, especially the pattern of light and dark fringes in the photo with the narrowest slit (on the far right) is very mysterious; it doesn't fit at all naturally with the ray model. You may think that the spreading out of the light could be explained within the framework of the ray model by reflection or scattering of the light in some way from the edges of the slit; scientists, including Newton, indeed used the ray model to construct such explanations. However, any explanations based on the ray model simply cannot explain the pattern of dark and light fringes that appear as the slit gets narrow. In fact, the narrower the slit, the wider and more pronounced the fringes become, and there isn't any way to combine rays of light in such a way as to cancel themselves and thus produce a dark fringe.

On the other hand, this phenomena is very reminiscent of what we observed with waves in Chapter 6: waves naturally spread out or diffract when they pass through narrow openings (Section 6.4); furthermore, waves can easily cancel one another, as in destructive interference (Section 6.2, Fig. 6.11). In addition, the two-hole interference pattern (Figure 6.21) had certain locations where the waves always cancelled one another out; this would be a natural way to explain the dark fringes. Finally, Huygens' Principle (Section 6.3) applied to waves striking a diffraction grating (Section 6.4) predicted that, as the angle of diffraction changed, the waves cancelled and reinforced and cancelled and reinforced (Figure 6.24); this would seem likely to produce a pattern of dark and light fringes.

Limitation of the ray model. Evidently the ray model is limited. When experiments are pushed beyond the limits of this model, it breaks down. The wave model is suggested by the diffraction of the single slit (Fig. 7.7), and by the interference pattern seen through the handkerchief (Fig. 7.2). As we will show below, it is a better model for light than the ray model. In other words, light beams are better represented as packets of light waves than as bundles of rays. The diffraction visible in Fig. 7.7 may be considered a consequence of the uncertainty principle for waves (Section 6.2). Attempts to localize the wave packet (that is, make it thinner from side to side) require a mixture of a broader range of wave numbers and wavelengths (that is, some of the light has a larger wavelength and thus spreads out sideways after leaving the slit). Within the limits of the uncertainty principle, or in the absence of diffraction, the ray model is satisfactory.

Adequacy of the wave model. Does the propagation of wave packets correctly explain Assumptions 1 to 5 (Fig. 5.16) about light rays? The answer is that it does, in view of Huygens' principle and the laws of reflection and refraction of waves (Eqs. 6.14 and 6.17b). The observed reflection and refraction of light corresponds directly with the observed behavior of water waves. In fact, the index of refraction of a material (Eq. 5.2) acquires a dramatic new significance in the wave model: it is the ratio of the speed of light waves in air to the speed of light waves in that medium (Example 7.1). With this new insight, a table of light speeds in various media can be constructed from that of indices of refraction (Table 5.1), with no measurement other than the speed of light in air (Table 7.5).

However, there is a contradiction lurking in the background: the speed of sound is *greater* in water and glass than in air, but the speed of light is *less* in water and glass than in air. This may seem like a small detail, and in the 1700s it was. But much later, in the late 1800s, after the wave model for light had become very well accepted, scientists recognized that this contrast between sound and light pointed to a very serious limitation of the wave model for light. In fact, scientists' attempts to use their experience with sound (and other) waves to identify the *medium* for light waves generated many other contradictions and problems that were only resolved with Einstein's revolutionary theory of relativity. We will explain this more fully below in Section 7.3.

Early history of models for light. The role of Isaac Newton in the development of models for light makes a remarkable chapter in the history of science. It is clear from Newton's writings that he understood Huygens' wave theory and that he was informed, through his own experiments and those of others, of the properties of light known in his day. These properties included the speed of light as measured by Roemer, the diffraction of light by a thin slit, the interference of light to form colors by multiple reflection from thin films (for example, soap bubbles), the association of color with wavelength (ultraviolet the shortest and red the longest), as well as the phenomena on which Newton based his formulation of the ray model.

Newton's rejection of the wave model. Newton summarized all these data in a set of rhetorical questions that defined the wave model for light. Included in his reasoning was the existence of a medium (aether) whose properties he estimated by assuming that the light waves were pressure waves in aether analogous to sound waves in air. Newton gave three principal reasons for rejecting the wave model.

First, Newton expected that light waves would be diffracted more ex-

EXAMPLE 7.1

From Equation 5.2:
$$\frac{sine\ \theta_i}{sine\ \theta_r} = \frac{n_r}{n_i} \quad (1)$$
and from Equation 6.17(b)
$$\frac{sine\ \theta_i}{sine\ \theta_r} = \frac{v_i}{v_r} \quad (2)$$

Let the incident medium be air. Then $n_i = 1$ (from Table 5.1), and $v_i = v_{air}$, $n_r = n_{medium}$, and $v_r = v_{medium}$.

Putting together Eqs. (1) and (2):
$$n_{medium} = \frac{v_{air}}{v_{medium}} \quad (2)$$

"Are not all Hypotheses erroneous, in which Light is supposed to consist in Pression or Motion, propagated through a fluid Medium?"

Isaac Newton
Opticks, 1704

tensively than observations showed. He dismissed the diffraction that had been observed as being too small to arise from the interference of waves and ascribed it instead to a repulsive interaction with the edges

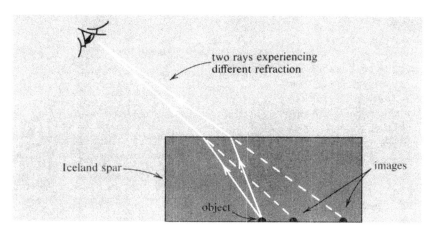

two rays experiencing different refraction

Iceland spar

images

object

Figure 7.8 Double refraction of light by the mineral Iceland spar. Two images are seen. The distance between images depends on the viewing angle and the thickness of the mineral specimen.

of the slit.

Second, Newton found that the minerals Iceland spar and crystal quartz could split a light beam into two beams refracted through different angles (double refraction, Fig. 7.8). This was incompatible with Newton's concept of a wave unless the crystal itself were to modify the light wave. But one of Newton's fundamental assumptions was that the properties of light other than speed and direction were determined by the source and not by the media traversed.

Third, Newton rejected the aether concept because the very large speed of light required properties (extremely low inertia, strong interaction) that seemed unphysical. Furthermore, Newton was unable to find any reason for its existence other than that it could serve as a medium for the propagation of light.

Newton's corpuscular model. Having thus demolished the wave model, Newton proposed his own. Upon his belief that light consisted of "very small Bodies," or corpuscles, Newton built the corpuscular model for light. Besides particle structure, other features of this model were an association of particle size with color, an interaction that accelerated and deflected corpuscles falling on a dense medium (refraction), and an MIP model for matter in which the light corpuscles were easily emitted and absorbed by matter particles. The corpuscular model for light rays answered Newton's three objections to the wave model. The corpuscles can interact-at-a-distance and be deflected, but are not diffracted. The double refraction by Iceland spar and quartz crystals was explained by ascribing a shape to the corpuscles; depending on how the corpuscles were aligned relative to the micro-domain structure of the crystals, they would be deflected through different angles (Fig. 7.9). And the aether was unnecessary.

These compelling arguments show how, in the formulation of scientific models, it is usually necessary to make a compromise: some parts are more satisfactory and other parts are less satisfactory. The greatest

"Are not the rays of Light very small Bodies emitted from shining Substances? For such Bodies will pass through uniform Mediums in right Lines...."

Isaac Newton
Opticks, 1704

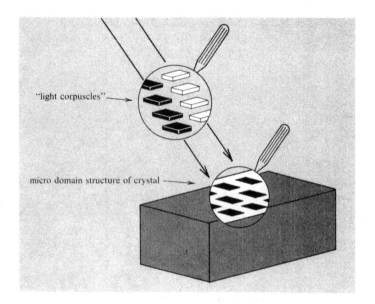

Figure 7.9 Working model for Iceland spar to illustrate Newton's interpretation of double refraction. Newton ascribed double refraction to the shape of the "light corpuscles," which could be oriented parallel or at right angles to structures in the crystal. (The details of this model were invented by the author.)

triumph of wave theory was the law of refraction, purchased at the cost of an aether. Partly because of his idea that refraction of rays was deflection of corpuscles, Newton was not willing to pay the price.

Huygens' preference for the wave model. Huygens, however, was willing to pay the price. He believed in the wave model for light. His reasoning overlaps Newton's but comes to different conclusions. He was aware of the very great speed of light, but felt that this was incompatible with particles of matter being shot from the source to the eye. Instead, a wave motion propagating through an intervening medium was the more attractive model to him. Huygens was aware also that rays of light could cross one another without disturbing each other. This observation led him to reject particles, which might collide, and made him favor waves that obey a superposition principle. In Huygens' theory, pulses reinforce one another at their common tangent line (Section 6.3). Huygens, unfortunately, did not know about the interference of wave trains and the resulting standing waves, nor about the double-slit interference pattern (Figure 6.21). These phenomena, which strongly confirmed the wave model, were discovered later and would have substantially bolstered Huygens' arguments.

Particle versus wave theory of refraction. Of the two men, Newton had the greater reputation, and his model was accepted by most of his contemporaries. A clear-cut detectable difference between the two models lay in their prediction of the speed of light in dense media like water and glass. According to Newton, the speed was *increased* by the attractive interaction that deflected (refracted) the corpuscles at the surface. According to Huygens, the observed refraction required a decrease in the speed of light (Table 5.1 and Example 7.1). As mentioned above, it was only much later, in the middle 1800s, that Foucault determined that light in fact traveled slower in water than in air. From the modern point of view, both models have weaknesses, but these were only resolved in the twentieth century, after development of the theory

Thomas Young (1773-1829) was an astonishingly versatile figure: physician, linguist, and scientist. While a medical student, Young made original studies of the eye and later developed the first version of the three-primary-color theory of vision. A large inheritance in 1797 enabled him to devote himself primarily to science. After becoming professor at the Royal Institute in 1801, he turned to physical optics and discovered that Newton's work was explainable in terms of waves. Young was also a pioneer in Egyptology and was among the first to try to decipher the Rosetta Stone.

"Those who are attached to the Newtonian Theory of light. .. would do well ... to imagine anything like an explanation of these experiments derived from their own doctrines ..."

Thomas Young
Philosophical Transactions,
1804

To Newton, the downfall of his corpuscular theory would not have been entirely unexpected. Unlike many of his predecessors and followers, Newton did not confuse theory with doctrine: "Tis true, that from my theory I argue the corporeity of light: but I do it without any absolute positiveness ... I knew, that the properties, *which I declared of light, were in some measure capable of being explicated not only by that, but by many other mechanical hypotheses."*

of relativity and quantum mechanics. (A dual theory is now in use, with a wave packet (Section 6.2) playing a central role. Propagation of light is determined by the wave character of the packet, while emission and absorption are determined by the corpuscular aspect of the wave packet as a "chunk" of light.)

Discovery of interference. The next development in the physics of light took place 100 years after the time of Newton and Huygens. Thomas Young conducted experiments on the diffraction of light by two slits; he also observed the patterns we showed for water waves in Fig. 6.2, and he introduced the concepts of superposition and interference of waves in a series of three papers. The two-slit interference experiment was very much more suggestive than the much earlier single-slit diffraction experiments and served to revive the wave model for light. Curiously, Young felt compelled in his first publication to ascribe the original wave model for light more to Newton (who considered it in his treatise) than to Huygens. Young played down Newton's own complete rejection of the wave model.

Young went considerably further in his second and third papers, in which he described the conditions for constructive and destructive interference in terms of the wave path difference measured in "breadths" (wavelengths) of the supposed "undulations," (waves) which differed for different colors. Young was then no longer so respectful of Newton; he used Newton's data to illustrate his own ideas, and he completely rejected Newton's interpretation.

Acceptance of the wave model. The last blow to Newton's corpuscular model was delivered in the middle of the nineteenth century by Foucault's measurements of the speed of light in water. We have already reported that this speed was found to be less than the speed in air, as required by the wave model's explanation of refraction and in contradiction to Newton's prediction based on the corpuscular model. With this achievement, the wave model was unanimously accepted and physicists' attention could turn to new questions: What is the aether? What kind of waves are light waves? How is light emitted and absorbed by matter?

Emission, reflection, and absorption spectra. The emission of light by a hot source means that there is energy transfer from thermal energy of the glowing material to radiant energy of light, which travels to a distant energy receiver. The reverse energy transfer occurs when light is absorbed; then the radiant energy of the light is transferred to a form of energy in the receiver. This is usually thermal energy, but some of it may be chemical energy also (as in photosynthesis, photography, and sunburn).

Emission and absorption by gases. The diffraction grating has been used to decompose the light from many sources into its monochromatic parts. Once the wave nature of light was generally accepted, such studies were used to gain information about the light source itself. It was

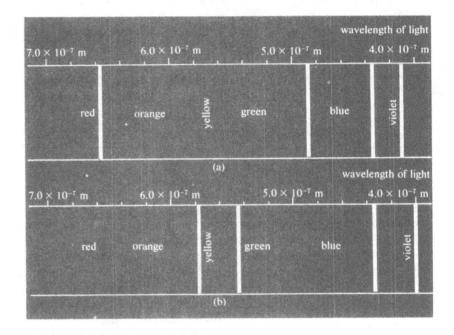

*Figure 7.10 Line emission spectra of gases.
(a) Hydrogen gas.
(b) Mercury gas.*

TABLE 7.7 EMISSION SPECTRAL LINES OF SELECTED ELEMENTS

Element	Wavelength (m)
calcium	4.9×10^{-7}
	6.1×10^{-7}
	6.4×10^{-7}
copper	4.6×10^{-7}
helium	5.3×10^{-7}
mercury	4.4×10^{-7}
	5.5×10^{-7}
	5.8×10^{-7}
neon	5.4×10^{-7}
	6.4×10^{-7}
	6.5×10^{-7}
sodium	5.9×10^{-7}

---

found that many glowing gases emitted characteristic line spectra (Fig. 7.10). That is, the spectrum of light they produced did not include all colors, but only certain colors in very narrow bands of wavelengths called spectral lines. These emission line spectra are unique for each element and can be used for identification, just as fingerprints can be used to identify a person (Table 7.7).

Absorption of light by gases, like emission, is selective. That is, gases absorb light only at certain wavelengths or absorption lines. Most of the absorption lines have the same wavelength as emission lines and can be used to identify the presence of a chemical element (Fig. 7.11). Virtually all information about the chemical composition of the sun and stars comes from the analysis of spectral lines. Astronomers have now recorded and analyzed spectra from essentially all of the stars and other objects they have found in the sky; this huge body of evidence leads to a conclusion that may seem disappointing to sci-fi fans: the entire known universe is composed of the same chemical elements as the earth, though in different proportions than found on the earth.

Emission and absorption by solids. Glowing solids, such as light bulb filaments or hot coals, emit continuous spectra (Fig. 7.6). That is, all wavelengths are represented, and not only selected ones as in the

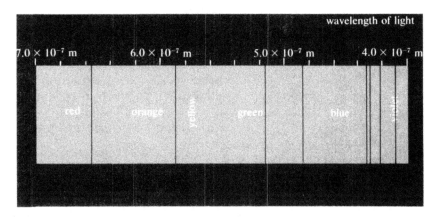

Figure 7.11 Light emitted by the sun. Dark lines (called Fraunhofer lines after their discoverer) are evidence of absorption of light by chemical elements in the gases at the surface of the sun.

line spectrum emitted by a glowing gas. By contrast with line spectra from gases, the continuous spectra from different glowing solid materials are very similar to one another and depend on the temperature but not on the composition of the material.

We have already described the selective reflection and absorption of light that is responsible for the relative brightness and the color of a reflecting surface (Section 5.2). By analyzing the reflected light with a diffraction grating, you can find out whether a color is pure (monochromatic) or mixed. By inference from the colors of the incident and reflected light, you can find which colors (wavelengths) are absorbed.

Micro-domain model for emission and absorption. Are the sources of light macro-domain systems? Since the wavelength of light is at the borderline of the two domains, the sources are in all likelihood very small by macro-domain standards. Another clue comes from the spectra of gases. According to the MIP model, the particles in gases are far apart and do not interact with one another appreciably. Since gases nevertheless emit and absorb light, the source must be a gas particle acting alone. Furthermore, since gases emit line spectra (isolated wavelengths and frequencies), you may conclude that a gas particle acts like a tuned system with several oscillators, one frequency corresponding to each spectral line emitted. In this interpretation of spectral lines, therefore, their frequency is a more significant property than their wavelength.

This model helps to explain why solid materials emit continuous spectra rather than line spectra. In the solid phase, the particles of the MIP model interact strongly with one another. The interaction shifts the frequencies of the individual oscillators so all possible frequencies are represented. Each of these oscillators emits light of its own frequency, but all oscillators together give rise to a continuous spectrum of light.

The frequency of an oscillator responsible for the emission of light

Equation 7.5

speed of light $= v$

wavelength $= \lambda$

frequency $= f$

$$f = \frac{v}{\lambda} = \frac{3 \times 10^8 \; m/sec}{5.9 \times 10^{-7} \; m}$$

$$= 5 \times 10^{14} \; /sec$$

James Clerk *Maxwell (1831-1879) was born into a wealthy Scottish family in Edinburgh. After education at Edinburgh and Cambridge, Maxwell was Professor of Physics at Marischal College, Aberdeen, and Kings College, London. He published important papers in 1859-1860 on Saturn's rings and on the kinetic theory of gases. His greatest work, however, was in electromagnetism. Adopting Faraday's theory of fields, Maxwell set out to establish a unified mathematical description of electric and magnetic phenomena. Maxwell's findings, published in 1865 and 1873, are a landmark in theoretical physics.*

can be calculated from the known speed of light and the measured wavelength of one spectral line. To take an example, the frequency of sodium light is found to have a value of enormous magnitude (Eq. 7.5). Since the wavelengths of all visible light do not differ greatly, the frequencies of all the oscillators in the particle model are of similar magnitude. What could be oscillating with such high frequency?

7.3 The electromagnetic theory of light

As we explained in the previous section, the model of light as displacement waves propagating in the aether was generally accepted by the middle of the nineteenth century. In spite of the satisfactory state of affairs, however, there were loose ends yet to be explained. These problems had been pointed out by Newton in his critique of the wave model: no independent evidence of the aether's existence and no mechanism for the emission and absorption of light. Soon, however, there were several developments that led ultimately to a brilliant confirmation of the wave model's applicability to the propagation of light; however, they also required considerable adjustment in the models that scientists had for light waves, and indeed for other physical phenomena.

Maxwell's theory. The first step was a theoretical synthesis, by J. Clerk Maxwell, of the discoveries regarding electric fields, magnetic fields, the magnetic effects of electric currents, and the electric effects of moving magnets, which had been made during the preceding decades. Maxwell came to the conclusion that rapidly vibrating electric charges would generate electric and magnetic fields whose intensity exhibits wavelike patterns, much as a vibrating violin string generates air pressure variations that exhibit wave patterns (Section 7.1).

Maxwell called his waves electromagnetic waves. He viewed them as oscillatory displacements of the aether, again in analogy to sound

Figure 7.12 Electric and magnetic fields in an electromagnetic wave.

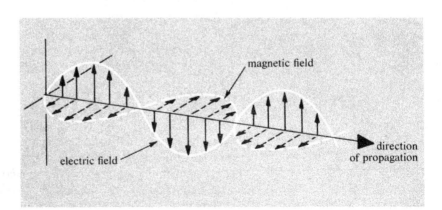

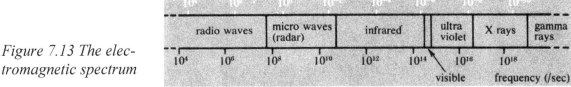

Figure 7.13 The electromagnetic spectrum

"The ... difficulties ... which are involved in the assumption of particles acting at a distance ... are such as to prevent me from considering this theory as an ultimate one ... I have therefore preferred to seek an explanation of the facts in another direction. ... The theory I propose may ... be called a theory of the Electromagnetic Field, because it has to do with the space in the neighbourhood of the electric or magnetic bodies."

James Clerk Maxwell
Philosophical Transactions,
1865

waves or water waves. A diagram indicating the electric and magnetic field patterns in such a wave is shown in Fig. 7.12. The electric and magnetic fields are at right angles to each other. At any particular point in space, the intensity of the fields oscillates in magnitude and/or direction with the frequency of the source. The entire pattern shown moves in the direction of propagation with the speed of light.

Maxwell's calculation indicated that the waves would propagate in aether with a speed of 3×10^8 meters per second. This speed was, to everyone's amazement, just equal to the speed of light measured a few years earlier (Table 7.6). Light waves were therefore identified as electromagnetic waves with a wavelength of about 5×10^{-7} meters. The frequency of these waves is related to their wavelength through the same equation that applies to sound and other waves ($v = f/\lambda$, Eq. 7.5); the frequency of light waves, however, is enormously high compared to that of sound waves. Maxwell's theory demonstrated conclusively that electric charges were the source of the vibrating electric and magnetic fields; thus these charges would have to exist within any object that is a source of light, and in order to generate light, they would have to be able to vibrate at the high frequencies of light waves. Furthermore, if the electric charges vibrate (oscillate) at such a high frequency, they must have an extremely small inertia and be subject to very strong interaction compared to the oscillators that are responsible for generating sound waves. This requirement will turn out to be extremely important in the search for understanding of the structure and constituents of matter.

The characteristic spectra of gases exhibit sharp frequencies and are evidence that gases contain "tuned" electrically charged systems capable of oscillating at well-defined high frequencies. In Chapter 8 we will explore the sources of light waves, which are now identified with the atoms and molecules composing all matter.

The electromagnetic spectrum. Maxwell's theory suggested strongly that electromagnetic waves should exist with frequencies different from those of light. One only had to arrange for electric charges to vibrate at a lower frequency. Heinrich Hertz (1857-1894) succeeded in generating waves with a wavelength of a few centimeters (rather than the 5×10^{-7} m wavelength of light) by making sparks in simple electric circuits, and Marconi (1874-1937) turned this discovery to practical use in his invention of the wireless telegraph (radio). Many other forms of electromagnetic (E-M) radiation have since been discovered. Such E-M waves are extremely useful; for example, radio and radar waves can transmit information over great distances, and x-rays and infrared can create images of objects that are otherwise invisible. The entire frequency range is called the electromagnetic spectrum (Fig. 7.13). It includes not only visible light and radio waves, but also X-rays, ultraviolet and infrared radiation, radar, and even the electric and magnetic fields associated with 60-cycle alternating house current (Section 12.4). Visible light actually spans only a minute portion of the spectrum.

Besides giving a clue about the sources of light and vastly expanding the spectrum, Maxwell's theory made it possible to associate energy with light waves, since energy is associated with electric and magnetic fields. Maxwell's great contribution, however, also called attention once again to the aether, the medium in which light waves, now viewed as wave patterns of electric and magnetic fields, were believed to propagate.

The aether mystery. You might think that the existence of the aether, so severely criticized by Newton, was now firmly established with the success of Maxwell's theory. Far from it! Now that aether had to be taken seriously, its properties were investigated more thoroughly.

The first question was asked by Maxwell himself: What about the motion of the aether? Clearly, he reasoned, light waves traveling at a certain speed relative to the aether would be observed to travel at a different speed relative to objects moving with respect to the aether. Nobody knew, of course, what objects moved with respect to the aether, but the planets' relative orbital motion at different rates made it impossible for the aether to be at rest with respect to all of them at the same time.

Many experiments were carried out to detect motion of the earth relative to the aether by comparing the speed of light measured under many different conditions: parallel to the earth's orbital motion around the sun, perpendicular to this motion, inside rapidly moving liquids, light generated on earth and coming from moving sources, and so on. The results were negative; the speed of light gave no evidence that the earth moved relative to the aether. Was the earth, then, really at rest in the aether, while the entire universe moved around the earth? More than 300 years after Copernicus this was not an acceptable hypothesis.

There were other mysteries about the aether; we pointed out above, in Section 7.2, that the behavior of light and sound was not consistent: light traveled slower in denser materials (water and glass) than air but sound traveled faster in such materials. This apparent contradiction raised doubts about the similarities between the medium for sound and the aether. Such concerns became more serious when scientists started actively investigating the properties of the aether, using Maxwell's theory of electromagnetism. For sound (and other waves), the faster the speed of the wave, the lower the inertia and the higher the interaction among the oscillators of the medium. However, applying this kind of thinking to light and its extremely high speed meant that the aether would have to be made up of oscillators with contradictory properties: an inertia that was much, much lower, yet a strength of interaction that was much, much stronger, than in any existing substance!

The theory of relativity. Attempts to resolve these contradictions made little progress until Albert Einstein, early in the twentieth century,

---

"It appears therefore that certain phenomena in electricity and magnetism lead to the same conclusion as those of optics, namely, that there is an aethereal medium pervading all bodies, and modified only in degree by their presence ..."

James Clerk Maxwell
Philosophical Transactions, 1865

Albert Einstein (1879-1955), perhaps the greatest theoretical physicist since Isaac Newton, was born in Ulm, Germany and educated in Munich and Switzerland. After graduation he could not obtain a university teaching position and had to accept the obscurity of a minor post at the Berne Patent Office. This obscurity ended dramatically in 1905 when Einstein published five important papers, including two that shook the scientific world to its foundations – one on the photoelectric effect, and the other on the special theory of relativity. In 1933, Einstein resigned as Director of the Kaiser Wilhelm Institute of Physics in Berlin as a protest against Hitler's fascist policies. He emigrated to the United States, where he spent the rest of his life working at the Institute for Advanced Study in Princeton.

approached the subject from a new, apparently unrelated, direction; Einstein seriously pursued the problem of how the interaction of electrically charged bodies and the light waves they generate would appear to two observers in relative motion. In particular, how would a light wave in vacuum appear to an observer who moves alongside it, with the same speed? Maxwell's theory excluded the possibility of a stationary light pulse.

Einstein took the viewpoint that the interaction of two electrically charged objects should depend only on their motion relative to one another and not on their common motion relative to some outside reference frame. To reconcile this requirement with the known laws governing electric and magnetic interactions, Einstein found it necessary to abandon the aether and to modify the commonsense concepts of space and time.

At the basis of Einstein's reasoning was an operational approach to the methods by which observers in relative motion can communicate their

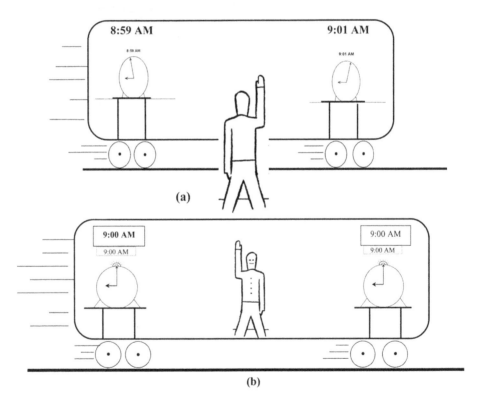

(a)

(b)

Figure 7.14 Two clocks on a train, as observed by two observers in relative motion. The effect is vastly exaggerated and would not be observable in a train.
(a) To the observer opposite the car's center, but off the train, the clock at the rear of the train would appear to be running two minutes behind the clock at the front of the train.
(b) To the observer at the car's center on the train, the two clocks would appear perfectly synchronized.
This paradoxical behavior is the direct result of the assumption that the speed of light is the same for all observers, which also seems to contradict our experience but is extremely well documented. It would seem difficult to build a reliable theory of physics based on such anti-intuitive ideas, but Einstein did exactly that. His theory of relativity provides a solid basis of understanding for all situations in which velocities near the velocity of light are involved.

observations to one another. Instead of assuming that this could be done, as had Galileo and Newton, he described how observers must use light signals (the fastest known method of communication over a distance) to standardize their instruments for measuring distances, time intervals, and so on. Built into the scheme were the experimental results, which indicated that the observed speed of light was not influenced by motion of neither the light source nor the light detector.

Einstein's results, embodied in his theory of relativity, have substantially influenced both science and philosophy. Einstein demonstrated that several of our apparently "intuitive" ideas about the physical world had to be discarded. The theory of relativity was both consistent and comprehensive, and it required us to accept several new ways of thinking that seemed to conflict with intuition: First, two events that are simultaneous for one observer are not simultaneous for a second observer in motion relative to the first (Fig. 7.14). Second, if observer A is moving with respect to observer B and they both measure the length of a given object and the duration of a given event, (using standard rulers and clocks moving with them), they will *not* find the same results. Third, an automobile moving at exactly 75 miles per hour passing another at exactly 55 miles per hour is not traveling at exactly 20 miles per hour relative to the second car (Section 2.2), though the difference is insignificant for such slowly moving objects. Finally, the answer to Einstein's original question about the observer "catching up" with the light wave in vacuum is deceptively simple: the properties of space and time are such that this can never happen!

Summary

Sound and light are intermediaries in interaction-at-a-distance between a source and a receiver. The modern understanding of both phenomena is achieved with the help of a wave model.

Sound waves are pressure waves in solids, liquids, or gases. Associated with the pressure variations are displacements of micro-domain particles making up the material. The tone or pitch of sound is associated with the wave frequency, the intensity with the wave amplitude. The speed of sound in air is 344 meters per second.

The human ear can detect sound waves whose frequencies fall between about 20 vibrations per second and 20,000 vibrations per second, with most speech using waves from 200 to 2000 per second. The wavelength of sound waves in speech, therefore, is comparable in size to the human body and to many environmental objects in the macro domain. These sound waves are so strongly diffracted by objects of this size that no sharp acoustic images can be formed.

Light waves are electromagnetic waves that propagate in vacuum and in various media. Visible light is only a very narrow portion of the electromagnetic spectrum, which also includes radio, microwaves (radar), infrared, ultraviolet, X-ray, and nuclear radiation. The color of light is associated with the wave frequency, the intensity with the wave amplitude. The speed of electromagnetic waves in vacuum is 3×10^8 meters per second.

The wavelength of visible light in air is very small, about 5×10^{-7} meters. Evidence of the wave nature of light is therefore difficult to obtain through experiments in the macro domain. The ray model describes light very well until it interacts with objects at the lower limit of the macro domain. Then the effects of diffraction and interference can be observed and the inadequacies of the ray model are revealed.

As a matter of fact, light spans the micro, macro, and cosmic domains. The speed of light is so great that it traverses the cosmic distance from the earth to the moon in about 1 second. The wavelength of light is in the micro domain. And the light rays (pencils of light) that make possible human vision are in the macro domain. No wonder that so much controversy surrounded the models for light! For this very reason, however, light has been a powerful tool in the study of systems in the micro and cosmic domains.

One of the most revolutionary consequences of the electromagnetic theory, as applied by Einstein, was to eliminate the need for the existence of the aether. Light waves propagate in vacuum, without a medium. There is no medium, therefore, that could serve as a special reference frame for measurements of the speed of light. The startling consequences of this conclusion have changed out view of space and time.

List of new terms

octave	shock waves	infrared
ultrasonics	spectrum	double refraction
sonar	absorption	line spectrum
sub-audible waves	emission	electromagnetic waves
seismic waves	ultraviolet	aether
theory of relativity		

List of symbols

v	wave speed	Δs	distance traversed
λ	wavelength	Δt	time interval
f	frequency	d	distance between grating slits
c	speed of light in vacuum	l, L	distances in experimental
θ	angle		arrangements
n	index of refraction		

Problems

1. Prepare several more or less well-tuned systems that act as sound sources (rubber band, stretched wire, air in a bottle, glass of water) and experiment to produce musical notes with them.
    (a) Describe how you can change the pitch of the note (i.e., tune the system) in terms of the wave model for sound.
    (b) Relate the pitch of the sound to the wave speed in the medium and the dimensions of the tuned system.

2. Mersenne's shorter brass wire was 22 centimeters long (Section 7.1). Estimate the wave speed on the wire in his experiment. (Note: the

speed of a mechanical wave on a wire is usually *not* equal to the speed of a sound wave in the wire material. Mechanical waves of the wire as a whole involve macro-domain displacements; whereas sound waves in the material, set up, for example, by scraping the wire with a file, involve micro-domain displacements of the many interacting particles making up the wire.)

3. Explain the following three statements with the help of the gasbag model for sound waves and the conditions for wave propagation. If necessary, make hypotheses about the materials to explain the observations. (Note: a successful explanation is evidence that the hypothesis is valid.)

(a) The speed of sound in air at room temperature does not change if the air pressure is increased or decreased.

(b) The speed of sound in air increases with the temperature. At the boiling temperature of water it is 386 meters per second.

(c) The speed of sound in carbon dioxide is less, and in hydrogen gas is more, than in air at the same temperature.

4. Complete the wavelength column in Table 7.2.

5. Calculate the frequencies and wavelengths of the highest and lowest notes on an 88-key piano.

6. An organ pipe that plays a B note is filled with hydrogen gas. What note will it play then?

7. Look for "sound shadows" produced by buildings or other large obstacles. Describe the relative position of the sound, source, obstacles, and receiver for a significant shadowing to be observable. (Note: you may use your ear as receiver, but you will need a reliable and cooperative sound source, such as a friend with a musical instrument.)

8. Explain the operation of sonar with respect to the following. (a) What size underwater objects can be reliably detected by sonar of 20,000 vibrations per second? Justify your estimate. (b) The echo from an underwater object is received 5 seconds after the sonar pulse was emitted. How distant is the object?

9. Find the wavelength of bat "sonar" pulses. (frequency about 10^5 /sec)

10. (a) Estimate the wavelength of seismic waves.

(b) Would you recommend that geologists use a ray model or a wave model for seismic waves? Explain your answer and any limitations it may have.

11. Use a fine, regularly woven fabric to measure the wavelength of light approximately (Figs. 7.2, 7.3, Problem 5.9).

12. Look at the glancing reflection of a bright, medium-distant light source in a phonograph record, compact disc (CD) or digital videodisc (DVD). (See figure at left.)

(a) Describe your observations and explain them in terms of the wave theory of light.

(b) Make appropriate measurements and calculate the approximate spacing of the tracks on the record, CD, or DVD.

Diagram for Problem 12

Problems 13-16. To observe interference of light, you will need a narrow light source and one or more very thin slits. The simple slit holder shown in Fig. 7.15 below gives satisfactory results. A "source slit" aimed at a lamp acts as a narrow light source. A diffraction grating, a single slit, or a double slit can be attached to the opening in near your eye so as to view the light from the source slit.

Figure 7.15 (to right) Slit holder and slits for Problems 13-16. The slits shown in the detailed drawings above and below the slit holder may be attached to the slit holder or may be built on separate pieces of cardboard that are attached to the slit holder with paper clips.

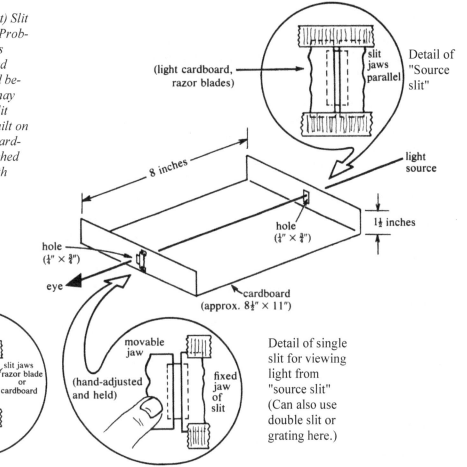

Figure 7.16 (above) Construction of a double slit. The slits should be as narrow as possible, with a hair centered in a cardboard slit (Fig. 7.15) to make it into a double slit. Keep the slit jaws parallel. You may need to work with a magnifier

13. Make a variable-width slit (Fig. 7.15) and attach it to the slit holder. Look at the light bulb through the holder and observe the single-slit diffraction pattern. Describe the colors, the number of bright fringes, and the spacing of the fringes while you vary the slit width.

14. Make a double slit (Fig. 7.16) and attach it to the slit holder. Look at a light bulb through the holder and observe the double-slit diffraction pattern.

   (a) Describe your observations (color, number of bright fringes, position of dark fringes).

   (b) Measure the position of the dark fringes and the slit dimensions as well as you can. Use these to calculate (roughly) the wavelength of light (Section 6.4, Problem 6.12).

15. Attach a plastic diffraction grating to the holder and look at various light sources.

(a) Describe what you observe when you look at an ordinary light bulb, a fluorescent bulb, a red "neon" sign, a green "neon" sign, a mercury arc lamp and a sodium vapor lamp. Sodium vapor lamps are often used in street lights because of their high efficiency; their light is very strongly yellow and seems harsh.

(b) Measure the displacement of the diffraction images (distance labeled "l" in Fig. 7.5a) for the spectral lines of one element.

(c) Calculate the distance between slits in the grating from your measurements using Eq. 7.3 and the known wavelength of the light from that element (see Table 7.7).

(d) Measure the displacement of the diffraction images for the longest- and shortest-wavelength light you can see from an ordinary light bulb.

(e) Calculate the shortest and longest wavelengths of light you can see.

(f) Measure the wavelengths of the various colors as you see them in the spectrum of an ordinary light bulb. Compare your results with Table 7.5.

16. Obtain colored cloth or paper (preferably not glossy) and place it under a bright lamp. Look at this colored material through a diffraction grating on your slit holder (Fig. 7.15). Describe the light that is reflected by the material. (You may wish to compare the result of this experiment carried out in bright sunlight with that obtained when you perform it under artificial light.) Why should you avoid glossy material?

17. Explain why it is more accurate to say that the color of light is associated with its frequency rather than with its wavelength.

18. Describe your evaluation of the disagreement between Huygens and Newton. Optional: Do additional reading on the subject.

19. (a) Calculate the wavelength range of the standard AM broadcast band (frequency 5.6 –1600 kHz, 1 kHz = 1,000 /sec)

(b) Calculate the wavelength range of FM broadcasts. (frequency 88–106 MHz, 1 MHz = 1,000,000 /sec)

(c) Explain why hilly terrain interferes with FM broadcast reception much more seriously than with AM radio reception.

20. Identify one or more explanations or discussions in this chapter that you find inadequate. Describe the general reasons for your dissatisfaction (conclusions contradict your ideas, or steps in the reasoning have been omitted; words or phrases are meaningless; equations are hard to follow; and so on) and pinpoint your criticism as well as you can.

Bibliography

L. Barnett, *The Universe and Dr. Einstein*, Harper and Row, New York, 1948.

A. H. Benade, *Horns, Strings, and Harmony*, Doubleday, Garden City, New York, 1960.

R. E. Berg and Stork, D. G., *The Physics of Sound*, 2nd Ed., Prentice Hall, 1995.

A. Einstein, *Relativity*, Doubleday, Garden City, New York, 1947.

M. Gardner, *Relativity for the Millions*, Macmillan, New York, 1962.

D. E. Hall, *Musical Acoustics*, 2nd Ed, Brooks/Cole Publishing, 1991.

C. M. Hutchins and F. L. Fielding, "Acoustical Measurement of Violins," *Physics Today* (July 1968).

C. Huygens, *Treatise on Light*, Dover Publications, New York, 1912.

W. E. Kock, *Sound Waves and Light Waves*, Doubleday (Anchor), Garden City, New York, 1965.

E. Mach, *The Principles of Physical Optics: A Historical and Philosophical Treatment*, Dover Publications, New York, 1952.

G. Murchie, *Music of the Spheres*, Houghton-Mifflin, Cambridge, Massachusetts. See Chapters 11 and 12.

J. Rigden, *Physics and the Sound of Music*, Wiley, 1977.

T. D. Rossing, *The Science of Sound*, 2nd Ed, Addison-Wesley 1990.

B. Russell, *The ABC of Relativity*, New American Library, New York, 1959.

W. A. M. Bergeijk and others, *Waves and the Ear*, Doubleday, Garden City, New York, 1960.

E. Wood, *Experiments with Crystals and Light*, Bell Telephone Laboratories, Murray Hill, New Jersey, 1964.

Articles from Scientific American. Some or all of these, plus many others, can be obtained on the Internet at http://www.sciamarchive.org/.

E. D. Blackham, "The Physics of the Piano" (December 1965).

E. Bullard, "The Detection of Underground Explosions" (July 1966).

E. E. Helm, "The Vibrating String of the Pythagoreans" (December 1967).

E. N. Leith and J. Upatnieks, "Photography by Laser" (June 1965).

The entire September 1968 issue of *Scientific American* is devoted to "light." The following articles are of particular interest:

P. Connes, "How Light is Analyzed."

G. Feinberg, "Light."

D. R. Herriott, "Applications of Laser Light."

U. Neisser, "The Processes of Vision."

A. L. Schawlow, "Laser Light."

V. F. Weisskopf, "How Light Interacts with Matter."

*Hitherto philosophy has been
chiefly conversant about
the more sensible properties of
bodies: electricity,
together with chemistry,
and the doctrine of light
and colours, seems to be giving
us an inlet into their internal
structure, on which all
their sensible properties depend . . . .
New worlds may open to our
view, and the glory of the great
Sir Isaac Newton himself,
and all his contemporaries,
be eclipsed, by a new set of
philosophers, in quite a new
field of speculation.*

JOSEPH PRIESTLEY
History of Electricity,
1767

Models for atoms

In the previous chapters we have had frequent occasion to contrast direct observations of nature with the models scientists have made for interrelating these observations. The many-interacting-particles (MIP) model for matter has been a powerful tool to this end. In this model, matter is composed of micro-domain particles that are in motion and that interact-at-a-distance with one another by means of an intermediary field. During the nineteenth century the various types of energy—kinetic, thermal, elastic, chemical, phase, electromagnetic field (including radiant), and gravitational field—were investigated intensively. As we pointed out in Section 4.5, and as we will describe in greater detail in later chapters, it is possible to explain thermal, elastic, chemical, and phase energies in terms of the kinetic energies of all the particles and the field energy arising from their interaction.

During the twentieth century, physicists have set themselves the goal of explaining all macro-domain phenomena, including perhaps even life, in terms of these particles, which are called atoms or molecules. Great strides in formulating these explanations have been made, except in the case of phenomena involving the gravitational field. Even Einstein, who reformulated the gravitational interaction in a very novel and general way in his general theory of relativity, was not able to relate gravitation effectively to the electric, magnetic and other fields.

One of the most notable areas of progress has been in the invention of models for atoms themselves. Instead of being conceived of as simple point-like particles with no internal structure, atoms are now viewed as complex systems composed of simpler constituents. The properties of atoms are explained in terms of the arrangement and motion of the constituents. In this chapter we will review some of the studies of the current century that have led to the presently accepted models for atoms.

8.1 The electrical nature of matter

Dalton's atomic theory. When John Dalton proposed his atomic theory of chemical reactions, the particles were called *atoms* (from the Greek word for "indivisible") because they were conceived of as being ultimate constituents that would permit no further subdivision. In this view, which gradually became accepted during the first half of the nineteenth century, each chemical element is composed of a different kind of atom. There were about 90 kinds of atoms, each with its own characteristics (Fig. 8.1). The atoms were believed to have properties that could account for the chemical activity and various other macro-domain properties (hardness, appearance, melting and boiling temperatures, and so on) of specimens of the element.

Electric conduction in solid and liquid materials. As soon as the existence of Dalton's atoms became non-controversial, questions arose about the "intrinsic" properties of the atoms and whether models could be constructed to account for them. In other words, scientists asked in what way an oxygen atom differed from a nitrogen atom, why solid copper conducts an electric current while solid sulfur does not, and how

"... the existence of these ultimate particles of matter can scarcely be doubted, though they are probably much too small ever to be exhibited by microscopic improvement. I have chosen the word atom to signify these ultimate particles...."
John Dalton
A New System of Chemical Philosophy, 1808

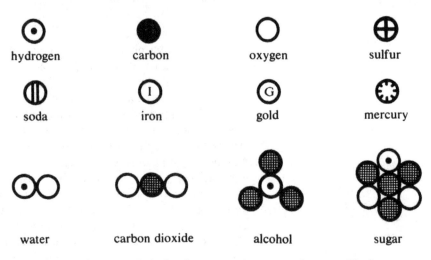

Figure 8.1 Dalton's symbols for the atoms of common elements. The bottom row shows his models for the more complex "atoms" of common compounds.

the metal sodium and the gas chlorine could interact to produce the white crystalline substance sodium chloride (ordinary table salt). Furthermore, a solution of sodium chloride and water conducts an electric current, but solid sodium chloride and pure water separately do not.

Electrolysis. Sir Humphry Davy and Michael Faraday found another effect: melted table salt (sodium chloride) can be decomposed into sodium and chlorine by the passage of an electric current, but the element copper is not modified on the macro level by an electric current. The decomposition of sodium chloride is an example of *electrolysis* (using electricity to divide something into its parts). An electric current can

Figure 8.2 (to the right)
The electrolysis of water to form hydrogen and oxygen gases. Note the ratio of the volumes of gases produced. What do you conclude about the composition of water? On this basis, can you suggest a modification of Dalton's symbol for water above in Figure 8.1?

It is interesting that Dalton knew about this experiment, and others giving similar results about the simple whole number ratios of the volumes of combining gases, but he did not accept these findings, possibly because the particular model of atoms he had developed couldn't explain it!

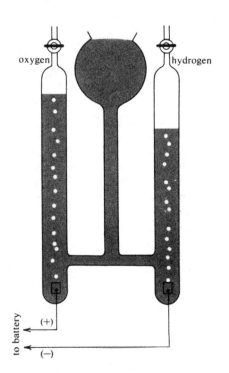

The quantity of electric charge is measured by the mass of hydrogen it liberates in the electrolysis of water. The unit of electric charge, called the Faraday (symbol F), is the quantity of electricity associated with 1 gram of hydrogen.

Sir Joseph John Thomson (1856-1940) was born in Manchester and studied at Cambridge, England. In 1884 he was appointed to what was then the most prestigious position in physics, the Cavendish chair of physics at Cambridge. His discovery of the electron and its properties in 1897 won him a Nobel Prize in 1906.

"I hope I may be allowed to record some theoretical speculations ... I put them forward only as working hypotheses, ... to be retained as long as they are of assistance ... The phenomena in these exhausted tubes reveal to physical science a new world— a world where matter may exist in a fourth state, where the corpuscular theory of light may be true, and where light does not always move in straight lines, but where we can never enter, and with which we must be content to observe and experiment from the outside."
                        Sir William Crookes
        Philosophical Transact., 1879

also break up (electrolyse) water, producing hydrogen gas and oxygen gas in a volume ratio of two to one (Fig. 8.2). Faraday concluded from these observations that atoms of matter must be endowed with electrical charges. In fact, the mass of material produced in electrolysis can be used to formulate an operational definition of the quantity of electric charge; the details of the definition are in the left margin.

Franklin's electric fluid. Even air permits the passage of an electric current, as in lightning or an electric spark. Certain materials can be given an electric charge by rubbing. Benjamin Franklin studied these phenomena and concluded, as explained in Section 3.5, that there was one electric fluid whose presence in greater or lesser amounts showed up as positive and negative charges. According to Faraday, these electric charges had to originate within the atom. The generation of electromagnetic waves in association with sparks (Section 7.3) was further evidence that the constituents of matter had electrical properties. Arrayed against these conclusions was the lack of electric effects in ordinary pieces of matter such as a glass of water, a coin, or the air we breathe.

Electric conduction in air. William Crookes (1832-1919) and J. J. Thomson studied in detail the conduction of electric current through air in a glass tube. Fig. 8.3 shows their apparatus. Electric terminals were connected to metal pieces (called electrodes), which were sealed through the ends of the tube. The gas in the tube could be pumped out so as to produce a partial vacuum inside. At ordinary atmospheric pressure, sparks jumped from one electrode to the other. At low pressure, the gas became luminous. But at very low pressure, the glass of the tube itself glowed—evidence of interaction-at-a-distance between the electrodes and the glass. When a metal screen with a fluorescent covering was placed in the tube, it glowed and cast a "shadow" on the glass tube, presumably by intercepting the "rays" that were passing between the electrodes and the glass. In this way it was possible to show that the negative electrode, called the *cathode,* was the source of the rays, which were called cathode rays, and the apparatus became known as a "cathode ray tube." Whether the rays were a stream of particles or a wave phenomenon was not yet known.

Figure 8.3 (to right) An electric current can pass through low-pressure air in a glass tube, giving rise to many interesting luminous effects. This simple "cathode ray tube" stimulated many fruitful investigations, giving birth to the electron, electronics, the TV tube, X-rays, and other discoveries.

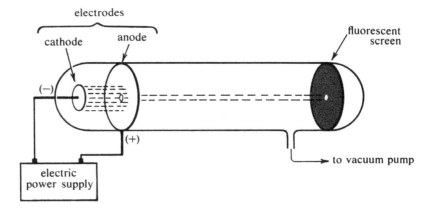

Figure 8.4 (to right).
Deflection of cathode rays by
magnetic and electric fields.
(a) Deflection by a magnetic field.
(b) Deflection by an electric field.
(c) Deflection by both magnetic
and electric fields.

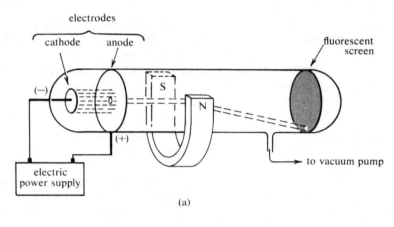

(a)

"... The most diverse opinions
are held as to [cathode] rays;
according to the almost unani-
mous opinion of German physi-
cists they are due to some
[wave] process in the aether to
which—inasmuch as in a uni-
form magnetic field their course
is circular and not rectilinear—
no phenomenon hitherto ob-
served is analogous; another
view of these rays is that, so far
from being wholly aetherial
[wave-like], they are in fact
wholly material, and that they
mark the paths of particles of
matter charged with negative
electricity . . . I can see no es-
cape from the conclusion that
they are charges of negative
electricity carried by particles
of matter. The question next
arises, What are these parti-
cles? are they atoms, or mole-
cules, or matter in a still finer
state of subdivision?"
J. J. Thomson
Philosophical Magazine, 1897

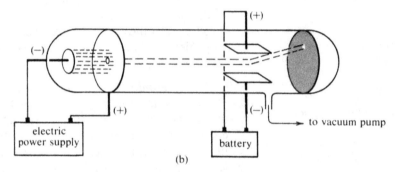

(b)

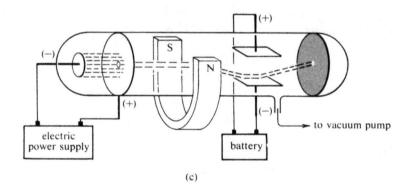

(c)

Electrons. By collecting the rays, letting them interact with electrically charged bodies, deflecting them with magnets (Fig. 8.4), and quantitatively measuring the effects, J. J. Thomson was able to show that the cathode rays carried electric charge and inertial mass in a fixed ratio. He therefore proposed a particle model to explain the interaction-at-a-distance between the cathode and glass: many tiny, identical micro-domain particles with a definite mass and negative charge are emitted by the cathode and acquire kinetic energy on being attracted by the positive electrode (the *anode*). Thomson adopted the term *electron* for these particles. Robert Millikan later measured the electric charge of a single electron in 1909 by means of his ingenious oil drop experiment, the first measurement of a physical

Figure 8 5 Color television tubes in an assembly line are being baked over heat lamps to bond a special phosphorescent material to the inside of the glass screen. TV tubes are essentially cathode ray tubes; the electrons are ejected by the cathode (at top of photograph), formed into a beam, and allowed to strike the screen (at bottom), transferring some of their energy to the phosphorescent material, which emits light of specific colors. Magnetic fields are used to direct the electron beam to specific points on the screen and thus form the image we see.

Robert A. Millikan (1868-1953), a physicist at the University of Chicago and an experimentalist of extraordinary talent and patience, developed his "oil drop experiment" by which changes of one electron charge could be detected. He was the first to carry out a direct measurement of the charge of the electron. He also verified Einstein's hypothesis of light quanta. His painstaking work on the charge of the electron and the photoelectric effect won him the Nobel Prize in 1923.

Millikan found that the charge of one electron was 1.7×10^{-24} faraday. Hence 1 gram of hydrogen has 6.0×10^{23} electrons associated with it.

quantity in the micro domain. Millikan's value of the electric charge plus Thomson's value for the charge-to-mass ratio yielded the actual mass of the electron. Modern application of the cathode ray tube has created the electronics industry and revolutionized communications (Fig. 8.5).

Electron waves. According to J. J. Thomson in 1897, cathode rays were minute, electrically-charged "corpuscles." However, physicists in Germany, where Crookes had originally discovered cathode rays, thought of the cathode rays as similar to light and other types of electromagnetic waves (including the recently discovered X-rays), and they attempted to explain the cathode rays on the basis of a wave motion of the "aether" (see quote from Thomson on previous page). Thomson overcame these arguments by the turn of the century with a masterful series of experiments, volumes of data and many outstanding papers demonstrating how his data confirmed the predictions of the particle model and conflicted with those of the wave model.

As a matter of fact, no one at the turn of the century actually tested the cathode rays, directly, for wave properties. One reason for this was Thomson's experiments and tightly-woven arguments; another was the fact that electric charges had always been associated with matter, and no one had ever observed charges transported from place to place by waves. In any case, no one pursued this for almost 30 years, until, as we explain below in Section 8.4, a wave model for all matter was proposed on theoretical grounds. As a result, in 1926, Davisson and Germer performed an experiment in which cathode rays were intercepted by a nickel crystal, which had regularly arranged atoms that functioned like a diffraction grating. Davisson and Germer found a diffraction pattern, as predicted by a wave model for the cathode rays! However, the wavelength of the electrons was extremely short, much shorter than the wavelength of visible light.

The present view, therefore, is that the propagation of electrons is best described as a wave phenomenon. In most situations, the wavelength is very short, however, so that diffraction and other wave-like behavior appear only with micro-domain-sized slits, as in the nickel crystal. Thus Thomson's experiments, with macro-domain size slits, did not reveal wave effects. The modern theory includes the possibility of forming micro-domain wave packets from electron waves. These tiny wave packets, which carry mass and electric charge, behave like particles in macro-domain experiments.

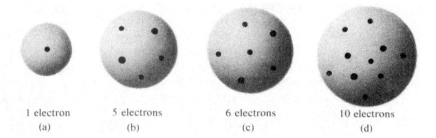

1 electron	5 electrons	6 electrons	10 electrons
(a)	(b)	(c)	(d)

Figure 8.6 (to right). J. J. Thomson's "plum pudding" model for the atom. The tiny electrons (the raisins) repel one another but are held in place (in mechanical equilibrium) by a much larger "pudding" of positive charge. The positive charge has most of the mass of the atom but, unlike the electrons, is like a diffuse "cloud" spread throughout the volume of the atom. The electrons, if stimulated, can vibrate in place to produce light waves.

Figure 8.7 (below). Ernest Rutherford's scattering experiment. The alpha particles, which were known to have much more mass than an electron, emerge from the source in a narrow beam at high speed and strike the gold foil target. Most of the alphas continue undeflected through the foil in a straight line, but some are deflected at various angles. The physicist sits in darkness for long periods, counting the bright flashes on the circular screen created by the impacts of the deflected alpha particles.

Rutherford found that the number of alphas he counted at various angles did not match the predictions of the Thomson model (above). In particular, Rutherford was astounded (see quote on next page) to find a very few flashes showing that the alphas were occasionally bouncing <u>backwards</u>. He initially thought that such flashes could not be genuine.

It is extraordinarily unlikely that Thomson's plum pudding model (above) would bounce a heavy alpha particle backwards. Thus Rutherford invented another model, with the positive charge concentrated in a tiny but heavy <u>nucleus.</u>

The MIP model. At the beginning of the twentieth century it was definitely established by experiment (electrolysis, the ability of matter to emit electromagnetic waves, and cathode ray studies) that matter could be divided into electrically charged components. Presumably, therefore, each electrically neutral atom had positive and negative parts. At last it was possible to identify how the particles interacted in the MIP model: the interactions were electric, transmitted by means of an electric field. Macro-domain chemical, phase and elastic energy were, therefore, really just electric field energy of the atoms and molecules. The mutual attraction of the positive and negative parts in adjacent atoms could explain the structure of molecules, crystals, and liquids. The relative motion of the electrically charged parts could explain the emission of electromagnetic waves (Section 7.3), including the spectra of visible light.

The negative constituents of atoms were the very light electrons, which Thomson had measured as less than one thousandth the mass of the atom as a whole. What was the positive part, which appeared to have almost all the mass of the atom?

8.2 Early models for atoms

Thomson's model. J. J. Thomson himself was one of the first to propose an atomic model that took into account the atom's electrical nature (Figure 8.6). He suggested that an atom consists of a positive electric charge that is uniformly distributed within a spherical region of about

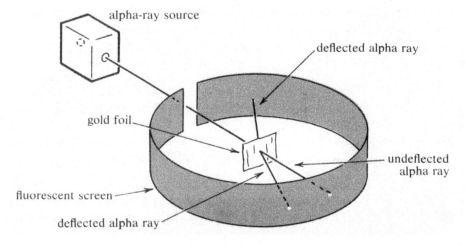

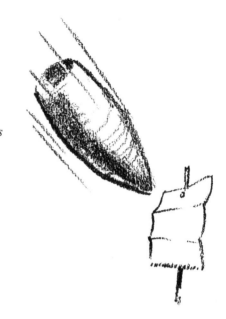

Figure 8.8 Analogue model for the collision of an alpha ray with an electron: a 15-inch shell hits tissue paper, as suggested by Ernest Rutherford in quote below to left. Would you expect the shell (or the alpha ray) to bounce back? Estimate the ratio of masses - this gives a rough estimate of the probability that the shell would bounce back.

Ernest Rutherford (1871-1937) was a New Zealander who came to England on a scholarship in 1895. After three years as a student of J. J. Thomson at Cambridge, he accepted a professorship at McGill University, Canada. Rutherford showed such virtuosity as an experimentalist that he was brought to Manchester as director of a research laboratory in 1907. At Manchester, Rutherford developed his nuclear model of the atom. An extraordinarily warm, perceptive, and stimulating man, Rutherford fostered and influenced an amazingly gifted group of young men, including Niels Bohr.

"It was quite the most incredible event that has ever happened to me in my life. It was almost as incredible as if you had fired a fifteen-inch shell at a piece of tissue paper and it came back and hit you." [see Figure 8.8 above]
Ernest Rutherford

the same size as an atom. The small, relatively light electrons were sprinkled throughout this region, somewhat like seeds in a watermelon or raisins in a plum pudding (Fig. 8.6). Since the negatively charged electrons repel one another, they tend to spread apart. The attraction of the positive charge, however, kept them from separating completely.

Thomson concluded that the electrons in this model should arrange themselves in ring-like layers. He also thought that atoms were held together in molecules and crystals by electrical forces between the unbalanced charges that are created when electrons are lost by one atom (which is then electrically positive) and are acquired by another atom (which is then electrically negative). In addition, Thomson tried, only partially successfully, to estimate how the electrons might vibrate so he could predict the frequencies of the emitted electromagnetic radiation.

Rutherford's experiment. An investigation conducted in Ernest Rutherford's laboratory in 1908 yielded a surprising result that could not be interpreted with the Thomson model for the atom. An extremely thin gold foil was bombarded with alpha rays, which are the disintegration products of the radioactive element radium (Fig. 8.7). From their interaction with magnets and electric charges, it was known that alpha rays are positively charged and are about 8000 times as massive as electrons. Most of the alpha rays that impinged on the foil went through without a measurable deflection, but a very small number were deflected by more than 90°, that is, back towards the source. Apparently, the deflected alpha rays hit something that was present here and there in the gold foil

In the Thomson model, the electrons were present "here and there." Could they have deflected the alpha rays? The very low mass of the electrons compared to that of the alpha rays meant that *none* of the alphas should be deflected by more than 90°, or backwards. *All* deflections from electrons should be small and none large (Fig. 8.8 and Figure 8.9a). But Rutherford's many experiments clearly showed that this was not true; therefore, he had to devise his own model for the way the electrons and the positive charge were arranged in the atom.

*Figure 8.9 Deflection of alpha rays by a gold foil.
(a) As predicted by Thomson's "plum pudding model," most alphas would be deflected by small angles, but none would be deflected by more than 90°.
(b) As predicted by Rutherford's nuclear model, most alphas would be undeflected, but a few would actually strike the nucleus and thus be deflected by substantial angles. In addition, among alphas actually striking the nucleus, there would be a very few that would have "head-on" collisions. These alphas represent the very small, but not zero, number deflected by more than 90°*

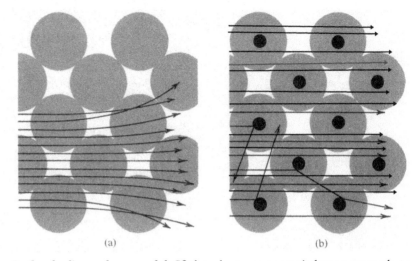

(a) (b)

Rutherford's nuclear model. If the electrons weren't heavy enough to bounce the alpha rays back, what else in the atom could do it? Rutherford therefore thought further about the positive charge in the atom, which Thomson's model represented as being spread out like a cloud, but which included almost all the mass of the atom. Perhaps it was not spread out; perhaps it was actually concentrated with all its mass in a very small region of space. If alpha rays strike such a small, massive object head on, they would recoil in the direction from which they came or be deflected by large angles, as shown in Figure 8.9b. The alpha rays are like random "bullets" striking a target; only the ones that happen to hit the "bull's-eye" (the positive charge) would be deflected by large angles. Most of the alpha rays would simply continue in a straight line. The percentage of deflected alphas should be directly related to the size of the bull's-eye.

On the basis of this relationship Rutherford estimated the size of "bull's-eye," that is, the region occupied by the positive charge within the atom. Rutherford found that this region was much, much smaller than the size of a gold atom. Rutherford called this small, massive, positive charge the *nucleus* of the atom.

The atomic model invented by Rutherford, called the *nuclear atom,* has stood the test of time and, with many refinements, is still accepted today. However, when Rutherford proposed the nuclear model, there were many serious difficulties and various "mysteries" associated with this new model. We will now explain these problems and how they were resolved, which laid the groundwork for quantum theory.

Difficulties of the nuclear model. One of the mysteries was to explain how all the positive electric charge could remain clustered in a small region of space in spite of the enormous mutual repulsion of all the parts of the charge for one another. To solve this problem, it is necessary to make a model for the atomic nucleus itself, showing the parts it is composed of and how they interact (Section 8.6).

The problem of atomic stability. A second mystery was why the negative electrons, which were strongly attracted by the positive nucleus, remained outside the nucleus and were not pulled into it. The first attempt to resolve this mystery was not successful. Rutherford suggested that the atom is like a miniature solar system, with the light, negative electrons orbiting around the heavy, positive nucleus: the planetary model for the atom. Even though such an atom would not collapse immediately, the relative motion of negative and positive charges was known to lead to the emission of electromagnetic waves, as in Hertz's spark gap experiments (Section 7.3). These waves should be observable as light emitted by atoms and would gradually rob the atoms of energy until the electrons eventually collapsed into the nucleus (Fig. 8.10). Since atoms are not observed to emit light until they collapse (in fact, they are stable and do not collapse), the planetary version of the nuclear atom did not seem satisfactory.

The problem of line spectra. A third mystery was the line spectrum of light emitted by hot gases (Section 7.2). Hot gases emit light of selected frequencies, and the atomic models should give an explanation of the spectrum. The planetary nuclear atom led to the emission of light, as we have just explained, but the emission never stopped, and the frequency varied as the electron spiraled with an increasing frequency on its way toward the nucleus.

In spite of its difficulties, the nuclear atom also enjoyed successes. By studying the deflected alpha rays, Rutherford was able to determine the relative amounts of positive electric charge on various nuclei. He found that the positive charges on nuclei of different elements, such as carbon, aluminum, copper, and gold, varied in the same ratio as the atomic number of the element (Section 4.5). Since the negatively charged electrons in an atom must neutralize the positive charge on the nucleus, the number of electrons is just equal to the atomic number and varied from one element to the next in the periodic table. This startling result explained how the atoms of various elements differed from one another; furthermore, it explained the significance of the chemists' atomic number in terms of the structure of atoms.

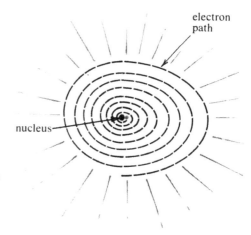

electron
path

nucleus

Figure 8.10 In the planetary model, the electron emits light and spirals into the nucleus as it loses more and more energy.

8.3 Bohr's model for the atom

Equation 8.1

energy transferred	$= \Delta E$
frequency of light	$= f$
Planck's constant	$= h$

$$\Delta E = hf$$

Planck's constant h has a numerical value in the micro domain,

$$h = 6.6 \times 10^{-34} \text{ kg-m}^2/\text{sec}$$

It therefore does not influence macro-domain phenomena directly.

" ... an electron of great velocity in passing through an atom and colliding with the [bound] electrons will lose energy in distinct finite quanta ... very different from what we might expect if the result of the collisions was governed by the usual mechanical laws."

Niels Bohr
Philosophical Magazine,
1913

Bohr's theory. Niels Bohr, a young Danish theoretical physicist temporarily working at Rutherford's laboratory, initiated a completely new approach to atomic models and achieved a decisive breakthrough in 1912. He accepted the findings of Thomson and Rutherford, but he did not submit their models merely to the laws of motion and electromagnetism as formulated by Newton and Maxwell, respectively. In addition, he was stimulated by the inconsistencies of the existing models to introduce new principles called *quantum rules* (stated to the left), and he found that these solved all the difficulties associated with the electron motion and light emission in the nuclear atom. Equation 8.1 is one of the quantum rules, called *Bohr's frequency condition*. The quantum rules and the procedures for using them make up Bohr's theory. When Bohr applied the quantum rules to the nuclear model for the hydrogen atom, which is the simplest of all atoms in that it has only one electron, he obtained excellent quantitative agreement with observations of the hydrogen spectrum. This explanation of the frequencies of the light emitted and absorbed by the hydrogen atom was a startling success of Bohr's theory.

A little later we will explain the details of the quantum rules and how to apply them. Now we will describe how they eliminate the critical difficulties of the nuclear model.

States of an atom. In any model for atoms, the state of an atom is described in terms of the arrangement and motion of its parts. Before Bohr, the radius of an electron's orbit around the nucleus was believed to become smaller and smaller as the electron lost energy by radiation and spiraled into the nucleus. The energy of the atom in any state was calculated from the speeds of the electron (kinetic energy) and the electric interaction among the electrons and the nucleus (electric field energy). With Bohr's rules, certain electron orbits and therefore certain states of atoms were selected, and all others were eliminated. The energy of each allowed state, however, was calculated from the kinetic and electric field energies as before. Each allowed state of an atom is called an *energy level* (Fig. 8.11).

Two important consequences follow from the existence of energy levels. First, there is an atomic state with the lowest permitted energy, called the *ground state*. An atom in the ground state cannot emit light, because emission of light would reduce its energy below the lowest permitted value. The ground state is, therefore, safe against collapse of the electrons into the nucleus. Second, when the atom is not in the ground state, it can emit light as the electron "jumps" from an orbit of higher energy to another of lower energy. The energy transferred to the light is a specific amount equal to the energy difference between the two states of the atom (Fig. 8.12). The frequency of the light can be calculated from the energy transfer according to the formula in Bohr's second rule and his frequency condition (Eq. 8.1). Hence the emission spectrum will consist only of the spectral lines arising from all the allowed orbital "jumps" of the electrons; the spectrum will not be a complete rainbow including all frequencies.

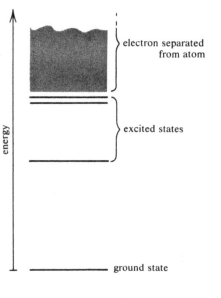

Niels Bohr (1885-1962) had just received his doctorate in Denmark when he went to England in 1911. He joined Rutherford's research group in Manchester, the "home" of the nuclear model of the atom. Bohr was intimately familiar with the latest physics theory (mostly developed in continental Europe, especially Germany, England's archrival) which revealed the stubborn contradictions and difficulties inherent in the nuclear model, and he had a remarkable capacity to connect with others as well as boundless enthusiasm for discussing the details of apparently conflicting ideas. The 26 year-old Danish theorist responded to the stimulation of Rutherford's lab with an extraordinary burst of creativity: audaciously inventing his own new "quantum rules" and using them, together with the latest theory, to resolve the well-known problems delaying acceptance of the nuclear atom. For this work, published in 1913 and 1915 (while Europe was entering World War I), Bohr was awarded the Nobel Prize in 1922.

Bohr's scientific genius was equaled by his courage, compassion, and wisdom. During World War II he helped rescue thousands of Jewish refugees by smuggling them out of Nazi-occupied Denmark to the safety of Sweden. Bohr's last years as Director of the Copenhagen Institute for Theoretical Physics were distinguished by unceasing efforts to secure international peace in an age threatened by nuclear holocaust.

Figure 8.11 Diagram to represent the allowed states of an atom. Each line represents one state allowed by Bohr's quantum rules. The distance between lines represents the energy difference between the states. The diagram is called an energy level diagram.

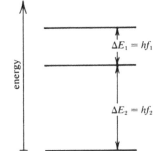

Figure 8.12 The frequencies of emitted spectral lines are found from the energy differences on an energy level diagram.

Revolutionary aspect of Bohr's rules. Bohr's theory provided a breakthrough because it added to the accepted laws and principles of physics. It did not merely combine the accepted laws to obtain a new conclusion. This is characteristic of a scientific revolution.

Bohr's rules do seem arbitrary and ad hoc, and they appear to conflict with other, well-established laws of physics. The question naturally arises why should Bohr's rules hold? In 1912 many physicists asked this question. However, a "why" question of this type is more a philosophical than a scientific question. The quantum rules are an integral part of Bohr's model: the rules must be accepted if the model is successful and rejected, or modified, if it fails. If a second successful model is available, we could choose the one that better satisfies our curiosity or preferences. But, in 1912, calculations based on Bohr's theory, and no other, produced sensible results that agreed precisely with the observed frequencies of light emitted and absorbed by the hydrogen atom. With only one model, we do not have a choice. Nevertheless, many of Bohr's contemporaries questioned the new theory, and several years passed before the quantum rules were accepted for their value.

Quantum ideas before Bohr. The idea of the quantum (and the associated idea that micro-domain systems had discrete, not continuous,

Equation 8.2

energy of oscillator   = E
frequency of oscillator   = f
Planck's constant   = h
   $E = hf$

"While this constant was absolutely indispensable to the attainment of a correct expression ... it obstinately withstood all attempts at fitting it, in any suitable form, into the frame of a classical theory."

          Max Planck
    Nobel Prize address, 1920

Equation 8.3
energy of 1 quantum   = E
   $E = hf$

Equation 8.4
electron momentum   = $\mathcal{M}$
orbit circumference   = $2\pi r$
quantum number   = n
   $2\pi r \mathcal{M} = nh$
   $(n = 1, 2, 3, 4 ...)$

Equation 8.4 restricts the radius and momentum of the electron to only certain allowed values. The radius r and the momentum $\mathcal{M}$ must have values that fit into one of the following:
$2\pi r \mathcal{M} = h$, or
$2\pi r \mathcal{M} = 2h$, or
$2\pi r \mathcal{M} = 3h$, and so on....

Figure 8.13 (to the right) The photoelectric effect. When ultraviolet light illuminates copper plate A, electrons are emitted by the copper plate, leaving it with a positive electric charge. The electrons are absorbed by copper plate B, giving it a negative electric charge.

energy levels, as shown in Figure 8.12) did not originate with Bohr. Max Planck (1858-1947) had introduced it into the theory of light emission from glowing solids (Section 7.2). Planck used the microdomain model that represented a solid as a system of electrically charged oscillators (Section 7-3). To explain the data, Planck assumed that only selected states of the oscillators could radiate their energy and that the energies of these states were directly proportional to the oscillator frequency (Eq. 8.2). The formula relating the energy to the frequency depended on a new number, which Planck called the *quantum of action* (symbol h). This number (now called Planck's constant) and formula were similar to the ones introduced later by Bohr.

A few years after Planck's work, Einstein applied the quantum concept to the photoelectric effect (Fig. 8.13). When light is absorbed by certain metals, electrons may be ejected from the metal surface. The frequency of the light and the kinetic energy of the electrons were measured and compared. Below a certain frequency, no electrons were emitted. Above that frequency, the kinetic energy increased in a regular fashion when the light frequency was increased. Einstein found he could explain the experimental measurements in terms of energy transfer from the light to each ejected electron with the following assumption: radiant energy can be transferred to one electron only in certain definite amounts, called a quantum. The energy of one quantum of light is directly proportional to its frequency, with Planck's constant again making its appearance (Eq. 8.3).

Quantum number—Applications of Bohr's quantum rules. Bohr used Planck's constant and the electron momentum for the application of his first quantum rule to specific cases. You will recall that the momentum of a particle is its inertial mass multiplied by its speed (Eq. 3.1). The allowed circular orbits (quantum states) of Bohr's first rule are selected with the requirement that the electron momentum multiplied by the circumference of the orbit is equal to a whole number times Planck's constant (Fig. 8.14 and Eq. 8.4). The whole number n that appears in Eq. 8.4 is called a quantum number. This is known as quantization in which some quantity (in this case, $2\pi r \mathcal{M}$) can have only certain restricted values that are multiples of a basic quantity (in this case h).

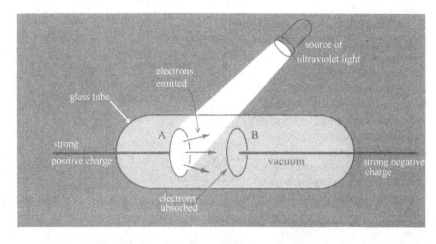

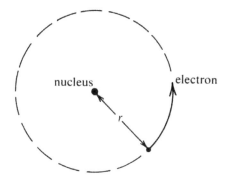

Figure 8.14 Electron in circular orbit around the nucleus. The radius of the orbit is r, the circumference is 2πr.

In Bohr's second rule, the energy transferred when an electron jumps from one orbit to another is directly proportional to the light frequency (in (Eq. 8.1) in the same relation with Planck's constant as found by Planck and Einstein (Eqs. 8.2 and 8.3). The quantity of energy transferred can be expressed by means of the quantum numbers of the two orbits. As we stated above, this model predicts quite accurately the observed spectrum of the hydrogen atom. This outstanding success of Bohr's theory made it very attractive to many physicists in spite of its revolutionary nature, and intensive research to test its implications was undertaken.

Limitations of Bohr's theory. With the passage of time and accumulation of new, more precise data, Bohr's theory had to be modified in minor ways. However, more serious inadequacies were revealed in its applications to complex atoms containing many electrons. Bohr's theory led to the conclusion that all the electrons in the atom's ground state would occupy the same orbit close to the nucleus. Both the spectra and the sizes of atoms, however, indicated that this did not happen, but that electrons arranged themselves in different orbits, some close to the nucleus, some farther away. Furthermore, the repulsive interaction of the electrons with one another could not be included satisfactorily in the theory. The difficulty of treating systems containing many electrons made it impossible to apply the model to the formation of molecules out of atoms, because all molecules do include many electrons.

In addition, the model suffered from the philosophical weakness that it combined the old laws of Newton and Maxwell with the new quantum rules. This rather capricious combination was unattractive for some physicists, who could not see clearly just when the old and when the new was to be employed. Nevertheless, Bohr's theory was the first step, the breakthrough, of a scientific revolution in the theories and models for micro-domain physical systems. This revolution, which reached its climax in the nineteen twenties, was throughout inspired and guided by Bohr.

8.4 The wave mechanical atom

In Bohr's mysterious quantization, the electron momentum multiplied by the orbit's circumference is equal to a whole number times Planck's constant (Eq. 8.4). Another way to state this condition is that the length of the path of the electron in its orbit (2πr) is a whole number times Planck's constant divided by the electron momentum (Eq. 8.5)

Electron waves. Now, this relation reminded the French physicist Louis de Broglie of the condition for the wavelength of waves permitted in a tuned system (Eq. 8.6; see Eq. 6.4). Using Bohr's model for the atom, de Broglie interpreted the circumference of the orbit as the length of the tuned system. By comparing Eqs. 8.5 and 8.6 and identifying the circumference, 2πr, with L, we can see the close similarity between $h/\mathcal{M}$ and the wavelength, λ. De Broglie therefore introduced the idea that the motion of electrons is governed by a wave whose wavelength is inversely proportional to the electron momentum (Eq. 8.7a) and whose wave number (k = 1/λ) is therefore directly proportional to the momentum (Eq. 8.7b). An atom is thus viewed as a tuned system that contains standing waves of one or more electrons. The electrons are refracted around the positively charged nucleus by its electrical field, which affects electron waves in the same way that a change of index of refraction affects light waves (Fig. 8.5).

Free electrons. If the motion of electrons is really governed by waves while they orbit the nucleus, the same should be true for electrons in a cathode-ray beam. Such electrons should therefore be diffracted by a suitable "grating" whose slits are separated by a distance comparable to the wavelength as given in Eq. 6.13. A nickel crystal provides such a grating. And, as we pointed out in Section 8.1, Davisson and Germer accurately verified this prediction of de Broglie's model.

Wave mechanics. The branch of physics in which the motion of electrons is represented as wave propagation is called wave mechanics. Electron waves are diffracted by obstacles. They are refracted when passing through the electric fields of the nucleus and of other electrons. The computational method for finding electron standing wave patterns was developed by the Austrian physicist Erwin Schrödinger and has been successfully applied to simple atoms and molecules. In principle, wave mechanical calculations make it possible to compute the physical and chemical properties of all substances. In practice, the calculations could be done for only the simplest atoms. At that time, the calculations for atoms with more than 2 or 3 electrons were much too complicated; now high-speed computers have successfully calculated the properties

Equation 8.5

$2\,\pi r = n\,(h/\mathcal{M})$

(n = 1, 2, 3, ...)

Equation 8.6

length of tuned system = L
wavelength = λ
number of waves = n

$L = n\lambda$

Equation 8.7

wave number = k

$\lambda = h/\mathcal{M}$ (a)

$k = \mathcal{M}/h$ (b)

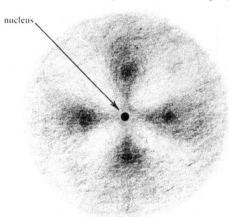
nucleus

Figure 8.15 Diagram showing electron waves refracted around the nucleus. Four wave maxima and four minima are shown. Compare with Fig. 8.14, which represents the orbit of electron particles.

Louis Victor de Broglie (1892-1987) is unique among great physicists because he became interested in physics at an age when most physicists have completed their major work, and even this late start was interrupted by service in the army during World War I. After the war, in his brother's scientific laboratory, the idea that electrons may be considered to have wave properties occurred to him. This theory, delivered in his doctoral thesis, Recherches sur la Theorie des Quanta, in 1924, won de Broglie the Nobel Prize in 1929.

Erwin Schrödinger (1887-1960) was born and educated in Vienna. Violently anti-Nazi, Schrödinger barely escaped a concentration camp by fleeing Austria in 1940. Most of his last years were spent as director of the Institute for Advanced Studies in Dublin, Ireland, where his interest was attracted to questions related to the nature of life. In 1924 and 1925 Schrödinger developed wave mechanics using de Broglie's idea that stationary states in atoms correspond to standing matter waves. For this distinguished work, Schrödinger shared the Nobel Prize with P. A. M. Dirac in 1933.

of many atoms and molecules; thus confirming the validity of the wave mechanical model to extremely high accuracy.

Electron waves in atoms and molecules. According to the wave mechanical model, therefore, an atom consists of a small nucleus surrounded by a standing wave of one or more electrons. The atom is very much larger than the nucleus, with most of its space occupied by the electron waves. The size of the atom is approximately equal to one or a few wavelengths of the electron waves forming a standing wave pattern. The atomic collapse predicted in the particle model for the electron does not occur because the particle description is not applicable. The separation of adjacent atoms in a molecule or crystal is maintained by the mutual repulsion of the negatively charged electron waves surrounding the nuclei of adjacent atoms. The formation of molecules is explained by the partial merging of certain electron waves, called valence electrons, for which the attraction by the adjacent nucleus is stronger than the repulsion by the adjacent electron wave.

Interaction of atoms with light. The wave mechanical model retains the concept of quantum jumps in which the electron gains or loses energy when the atom absorbs or emits light. In a quantum jump, the electron standing wave patterns are transformed from those of the initial state to those of the final state. The frequency of the light, still determined by Bohr's frequency condition (Eq. 8.1), is directly proportional to the energy gain or loss. Schrödinger's mathematical formulation, however, makes it possible not only to calculate the frequency but also to relate the intensity of the emitted light to the standing wave patterns in the atom's initial and final states.

8.5 Wave particle duality and the uncertainty principle

You may wonder how it is possible to reconcile the particle and wave models for the electron. Cathode rays were originally invented as the intermediaries in an interaction-at-a-distance between the cathode and the glass of the discharge tube. When it was found that they possessed electric charge and inertial mass, a particle model became accepted; the particles were called electrons. Later, the diffraction experiments gave evidence of the electron's wave nature.

The answer to the question "Where is the electric charge of the electron located?" illustrates the apparent contradictions between the two models. In the particle model, the charge is located at the position of the particle. In the wave model, the charge is spread throughout the region occupied by the wave; the electron is found more often (and the charge is more concentrated) where the wave has a large amplitude and the electron is found less often (and the charge is less concentrated) where the wave has a small amplitude. It should be possible, you may say, to find the correct model by means of an experiment. Unfortunately, as we will explain below, the experiments that can be carried out do not help to distinguish between the two models. As a matter of fact, there was a famous years-long debate between Einstein and Bohr, in which Einstein tried to find weaknesses in the quantum theory, but in which (in the opinion of most physicists today) Bohr actually found the weaknesses of Einstein's objections.

The effect of measurement on the state of a system. Consider the following simple example. A blind person tries to find the location of a glass marble on a table. He feels with his hands, measuring from the edges of the table, until his fingers touch the marble. Then he knows where he found it, but he will also have bumped it slightly and nudged it to a new location.

In any real experiment, some "nudging" of the system under study always takes place. What we tend to do, however, is to idealize by imagining a very, very gentle "nudge"—just enough, as it were, to tell the blind person that he has found the marble, but not enough to displace it. You can readily see that this ideal does not really exist, that finding the marble will always influence the marble, and observing a system will always influence or disturb the system in some way.

Even though the scientist has much more delicate instruments than the blind man's hands, the quantization of energy limits the "gentleness" with which he can operate. When he detects the position of the electron, there has to be some interaction and energy transfer between the electron and the measuring instrument. The smallest amount of energy that can be transferred is 1 quantum. Since the electron's inertial mass is very small, even the transfer of a very small amount of energy greatly affects its state.

Probability interpretation. The structure of the quantum theory, in other words, makes it impossible to carry out operations that identify an electron definitely as a wave or a particle. When you treat the electron as a particle and direct it through a small opening, the resulting diffraction pattern reminds you of its wave nature. When you treat the electron as a wave and try to separate a portion of the wave pattern from the remainder, you find that you get either all of it or none of it, as you would for a particle. The question "Is the electron a wave or a particle?" therefore does not have an operational meaning within the theory.

Nevertheless, there is a relationship between the two views. The amplitude of the electron wave can be related to the probability for locating the charge of the electron. Suppose, for example, an investigator makes many measurements of the position of the electron's charge in the ground state of an atom. He then finds the charge frequently (that is, with high probability) in regions where the wave has a large amplitude (Fig. 8.15). He finds the charge rarely (that is, with low probability) in regions where the wave has a small amplitude. He has to make many observations, however, and each time he finds all or none of the charge of the electron at the position he is observing.

Electron wave packets. Another way to reconcile the wave and particle models is to represent the electron as a wave packet. A wave

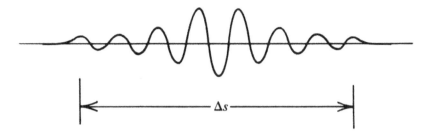

Figure 8·16 An electron wave packet of size Δs.

Equation 8.8

size of the wave packet
= Δs
uncertainty in wave number
= Δk

Δs Δk = 1

Equation 8.9

uncertainty in wave number
= Δk
uncertainty in momentum
= Δ$\mathcal{M}$

Δk = Δ$\mathcal{M}$ / h

Equation 8.10

Δs Δ$\mathcal{M}$ = h

You must remember that a particle is an idealized object used in the construction of models. A particle occupies a point in space and has inertial mass, speed, and momentum. A wave packet is an alternative, somewhat more complex, idealized object which we use instead of a particle in constructing models in the micro domain.

packet differs from a single wave train in that it does not extend throughout space. It differs from a particle in that it is not localized at one point. Wave packets are described by an uncertainty principle (Eq. 8.8), which was explained in Section 6.2. This principle applies to electron waves also.

We will now interpret the uncertainty principle for electron wave packets. The size Δs of the wave packet (Fig. 8.16) is interpreted as the uncertainty in the position of the electron. The uncertainty Δk in the wave number is interpreted as an uncertainty in the momentum of the electron (Eq. 8.9, derived from de Broglie's relation Eq. 8.7b). You can therefore infer an uncertainty relation between the position and the momentum of the electron (Eq. 8.10). The smaller the uncertainty in the position, the larger the uncertainty in momentum. And vice versa. It is not possible to find the precise position and momentum of an electron simultaneously. This property of an electron means that it cannot be described by a particle model.

Matter waves. Experiments with beams of alpha rays, atomic nuclei, and even entire atoms and molecules have shown that these, too, are diffracted; in other words, their motion is in accord with the theory of wave propagation. You may now ask, "Where does this end? Does all matter exhibit wave properties?" The answer is, in principle, yes. A wave packet is associated with every material system. In practice, however, the wavelengths of systems in the macro domain are so small that diffraction effects are undetectable for them. The uncertainty principle, for example, reveals that a speck of dust with a mass of 1 milligram can be treated as a "particle" with negligible uncertainty of position and momentum, even though an electron cannot be treated as such (Example 8.1). The reason is the micro-domain numerical value of Planck's constant.

EXAMPLE 8·1 Applications of the uncertainty principle using Eq. 8·10, $\Delta s \approx h/\Delta \mathcal{M}$.

(a) Electron:
mass $M_1 = 10^{-30}$ kg
speed $v = 10^7$ m/sec

momentum $\mathcal{M} = M_1 v = 10^{-30}$ kg × 10^7 m/sec = 10^{-23} kg-m/sec ten percent uncertainty in momentum:

$$\Delta \mathcal{M} = 10^{-24} \text{ kg-m/sec}$$

uncertainty in position:

$$\Delta s \approx \frac{h}{\Delta \mathcal{M}} = \frac{6.6 \times 10^{-34} \text{ kg-m}^2/\text{sec}}{10^{-24} \text{ kg-m/sec}} = 6.6 \times 10^{-10} \text{ m}$$

(approximately five times the diameter of a hydrogen atom)

(b) Dust particle:
mass $M_l = 10^{-6}$ kg
speed $v = 1$ m/sec

$$\text{momentum } \mathcal{M} = M_l v = 10^{-6} \text{ kg-m/sec}$$

one-tenth percent uncertainty in momentum:

$$\Delta \mathcal{M} = 10^{-9} \text{ kg-m/sec}$$

uncertainty in position:

$$\Delta s = \frac{h}{\Delta \mathcal{M}} = \frac{6.6 \times 10^{-34} \text{ kg-m}^2/\text{sec}}{10^{-9} \text{ kg-m/sec}} = 6.6 \times 10^{-25} \text{ m}$$

(a completely negligible uncertainty)

8.6 The atomic nucleus

In the preceding section, the atomic nucleus was described as a massive, positively charged particle whose electric field refracted the electrons into a standing wave around the nucleus. As a matter of fact, scientists have formulated models for the nucleus itself as a complex system of interacting parts, a system capable of acting as energy source or energy receiver by changing its state.

Radioactivity. Already before Rutherford's invention of the nuclear atom, mysterious rays (including Rutherford's alpha rays) had been observed by Henri Becquerel to affect photographic plates near certain materials, especially compounds containing uranium, and had been given the name radioactivity. These rays were soon identified by their interaction with a magnetic field as having three components (Fig. 8.17): massive, positively charged alpha rays (Rutherford's tool for discovering the nucleus); light, negatively charged beta rays (later identified as electrons); and electrically uncharged gamma rays, which acted similarly to X rays (later identified as electromagnetic radiation). Scientists studied the rays and the sources intensively, finding that certain elements must be the source of the rays and that, amazingly enough, these elements seemed to be "decaying" into new elements (the objective of the alchemists' centuries-old, but unsuccessful, quest). Initially, because the rate of all known chemical reactions depended on temperature, it also seemed obvious that the rate at which the rays were emitted should depend on temperature, or on the specific other elements with which the radioactive elements were combined. However, many studies established conclusively the opposite: the rate of radioactivity depended only on the amount of the particular radioactive elements present, and neither on the temperature nor anything else.

Radioactive elements each have a characteristic half-life (the time it takes for half of a sample of the element to decay into another element, leaving one half of the original element). The half-life of radium is about 1600 years, and uranium's is almost five billion years! On the other hand, the half-life of other radioactive elements is a tiny fraction of a second.

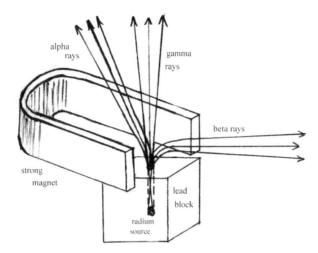

Figure 8.17 Becquerel's separation of alpha, beta, and gamma rays by using a magnetic field. Compare the deflections with those of cathode rays in Figure 8.4.

Henri Becquerel (1852-1908) studied at the Ecole Poly-technique, where he was appointed a demonstrator in 1875 and a professor in 1895. His accidental discovery of radioactivity in 1896 triggered an avalanche of research into radioactivity and its source in the atomic nucleus.

" I particularly insist on the following fact, which appears to me exceedingly important and not in accord with the phenomena which one might expect to observe: the same encrusted crystals placed with respect to the photographic plates in the same conditions and acting through the same screens, but protected from the excitation of incident rays and kept in the dark, still produce the same photographic effects."
Henri Becquerel
Comptes Rendus, 1896

The discovery of helium in radioactive ores and the experimental similarity of alpha rays to electrically charged helium atoms led Rutherford in 1903 to the conclusion that radioactivity was a breakdown of an atom of one element into an atom of another element. Uranium was gradually converted to radium and finally to lead, with alpha rays being emitted. The alpha rays then form helium atoms, which accumulate in the material.

This was the first serious attack on the concept that atoms were indivisible and immutable. (The next major step was Rutherford's invention of the nuclear atom.) The electric charges, masses, and energies observed in radioactivity quickly led Rutherford to the further conviction that radioactivity was a breakdown of the nucleus itself and did not involve the atomic electrons in a significant way. From the very beginning of the nuclear model it was clear that the nucleus was divisible and changeable. It was natural, therefore, for Rutherford to be puzzled that the positive charges in the nucleus held together as well as they did. Because of the mutual electrical repulsion of the positive charges, he wondered why every atomic nucleus did not disintegrate, as radioactive nuclei in fact do when they emit alpha rays.

Modern nuclear models. According to the currently accepted model, the atomic nucleus consists of positively charged constituents called *protons* and electrically neutral constituents called *neutrons,* all refracted into a standing wave pattern by their interaction. Neutrons and protons have approximately the same mass, but each is about 2000 times as massive as an electron. The nucleus of a hydrogen atom is one proton, an alpha ray consists of two protons and two neutrons, an oxygen nucleus includes eight protons and eight neutrons, and a uranium nucleus contains 92 protons and 143 to 146 neutrons.

Atomic number. The positive electric charge of the proton is

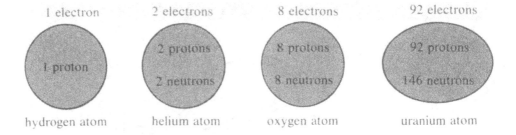

| 1 electron | 2 electrons | 8 electrons | 92 electrons |

| 1 proton | 2 protons
2 neutrons | 8 protons
8 neutrons | 92 protons
146 neutrons |

| hydrogen atom | helium atom | oxygen atom | uranium atom |

Figure 8.18 (above) The composition of four electrically neutral atoms. The numbers of electrons and protons are always equal.

equal in magnitude to the negative electric charge of the electron. Therefore, the hydrogen atom composed of one proton and one electron is electrically neutral. The number of protons determines the total positive electric charge of the nucleus, and this, in turn, determines the number of electrons that are bound in the neutral atom (Fig. 8.18). The number of electrons, finally, determines the chemical properties of an element (Section 8.2). The chemical properties, in this model, are therefore traceable to the number of protons in an atomic nucleus of the element. The number of neutrons in the nucleus does not play a role in an element's chemical properties but it does contribute to the mass of the nucleus and to its stability.

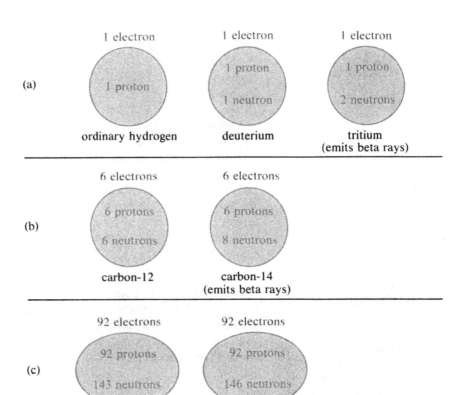

Figure 8.19 Diagrammatic representation of isotopes.
(a) Isotopes of hydrogen.
(b) Isotopes of carbon.
(c) Isotopes of uranium.

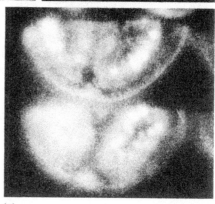

Figure 8.20 Application of radioactive isotopes.
(a) Radioactive carbon is being used in this drug metabolism study. Drugs labeled with radioactive materials yield information about concentration of compounds in tissue and about molecular structures, and provide clues to biosynthesis and interactions of normal body chemicals.
(b) The first "radio autograph" made in the dark with an artificial isotope of phosphorus. The isotope, produced in E. O. Lawrence's cyclotron, shows its presence in the leaf by affecting the photographic plate on which the leaf was placed.
(c) Radio autograph of a sliced tomato fed with a trace of radioactive zinc. The lightest spots show where the radioactivity is strongest. What can you infer about the role of zinc in growing tomatoes?

In the high temperatures at the center of stars atomic nuclei become separated from their electrons. Thus, the nuclei are not surrounded by mutually repelling electron waves; therefore, the nuclei can interact with each other when they collide. In fact, such collisions result in the large nuclear energy release that, in turn, maintains the high temperature of the sun and stars.

Isotopes. The name "isotope" is given to atoms of one element whose nuclei differ only in their number of neutrons (Fig. 8.19). Most elements have one or a few known stable isotopes and a few radioactive isotopes, which emit beta rays. The radioactive isotopes are thereby transformed into stable isotopes of another element. Isotopes have found many uses in industry, science, and medicine (Fig. 8.20).

Nuclear stability. Like electrons, both neutrons and protons exhibit wave-particle duality and obey the uncertainty principle. Most important, protons and neutrons participate in a non-electrical attractive interaction-at-a-distance, which binds them to form stable nuclei. The interaction, called the *nuclear interaction,* is much stronger than the electrical repulsion for inter-particle distances less than about 10^{-15} meters, but it becomes exceedingly weak for larger distances. The nuclear interaction acts over such extremely small distances that not even the nuclei of adjacent atoms in solids or liquids are affected. Consequently, there are no macro-domain manifestations of the nuclear interaction and it does not play a role in everyday phenomena.

The role of the neutrons in the nucleus appears to be one of stabilizing the nuclear system. They participate in the strong attractive

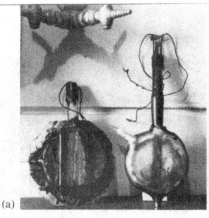

(a)

(b)

Figure 8.21 The cyclotron.
(a) Lawrence's first cyclotron
chambers, 1930.
(b) The 184-inch cyclotron at the
Lawrence Radiation Laboratory,
Berkeley, in about 1969.
Note the difference in size!

Ernest O. Lawrence (1902-1958), the father of the modern accelerator, was professor of physics at the University of California at Berkeley and the first director of the university's famed Radiation Laboratory. For his researches in atomic structure, development of the cyclotron, and its use in artificially induced radioactivity, Lawrence won the Nobel Prize in 1939. During World War II, he was one of the chief participants in the race to develop the atomic bomb before the Germans

nuclear interaction and not in the mutual electrical repulsion of the protons. Stable nuclei contain a number of neutrons about equal to or somewhat larger than the number of protons (Figs. 8.19 and 8.20). Nuclei that deviate from the ideal are radioactive and disintegrate by the emission of alpha or beta rays to form more stable nuclei.

Nuclear reactions. *The accelerator* is a research tool that has made possible the systematic study of atomic nuclei and their properties. An accelerator produces a beam of protons, electrons, or alpha rays with very high kinetic energy. The first of these machines was the *cyclotron* invented by E. O. Lawrence in 1931 (Fig. 8.21). Since then, new families of accelerators have been designed and built to study the properties of nuclei during highly energetic collisions.

When the energetic rays produced by an accelerator interact with the nuclei in a target such as thin aluminum foil, they set off a series of changes in the state of the target nucleus. These changes are called *nuclear reactions* (analogous to chemical reactions). The result is the formation of new nuclei with a changed number of neutrons and protons. Frequently these nuclei are radioactive isotopes and are useful in scientific research and in medicine.

Energy transfer. One of the most important practical outcomes of the study of radioactivity and nuclear reactions was the discovery that nuclear transformations involve very a large energy transfer from nuclear field energy to kinetic energy of the reaction products. This is easy to understand in the light of the model we have presented above. The nucleus consists of very strongly interacting neutrons and protons. When the standing wave pattern of these is changed, the nuclear energy stored in the system is changed, with the energy difference being transferred

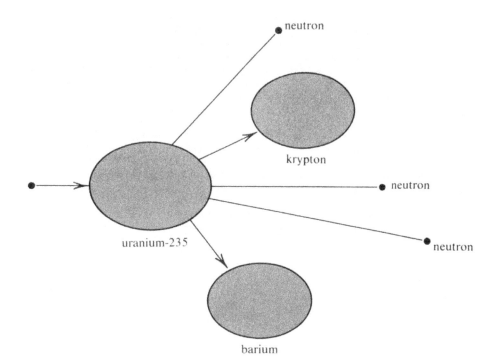

neutron

krypton

neutron

uranium-235

neutron

barium

Figure 8 22 Diagram of the nuclear fission process. The isotope uranium-235 is especially susceptible to fission after being struck by a neutron.

to other forms, such as kinetic energy of the nuclear reaction products.

Interestingly enough, the change in energy of a nucleus (as determined by the energy transfer to other forms) also manifests itself as a change in the mass of the nucleus. The energy transfer and the mass change are directly proportional (Eq. 8-11), as predicted by Einstein in his special theory of relativity in 1905.

Nuclear fission. *A* nuclear reaction in which an especially large amount of nuclear energy is released is the process of *fission.* Very large nuclei, such as uranium, are capable of fission when they are bombarded by neutrons. In the fission process, a neutron interacts with the uranium nucleus, which then disintegrates into two very energetic nuclear fragments (of about half the mass of uranium) and two or three neutrons (Fig. 8.22). The nuclear fragments transfer their kinetic energy to atoms with which they collide and thereby increase the temperature of the material in which the fissioning nucleus was embedded. In this process, nuclear energy is transformed into thermal energy, with the fission fragments acting as intermediate energy receivers and sources.

Nuclear chain reactions. The utilization of nuclear fission for macroscopic energy transfer requires the fissioning of large numbers of uranium nuclei. This has been accomplished successfully by using the principle of the *chain reaction.* One neutron is required to trigger the fission process, and two or three neutrons are produced. If one of the latter neutrons is allowed to fission a second uranium nucleus,

Equation 8.11
energy transfer	$= \Delta E$
change in mass	$= \Delta M_1$
speed of light	$= c$

$$\Delta E = \Delta M_1 \, c^2$$

"If a body gives off the energy E in the form of radiation, its mass diminishes by E/ c^2. The fact that the energy withdrawn from the body becomes energy of radiation evidently makes no difference, so that we are led to the more general conclusion that the mass of a body is a measure of its energy-content."

Albert Einstein
Annalen der Physik, 1905

Figure 8·23 Self-sustaining chain reaction. The neutrons are represented by the small black dots. After a neutron strikes a uranium nucleus, the nucleus is likely to split (fission) into two smaller nuclei and emit other neutrons. These neutrons can be lost (leave the sample or be absorbed by impurities), or they can strike another uranium nucleus.

(a) Non-explosive chain reaction, as in a nuclear power reactor. If some neutrons are lost, so that exactly one neutron from each fission causes another fission, the number of chain reactions is maintained, and the amount of energy released stays constant.

(b) Explosive chain reaction, as in a bomb. If the loss of neutrons is minimized, so that more than one neutron from each fission causes another fission, the number of fission reactions multiplies, and the energy released grows rapidly without stopping. In this diagram, two neutrons from each fission are shown, each of which causes another fission reaction. Thus the number of reactions doubles in each "generation." To see how extraordinarily fast such growth can be, calculate how much money you would have after a month if you earned one penny on the first day, two pennies on the second day, and so on.

and one of those produced then fissions a third, and so on, the process continues with the result that much nuclear energy is converted to thermal energy (Fig. 8.23a).

The chain reaction can lead to a nuclear explosion, as in an atomic bomb, if two of the neutrons produced during a fission process trigger another fission process (Fig. 8.23b). Then one fission is followed by

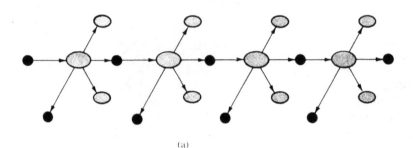

(a)

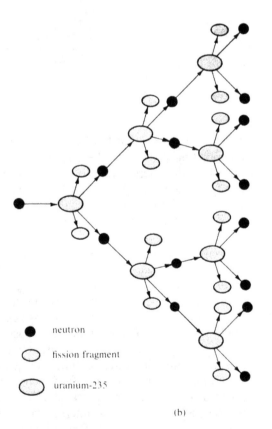

● neutron

◯ fission fragment

⬭ uranium-235

(b)

Figure 8 25 In this nuclear reactor for laboratory research, the uranium-containing reactor core is at the bottom of a 24-foot deep tank filled with water. Control rods and measuring devices extend down from the top.

Figure 8 24 Nuclear explosion, Nevada, 1955.

two, the two by four, the four by eight, and so on, until (after about 70 steps) all remaining nuclei fission at once with an enormous energy release equivalent to thousands of tons of TNT (Fig. 8.24).

Nuclear reactors. To harness the chain reaction, it is necessary to have exactly one neutron propagate the process, not more and not less. An excess of neutrons leads to an explosion, a deficiency to extinction of the chain reaction. *A nuclear reactor* is a system in which a steady chain reaction is achieved by careful control of the neutron economy (Fig. 8.25). Some neutrons escape from the system, some neutrons are absorbed by structural members of the reactor, and some neutrons are absorbed without causing fission by cadmium or boron *control rods.* Everything is placed so that exactly one neutron per fission process sustains the reaction. As uranium in the reactor is depleted, the control rods are gradually withdrawn to compensate for the remaining uranium's lowered efficiency. When the reactor is to be shut down, the control rods are inserted more deeply. A control mechanism that carefully monitors the level of neutron production supplies negative feedback via the control rods to counteract any deviations from the desired operating level.

Summary

The many-interacting-particle model for matter had been proposed, discarded, and resurrected several times since the days of the Greek philosophers, when Dalton directed attention at the ratios of weights and volumes in which elements combine to form compounds. Dalton's model was so useful in correlating chemical data that it was accepted after a few decades. Scientists could then turn to the question of the structure of the atoms themselves.

The investigations of electrolysis and spectra gave evidence that matter had an electrical nature and that atoms themselves must be complex systems rather than indivisible entities. The building of models for atoms has revolutionized the physics of the twentieth century. Early attempts involved models made up of particles (electrically negative electrons and the positive nucleus) in electrical interaction with one another. Bohr concluded that the known laws of physics did not apply to atoms, because systems of electrically charged particles could not have the permanence that atoms obviously did have. He therefore assumed new laws, called quantum rules, to supplement the laws of Newton and Maxwell. This approach, while partially successful, soon had to be replaced by a completely new point of view, in which the particle model for matter was abandoned.

The new concept, introduced by de Broglie and Schrödinger, was to represent the constituents of matter as wave packets instead of as particles. The matter waves are refracted by their interactions with one another, just as particles are deflected by their mutual interactions. Stable atoms are tuned systems in which the electron wave packets oscillate with a characteristic frequency. When the state of the atom changes because radiation is emitted or absorbed, the frequency of the radiation is related to the energy transfer by Bohr's frequency condition (Eq. 8.1).

Further developments have led to the recognition that the nucleus of the atom is not indivisible but can undergo spontaneous disintegration in the process called radioactivity. In presently accepted models for the nucleus, protons and neutrons are refracted into a stable standing wave pattern of exceedingly minute physical dimensions by an enormously strong nuclear interaction of very short range. When the state of the nucleus changes, something that happens only during radioactive decay under ordinary conditions on earth, the energy transfer is very much larger than during a change in the electron standing wave pattern in an atom. The technological exploitation of nuclear energy release has led to nuclear reactors for power production as well as to "atomic" (nuclear) bombs.

Equation 8.1
energy transferred $= \Delta E$
frequency of light $= f$
Planck's constant $= h$

$$\Delta E = hf$$

"The new discoveries made in physics in the last few years, and the ideas and potentialities suggested by them, have had an effect upon the workers in that subject akin to that produced in literature by the Renaissance... In the distance tower still higher peaks, which will yield to those who ascend them still wider prospects, and deepen the feeling, whose truth is emphasized by every advance in science, that `Great are the Works of the Lord.'"

J. J. Thomson, 1909

List of new terms

electrolysis	Planck's constant	gamma rays
electrode	energy level	proton
cathode	ground state	neutron
anode	quantum	atomic number
cathode rays	photoelectric effect	isotope
electron	quantum number	nuclear interaction
electron waves	electron diffraction	nuclear reaction
Thomson's model	wave mechanics	accelerator

alpha rays   valence electron   cyclotron
nucleus   probability   nuclear fission
nuclear model   electron wave packet chain reaction
planetary model   radioactivity   nuclear reactor
quantum rules   beta rays   control rod

List of symbols

E	energy	$\mathcal{M}$	momentum
ΔE	energy transfer	k	wave number
f	frequency	M_I	inertial mass
h	Planck's constant	v	speed
Δs	size of wave packet	$\Delta\mathcal{M}$	uncertainty of momentum
	(or uncertainty of	Δk	uncertainty of wave number
	position)		

Problems

1. Compare the many-interacting-particles models for matter proposed by Greek philosophers, eighteenth-century scientists, and John Dalton. Make reference to: (a) variety in the kinds of particles; (b) interaction among the particles; (c) quantitative relations among the particles.

2. Review and comment upon the evidence that matter has electrically charged constitutents. Which piece of evidence do you find most compelling?

3. Electrons have been described as particles and as waves. Explain briefly what you understand by these terms and how they were appropriate and/or inappropriate.

4. What were some of the problems Bohr was trying to resolve when he introduced his quantum rules? Appraise his success in solving them. (Refer to the Bibliography for additional reading on this subject.)

5. Electron waves are used to study the structure of molecules in a technique called electron diffraction. The inertial mass of electrons is approximately 1×10^{-30} kilogram. Find the wavelength of the electron waves in an electron beam when the electron speed is (a) 1.0×10^6 meters per second; (b) 4.5×10^6 meters per second.

6. Would you expect the diffraction of the electron waves in Problem 5 to give evidence of structure on a scale of sizes equal to, larger than, or smaller than visible light? Explain.

7. George Gamow's *Mr. Tompkins Explores the Atom* (published in 1993 as part of *Mr. Tompkins in Paperback*). describes a world where Planck's constant has a numerical value in the macro domain. Read and comment on this science fiction story.

8. Suppose that Planck's constant had the numerical value of 1 kilogram-(meter)^2 per second, which is a numerical value in the macro domain. Estimate the consequences of this supposition for the behavior of a few macro-domain objects.

9. Estimate the largest possible value of Planck's constant that is compatible with your everyday experience.

10. Prepare a chronology of events in the development of models for atoms and atomic nuclei between 1900 and 1930.

11. Identify one or more explanations or discussions in this chapter that you find inadequate. Describe the general reasons for your judgment (conclusions contradict your ideas, steps in the reasoning have been omitted, words or phrases are meaningless, equations are hard to follow, . . .), and make your criticism as specific as you can.

Bibliography

I. Adler, *Inside the Nucleus,* Day, New York, 1963.

D. L. Anderson, *The Discovery of the Electron,* Van No strand, Princeton, New Jersey, 1964.

J. Bernstein, *Hitler's Uranium Club, The Secret Recordings at Farm Hall,* American Institute of Physics, Woodbury, New York, 1996.

R. T. Beyer, Ed., *Foundations of Nuclear Physics,* Dover Publications, New York, 1949.

D. Bohm, *Causality and Chance in Modern Physics,* Van Nostrand, Princeton, New Jersey, 1957.

M. Born, *The Restless Universe,* Dover Publications, New York, 1957.

R. Calder, *Living With the Atom,* University of Chicago Press, Chicago, Illinois, 1962.

H. Childs, *An American Genius- The Life of Ernest Orlando Lawrence,* E. P. Dutton, New York, 1968.

C. S. Cook, *Structure of Atomic Nuclei,* Van Nostrand, Princeton, New Jersey, 1964.

E. Curie, *Madame Curie,* Doubleday, Garden City, New York, 1937.

W. C. Dampier and M. Dampier, Editors, *Readings in the Literature of Science,* Harper and Row, New York, 1959.

L. de Broglie, *Matter and Light—The New Physics,* Dover Publications, New York, 1939.

L. de Broglie, *The Revolution in Physics,* Noonday Press, New York, 1953.

A. Eddington, *The Nature of the Physical World,* University of Michigan, Ann Arbor, Michigan, 1958.

Editors of *Scientific American, Atomic Power,* Simon and Schuster, New York, 1956.

A. S. Eve, *Rutherford,* Macmillan, New York, 1939.

L. Fermi, *Atoms in the Family: My Life With Enrico Fermi,* University of Chicago Press, Chicago, 1954.

K. W. Ford, *The World of Elementary Particles,* Blaisdell, Waltham, Massachusetts, 1963.

M. Frayn, *Copenhagen* (play), Methuen, London, 1998.

D. H. Frisch and A. M. Thorndike, *Elementary Particles,* Van Nostrand, Princeton, New Jersey, 1964.

G. Gamow, *Atom and Its Nucleus,* Prentice-Hall, Englewood Cliffs, New Jersey, 1961.

G. Gamow, *Mr. Tompkins Explores the Atom,* Cambridge University Press, New York, 1945.

G. Gamow, *Thirty Years That Shook Physics,* Doubleday, Garden City, New York, 1966.

S. Glasstone, *Source Book on Atomic Energy,* Van Nostrand, Princeton, New Jersey, 1958.

B. Greene, *The Elegant Universe—Superstrings, Hidden Dimensions, and the Quest for the Ultimate Theory,* Vintage Books, New York, 1999.

O. Hahn, *Otto Hahn, A Scientific Autobiography,* Scribner's, New York, 1966.

S. Hecht, *Explaining the Atom,* Viking, New York, 1954.

P. L. Rose, *Heisenberg and the Nazi Atomic Bomb Project, A Study in German Culture*, University of California Press, Berkeley, 1998.

B. Hoffman, *The Strange Story of the Quantum,* Dover Publications, New York, 1959.

J. Irvine, *The Basis of Modern Physics,* Oliver and Boyd, Edinburgh and London, 1968.

G. Johnson, *Strange Beauty—Murray Gell-Mann and the Revolution in Twentieth-Century Physics*, Vintage Books, New York, 1999.

R. Jungk, *Brighter Than a Thousand Suns,* Harcourt Brace, New York, 1958.

W. B. Mann and S. B. Garfinkel, *Radioactivity and Its Measurement,* Van Nostrand, Princeton, New Jersey, 1966.

A. I. Miller, *Einstein, Picasso—Space, Time and the Beauty That Causes Havoc*, Basic Books, New York, 2001.

R. Moore, *Niels Bohr,* Alfred A. Knopf, New York, 1966.

J. Needham and W. Pagel, *Backgrounds to Modern Science,* Macmillan, New York, 1953.

O. Oldenberg, *Introduction to Atomic Physics,* McGraw-Hill, New York, 1954.

H. Smyth, *Atomic Energy for Military Purposes,* Princeton University Press, Princeton, New Jersey, 1945.

A. K. Solomon, *Why Smash Atoms?,* Penguin Books, Harmondsworth, Middlesex, England, 1959.

J. Tyndall, *Faraday as a Discoverer,* Crowell, New York, 1961.

E. Zimmer, *The Revolution in Physics,* Faber and Faber, London, 1936.

Articles from Scientific American. Some or all of these, plus many others, can be obtained on the Internet at http://www.sciamarchive.org/.

H. Alfven, "Antimatter and Cosmology" (April 1967).

L. Badash, "How the `Newer Alchemy' Was Received" (August 1966).

V. D. Barger and D. B. Cline, "High-Energy Scattering" (December 1967).

H. R. Crane, "The g Factor of the Electron" (January 1968).

F. J. Dyson, "Mathematics in the Physical Sciences" (September 1964).

G. Feinberg, "Ordinary Matter" (May 1967).

J. F. Hogerton, "The Arrival of Nuclear Power" (February 1968).

R. B. Leachman, "Nuclear Fission" (August 1965).

Part three: energy

JAMES PRESCOTT JOULE
(1818–1889)

Operational definitions of energy

Your common-sense notion of energy is derived from certain systems (bent bows, candle-oxygen systems) that can act as energy sources. In Chapter 4, we identified several forms of energy storage, such as the elastic energy of the bow and the chemical energy of the candle-oxygen system. We also pointed out that a system stores a quantity of energy that depends on the state of the system.

In this chapter, we will introduce two operational definitions of energy. To do so, we will describe standard units of measurement and procedures for comparing the standard units with the system whose energy we wish to measure. You can apply these operational definitions to real systems in experiments; you can also use the operational definitions to estimate the energy stored in systems by making working models for such systems and carrying out thought experiments on the models. We will use these operational definitions extensively in later chapters of this text to find mathematical models for the energy stored in a wide variety of systems. In fact, energy is the quantity that connects all of physics, a "natural currency" or a "medium of exchange." We will use the concept of energy and our operational definitions as the basis for understanding the remaining topics in this book: temperature, work, force, electricity, motion, Newton's laws, periodic motion and the solar system, heat engines and refrigerators, and kinetic theory of gases.

9.1 Measurement of energy

Unlike length or mass, the energy stored in a system cannot be compared directly with a standard unit of energy. For instance, when you compare two bows that have been bent to a different degree, it is not possible to arrange them side by side and infer by direct comparison that the one has, say, three times the stored energy of the other one. It is not even possible to place one standard system beside the two bows and to read off their energies, as you would use a ruler to compare the heights of two children. Instead of a direct comparison, you have to make an indirect comparison by transferring the energy from each bow in turn to a third system (perhaps a third bow, a spring, or the gravitational field of a weight-earth system, see Fig. 9.1). Suitable coupling elements have to be provided, a procedure that is easy in thought experiments but difficult in real experiments. In the end, the third system permits the energy comparison if a suitable energy scale is available.

A burning candle. Think of a candle, and how you could determine the energy stored in the candle-air system, energy that is transferred to other systems when the candle burns. The best way to determine this energy is to burn the candle and let the energy be transferred to a system that shows the effect of the added energy in a measurable way: either cold water that is heated, or a block of ice that is partially melted will do. The result of your measurement is expressed as the temperature rise of the water or as the mass of ice melted. Now you know the

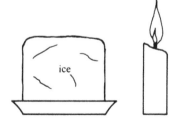

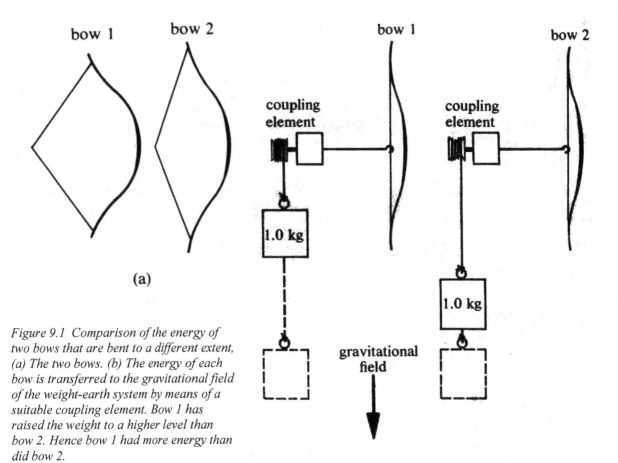

Figure 9.1 Comparison of the energy of two bows that are bent to a different extent, (a) The two bows. (b) The energy of each bow is transferred to the gravitational field of the weight-earth system by means of a suitable coupling element. Bow 1 has raised the weight to a higher level than bow 2. Hence bow 1 had more energy than did bow 2.

Figure 9.2 Measurement of the energy of an air-and-candle system.
(a) You need two identical candles, A and B.
(b) Candle B is burned and the energy is transferred to melting ice.
(c) In a thought experiment, candle A can transfer an equal quantity of energy to melting ice.

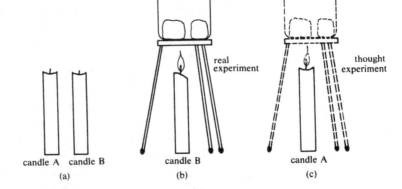

energy, but you no longer have the candle. If you could only find the energy stored without burning the candle!

This is where a thought experiment comes in. Instead of burning your candle A, take another candle B, as similar to candle A as possible,

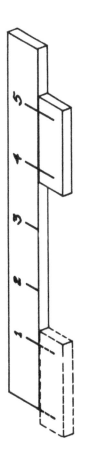

Rumford was an unusual individual who made key contributions to the study of energy and heat. We will explain his work below in Section 10.5.

and burn it to measure the energy transferred (Fig. 9.2). Now imagine that you burn candle A. Since the two candles were very much alike, candle A will heat the water or melt the ice to the same extent as did candle B. The energy of the air and candle A system, therefore, is equal to the energy transferred during the burning of candle B. You now have a measure of the energy of the air and candle A system, and you still have candle A. In fact, you can apply your knowledge to all other similar candles after you have burned the one candle and measured the energy transfer.

The foregoing description has several points that should be noted. First of all, it shows that energy can be measured directly only by being transferred, as was true also in the bow example. Second, it suggests two operational definitions of energy similar to those used over 200 years ago by Count Rumford: energy may be measured by the temperature rise of water or by the mass of melted ice. Third, it shows that a standard system (the cold water, the ice) must be chosen for the definition of the unit of energy. Fourth, it illustrates how the energy of one system, such as candle A, may be estimated through a thought experiment that makes use of the operational definition and of observations on another, but very similar, system.

9.2 Energy scales

Before we describe operational definitions of energy, we will define units of energy and construct scales on which energy may be measured. This task is analogous to the selection of a unit of length (the meter, defined as the distance between two scratches on a certain platinum-iridium bar, Fig. 1.9), and the construction of a ruler, by making equidistant marks using the unit of length. Just as the ruler has a scale for measuring lengths or distances, so we now need a scale for measuring energy.

Thermal energy scales. The energy scale consists of a set of systems that have 0 units, 1 unit, 2 units,... of energy, just as the ruler is marked for 1, 2, . . . units of length. The thermal unit of energy may be selected in any one of several ways. For instance, it may be defined as the energy required to bring 1 kilogram of water from its freezing temperature to its boiling temperature. This unit of energy is inconveniently large, and the generally accepted thermal unit of energy, called the *Calorie,* is capable of raising 0.010 kilogram of water from its freezing temperature to its boiling temperature (Fig. 9.3). Other thermal units of energy could be defined in the same way as the Calorie, but using substances such as aluminum, mercury, or alcohol instead of water. Water, however, is the material most conveniently available and generally used for energy scales.

Still another thermal unit of energy may be defined as the energy required to melt 1 kilogram of ice (solid water) at its melting temperature. A system for constructing a thermal energy scale based on this unit is shown in Fig. 9.4.

Figure 9.3 A system that provides a thermal energy scale consists of four (or more) 0.010 kilogram samples of water.

(a) All samples are at freezing temperature. The system has 0 Calories of energy,

(b) One sample has been heated to boiling temperature; the three others remain at freezing temperature. The system now has 1 Calorie of energy,

(c) Two samples have been heated to boiling temperature; two remain at freezing temperature. The system now has 2 Calories of energy,

(d) Three samples have been heated to boiling temperature; one remains at freezing temperature. The system now has 3 Calories of energy.

Equation 9.1 (Thermal energy)

mass of ice melted* (kg) $= M_G$
energy increase of system (Cal) $= E$

$$E = 80 M_G$$

(*M_G is a negative number if water freezes)

EXAMPLES
a) mass of ice melted = 2.6 kg
  $M_G = 2.6\ kg$
  $E = 80 \times 2.6\ Cal$
    $= 208\ Cal$
b) mass of ice melted = 0.3 kg
  $M_G = 0.3\ kg$
  $E = 80 \times 0.3\ Cal$
    $= 24\ Cal$

<u>Calorie, calorie and kilocalorie</u>
We will use the Calorie (with an uppercase "C") as the unit of thermal energy. The calorie (with a lowercase "c") is a much smaller unit: 1000 cal = 1 Cal. Many books also refer to the "kilocalorie" (=10^3 cal), which is simply another name for the Calorie.

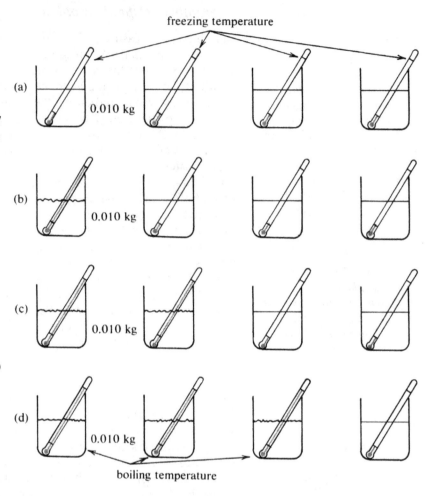

freezing temperature

(a) 0.010 kg

(b) 0.010 kg

(c) 0.010 kg

(d) 0.010 kg

boiling temperature

You may ask how the Calorie and the ice-melting unit of energy compare. An experiment to make the comparison is illustrated in Fig. 9.5. The result of this experiment is that 1 ice-melting unit equals 80 Calories. A second question is how the water-heating and ice-melting energy scales compare. In other words, if 1 ice-melting unit equals 80 Calories, are 2 ice-melting units equal to 160 Calories, and so on? The experimental result is that the two energy scales are equivalent, that quantities of energy can be measured on either scale and converted to the other scale, at 80 Calories per ice-melting unit (Eq. 9.1). One Calorie, therefore, is the energy required to melt 1/80 kilogram (0.0125 kg) of ice.

Mechanical energy scales. Mechanical units of energy may be defined as the elastic energy of a standard spring that is wound up a specified amount, or of a standard rubber band that is stretched a specified amount. Another conceivable mechanical unit of energy is the kinetic energy of an object with an inertial mass of 1 kilogram, moving

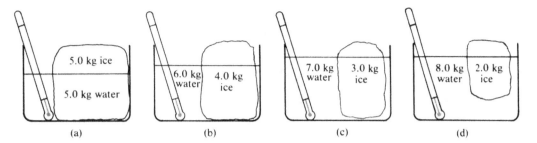

Figure 9·4 *A system that provides a thermal energy scale based on the ice-melting unit consists of 5 kilograms of liquid water and 5 kilograms of ice at equilibrium.*
(a) The system has zero energy.
(b) The system has 1 unit of energy.
(c) The system has 2 units of energy.
(d) The system has 3 units of energy.

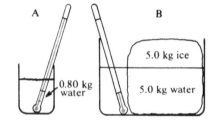

Figure 9·5 *Comparison of the Calorie and ice-melting units of energy. (a) The initial state of the system. Subsystem A at the boiling temperature has an energy of 80 Calories. Subsystem B at the freezing temperature has an energy of 0 ice-melting units. (b) The final state of the system. The two subsystems have come to equilibrium at the freezing temperature. Subsystem A has an energy of 0 Calories, subsystem B has an energy of 1.0 ice-melting units.*

with a speed of 1 meter per second. Undoubtedly other possibilities will also occur to you. Energy scales based on two of the examples mentioned are illustrated in Figs. 9.6 and 9.7.

The mechanical energy scale we will use is derived from energy stored in the gravitational field of the earth interacting with certain standard weights. The unit is defined as the energy required to raise a weight with a 0.10 kilogram gravitational mass through a height of 1 meter. This unit of energy is called the *joule*.

The 0.10 kg mass (more accurately, 0.102 kg) is chosen for reasons that will be explained in Section 14.4.

When you come to make an energy scale based on the joule, you would seem to have two alternatives: a) to raise the standard weight through various heights (1 meter for 1 joule, 2 meters for 2 joules, and so on, Fig. 9.8a), or b) to use more weights and always raise them through 1 meter (0.20 kilogram for 2 joules, 0.30 kilogram for 3 joules, 0.05

Figure 9.6 A system that provides a mechanical energy scale consists of a set of similar springs.
(a) The system has 0 "spring units" of energy.
(b) The system has 1 "spring unit" of energy.
(c) The system has 2 "spring units" of energy.

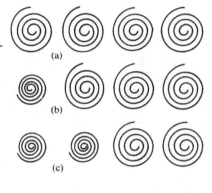

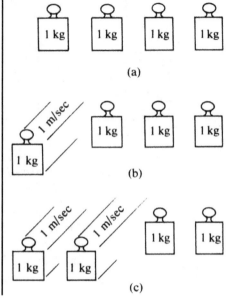

Figure 9.7 (above) A system that provides a mechanical energy scale consists of a set of 1 kilogram weights that can be put in motion at 1 meter per second.
(a) The system has 0 "kinetic units" of energy.
(b) The system has 1 "kinetic unit" of energy.
(c) The system has 2 "kinetic units" of energy.

Figure 9.8 Energy scales in joules.
(a) (to left) One weight is raised or lowered by various heights.
(b) (below) Various weights are raised or lowered through 1 meter.

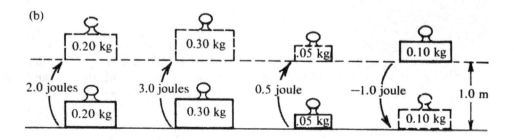

gravitational mass of
  the weight (kg) $= M_G$
height the weight is
  raised above
  starting level (m) $= h$
energy (joules) $= E$

$$E = \left(\frac{M_G}{0.10}\right) h = 10 M_G h$$

*(h is a negative number
if the weight is lowered
below starting level.)*

EXAMPLES

$M_G = 2.0 \, kg; \ h = 10 \, m$

$E = 10 \, M_G h$

  $= 10 \times 2.0 \times 10$
  $= 200 \, joules$

$M_G = 1.0 \, kg;$
$h \ \ = 0.30 \, m$

$E = 10 \, M_G h$

  $= 10 \times 1.0 \times 0.30$
  $= 3.0 \, joules$

OPERATIONAL
DEFINITION
*Thermal energy of a system
is measured by the mass of
ice melted as the system
comes to equilibrium with a
mixture of water and ice.*

kilogram for 0.5 joule, and so on, as shown in Fig. 9.8b). As in the examples of thermal energy scales, experiments on the energy transfer among various systems of weights interacting with the earth lead to the conclusion that the two alternatives provide equal energy scales. A single mathematical model relates the energy in joules to the mass and height: E = 10 M_Gh (Eq. 9.2). We will use this model and whichever of the two scales in Fig. 9.8 is more convenient. The relationship between the Calorie and the joule will be explained below (Section 9.4).

9.3 The definitions of energy

In this section we will introduce two operational definitions of energy that are easy to understand, can be used in practice with only moderate difficulty, and can be used easily in thought experiments. The comparison operation in each definition requires that the energy be transferred to a standard system that provides an energy scale.

Using the thermal energy scale. The first operational definition of energy makes use of the thermal energy scale derived from ice melting (Fig. 9.4). We have found that this scale is more convenient than the scale that derives from heating water to its boiling temperature, for this reason: systems at ordinary temperatures can transfer energy to ice and cause it to melt. Such systems, however, could not bring water to its boiling temperature, as would be required for use of the water-heating temperature scale (Fig. 9.3).

Applications of this operational definition are illustrated in Fig. 9.9. Because the water-ice mixture mentioned in the definition is at the freezing temperature of water, any system that is also at this temperature is already at equilibrium with the water-ice mixture and will not melt any ice at all. Systems at this temperature, therefore, have 0 Calories of energy.

Negative energies. What is the energy of a system whose initial temperature is *below* the freezing temperature? Fig. 9.9c shows this situation: a very cold chicken from a deep freeze interacts with the water-ice mixture. After the chicken comes to equilibrium with the water-ice mixture, their temperatures are the same; the temperature of the chicken has *increased.* The chicken has gained energy; the water-ice mixture has lost it, and some of the water has frozen and turned to more ice. Thus this process has transferred energy *from* the freezing water *to* the meat. You can weigh the amount of additional ice (say, 0.25 kilogram) and compute the number of Calories transferred from the water-ice mixture to the meat: 0.25 kg x 80 Cal/kg = 20 Calories.

Now consider the following reasoning. When the meat finally is at the freezing temperature of water, it has 0 Calories. While the meat was warming up and some of the water in the mixture was freezing, energy was being transferred from the water (which was losing energy) to the meat (which was gaining energy). Originally, therefore, the meat must have had *less than* 0 Calories of energy; in other words, its energy was <u>negative</u>. Hence we say that the frozen meat had -20 Calories of energy. More generally, the energy of systems below 0° Celsius is described by a *negative* number of Calories according to our definition.

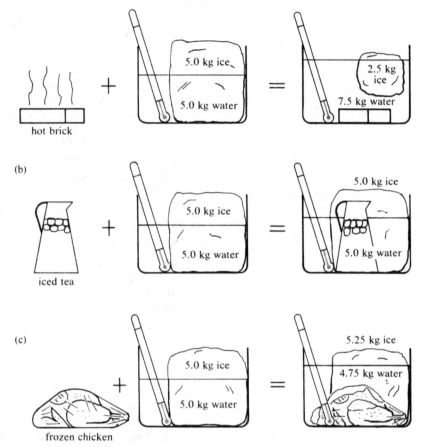

Figure 9.9 Application of the operational definition of energy based on ice melting. Sufficient water and ice must be in the system initially so that some of each remains in the system's final state. 1 Cal = energy to melt 0.0125 kg of ice.
(a) The brick melts 2.5 kilograms of ice. Its energy was 200 Calories.
(b) The iced tea melts no ice. Its energy was 0 Calories.
(c) The frozen chicken causes 0.25 kilogram of water to freeze. Therefore, the original energy of the chicken must have been <u>less than zero (negative)</u>, or, more precisely, according to Eq. 9.1, -0.25 kg x 80 Cal/kg = -20 Calories.

OPERATONAL DEFINITION
Mechanical energy of a system is measured by the height to which the system can raise a standard weight in gravitational interaction with the earth.

Using the mechanical energy scale. The second operational definition of energy makes use of the mechanical energy scale derived from the gravitational field and illustrated in Fig. 9.8. In applying this definition, a suitable coupling element has to be provided so that the energy of the system under study can be transferred completely to the standard weight with a 0.10 kilogram gravitational mass. In conducting real experiments that measure energy, you must minimize the limitations of the actual coupling element. Usually you will be carrying out thought experiments about energy transfer. Then you may assume that the coupling element is perfectly capable of transferring the energy in the desired way. Applications of this operational definition and the mathematical model of Eq. 9.2 are illustrated in Fig. 9.10.

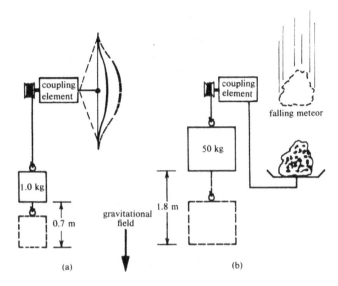

*Figure 9.10 Applications of the operational definition of energy.
(a) The bow lifts 1 kilogram through 0.7 meter. Its energy was 7 joules.
(b) The falling meteor raises 50 kilograms through 1.8 meters. Its energy
was 900 joules.*

9.4 Comparison of the joule and the Calorie

Since we have defined two energy scales, you may well ask how they
are related. If it were possible to transfer energy from a raised weight to
a water-ice mixture or vice versa, you could answer this question. Fortunately, the frictional interaction (Fig. 9.11) makes the necessary energy transfer possible. In particular, we can arrange the apparatus
shown in Fig. 9.12 to achieve this energy transfer.

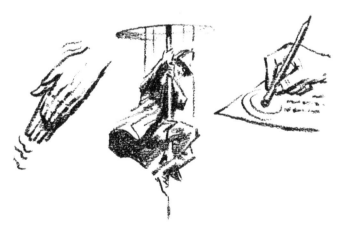

*Figure 9.11 Examples of the frictional interaction. The temperature increase in these
situations is evidence of friction.*

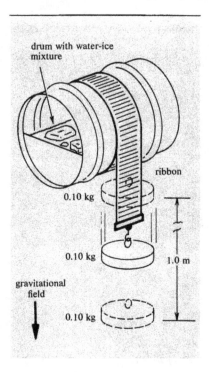

Figure 9.12 Apparatus for a
thought experiment to find the
number of joules in one Calorie.

The water-ice mixture is contained in a metal drum. A nylon ribbon is wrapped around the drum and has a standard weight hanging from it. The ribbon is pressed against the drum so that the frictional interaction is large enough to let the weight drop only very slowly. As the weight slowly drops through 1 meter, the ribbon rubs against the drum, transfers 1 joule of energy to the ice and water, and causes some ice to melt. This process is repeated until 0.0125 kilogram of ice has been melted. The number of times the process has to be repeated is the number of joules per Calorie.

Because of practical limitations, this specific experiment has never, to our knowledge, actually been carried out just as we have described it. The problem with the method we have described is that the joule turns out to be a very small unit of energy compared to the Calorie; thus, in the experiment above very little ice would be melted.

Nevertheless, over the last 100 years Joule and others have done many very similar experiments with specially designed equipment in which a measured quantity of mechanical energy is transferred, usually via friction, to a measured quantity of heat. Such experiments have established to a high degree of accuracy that the mechanical and thermal energy scales are equivalent with 4186 joules = 1 Calorie. In this text, we will use the approximate value of 4000 joules = 1 Calorie.

Equation 9.3
1 Cal = 4000 joules

Summary

To measure how much energy is stored in a system, you must conduct an experiment in which the energy is transferred to a standard system that has been chosen to provide the energy scale. Two such standard systems have been selected for the formulation of two operational definitions of energy. In one definition, the standard system is a mixture of water and ice at its equilibrium temperature of 0° Celsius. When

Equation 9.1 (Thermal energy)

mass of ice melted (kg) $= M_G$
*energy increase of
    system (Cal)* $= E$

$$E = 80 M_G$$

*(M_G is a negative number
   if water freezes)*

Equation 9.2 (Mechanical energy)

*gravitational mass of
  the weight (kg)* $= M_G$
*height the weight is
  raised above starting
  level (m)* $= h$
energy (joules) $= E$

$$E = \left(\frac{M_G}{0.10} \right) h = 10 M_G h$$

*(h is a negative number if the
  weight is lowered below
  starting level.)*

energy is transferred to the system, some ice melts. When energy is transferred from the system, some water freezes. The unit of energy is the Calorie; 1 Calorie is the energy transferred when 0.0125 kilogram of ice melts or 0.0125 kilogram of water freezes (Eq. 9.1).

In the second definition, the standard system includes the earth, any convenient macro-domain object, and their gravitational field. When energy is transferred to this system, the object is raised to a higher level. When energy is transferred from the system, the object is lowered. The unit of energy is the joule. One joule is the energy transferred when an object with gravitational mass of 0.10 kilogram is raised or lowered by 1 meter (Eq. 9.2). Approximately 4000 joules of energy are equal to 1 Calorie.

List of new terms

energy scale
Calorie
joule

List of symbols

E energy
h height
M_G gravitational mass

Problems

1. A 0.01 kilogram candle that burns up completely is observed to melt 1.4 kilograms of ice. Make a mathematical model that relates the energy stored in a candle-air system to the mass of the candle. Compare the use of this model to the procedure for finding the energy illustrated in Fig. 9.2.

2. Compare practical advantages and disadvantages of the energy scales illustrated in Fig. 9.3 and Fig. 9.4.

3. Explain why the legend in Fig. 9.5 can claim that subsystem A has 80 Calories initially.

4. Comment on the use of the gravitational mass (rather than inertial mass) in Eq. 9.1.

5. The energy scale in Fig. 9.4 is expanded to include these states of the water-ice system:
    (a) 1 kilogram of ice, 9 kilograms of water;
    (b) 9 kilograms of ice, 1 kilogram of water;
    (c) 7.5 kilograms of ice, 2.5 kilograms of water;
    (d) 3.7 kilograms of ice, 6.3 kilograms of water.
  What is the energy of the system in each state?

6. One kilogram of water at the boiling temperature is poured over a 1 kilogram block of ice. Describe the final state of this system, whose mass is 2 kilograms. What is the energy of this system?

7. Point out some advantages and/or limitations of the energy scale illustrated in Figs. 9.6 and 9.7 when compared to that in Fig. 9.8.

8. Invent and describe two mechanical energy scales other than the ones illustrated in the text.

9. Comment on the possibility of obtaining the mathematical model in Eq. 9.2 by thought experiments and/or by real experiments.

10. Apply the concept of "negative stored energy" (Fig. 9.8) to the measurement of mechanical energy. Describe a situation in which this concept might be used.

11. Describe and explain the advantages and limitations of using the three energy scales below:
    (a) The unit of energy is the energy of a fresh flashlight battery. Two batteries have 2 units of energy, and so on.
    (b) The unit of energy is the energy of a paper match-air system. Two matches (plus air) have 2 units of energy, and so on.
    (c) The unit of energy is the energy of 1 kilogram of matter according to Einstein's relation (Eq. 8.11). First, calculate the magnitude of this unit when expressed in joules.

12. Identify one or more explanations or discussions in this chapter that you find inadequate. Describe the general reasons for your judgment (conclusions contradict your ideas, steps in the reasoning have been omitted, words or phrases are meaningless, equations are hard to follow, . . .), and make your criticism as specific as you can.

Equation 8.11

energy (joules) $= \Delta E$
change in mass (kg) $= M_I$
speed of light (m/sec) $= c$

$$\Delta E = M_I c^2$$

When dining, I had often observed that some particular dishes retained their Heat much longer than others; and that apple pies . . . remained hot a surprising length of time. Much struck with this extraordinary quality of retaining Heat, which apples appear to possess, it frequently occurred to my recollection; and I never burnt my mouth with them . . . without endeavouring . . . to find out . . . some way of accounting . . . for this surprising phenomenon.

BENJAMIN THOMPSON,
COUNT RUMFORD
Collected Works

Temperature and energy

10

Have you ever burned your tongue on hot apple pie? If so, you have probably also noticed that, while you must be careful about the apple filling, the crust is not dangerous even when the pie has just come out of the oven. Visitors to the San Francisco Bay region know the remarkably mild climate enjoyed by the environs of the Bay, where the summer temperature is often 20 or more degrees less than it is only a few miles inland. Both the apple pie effect and the San Francisco climate are related to the fact that water is different from most other materials in its capacity to store thermal energy. All substances become colder or hotter as they lose or gain thermal energy (assuming they are not changing phase, such as in melting or boiling). For water, however, the temperature change is relatively small for a given energy change. The wet pie filling, therefore, transfers energy to the tongue and still remains hot, while the crust cools off substantially when it interacts with the tongue. Conversely, both land and water receive similar amounts of energy from the sun in the summer, but San Francisco Bay water rises in temperature much less than does the adjacent land.

These two explanations rest on your awareness that thermal energy and temperature are distinct factors, though they usually increase or decrease together. The definition of the Calorie in Section 9.2 makes it possible for you to measure the thermal energy stored in any system in which you may be interested. We still must explain the measurement of temperature. In this chapter, therefore, we will explore the relationship of thermal energy to the temperature of a system and make mathematical models for the relationship. By using these models and thought experiments, you will be able to calculate the equilibrium temperature of a system whose subsystems were initially at different temperatures.

Chemical energy, phase energy (the energy of liquid versus solid or gas versus liquid at the same temperature), and thermal energy are easily transformed into one another, as you know from burning a candle and making iced tea. We will therefore include all three forms of energy storage in our discussion. Finally, we will describe how the many-interacting-particles (MIP) model for matter explains the thermal energy transfer that is observable in the macro domain.

10.1 Operational definition of temperature

An instrument for measuring temperature is called a *thermometer.* You can make a thermometer out of any system that gives easily observed evidence of a change in its temperature (Fig. 10.1). Such a system is called a *thermometric element.* Thermal expansion of solids or liquids is the property of matter most commonly used in the construction of thermometers, but Galileo used air in his "thermoscope" (Fig. 10.2). Since the expansion effects are small, one practical procedure is to confine an amount of liquid (mercury, red-dyed alcohol, or kerosene) in a bulb and to let the liquid extend into a very fine tube. Even a small change in volume brought about by warming or cooling of the liquid will then result in a visible change in length of the liquid in the tube

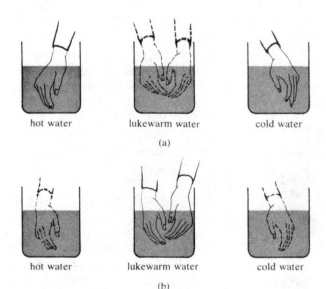

(a)

(b)

Figure 10.1 (above) Is your hand a good thermometer? To find out, proceed as follows:
(1) Place both hands in the lukewarm water. Do both hands sense the same temperature?
(2) Place one hand in the hot water, and one in cold, for a minute.
(3) Place both hands in the lukewarm water. Do both hands still sense the same temperature?
Conclusion: Your hand _is_ a thermometric element, but it makes a poor thermometer!

Figure l0.3 (below)
As the liquid in the bulb expands or contracts because its temperature changes, the liquid level in the narrow tube rises or falls.

Figure l0.4 (below)   A bimetallic strip; the two metals (brass and iron) are tightly bound together with rivets and/or strong cement. Brass expands (or contracts) more when heated (or cooled) than does iron, and so the strip bends as shown. The longer metal is on the _outside_ of the curve. Why is this?

Figure 10.2 (above)
Galileo's thermoscope. A glass bulb filled with air has a long tube whose end is dipped into water. When the bulb cools, the water rises in the tube; when the bulb is warmed, the water level drops.

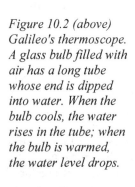

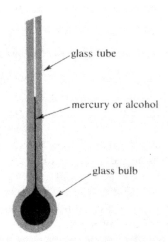

glass tube

mercury or alcohol

glass bulb

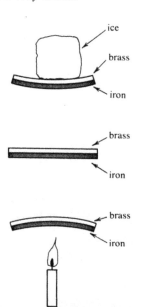

ice

brass

iron

brass

iron

brass

iron

TABLE 10.1 A SCALE OF THE DEGREES OF HEAT*

0	*The heat of the air in winter at which water begins to freeze. This heat is determined by placing a thermometer in packed snow while the snow is melting.*
0, 1, 2	*Heats of the air in winter,*
2, 3, 4	*Heats of the air in spring and in autumn.*
4, 5, 6	*Heats of the air in summer*
6	*Heat of the air in the middle of the day in the month of July.*
12	*The greatest heat that a thermometer takes up when in contact with the human body. This is about the heat of a bird hatching its eggs.*
17	*The greatest heat of a bath that can be endured for some time when the hand is dipped in it and is kept still.*
34	*The heat at which water boils violently...*
48	*The lowest heat at which a mixture of equal parts of tin and bismuth liquifies...*
96	*The lowest heat at which lead melts...*
136	*The heat at which glowing bodies shine at night but hardly in the twilight. . .*
161	*The heat at which glowing bodies in the twilight, just before the rising of the sun or after its setting, plainly shine, but in the clear light of day not at all or only very slightly.*
192	*The heat of coals in a little kitchen fire made from bituminous coal and excited by the use of bellows . . .*

*From Isaac Newton, *Opuscule,* 1701

Newton used two distinct operational definitions to define temperature (or "heat" as he called it). One made use of a linseed oil thermometer marked 0 at the freezing temperature of water and 12 at human body temperature. The other (used at higher temperatures, where the linseed oil thermometer failed) measured temperature by the time required for a particular hot block of iron to cool down to body temperature in a uniformly blowing cold wind. Newton conducted experiments that enabled him to convert from one scale to the other.

(Fig. 10.3). A second common procedure is to make a so-called bi-metallic strip, which bends when its temperature is changed (Fig. 10.4). All thermometric elements have limitations, because the liquids solidify at very low temperatures and the solids melt at high temperatures.

Temperature scales. Even though many temperature scales have been used in the past (see Table 10.1 for an example), and the Fahrenheit scale is commonly used in the United States today, we will describe the Celsius temperature scale, which is used for all scientific purposes. As we explained in Section 4.6, the temperatures of 0° and 100° on this scale are equal, respectively, to the temperatures of a water-ice system (liquid water and ice at equilibrium) and a boiling water system (liquid water and steam at equilibrium) at ordinary atmospheric pressure. The question that remains is how to subdivide the interval between these two temperatures, and how to extend the scale above and below these temperatures. Each procedure leads to an operational definition of temperature.

Perhaps the simplest approach is to make a liquid-in-tube thermometer (Fig. 10.3), mark the stem for 0° and 100°, and then divide the interval between these marks into 100 equal spaces, each being one degree. Mercury, alcohol, solvents, and linseed oil have been employed as the liquid in thermometers. Since the scale constructed in this way depends on which liquid is used, one liquid has to be chosen as standard.

There are many other procedures that could be used. For example, a bimetallic strip (Fig. 10.4) could be used to operate a pointer that sweeps over a dial (Fig. 10.5). Marks are made on the dial for 0° and 100° as described above; then the interval is divided into 100 equal spaces. Or an energy-based scale could be constructed (Fig. 10.6).

The Celsius scale. The presently accepted scientific temperature scale has a formal definition remote from the simple operations described above. It evolved from the simple operational definition as the theories regarding temperature and energy transfer were developed during the nineteenth century.

We will be content with an operational definition (in box to left) making use of the mercury thermometer. This scale is very close to the energy-based scale (Fig. 10.5) as well as to the scientist's scale, which differ from one another by less than 1% in the magnitude of a degree interval. By contrast, the degree intervals on a Stoddard's solvent thermometer, which has been marked to read the Celsius scale defined above, are significantly non-uniform, being much larger near 100° Celsius than near 0° Celsius (Fig. 10.7).

The reference to the water-ice and water-steam equilibrium temperatures in the operational definition enables anyone to check the performance of a thermometer by testing it under these conditions. This is good practice because inexpensive commercial thermometers are not always reliable.

Heat. When we discussed energy degradation in Section 4.1, the spontaneous tendency of systems at different temperatures to come to equilibrium at an intermediate temperature was mentioned. As these systems come to equilibrium, there is energy transfer from the hotter ones (energy sources) to the colder ones (energy receivers). As long as the systems interact and there is a temperature difference, there is energy transfer. Energy transferred because of a temperature difference has been given the special name *heat*. Energy sources, such as furnaces and candles, that function because of temperature differences, are called *heat sources*. For example, a lit match is a heat source for

OPERATIONAL DEFINITION
Temperature is measured by the mercury level in a mercury thermometer, marked to read 0° in equilibruim with a water-ice system and 100° in equilibrium with a boiling water system (also called a water-steam system). The water-ice and water-steam systems must be at atmospheric pressure.

The symbol of temperature is T.

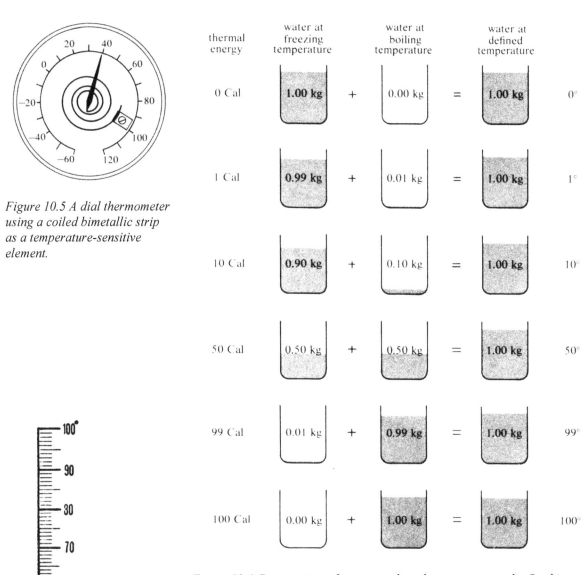

Figure 10.5 A dial thermometer using a coiled bimetallic strip as a temperature-sensitive element.

thermal energy	water at freezing temperature		water at boiling temperature		water at defined temperature	
0 Cal	1.00 kg	+	0.00 kg	=	1.00 kg	0°
1 Cal	0.99 kg	+	0.01 kg	=	1.00 kg	1°
10 Cal	0.90 kg	+	0.10 kg	=	1.00 kg	10°
50 Cal	0.50 kg	+	0.50 kg	=	1.00 kg	50°
99 Cal	0.01 kg	+	0.99 kg	=	1.00 kg	99°
100 Cal	0.00 kg	+	1.00 kg	=	1.00 kg	100°

Figure 10.6 Construction of an energy-based temperature scale. On this scale, the temperature (in degrees) of 1 kilogram of water is equal to its thermal energy in Calories.

Figure 10.7 The Celsius scale of a thermometer containing Stoddard's solvent.

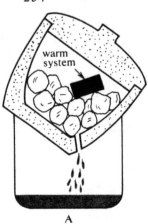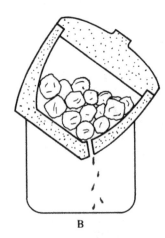

A   B

Figure 10.8 Apparatus for measuring the thermal energy of a warm or hot system. Insulated containers A and B are filled with ice at 0° Celsius. A warm system interacting with the ice in A melts some of it, in proportion to the system's thermal energy. Container B acts as a control; the amount of water melted in B is due to heat from the environment and must be subtracted from the amount melted in A.

an unlit candle, a glowing toaster coil is a heat source for the toast, human beings are heat sources in a cold room in the winter, and hot tea is a heat source for the ice cubes that convert it to iced tea. The functioning of a heat source depends on the temperature of the objects with which it interacts. Thus, the hot tea is a heat source with respect to ice cubes, but it is not a heat source with respect to the still hotter burner on the kitchen stove. For obvious reasons, heat input and heat output have an important effect on the temperature of systems.

10.2 Thermal energy and specific heat

You are now in a position to relate the thermal energy stored in a system to its temperature. According to the operational definition of temperature, you can find the temperature of the system with a Celsius thermometer. According to the operational definition of energy, you can find the energy of the system (in Calories) by letting it come to equilibrium with a water-ice system and weighing the additional melted water or solidified ice that is produced.

Measurement of thermal energy. The apparatus we will use for measuring the thermal energy of a system above 0° Celsius is shown in Fig. 10.8. It consists of two large, insulated containers of ice, each with a hole in the bottom through which water from melted ice can flow out. No water is included at the beginning. The system to be investigated is placed in one container A, and the other container B is used as "control" to determine the ice melted by interaction with room air in the absence of the system. To find the water produced by interaction of the system with ice, you subtract the mass of water coming out of the control container B from the mass of water coming out of the container with the system. Enough ice must be placed in the containers so not all of it is melted in the experiment. The time allowed must be adequate to cool the system to 0° Celsius. Thereafter, the water coming out of

TABLE 10.2 THERMAL ENERGY OF 1 KG OF WATER AT VARIOUS TEMPERATURES

Water temperature (Celsius)	Ice melted (kg)	Ice melted: control (kg)	Ice melted: net (kg)	Thermal energy (Cal)
10°	0.131	0.005	0.126	10
50°	0.625	0.003	0.622	50
75°	0.942	0.004	0.938	75

both containers is due to the melting of ice by its interaction with warm air in the room.

Results of using this apparatus to find the thermal energy of 1 kilogram of water and 1 kilogram of aluminum are listed in Tables 10.2 and 10.3 and are shown in Figs. 10.9 and 10.10. Two aspects of the results stand out: first, the graph of the experimental results is very close to a straight line; second, the thermal energy of 1 kilogram of water is between four and five times as large as the thermal energy of 1 kilogram of aluminum at the same temperature.

Specific heat. The observed straight-line relationship between thermal energy and temperature suggests a simple mathematical model (Section 1.3). The thermal energy of 1 kilogram of material is equal to the temperature multiplied by a *constant* number characteristic of the material (Eq. 10.1). This constant number is called the *specific heat* of the material (symbol C). The specific heat is therefore the energy transferred when 1 kilogram of material is raised or lowered in temperature by 1° Celsius. To find the specific heat of water, we use the data in the bottom line of Table 10.2: $C_{water} = (75 \text{ Cal})/(75°) = 1 \text{ Cal}/°C$. To find the specific heat of aluminum, we use the data from the bottom line of Table 10.3: $C_{aluminum} = (21.7 \text{ Cal})/(100°) = 0.22 \text{ Cal}/°C$. To obtain a more reliable value for the specific heat, we could make a large number of measurements and average the results.

So far we have discussed the thermal energy of 1 kilogram of material. With the help of a thought experiment in which you imagine 2 kilograms of a material interacting with the ice in a container, you can easily conclude that 2 kilograms contain two times the energy of 1 kilogram at the same temperature. The same holds true for larger and smaller amounts. In other words, the thermal energy is proportional to

Equation 10.1

thermal energy of 1 kg = E
temperature $\quad = T$
specific heat $\quad = C$

$$E = CT$$

(for water
$\quad C = 1.0 \text{ Cal/deg/kg}$
(for aluminum
$\quad C = 0.21 \text{ Cal/deg/kg}$

(For an explanation of this mathematical model, which is a "direct proportion," see Section 1.3 and the Appendix—Mathematical Background, Section A.2.)

TABLE 10.3 THERMAL ENERGY OF 1 KG OF ALUMINUM AT VARIOUS TEMPERATURES

Aluminum temperature (Celsius)	Ice melted: net (kg)	Thermal energy (Cal)
24°	0.064	5.1
60°	0.163	13.0
75°	0.205	16.4
100°	0.271	21.7

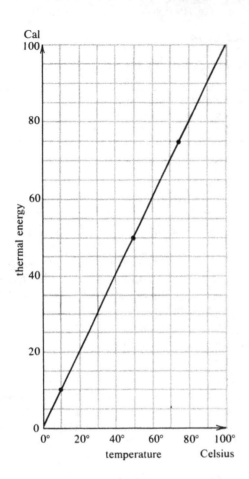

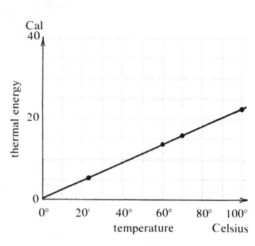

Figure 10.10 (above) Thermal energy of 1 kilogram of aluminum at various temperatures.

Figure 10.9 (to left) Thermal energy of 1 kilogram of water at various temperatures.

Equation 10.2 (thermal energy)

thermal energy(Cal) $= E$
mass of material (kg) $= M_G$
specific heat (Cal/deg/kg)
$\qquad = C$
temperature (°C) $= T$

$$E = CM_GT$$

the mass of the material. The combined proportionality to temperature and to mass is shown in Eq. 10.2. This mathematical model can be used for water, aluminum, or any other material for which the specific heat is known. The model is useful because it does apply to most materials at ordinary temperatures. If the model does not apply, and at extremes of temperature it does not, then additional measurements have to be made to find the thermal energy. Table 10.4, which contains specific heats of common materials, may be used for calculations of thermal energy according to the model of Eq. 10.2.

Applications. Most common systems are not made of one material. A glass of hot tea, for instance, is composed of hot water, a little dissolved tea material, and glass. A good model for the thermal energy of such a system is obtained by adding the energies of the various parts. This is done for a hot tea and glass system with the additional simplification that the energy of the small amount of dissolved tea is neglected (Example 10.1).

Subsystems at different temperatures. Sometimes a composite system includes subsystems that are initially at different temperatures. A glass of water with ice cubes is one example of such a system. Hot coffee and cold cream (initially in separate containers and not interacting) are another (Example 10.2). Boiling hot water and cold spaghetti are a third.

TABLE 10.4 SPECIFIC HEATS OF COMMON SUBSTANCES

Substance	Temperature* (Celsius)	Specific heat (Cal/deg C/kg)
air (gas)	0-100°	0.18
aluminum (solid)	-200°	0.08
	-100°	0.17
	0-100°	0.21
	100-600°	0.25
brass (solid)	0-100°	0.092
copper (solid)	0-100°	0.093
glass (solid)	20-100°	0.20
granite (solid)	10-100°	0.19
hydrogen (gas)	0-1000°	2.50
(liquid)	-252°	0.23
(solid)	-260°	0.57
iron (solid)	0-100°	0.110
lead (solid)	0-100°	0.031
oxygen (gas)	0-1000°	0.16
(liquid)	-200°	0.39
(solid)	-222°	0.34
sugar (solid)	20°	0.27
water (gas)	100-500°	0.37
(liquid)	0-100°	1.00
(solid)	-150°	0.25
	-100°	0.33
	-60°	0.39
	-20°	0.48
	-10-0°	0.50
wood	?	0.42
Wood's alloy (solid)	5-69°	0.035 to 0.05 (varies**)
(liquid)	70-150°	0.10 (approx.)

*Specific heats apply near given single temp. or throughout range.
**Varies depending upon composition of alloy.

Though the subsystems are initially at different temperatures, interaction among them will result in energy transfer, with the hotter ones acting as energy sources and the colder ones acting as energy receivers. The state of the system will change until the equilibrium state is reached with all subsystems at the same temperature. The mathematical model in Eq. 10.2 can be used to solve various problems associated with such an equilibrium state. For instance, what is the temperature of the cream and coffee system after the two subsystems are mixed?

You can estimate this temperature by treating the cup of coffee and the cream as an isolated system (Section 4.6). Therefore, the energy of the initial state of the system (before mixing) is equal to the energy of the final state (after mixing). Example 10.2 shows how to find the initial energy in joules. After mixing, the system comes to equilibrium. The coffee and cream are still sharing the same amount of energy as before, but they are now at the *same temperature*. Example 10.3 shows how to use this idea to calculate the final temperature of the mixture.

EXAMPLE 10·1. Thermal energy of tea–glass system.
Thermal energy of tea (assuming tea is water):

$$\text{tea } M_G = 0.25 \text{ kg}$$
specific heat of tea $C = 1.0$ Cal/deg/kg
temperature of tea $T = 70°$ Celsius

Energy of tea $= CM_GT$

$$= 1.0 \frac{\text{Cal}}{\text{deg-kg}} \times 0.25 \text{ kg} \times T$$

$$= 0.25 \frac{\text{Cal}}{\text{deg}} T$$

Thermal energy of glass:

$$\text{glass } M_G = 0.10 \text{ kg}$$
specific heat of glass $C = 0.20$ Cal/deg/kg
temperature of glass $T = 70°$ Celsius

Energy of glass $= CM_GT$

$$= 0.20 \frac{\text{Cal}}{\text{deg-kg}} \times 0.10 \text{ kg} \times T$$

$$= 0.02 \frac{\text{Cal}}{\text{deg}} T$$

Total thermal energy of system (E_{total}):

$$E_{\text{total}} = 0.25T + 0.02T$$

$$= 0.27 \frac{\text{Cal}}{\text{deg}} T$$

$$= 0.27 \frac{\text{Cal}}{\text{deg}} \times 70 \text{ deg}$$

$$= 18.9 \text{ Cal}$$

EXAMPLE 10·2. Thermal energy of hot coffee–cold cream system.
Thermal energy of coffee (assuming coffee is water):

$$\text{coffee } M_G = 0.30 \text{ kg}$$
specific heat of coffee $C = 1.0$ Cal/deg/kg
temperature of coffee $T = 60°$ Celsius

Energy of coffee $= CM_GT$

$$= 1.0 \frac{\text{Cal}}{\text{deg-kg}} \times 0.30 \text{ kg} \times 60 \text{ deg}$$

$$= 18 \text{ Cal}$$

Thermal energy of cream:

$$\text{cream } M_G = 0.05$$
$$\text{specific heat of cream } C = 1 \text{ Cal/deg/kg}$$
(assumed to be like that of water)
$$\text{temperature of cream } T = 5° \text{ Celsius}$$

Energy of cream $= CM_GT$

$$= 1 \frac{\text{Cal}}{\text{deg-kg}} \times 0.05 \text{ kg} \times 5 \text{ deg}$$

$$= 0.25 \text{ Cal}$$

Thermal energy of system (E_{total}):

$$E_{\text{total}} = 18 \text{ Cal} + 0.25 \text{ Cal} = 18.25 \text{ Cal}$$

EXAMPLE 10·3. Thermal equilibrium of cream and coffee.
Thermal energy of coffee:

$$\text{coffee } M_G = 0.30 \text{ kg}$$
$$\text{specific heat of coffee } C = 1 \text{ Cal/deg/kg}$$
$$\text{coffee temperature } T = \text{(to be found)}$$

Energy of coffee $= CM_GT$

$$= 1 \frac{\text{Cal}}{\text{deg-kg}} \times 0.30 \text{ kg} \times T \text{ deg}$$

$$= 0.30 \, T \text{ Cal}$$

Thermal energy of cream:

$$\text{cream } M_G = 0.05 \text{ kg}$$
$$\text{specific heat of cream } C = 1.0 \text{ Cal/deg/kg}$$
$$\text{cream temperature } T = \text{(to be found)}$$

Energy of cream $= CM_GT$

$$= 1.0 \frac{\text{Cal}}{\text{deg-kg}} \times 0.05 \text{ kg} \times T \text{ deg}$$

$$= 0.05T \text{ Cal}$$

Thermal energy of system (E_{total}):

$$E_{\text{total}} = 0.30T + 0.05T = 0.35T \text{ Cal}$$

From Example 10·2 you know that $E_{\text{total}} \approx 18$ Cal.
Therefore:

$$0.35T = 18 \text{ Cal}$$

$$T = \frac{18}{0.35}$$

$$T \approx 50°$$

Other uses of the model. The above model for thermal energy (Eq. 10.2, $E = CM_GT$) can be used in many other ways, too. One way is to find the specific heat of a material, such as the spaghetti dropped into hot water, from the cooling effect the spaghetti has on the water. Another is to find the rate at which the sun radiates energy to the surface of the earth from the heating effect of the sun on bodies of water. A third is to calculate how much cream must be added to coffee to achieve a desired temperature for the beverage. All of these examples illustrate how the mathematical model enables you to make many applications of data gathered in one standard situation, the determination of the specific heat, as summarized in Table 10.4.

10.3 Energy accompanying phase change

In Section 9.3 we used the melting of ice to define the thermal unit of energy. That is, the quantity of energy transferred to melt 0.0125 kilogram of ice was named 1 Calorie. This transfer of energy does not change the temperature of the water-ice system, it only changes the relative amounts of the solid and liquid phases of the material called "water" and thereby increases the phase energy of the water-ice system.

We introduced the term "phase energy" in Section 4.4 to denote the energy stored in a system by virtue of its phase. A gaseous material has more phase energy than the same material in liquid form, and a liquid material has more phase energy than the same material in solid form.

To learn more about the energy accompanying phase changes such as melting, solidification, evaporation, and liquefaction, we investigated the relation between energy and temperature of Wood's alloy, a mixture of bismuth, lead, tin, and cadmium. Wood's alloy has an especially low melting temperature for a metal, 69° Celsius. It is therefore easily possible to study its liquid phase and its solid phase. Figure 10.11 shows the data for an experiment in which the energy of 1 kilogram of Wood's alloy is measured at various temperatures. We describe how we made the measurements in the box at left.

We carried out the experiment with Wood's alloy in the same way as the experiments described above in Section 10.2 using water and aluminum. We heated the 1 kg metal sample to various temperatures and then allowed it to come to equilibrium with ice at 0°C. We measured the mass of ice melted and calculated the thermal energy transferred using Eq. 9.1 (E = 80 M_G). Finally, we plotted the results in Fig. 10.11.

Heat of fusion. The data points in Fig. 10.11 do not fit a single straight line (as in Figs. 10.9 and 10.10), but they do fit *two* straight lines, one for the temperature range (0-69°) where the alloy is solid,

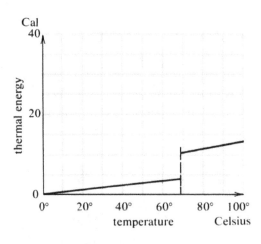

Figure 10.11 Apparent thermal energy of 1 kg Wood's alloy at various temperatures. The model of Eq. 10.1 does not fit the data points. The metal is solid below 69°C and liquid above that. At 69°C, you see a "step up" in the graph from 4 Cal to 11 Cal. Thus, the liquid phase at 69°C has 7 Cal more energy than the solid phase at the same temperature.

and the other for the temperatures (above 69°C) where it is liquid. At 69° Celsius there is a "step" in the graph. The significance of the step is that the liquid alloy at 69° has considerably more energy than the solid alloy at the same temperature. One kilogram of the liquid at 69° Celsius can melt 0.09 kilogram more ice than the solid alloy; this is simply another way of saying that it has about 7 Calories more energy than does the solid at the same temperature. The quantity of energy that is transferred when 1 kilogram of a material is melted or solidified without a temperature change is called the *heat of fusion.* The heat of fusion of Wood's alloy is 7 Calories per kilogram.

Modified form of Equation 10.2(including phase change, for energy of liquid Wood's alloy above melting temperature of 69°C).

(Eq. 10.2 in its original form, E = CMT, applies only to situations where there is no phase change.)

We therefore propose a mathematical model for the energy of Wood's alloy in which the alloy has thermal energy corresponding to a specific heat of about 0.05 Calorie per degree per kilogram in the solid phase (up to 69°C), an increase of phase energy equal to the heat of fusion of 7 Calories for each kilogram (at 69°C), and *additional* thermal energy corresponding to a specific heat of 0.10 Calorie per degree per kilogram in the liquid phase (which must be added if the sample is above 69°C). The algebraic form of this mathematical model is shown in the margin to the left. Calculations of the energy of Wood's alloy according to this model are in good agreement with the graph of the experimental results in Fig. 10.11 (Example 10.4).

Temperature(deg C) $= T$
Mass of liquid (kg) $= M$
specific heat of solid
    (Cal/deg C/kg) $= 0.05$
heat of fusion
    (Cal/kg) $= 7.0$
specific heat of liquid
    (Cal/deg C/kg) $= 0.10$

$E = (0.05)M(69°C)$
$\quad + (7.0)M$
$\quad + (0.10)M(T - 69)$

Energy of ice systems and water systems. The heat of fusion of water is 80 Calories per kilogram by definition of the Calorie in Section 9.2.

Comment:
The total energy (E) has three parts, each represented by a separate term in the above equation:

(0.05)M(69°C) is the thermal energy of the solid at 69°C.

(7.0)M is the phase energy of the liquid at 69°C.

(0.10)M(T – 69) is the additional thermal energy gained by the liquid when it is heated above 69°C.

EXAMPLE 10·4 The energy of 1 kilogram of Wood's alloy at 110° Celsius according to the model described in the text.
Data:

$\quad C$ (solid) $= 0.05$ Cal/deg/kg (up to 69°)
$\quad C$ (liquid $= 0.10$ Cal/deg/kg (above 69°)
heat of fusion $= 7.0$ Cal/kg (at 69°)

1 kilogram of solid at 40° Celsius:

$\quad E = CM_G T$

$\quad = 0.05 \dfrac{\text{Cal}}{\text{deg-kg}} \times 1.0 \text{ kg} \times 40 \text{ deg}$

$\quad = 2.0$ Cal

1 kilogram of solid at 69° Celsius:

$\quad E = CM_G T$

$\quad = 0.05 \dfrac{\text{Cal}}{\text{deg-kg}} \times 1.0 \text{ kg} \times 69 \text{ deg}$

$\quad = 3.5$ Cal

1 kilogram of liquid at 69° Celsius:

E = energy of solid + heat of fusion

$\quad$ = 3.5 + 7.0

$\quad$ = 10.5 Cal

1 kilogram of liquid at 110° Celsius:

E = energy of liquid at 69° + $CM_G(T - 69)$

$$= 10.5 \text{ Cal} + \left[0.10 \frac{\text{Cal}}{\text{deg-kg}} \times 1.0 \text{ kg} \times (110-69) \text{ deg}\right]$$

$$= 10.5 \text{ Cal} + \left[0.10 \frac{\text{Cal}}{\text{deg-kg}} \times 1.0 \text{ kg} \times 41 \text{ deg}\right]$$

$$= 10.5 \text{ Cal} + 4.1 \text{ Cal}$$

$$= 14.6 \text{ Cal}$$

". . . melting ice receives heat very fast, but the only effect of this heat is to change it into water, which is not in the least sensibly warmer than the ice was before. A thermometer, applied to the drops or small streams of water, immediately as it comes from the melting ice, will point to the same degree as when it is applied to the ice itself. . . A great quantity, therefore, of the heat, or of the matter of heat, which enters into the melting ice, produces no other effect but to give it fluidity . . . it appears to be absorbed and concealed within the water, so as not to be discoverable by the application of a thermometer."

Joseph Black
Lectures on the Elements of Chemistry, 1803

An energy of 80 Calories must be transferred to 1 kilogram of ice at 0° Celsius to melt it to water at 0° Celsius. Conversely, 80 Calories of energy must be transferred from 1 kilogram of liquid water at 0° Celsius to freeze it to 1 kilogram of ice at that temperature. It is clear, therefore, that 1 kilogram of ice and of water at 0° Celsius do not have the same energy from the viewpoint of energy conservation.

This conclusion is contrary to what is implied by the operational definition of energy in Section 9.2. Since no ice is melted while either ice or water at 0° Celsius comes to equilibrium with a mixture of water and ice, the energy of either system is 0 Calories according to our operational definition. The operational definition is therefore incompatible with energy conservation in its application to water and/or ice systems, and we must modify it. Therefore, we will now define the energy of liquid water at 0° Celsius to be 0 Calories. The energy of 1 kilogram of ice at 0° Celsius must then be -80 Calories (80 Calories less than zero). The energy of ice at temperatures below 0° Celsius is still less than this, as determined by the specific heat of ice (Table 10.4). The resulting model for the energy of water is similar to the model for Wood's alloy explained above and illustrated in Fig. 10.11. Can you sketch the graph or find an equation that gives the energy of water from below 0°C to above 100°C?

Heat of solution. Another important type of phase change, in addition to melting and vaporization, is the formation of a solution phase from a separate liquid phase and a solid phase, such as water and sodium chloride (table salt). This process is similar to the melting process in that the rigid solid material disappears and a new liquid phase is formed. There is a difference, however, in that the new liquid phase is a mixture of two substances rather than one pure substance. The energy of the solution phase is usually found (by application of the operational definition) to be different from the energy of the separate phases before mixing but at the same temperature. The energy difference is called the *heat of solution*. The heat of solution depends on both the amount of solid phase and the amount of liquid phase forming the solution. Since the solid undergoes the more drastic change, however, its amount has the more important influence; in the simplest mathematical model, the heat of solution is directly proportional to the mass of the solid material dissolved and does not depend on the mass of liquid used.

Applications. Values of the energy transfer accompanying phase changes at the listed temperatures are given in Table 10.5. You can see that the energy required for melting sodium thiosulfate (photographer's "hypo") is closely equal to that required for dissolving it in water. This observation supports the point of view that melting and dissolving are similar processes.

In all the instances except the solution of sulfuric acid, the process tabulated is associated with an increase in the energy of the system. In other words, energy has to be supplied from some heat source (perhaps a burning candle, warm water, room air, or any other system at a higher temperature) in order to bring about the change at the listed temperature. Energy has to be supplied to solid nitrogen at -210° Celsius to melt it. Energy has to be supplied to liquid alcohol at 78° Celsius to vaporize it. And energy has to be supplied to a system of solid sodium chloride and liquid water at 20° Celsius to produce a sodium chloride solution at 20° Celsius. Energy has to be removed from a sulfuric acid-water system at 20° Celsius when a solution is formed at the same temperature.

The reverse processes, of course, are accompanied by the opposite energy changes. In other words, liquid water freezing to form solid water at 0° Celsius acts as a heat source for colder systems (below 0° Celsius), gaseous water liquefying to liquid water at 100° Celsius acts as a heat source for colder systems (below 100° Celsius), and so on.

In all these examples, we have emphasized the temperature at which

TABLE 10.5 ENERGY ACCOMPANYING PHASE CHANGES

Substance	Process	Temperature (Celsius)	Energy increase (Cal/kg)
water	fusion	0°	80
water	vaporization	100°	540
mercury	fusion	-39°	2.82
mercury	vaporization	357°	65
nitrogen	fusion	-210°	6
nitrogen	vaporization	-195°	48
carbon dioxide	fusion	-56°	45
carbon dioxide	vaporization	-50°	85
alcohol	fusion	-114°	25
alcohol	vaporization	78°	204
sodium thiosulfate	fusion	46°	48
sodium thiosulfate	solution in water	20°	48
sodium chloride	solution in water	20°	22
sulfuric acid	solution in water	20°	-180

the phase change takes place. We have always compared two states of a system *at the same temperature* to avoid including thermal energy changes along with the phase energy changes. Table 10.5 lists the energy changes associated with a phase change (but no temperature change) of 1 kilogram of material.

Relation of phase change to temperature change. Frequently, phase changes and temperature changes occur together. For example, ice melts in a glass of warm water. The total energy of this system at the beginning of the process is approximately equal to the energy at equilibrium (some energy is exchanged with the room air), but the distribution of energy between the two subsystems has changed. Energy has been transferred from the warm water to the ice. An isolated system of 1 kilogram of ice at 0° Celsius in interaction with 1 kilogram of water at 80° Celsius will come to equilibrium in the form of 2 kilograms of water at 0° Celsius (Fig. 10.12). In this process heat flows from the warm water to the ice. The former cools down as energy is removed, the latter melts as energy is supplied.

The previous examples in which the melting of a solid substance (ice) was accompanied by a temperature drop in the isolated system certainly seemed natural to you. You may be a little surprised, however, when you mix water and sodium thiosulfate (both at room temperature, about 20° Celsius), shake the mixture to bring about phase equilibrium as the sodium thiosulfate dissolves, and then find that the temperature of the solution is only 10° Celsius (Fig. 10.13). How is it possible for two systems at the same temperature to come to equilibrium with one another

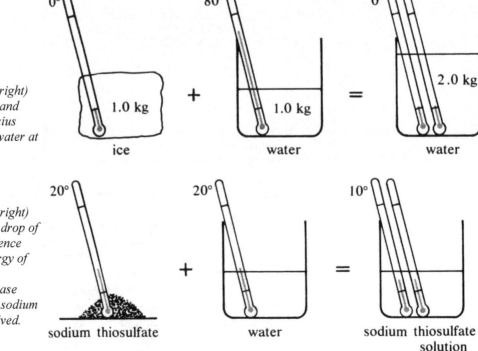

Figure 10.12 (to right) Ice at 0° Celsius and water at 80° Celsius interact to form water at 0° Celsius.

Figure 10.13 (to right) The temperature drop of the liquid is evidence that thermal energy of the system was transferred to phase energy while the sodium thiosulfate dissolved.

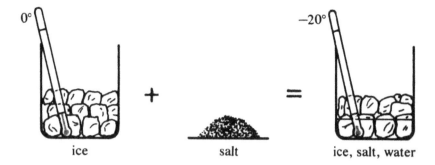

Figure 10.14 Salt (sodium chloride) sprinkled on ice interacts with the ice so that some of it melts. The temperature drop of the system is evidence that thermal energy was transferred to phase energy.

at a lower temperature? The answer is that the phase energy of the system increased as the sodium thiosulfate dissolved. Since the system in the jar was approximately isolated so that the total energy remained constant, most of the phase energy increase occurred at the expense of thermal energy stored in the system. Hence, the temperature decreased.

The same principle, transfer of thermal energy to phase energy, operates when ice cubes are dropped into cold salt water. The ice dissolves in salt water, its phase energy increases, and the thermal energy of the entire ice-water-salt system decreases, with a drop in temperature to as low as -20° Celsius. You can achieve even lower temperatures by sprinkling salt directly on ice cubes (Fig. 10.14). The salt dissolves in the water film on the surface of the ice to form a salt water phase, the ice then dissolves in the salt water phase with an increase in its phase energy, and the temperature of the entire system drops as its thermal energy is transferred to phase energy.

This phenomenon may startle you because you associate melting with an increase in temperature, freezing with a decrease. Here there was melting and decrease in temperature! The uncommon aspect of the experiments we have described was the *isolation* of the systems. Can you now explain why your commonsense expectation was misleading?

10.4 Chemical energy release

Chemical changes such as the burning of fuel are well known to result in the transfer of energy from the chemical form to the thermal form. The amount of energy can be measured by the melting of ice in a standard thermal system (water-ice mixture) and is found to be quite large, about 8000 Calories per kilogram for fuels such as gasoline, petroleum, candle wax, or coal.

Since burning fuel is an energy source, you can infer that the combustion products in the final state have a lower chemical energy than

TABLE 10.6 CHEMICAL ENERGY RELEASE

Fuel	Energy decrease of fuel-oxygen system (Cal/kg)
carbon	8,000
coal	5,000-8,000
wood	4,000
paraffin	10,000
alcohol	6,500
gasoline	11,500
kerosene	11,000
fuel oil	10,000
natural gas	11,000-13,000
hydrogen	30,000
iron	1,600
gunpowder	750
American cheese	4,500
egg	1,600
sirloin steak	2,500
rye bread	2,600
salad oil	9,000
sugar	4,000

did the fuel and oxygen in the initial state. The increase in temperature of the environment of the burning fuel is evidence of energy transfer. Table 10.6 lists the energy release accompanying combustion of various common fuels and foods.

One of the most significant technological achievements of the last 200 years is the transfer of chemical energy of fuel-oxygen systems to kinetic energy of steam engines, automobiles, airplanes, rockets, and other machinery. This process requires a suitably designed coupling element called a heat engine, which is described in Chapter 16.

10.5 Thermal energy and the MIP model for matter

The caloric theory. Greek philosophers, for whom fire was one of the elements of which all matter was composed, ascribed temperature differences to the presence of varying amounts of this element. Thus there originated the theory that heat (thermal energy) was a material substance, a "caloric" fluid.

This caloric theory, through the analogy of water flowing between reservoirs, enabled scientists to distinguish clearly between the temperature of a piece of material (represented by the level of the fluid) and the quantity of thermal energy (represented by the amount of fluid). The caloric fluid was endowed with properties not shared by water, air, or other physical fluids, in that the caloric could pass into and out of matter. The caloric theory could explain thermal equilibrium (the caloric fluid moves from one body to another until all interacting bodies contain the same level), various specific heats for different materials (some could contain more caloric for the same change in level than others, as a wide bowl can hold more water than a narrow vase), and conservation of thermal energy (the caloric could move from one body to another, but it was neither created nor destroyed).

Refinements in the theory helped it to accommodate new information. For instance, the transfer of thermal energy to phase energy when ice melts without a temperature increase gave rise to the dual concepts of "sensible" and "latent" caloric. Latent caloric was so closely tied to the particles of matter in a body that it could not flow from one body to another and heat it. Sensible caloric could do so. On melting, sensible caloric was converted to latent caloric; on combustion, latent caloric was converted to sensible caloric.

One weakness of the caloric theory was that efforts to weigh the fluid were in vain. All experimental evidence indicated that changes in temperature and/or phase changes did not change the gravitational mass of a system. Caloric therefore became identified as a fluid that had no mass. This may be contrary to common sense, but it is of no real consequence in the definition of a working model as long as no other aspects of the model are contradicted.

Joseph Black (1728-1799), a doctor of medicine at the Universities of Glasgow and Edinburgh, was one of the discoverers of carbon dioxide. However, his important work in thermal physics, including measurements of specific heat and heat of fusion, was published posthumously under the title Lectures on the Elements of Chemistry *(1803). Black was an advocate of the caloric theory of heat.*

Early forms of the kinetic theory. Some of the scientists in the seventeenth and eighteenth centuries did not accept the caloric theory. They took the view that thermal energy was a form of vibration or motion of the particles of material bodies. This theory is called the *kinetic theory*. In support of this theory, it was pointed out that thermal energy could be produced by motion, as through friction when you rub your hands together. Further support came from the related theory that gases exert pressure on their containers because they are made of rapidly moving particles that strike and recoil from the container walls.

The two theories, caloric and kinetic, competed until near the end of the eighteenth century. Then the experimental work of Joseph Black on the measurement of specific heats and other thermal properties of materials, which were so effectively explained by the caloric theory, caused that theory to be accepted almost universally.

Weaknesses of the caloric theory. We have already pointed out that the caloric fluid was thought to be mass less. Even though this was not a flaw in the logic of the caloric theory, it gave the fluid an aura of mystery not possessed by any ordinary material fluid. The amateur scientist Count Rumford therefore rejected the caloric theory.

A second strange quality convinced Count Rumford that the caloric theory was unacceptable: Rumford found that apparently unlimited amounts of caloric could be generated through friction. Rumford's observations were made during the boring of cannons, an activity he supervised at the military arsenal in Munich. He found that the cannon

Count Rumford (1753-1814). This amazing American grew up as Benjamin Thompson in Rumford, Massachusetts (now Concord, New Hampshire). He left his native land after the Revolution because of his Tory sympathies. Ambitious and aggressive, he was knighted by King George III and was employed by the Elector of Bavaria as Chamberlain, Minister of War, and Minister of Police. At the Bavarian court, Thompson was made a Lord, and it is ironic and touching that the title he chose was the name of the town to which he could never return. Somehow Rumford found the time to initiate extensive experiments designed to clarify the nature of heat.

"It will perhaps appear to some of you somewhat strange that a body apparently quiescent should in reality be the seat of motions of great rapidity; but you will observe that the bodies themselves, considered as wholes, are not supposed to be in motion. The constituent particles, or atoms of the bodies, are supposed to be in motion, without producing a gross motion of the whole mass. These particles or atoms, being far too small to be seen even by the help of the most powerful microscopes, it is no wonder that we cannot observe their motion. . ."

James Prescott Joule Lecture at St. Ann's Church, Manchester, England, 1847

barrels and the metal chips separated from them became hotter than boiling water, and this continued as long as the boring continued. He also measured the specific heat of the metal chips and the cannon barrels and found it unaffected by the boring process, during which a great deal of latent caloric had allegedly been converted to sensible caloric. Rumford concluded that anything that could be produced in unlimited quantity without an apparent source could not possibly be a material substance.

Rumford's experiments, which were carried out at the beginning of the nineteenth century, appear to us to be very convincing evidence against the caloric theory. His contemporaries, however, thought otherwise. Rumford did not provide a new theory that could explain as many observations as effectively and plausibly as the caloric theory. Hence the caloric theory remained in vogue for another 50 years. It is curious that this period of time coincided with the period following Thomas Young's experiments (Section 7.2), before the wave model for light triumphed over the corpuscular model.

The modern theory of thermal and phase energies. Dalton's atomic theory for chemical reactions (Section 4.5) established the model that matter was composed of many interacting particles in the micro domain. Quantitative properties such as mass, size, and specific interactions made the atoms of the mid-nineteenth century quite different objects from the material particles about which scientists had speculated previously. It was therefore no longer so difficult to add the ideas that these particles had kinetic energy and that energy could also be stored in the fields by which they interacted-at-a-distance.

Thermal energy. James Joule connected the MIP model with thermal energy in the way still accepted today. He proposed that macro-domain thermal energy was stored as micro-domain kinetic energy and field energy of the interacting particles. The transfer of thermal energy from hotter to colder bodies (macro-domain view) was represented by the transfer of kinetic energy through collisions of the particles (micro-domain view). Joule therefore associated temperature with kinetic energy of the particles: the hotter the body, the greater the average kinetic energy of the particles; the colder the body, the smaller the average kinetic energy of the particles. When a hot and a cold body interact, the more energetic particles in the hot body transfer energy to the less energetic particles in the cold body, until the average kinetic energies (and therefore the macro-domain temperatures) are equal. The random motion of the individual particles cannot be detected in the macro domain because the particles are too small.

Phase energy. Joule ascribed macro-domain phase energy changes to changes in the micro-domain field energy. He concluded that the particles are rearranged as a body is melted, that the fields by which they interact are altered, and that more energy is stored in the fields in the liquid phase than in the solid phase. When thermal energy is converted to phase energy (macro-domain view), kinetic energy is converted to field energy (micro-domain view). The converse occurs when a liquid material freezes to form a solid. Then phase energy is converted to thermal energy (macro-domain view), or field energy is converted to kinetic energy (micro-domain view).

Friction. The theory could also explain, in a way that Rumford had already anticipated, the increase of thermal energy through friction. During the relative motion of the hands rubbing against one another, the speeds of the particles and therefore their kinetic energies are increased. This topic is discussed more extensively in Section 11.8.

Gases. Some of the greatest quantitative successes of the new kinetic theory of thermal energy were achieved in its application to gases, for the following reason. We have already explained (Section 4.5) that gases in the MIP model are made of particles at such great distances from one another that they do not interact appreciably. Each particle is surrounded by the field it creates and is unaffected by the fields of the other particles. Consequently, the field energy stored in a gaseous system cannot change as long as the system remains a gas. All the thermal energy increases or decreases therefore coincide with increases or decreases of particle kinetic energy. This fact greatly simplifies the description of gases compared to solid and liquid materials, where complicated interactions do occur among the particles. In Chapter 16 we will describe a mathematical model that relates the temperature, pressure, thermal energy, and specific heat of gases to the properties of the particles in the MIP model. This achievement led to the universal acceptance of the kinetic theory and the abandonment of the caloric theory.

Summary

The most important message of this chapter is that thermal behavior of matter must be described by two variable factors: temperature and thermal energy. Temperature (in degrees Celsius) is defined operationally in terms of a mercury thermometer, which gives a visual indication by the expansion or contraction of the mercury in a slender glass tube. Temperature governs the direction of energy transfer, from hotter bodies, which act as heat sources, to colder bodies, which serve as energy receivers. If three bodies at different temperatures interact with one another, the one at the intermediate temperature receives energy from the hotter one and supplies energy to the colder one simultaneously. Two or more bodies at the same temperature are in thermal equilibrium, and no energy is transferred among them. Thermal energy may be transferred by physical contact of the interacting objects (when an ice cube is placed in hot water) or by radiation as an intermediary in interaction-at-a-distance (when the sun radiates to the earth).

Thermal energy of a system is defined operationally in terms of the

Equation 10.2

thermal energy (Cal) $= E$
mass of material (kg) $= M_G$
specific heat (Cal/deg C/kg)
$\qquad = C$
temperature (deg C) $= T$

$$E = C\, M_G\, T$$

*"We may conceive then, that
the communication of heat to
a body consists, in fact, in the
communication of (kinetic
energy) to its particles."*
  James Prescott Joule
  Lecture at St. Ann's Church
  Reading Room,
  Manchester, England
  1847

mass of ice *melted* when the system interacts and comes to equilibrium
with a mixture of water and ice. However, if the system was sufficiently cold that water is *frozen* (not melted) during this process, then
we must assign a *negative number* to the energy of the system. The unit
of energy is the Calorie, the energy required to melt 0.0125 kilogram of
ice.

The thermal energy of a system and its temperature are related: the
higher the temperature, the greater the thermal energy. For many systems made of one material and with all parts at the same temperature, a
simple mathematical model describes the relation adequately (Eq.
10.2). The specific heat, introduced as part of this model, is the thermal
energy transferred when 1 kilogram of the material is raised or lowered
1° Celsius in temperature. Phase energy or chemical energy of a system
can also be measured by the system's ability to melt ice.

The modern kinetic theory of thermal energy is based on the MIP
model for matter. What we have defined as thermal energy in the macro
domain is, according to the theory, stored as the total kinetic energy and
as some of the field energy of the interacting particles in the MIP model
for the system. What we have defined as phase energy in the macro
domain is stored as the remaining field energy of the particles in the
MIP model. What we have defined as temperature in the macro domain
is related to the average kinetic energy of the interacting particles in the
MIP model.

Additional examples

EXAMPLE 10.5. What is the specific heat of aluminum calculated
from the data point at 60° Celsius in Table 10.3?

Solution: $M_G = 1.0$ kg, T = 60° Celsius, E = 13.0 Cal, E = $CM_G T$ or

$$C = \frac{E}{M_G T} = \frac{13\,\text{Cal}}{1\,\text{kg} \times 60\,\text{deg}}$$
$$\approx 0.22 \text{ Cal/deg/kg}$$

EXAMPLE 10.6. What is the specific heat of liquid Wood's alloy from
the following data: 1.0 kilogram of hot aluminum at 200° Celsius interacts with 2.0 kilograms of liquid Wood's alloy at 81° Celsius until the
two metals come to equilibrium at a temperature of 144° Celsius?

Solution: Aluminum transfers energy as it cools from 200° to 144° Celsius (ΔT = 56° Celsius). The energy transfer ΔE and temperature drop
ΔT are related by the mathematical model $\Delta E = CM_G \Delta T$.
  For aluminum: $\Delta E = CM_G \Delta T = 0.22$ Cal/deg/kg x 1 kg x 56 deg
$$\approx 12.5 \text{ Cal}$$
  This energy is transferred to the Wood's alloy, which is warmed from
81° to 144° (ΔT = 63° Celsius).

For Wood's alloy: $\Delta E = CM_G \Delta T$ or

$$C = \frac{\Delta E}{M_G \Delta T} = \frac{12.5\,\text{Cal}}{2\,\text{kg} \times 63\,\text{deg}}$$

$$= \frac{12.5}{126} \approx 0.10\,\text{Cal/deg/kg}$$

EXAMPLE 10.7. One kilogram of copper at 80° Celsius is placed in 0.10 kilogram of water at 10° Celsius. What is the equilibrium temperature of the isolated copper and water system?

Solution 1 (Algebraic method.) The total initial energy of the system is the copper energy plus the water energy.

copper energy: $E = CM_G T$
     = 0.093 Cal/deg/kg x 1.0 kg x 80 deg = 7.4 Cal
water energy: $E = CM_G T$
     = 1.0 Cal/deg/kg x 0.10 kg x 10 deg = <u>1.0 Cal</u>
total energy: E = 8.4 Cal

The total final energy of the system is due to copper and water at the same final temperature (unknown).

copper energy: $E = CM_G T$
     = 0.093 Cal/deg/kg x 1.0 kg x T deg
                                 = $0.093\,T$ Cal
water energy: $E = CM_G T$
     = 1.0 x 0.10 kg x T deg   = <u>0.10 T Cal</u>
total energy: E = 0.193 T Cal
energy conservation: 0.193 T = 8.4
final temperature: $T = (8.4)/(0.193) = 43.5°$ Celsius

Solution 2 (Trial and error method). Energy will be transferred from the copper to the water until these two subsystems have come to the same temperatures. Use model

$$T = \frac{E}{CM_G}$$

Guess 1. Transfer 3.0 Calories from copper to water.
copper energy: 7.4 - 3.0 = 4.4 Cal

$$\text{copper temperature: } T = \frac{E}{CM_G}$$

$$= \frac{4.4\,\text{Cal}}{0.093\,\text{Cal/deg/kg} \times 1.0\,\text{kg}} = 47°\ \text{Celsius}$$

water energy: 1.0 + 3.0 = 4.0 Cal.

water temperature: $T = \dfrac{E}{CM_G}$

$$= \frac{4.0 \, Cal}{1.0 \, Cal/deg/kg \times 0.10 \, kg} = 40° \text{ Celsius}$$

More energy transfer is needed, but not much more.

Guess 2. Transfer 3.5 Calories from copper to water.
copper energy: 7.4 - 3.5 = 3.9 Cal

copper temperature: $T = \dfrac{E}{CM_G}$

$$= \frac{3.9 \, Cal}{0.093 \, Cal/deg/kg \times 1.0 \, kg} = 42° \text{ Celsius}$$

water energy: 1.0 + 3.5 = 4.5 Cal

water temperature: $T = \dfrac{E}{CM_G}$

$$= \frac{4.5 \, Cal}{1.0 \, Cal/deg/kg \times 0.10 \, kg} = 45° \text{ Celsius}$$

The copper temperature came out *below* the water temperature, which is physically impossible, indicating that Guess 2 was too high and that a little less energy transfer is needed. Since the water temperature is slightly too high and the copper temperature is slightly too low, we can estimate that the final equilibrium temperature will be about halfway between these two results, which is about 43.5°C. This technique of solving a problem by judicious use of estimating, plus what is called "successive approximation," may seem suspect, or even "unscientific." However, this method is quite powerful and generally useful, as it provides an independent check of a result found by use of an exact mathematical model. Most important, this method of successive approximation is what "real" scientists often use, because it works in circumstances where exact mathematical models do not exist or where finding an exact result just takes too long. In fact, most real-world problems do not have exact mathematical solutions, so making estimates and using successive approximation is the only method available.

List of new terms

thermal energy	heat	heat of solution
temperature	specific heat	chemical energy
thermometer	phase energy	caloric theory
thermometric element	heat of fusion	kinetic theory
Celsius temperature		

List of symbols

T	temperature (degrees Celsius)	ΔE	energy transfer
ΔT	temperature change	M_G	gravitational mass
E	energy	C	specific heat

Problems

1. (a) Carry out the experiment described in Fig. 10.1 and report your observations.
   (b) Describe an experiment designed to find out by how many degrees Celsius the sense of your hands can be "fooled."
   (c) Carry out the experiment described in (b) and report your result.

2. Galileo's thermoscope (Fig. 10.2) is sensitive to effects of temperature *and* of atmospheric pressure. Explain why this is the case. This fact was not recognized in Galileo's time.

3. The glass of the thermometer tube (Fig. 10.3) expands and contracts when it is heated or cooled. Describe the effect this circumstance has on the temperature readings of the thermometer.

4. Suppose the thermometer tube (Fig. 10.3) has a non-uniform inside diameter. What effect will this have on the temperature measurements?

5. Explain the functioning of a bimetallic strip as a thermometric element (Fig. 10.4).

6. Consider the device illustrated in Fig. 10.15. Describe how it could be used for the operational definition of temperature. Explain any precautions that must be taken.

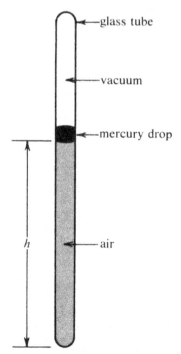

glass tube

vacuum

mercury drop

h

air

Figure 10.15 The sealed glass tube contains air on one side of a drop of mercury and a vacuum on the other. The mercury plugs the tube completely but can slide up and down.

7. Consider a thermometer constructed like Fig. 10.3, but using water as the thermometric element. Describe the limitations of this thermometer.

8. Find the thermal energy of each of these systems: (a) 1.5 kilograms of water at 66° Celsius; (b) 4.0 kilograms of wood at 45° Celsius; (c) 0.50 kilogram sugar at 52° Celsius; (d) 2.4 kilograms aluminum at negative 40 degrees Celsius (-40°C).

9. Find the temperature of each of these systems: (a) 1.8 kilograms of copper with a thermal energy of 11 Calories; (b) 6.0 kilograms of water with a thermal energy of 85 Calories; (c) 0.75 kilograms of wood with a thermal energy of -4.2 Calories; (d) 2.2 kilograms of Wood's Alloy with a thermal energy of 1.5 Calories. (Use values of specific heat from Table 10.4.)

10. Estimate the specific heat of the following materials from the data given.
    (a) 0.5 kilogram of ethyl alcohol initially at 60° Celsius melts 0.21 kilogram of ice at 0° Celsius.
    (b) 2.2 kilograms of hot stones at 300° Celsius melt 1.3 kilograms of ice at 0° Celsius.
    (c) 1.7 kilograms of paraffin at 45° Celsius melts 0.67 kilogram of ice at 0° Celsius.
    (d) When 1 kilogram of spaghetti at room temperature is dropped into 4 kilograms of boiling water, the water temperature drops to 90° Celsius.

11. Find the equilibrium temperatures of these systems:
    (a) 2.0 kilograms of water at 80° Celsius and 1.0 kilogram of water at 10° Celsius;
    (b) 1.0 kilogram of ice at 0° Celsius and 2.5 kilograms of water at 35° Celsius;
    (c) One dozen large eggs are taken from the refrigerator and put into 1 kilogram of boiling water. Assume that the flame under the boiling water is turned *off* immediately before you put the eggs into the water. Thus you can assume that the egg-water system is *isolated.* You will need to make and state your estimates for the temperature of the refrigerator and the mass of the eggs. *(Hint:* Treat the eggs as though they were made of water. To estimate the mass of the eggs, compare their apparent weight with something that you know, for example a quart of water or milk weighs about 1 kg. To estimate the temperature of the refrigerator, compare it with other temperatures that you know.)

12. Sprinkle salt on ice and report the lowest temperature you obtain (Fig. 10.14).

13. The specific heat of solid Wood's alloy found in the experiment reported in Fig. 10.11 is 0.050 Calorie per degree per kilogram, but that reported in Table 10.4 is 0.035 Calorie per degree per kilogram. Describe possible reasons for the discrepancy.

14. A family going on a camping trip was accustomed to taking along a gallon jug of cold water. When they filled it with 1 kilogram of ice at 0° Celsius and 3 kilograms of water at 10° Celsius, the ice lasted for 6 hours. How long should they expect the ice to last if they start out with 2 kilograms of ice and 2 kilograms of water? *(Hint:* Remember the energy transfer through the walls of the jug.)

15. Joseph Black, one of the physicists who contributed greatly to the concept of phase energy, recalled the following observation made by Fahrenheit in 1724: "He . . . exposed globes of water in frosty weather until . . . they were cooled down to the degree of the air, which was four or five degrees below the freezing point. The water . . . remained fluid so long as the glasses remained undisturbed, but on being taken up and shaken a little, a sudden freezing of a part of the water was instantly seen. . . . But the most remarkable fact is, that while this happens the mixture of ice and water suddenly becomes warmer, and makes a thermometer, immersed in it, rise to the freezing point." (Lectures on the Elements of Chemistry, Longman and Rees, London, 1803.) Explain Fahrenheit's observation in terms of phase energy.

Daniel Gabriel Fahrenheit (1686-1736) was the son of a German merchant. He was the first to introduce mercury as a thermometric substance and to succeed in making accurate thermometers. Fahrenheit is best remembered, of course, for the invention of the temperature scale that bears his name.

16. Explain how the following procedures or observations are related to phase energy.
(a) Salt is liberally mixed with the ice packed around an ice cream-making machine.
(b) A florist in Minneapolis wraps flowers to be delivered in very cold weather with soaking wet newspapers.
(c) A farmer puts a large tub of water into his cellar to keep the apples and potatoes stored there from freezing.
(d) You blow on hot soup to cool it more rapidly.
(e) Rubbing alcohol spilled on your hand feels cold.
(f) You feel cold when you step out of the swimming pool into a warm breeze.

17. Water is heated over a wood fire. How much tea can you make by burning 1 kilogram of wood under these conditions: you start with water at 10° Celsius; you want to heat the water to 80° Celsius only; only 10% of the energy from the fire is transferred to the water.

18. Warm-blooded animals maintain a steady state of body temperature generally above that of their environment. You can therefore represent a warm-blooded animal approximately as a passive coupling element, which converts chemical energy of a food-oxygen system to other forms of energy. Apply this concept to a Mr. X, who consumes food with an available energy of 2500 Calories per day.
(a) How much water evaporates from Mr. X's lungs and skin if all the energy becomes phase energy of gaseous water?
(b) How much hot air (warmed from room temperature of 20° Celsius to body temperature near 36° Celsius) can Mr. X produce in one day?

(c) How high must Mr. X climb if all the energy becomes gravitational field energy? His mass is 75 kilograms.

(d) Discuss qualitatively how the energy might be divided among these three and other possible forms. Comment whether you expect to find significant seasonal variations.

19. Which of the theories (caloric, kinetic) described in Section 10.5 seems most satisfactory to you for explaining the thermal phenomena described in Sections 10.1 to 10.4? Give reasons for your preference.

20. Interview four or more children (ages 8-14) to explore their concepts of thermal energy. Ask them to explain and/or predict their observations of some simple experiments, such as the mixing of two equal quantities of water of different temperatures or the melting of ice.

21. Interview four or more children (ages 8-14) to explore their concept of phase energy. Devise an appropriate demonstration and ask the children to explain and/or predict what happens.

22. Identify one or more explanations or discussions in this chapter that you find inadequate. Describe the general reasons for your judgment (conclusions contradict your ideas, steps in the reasoning have been omitted, words or phrases are meaningless, equations are hard to follow, . . .), and make your criticism as specific as you can.

Bibliography

S. C. Brown, *Count Rumford: Physicist Extraordinary,* Doubleday (Anchor), Garden City, New York, 1962.

S. G. Brush, *Kinetic Theory, Volume I: The Nature of Gases and Heat,* Pergamon, New York, 1965.

D. Cassidy, G. Holton, J. Rutherford, *Understanding Physics,* Springer Verlag, 2002.

C. A. Knight, *The Freezing of Supercooled Liquids,* Van Nostrand, Princeton, New Jersey, 1967.

E. Mach, *History and Root of the Principle of Conservation of Energy,* Open Court, La Salle, Illinois, 1911.

D. Roller, *The Early Development of the Concepts of Temperature and Heat: The Rise and Decline of the Caloric Theory,* Harvard University Press, Cambridge, Massachusetts, 1950.

J. Tyndall, *Heat, a Mode of Motion,* D. Appleton, New York, 1893.

A. Wood, *Joule and the Study of Energy,* G. Bell, London, 1925.

M. W. Zemansky, *Temperatures Very Low and Very High,* Van Nostrand, Princeton, New Jersey, 1964.

Articles from *Scientific American*

G. Y. Eastman, "The Heat Pipe" (May 1968).

W. Ehrenburg, "Maxwell's Demon" (November 1967).

D. M. Gates, "Heat Transfer in Plants" (December 1965).

D. Turnbull, "The Undercooling of Liquids" (January 1965).

Take then a quantity of
even-drawn wire . . . and coyl it . . .
into a helix of what length
or number of turns you please . . .
and hanging on several weights
observe exactly to what length each
of the weights do extend it
beyond the length that its own
weight doth stretch it to,
and you shall find that if one
ounce . . . doth lengthen it one line . . .
then two ounces . . . will
extend it two lines, . . . and three
ounces, . . . three lines . . .
And this is the rule or law
of nature upon which all manner
of restituent or springing
motion doth proceed . . . .

ROBERT HOOKE
De Potentia Restitutiva,
1678

Force, displacement, and energy transfer

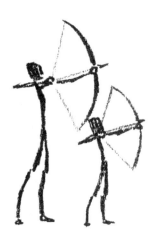

The archer bending a longbow, the ski tow pulling a skier, the fireman sliding down a brass pole, the housekeeper pulling a magnet off the front of a refrigerator, and the child hopping on a pogo stick are all examples of energy transfer from among the phenomena in our environment. The archer serves as the energy source for the elastic energy of the longbow, with the string acting as the coupling element. The gravitational field of the fireman and the earth serves as energy source for the brass pole and the fireman's hands, with the frictional interaction transferring the energy. And so on.

11.1 Factors in energy transfer

We will now analyze thought experiments of these two examples in more detail, with a view toward making a mathematical model of the energy transfer being accomplished by the interaction. It is clear that the more the bow is bent by displacement of the center of the string, the more elastic energy it stores (Fig. 9.1). The elastic energy can be measured roughly, though not in the joule unit, by the distance the arrow travels when it is released. It is also clear that a child's bow, which can be bent by a weaker arm, stores less energy even though it may be bent to the same extent. It appears, therefore, that a mathematical model for the energy of the bow and the energy transfer to the arrow must take into account two factors: the displacement of the string tied to the bow and the strength of interaction required to bend the bow.

An analysis of the fireman sliding down his brass pole leads to the same conclusion. The taller the pole, the more thermal energy will be created in the brass pole and the fireman's hands. Also, the more massive the fireman, the more tightly he will have to cling to the pole if he is to avoid breaking his legs on impact with the floor, and the hotter his

Figure 11.1 Examples of forces. The earth exerts a gravitational force on the falling apple. The bent fishing rod exerts an elastic force on the fishing line.

hands will become. Both the displacement of the fireman and the strength of his interaction with the pole must therefore be taken into account in a mathematical model for energy transfer in this example.

Force and work. We have mentioned the notion of interaction strength before. In Chapter 3 it came up as determining the magnitude of the response to interaction of an object or system that has inertia. We will introduce force as the measure of interaction strength that is appropriate to the archer bending his bow, the fireman sliding down his pole, the ski lift, and so on—that is, whenever energy transfer accompanies the displacement of an interacting object. This energy transfer has been given a name of its own; it is called work. In this chapter, therefore, we will make a mathematical model that relates work to force and displacement.

11.2 Interaction and force

The word "force" is used in everyday language to signify compulsion, either physical or mental. In scientific use, the word "force" has a narrow and quite specific meaning distinct from its everyday meaning. Rather than attempting to present a concise formal definition of force, we will use examples to show how the concept of force grows out of the more general concepts of interaction and energy.

The force concept. Isaac Newton (1642-1727) introduced the concept of force in his brilliant formulation of a theory for the motion of rigid bodies. In this theory, which we will present in Chapter 14, Newton ascribed the changes in motion of a body to a net (unbalanced) force acting on the body. Newton realized that several interactions might compensate for one another and produce no change of motion, as when the hands hold the bow and arrow ready for shooting. Only when the hand relaxes its hold does a net force exerted by the string on the arrow set the latter in motion.

Examples of force are illustrated in Fig. 11.1.You can see in Fig. 11.1 that a force has to be described not only by a magnitude (or strength) but also by a direction in space. Thus, the gravitational force exerted by the earth on an apple acts downward; the bent fishing rod exerts an upward-directed force on the line; the flowing creek water exerts a downstream force on the line; and so on. In this respect the force concept resembles the relative position and displacement concepts, which were described by a distance and a direction in space (Sections 2.1 and 2.3). Force will therefore be represented by a boldface letter $\mathbf{F}$ in print, by a letter F with an arrow over it in writing, and by an arrow in a diagram. The magnitude of the force is represented by the symbol $|\mathbf{F}|$, where the vertical bars symbolize that only the numerical strength of the force is important and that the direction of the force is to be disregarded. Forces can be combined (added, multiplied by numbers, and so on) in the same way that displacements are combined by arithmetic operations or by diagrammatic manipulation of the arrows representing them.

Force arrows will be drawn differently from position and displacement arrows.

displacement

force

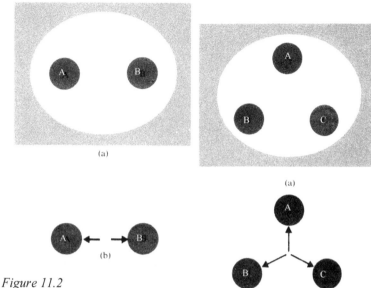

Figure 11.2
(a) Two interacting bodies
(b) The two forces of interaction,
represented by arrows. One force
acts on each body.

Figure 11.3
(a) Three interacting bodies
(b) The three forces of interaction,
represented by arrows. One force acts
on each body.

The Newtonian theory requires us to think about interaction between objects in a different way than we have so far. We have emphasized the *mutuality* of the interaction and have drawn your attention to the entire system of interacting objects, including the "field" which transmits the interaction. However, in the Newtonian point of view, we must focus our attention on one body at a time in order to study its motion. We think of interaction between two objects A and B as simply two forces: one net force acting on A (exerted by B), and a second net force acting on B (exerted by A). These two forces are indicated by arrows in Fig. 11.2. In a three-body system there are three forces, as shown in Fig. 11.3; one net force acts on A and is exerted by the subsystem composed of B and C, another net force acts on B and is exerted by the subsystem composed of A and C, and the third net force acts on C and is exerted by the subsystem A and B. In a four-body system there are four net forces, and so on. The net force acting on an object in a system of other objects is always exerted by a subsystem that includes all the *other* objects but not the object itself. In Newtonian theory, an object never exerts a net force on itself.

Combination of forces. You may find it difficult to visualize the net force exerted on body A by the subsystem composed of B and C in Fig. 11.3. However, you probably find it easier to think of the net force exerted by just one body B on A in Fig. 11.2. The sun-earth-moon system

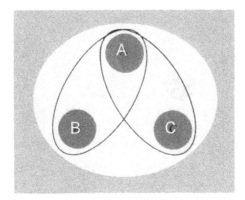

Figure 11.4 Two overlapping subsystems of the three-body system A-B-C.

is an example in which the moon is subject to the net force exerted by the sun-earth subsystem.

Now, the Newtonian method for dealing with such a three-body system is to simplify it by imagining the three-body system to be made up of three overlapping two-body subsystems (Fig. 11.4). There is an interaction within each two-body subsystem, and therefore a force exerted on A by B (in the A-B subsystem) and a force exerted on A by C (in the A-C subsystem) (Fig. 11.5). In the Newtonian theory it is assumed that each of these interactions is *unaffected by the presence of the third body*. In the astronomical example above, it is assumed that the earth-moon interaction is not affected by the sun, and the sun-moon interaction is not affected by the earth. This step may or may not seem appropriate to you, but it is central to grasping Newtonian theory.

The next step is to represent the two partial forces acting on A (exerted by B and by C) by arrows (Fig. 11.6a). The last step in finding the net force on A is to combine all the partial forces according to the procedure described in Section 2.3 for the addition of displacements.

Figure 11.5 Analysis of forces in a three-body system.
(a) The interaction of A and B in the three-body system (Fig. 11.4) is assumed equal to the interaction of A and B in the two-body subsystem.
(b) The interaction of A and C in the three-body system (Fig. 11.4) is assumed equal to the interaction of A and C in the two-body system.

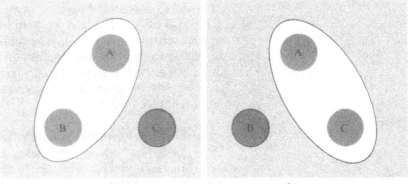

(a)   (b)

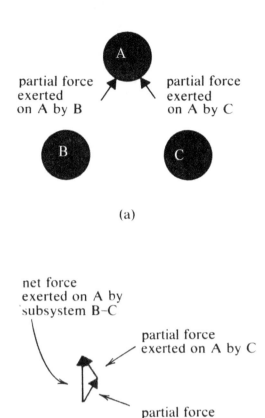

Figure 11.6 Partial and net forces in a three-body system.

(a, above) Newtonian diagram for partial forces exerted on A by B and by C separately.

(b, below) Addition of arrows that represent the partial forces acting on A to find the net force on A.

For an example of a body in equilibrium and in motion at the same time, think of your luncheon tray on a jet airliner moving at a constant 600 miles per hour relative to the ground. (What happens when the motion of the airliner changes suddenly?)

The tail of one arrow in the diagram is attached to the head of the other (Fig. 11.6b). The overall arrow from the tail of the first to the head of the second is the net force acting on body A. The net force is the sum of the partial forces, with directions and magnitudes of the partial forces being taken into account. The combination of forces is described further at the end of this section.

Mechanical equilibrium. If a body is not subject to a net force, then its motion is steady and does not change. Such a body is said to be in mechanical equilibrium, and the partial forces acting on it are called balanced. The apples on the tree, the fisherman's boots, and the steadily flowing brook water in Fig. 11.1 are in mechanical equilibrium. The falling apple, the turbulent brook water, and a boy swinging nearby are not. What about the tip of the fishing rod?

If a body in equilibrium is at rest, it remains at rest; if the body in equilibrium is in motion, it remains in steady motion with a constant speed in the same direction. This is not really a new result, because one kind of evidence of interaction described in Section 3.4 was the change in motion of interacting objects. In that section we pointed out that a body might not exhibit any change in motion, even though it is subject to interactions, if the interactions compensate. Alternate ways of describing this condition are mechanical equilibrium, a zero net force, and the balance of the partial forces acting on the body.

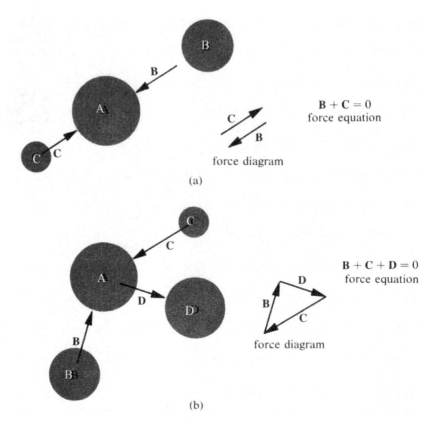

*Figure 11.7 Body A is in mechanical equilibrium. Symbols **B**, **C**, and **D** represent the forces exerted on A by the corresponding bodies,*
(a) Two balanced forces of repulsion,
 (b) Three balanced forces.

Addition of partial forces. The net force is absent (or equal to zero) if the arrow representing it in a diagram has zero length. This condition is achieved if the partial forces are represented by arrows that form a closed chain, with the head of the last arrow reaching the tail of the first arrow (Fig. 11.7). If only two partial forces are acting on a body in mechanical equilibrium, then the two must be represented by arrows of equal length and opposite direction (Fig. 11.7a). Three partial forces form a triangular diagram (Fig. 11.7b), and so on. Thus Fig. 11.7 illustrates sets of partial forces that are balanced. A body that is not in mechanical equilibrium (i.e., it is subject to a net force) can be brought to equilibrium by the application of an additional force that is equal in magnitude and opposite in direction to the net force.

Origin of the forces. This discussion of mechanical equilibrium has referred to a single body and the partial forces acting on it. The origin of the forces was not mentioned. Nevertheless, forces are a measure of interaction strength, and you should be aware of the interactions that operate even though you temporarily concentrate your attention on the motion or lack of motion of one particular body in the system. In particular, the interaction will affect the motion of the other bodies in the system.

Operational definition of force. The calculation of the net force acting on an object is based on measurements of the partial forces, and such measurements are in turn based on an operational definition of

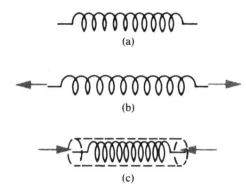

Figure 11.8 *Springs can be used to measure forces,*
(a) A spring not subject to forces,
(b) The same spring deformed by stretching,
(c) The same spring deformed by compression. (Compressed springs are usually kept inside a tube to prevent them from bulging to the side.)

force. We will now construct such an operational definition by selecting a force measuring device, calibrating it in standard units of force called *newtons*, and describing a procedure for using the device.

Any system that gives reliable and reproducible visual evidence of the action of a force can be used as force measuring device. A bow can be used, or a spring, or a rubber band. All these systems show a deformation when subject to a force, and return to their undeformed state when the force ceases to act. They are examples of elastic systems and will be discussed in greater detail in Section 11.6.

Standard spring scale. For convenience, we will select springs as force measuring devices. Springs may be deformed by stretching or by compression (Fig. 11.8). To calibrate a spring, we attach objects of various masses to the end of the spring and allow them to hang freely and vertically (no swinging) in the gravitational field of the earth. Such an object is subject to two partial forces: the downward force exerted by the earth via the gravitational field and the upward elastic force exerted by the spring. After such an object is allowed to come to mechanical equilibrium (that is, to stop bouncing), the spring is stretched just enough so that the force it exerts is equal in magnitude to the gravitational force.

It would be natural to choose a 1 kilogram mass as the standard object for one unit of force. For historical reasons we will not do this, but instead, we will choose a weight with a gravitational mass of 0.10 kilogram. This is the same standard object used to define the joule in Section 9.2. We will explain this apparently arbitrary choice in Chapter 14.

The unit of force on our scale is called a *newton*. The scale is marked 1 newton when a 0.10 kilogram weight hangs on the spring, 2 newtons when two such weights hang on the spring, and so on (Fig. 11.9). The spring with a scale is called a *spring scale*.

Definition. The operational definition of force employs a spring scale calibrated in newtons. When the spring scale is used to measure the net force acting on a body, it is attached to the body and used to hold it in mechanical equilibrium (Fig. 11.10). The magnitude of the net force is then equal to the scale reading. The direction of the net force is the direction in which the spring scale is extended. For the sake of simplifying the diagrams, we will henceforth represent spring scales by dials without numerical scale indications. The scale reading will be printed near the dial.

OPERATIONAL DEFINITION
Force is measured by a standard spring scale. The magnitude of the force is indicated by the scale reading. The direction of the force is indicated by the direction of the spring scale.

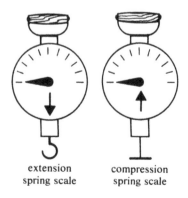

extension spring scale   compression spring scale

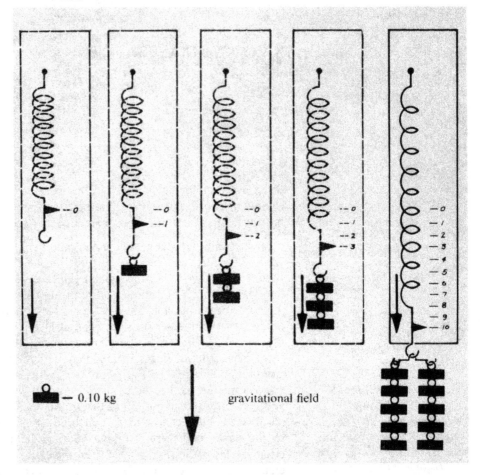

Figure 11.9 Calibration of a standard spring scale with weights having a mass of 0.10 kilogram. The arrow represents the direction of the force exerted on the spring scale.

Partial forces. The spring scale can also be used to measure partial forces acting on a body, like the force exerted on the fishing rod by the fisherman's left hand in Fig. 11.1. To do this, you only have to let the spring scale, instead of the man's left hand, hold the fishing rod in the same position. The force exerted by the spring scale then replaces and is equal to the force exerted by the hand. Other examples of the substitution method for measuring partial forces are shown in Fig. 11.11. Frequently these "measurements" are only carried out in thought experiments and the actual data come from the interpretation of indirect evidence.

Figure 11.10 The net force acting on the wagon in the absence of the spring scale is 2.1 newtons to the right. The partial forces acting on the wagon are exerted by the mouse (to right), the table (up), the earth (down), and the spring (to left).

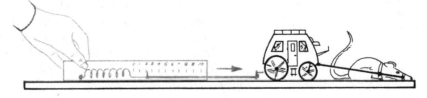

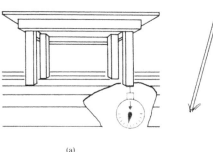

Figure 11.11 The substitution method for measuring partial forces.
(a) Compression spring scale is used to measure the partial force exerted by the floor on one table leg.
(b) Extension spring scale is used to measure the partial force exerted by the swing frame on the rope.

(a)   (b)

Addition of partial forces. Now that we have introduced a procedure for measuring forces, we can repeat some of the earlier explanations with quantitative illustrations. It is usually helpful to begin by making a force diagram, with an arrow for each partial force; the length of each arrow is scaled to the magnitude of the force (see Figs. 11.6 and 11.7). Since this is a force diagram and not a diagram or map of the position of the bodies on which the forces act, the force arrows can be drawn anywhere, as long as they have the correct magnitude and direction. To find the net force, draw one force, then draw the second with its tail attached to the head of the first, the third with its tail attached to the head of the second, and so on, until all forces have been added. Then draw an arrow from the *tail* of the first force to the *head* of the last one (Fig. 11.6b); this is the net force.

An alternative method of adding forces using arithmetic is to find and add the corresponding *rectangular components* of the forces. This approach requires the introduction of a rectangular coordinate frame (Fig. 11.12). The x component of the net force is then the sum of the x components, and the y component of the net force is the sum of the y components (Eq. 11.1). We recommend that you choose the x- or y-coordinate axes in the same direction as one or more of the forces (Fig. 11.13). If you do this, the components of these forces are especially easy to find. The components of forces not in the direction of the axis may be found graphically or with the sine and cosine functions.

It is also possible to use the method of addition of forces to find an

Equation 11.1
(Forces are defined in Fig. 11.12.)

$$F' + F'' + F''' = F$$

$$[1, 1] + [-2, 4] + [-3.5, 0] = [-4.5, 5.0]$$

$$F = [-4.5, 5.0] \text{ newtons}$$

Figure 11.12 Three partial forces F', F'', F''' and the net force F with the following rectangular components:
F' = [1.0, 1.0] newtons
F'' = [-2.0. 4.0] newtons
F''' = [-3.5, 0.0] newtons
F   = [-4.5, 5.0] newtons
(see Eq. 11.1).

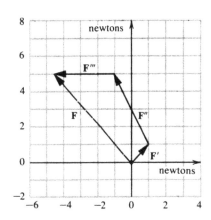

Figure 11.13 Examples of the addition of forces:

(a) partial forces:
[2.0, 0.0] newtons
[0.5, 5.0] newtons
[-1.0, -2.0] newtons
[-3.0, 2.0] newtons
net force:
[-1.5, 5.0] newtons

(b) partial forces:
[4.0, 0.0] newtons
[1.0, 1.5] newtons
[-1.0, 2.5] newtons
[-2.0, -1.0] newtons
[-2.0, -3.0] newtons
net force:
[0.0, 0.0] newtons

Note: The coordinate axes are purposely oriented at an angle, so that the x-axis is in the same direction as one of the forces. This is a good way to simplify the calculation. In fact, this technique of setting up the coordinate axes to fit the problem can be a powerful aide in problem solving.

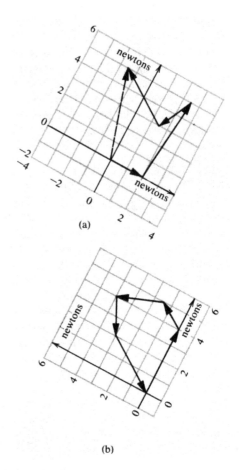

(a)

(b)

unknown partial force if the net force is known. This problem is similar to the earlier one in which, given the first part of a sailing trip and the destination, we showed how to use subtraction of displacements to find the displacement needed to complete the trip (Section 2.3, Fig. 2.19). A problem of this kind with forces is worked out below (Fig. 11.14).

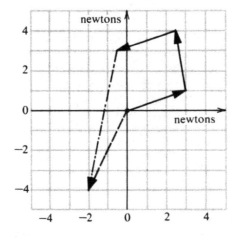

Figure 11.14 Example of how to find one partial force when the net force and the other partial forces are known.
net force: [-2.0, -4.0] newtons
partial forces:
*   [3.0, 1.0] newtons*
*   [-0.5, 3.0] newtons*
*   [-3.0, -1.0] newtons*
missing partial force :
*   [-7.5, -7.0] newtons*

11.3 The gravitational force

The gravitational interaction between the earth and objects near its surface is very important for all of us who dwell on the earth. You can measure the force of gravity exerted by the earth on any object free of other interactions by hanging the object from a spring scale. Then the force exerted by the spring scale compensates the force of gravity and holds the object in mechanical equilibrium.

We have already described how you may use a plumb line to define the direction of vertical (p. 15). In comparing this with the definition of force, you see that the vertical is precisely the direction of the force of gravity exerted by the earth on any nearby object. If you were to carry out the force measurement at various places on the earth, you would find that the direction is toward the center of the earth and that the magnitude is the same.

We will use the boldface symbol $\mathbf{F}_G$ for the force of gravity on an object. The directions of $\mathbf{F}_G$ at various places on the earth's surface are shown in Fig. 11.15. The magnitude of the force of gravity on an object with mass of 0.10 kilogram is, by definition, 1 newton. The force on other objects is described by the following mathematical model: $|F| = 10M_G$ (Eq. 11.2).

These spring scale measurements can be made with the object at rest *or in steady, or uniform, motion* relative to the earth's surface (Fig. 11.16). However, if the motion of the object changes, then a spring scale does not measure the net force (Fig. 11.17). These statements can be tested experimentally.

Gravitational intensity. The strength of the gravitational interaction is close to 10 newtons per kilogram, regardless of the shape, composition, or other properties of the object, and it is directed vertically downward. The symbol **g** is used to denote the gravitational force per unit mass, and it is called the *gravitational intensity* (Eq. 11.3). The relation of gravitational force, gravitational intensity, and mass is given by

Equation 11.2
force of gravity (at surface of earth) (magnitude, newtons)
$$= |\mathbf{F}_G|$$
gravitational mass (kg)
$$= M_G$$

$$|\mathbf{F}_G| = \frac{M_G}{0.10} = 10M_G$$

EXAMPLES :
(*a*) *Force of gravity on a person.*

$$M_G = 60 \ kg$$

$$|\mathbf{F}_G| = 10M_G$$
$$= 10 \times 60 \ newtons$$
$$= 600 \ newtons$$

(*b*) *Force of gravity on quarter pound (4-ounce) hamburger.*

$$M_G = 0.11 \ kg$$
$$|\mathbf{F}_G| = 10M_G$$
$$= 10 \times 0.11 \ newtons$$
$$= 1.1 \ newtons$$

Equation 11.3
gravitational intensity **g** *at the surface of the earth.*

$$\mathbf{g} = (10 \ newtons / kg,$$
$$downward)$$
$$|\mathbf{g}| = 10 \ newtons / kg$$

Figure 11.15 The force of gravity at various locations on the earth.

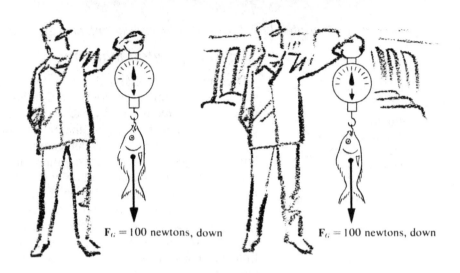

$F_G = 100$ newtons, down $F_G = 100$ newtons, down

Figure 11.16 (above) The force of gravity acting on the fish is measured, properly, as 100 newtons whether it is being weighed inside a train moving at constant speed in a straight line (at right) or at rest in the station (at left).. In both cases, the fish and scale are moving at <u>constant</u> velocity relative to a reference frame fixed on the surface of the earth. The fish is in mechanical equilibrium in both situations. The essential similarity of these two situations is the basis of Newton's first law and will be explained in more detail in Section 14.2.

Eq. 11.4, which is a restatement of Eq. 11.2. On the sun and the moon, the gravitational intensity has different values than it does at the earth's surface (Table 11.1).

You may wonder whether the fact that the force of gravity is proportional to the gravitational mass is a law of nature, or whether it is a result of our definition of the magnitude of the force. Actually, it is the latter. After all, we calibrated the spring scale by hanging objects on it while they were subject only to the earth's gravitational field and we marked force units in proportion to the mass, 1 newton for every 0.10 kilogram.

TABLE 11.1 GRAVITATIONAL INTENSITY (**g**)

Location	gavitational intensity (**g**, newtons/kg)
earth's surface	10
sun's surface	270
moon's surface	1.7

Figure 11.17 (to right) The force of gravity acting on the fish is <u>not</u> measured correctly by the spring scale under the conditions shown. As the carousel turns, the velocity of the fish is not constant but is <u>changing in direction</u> (though not magnitude). As a result, the fish is <u>accelerating and not</u> in equilibrium.

$F_G = 100$ newtons, down

11.4 Work

WORK IN PHYSICS
*Energy transfer resulting
from a force acting on an
object that is displaced in
the direction of the force.*

The concept of *work* represents a key connection between two different styles of thought and two different types of theory: 1) A "wholistic," "systems," or "field theory" approach based on concepts such as energy, waves, and fields (gravitational, electric and magnetic). These concepts all apply to a system as a whole or to an extended region of space. 2) A Newtonian, particle-based approach, based on breaking a system down into its parts (each piece reduced to a simple mass-point at a defined position in space), analyzing the interactions between the various particles, identifying all the forces acting on any individual particle, finding the net force on that particle, and repeating this procedure for each particle in the system so as to build up an understanding of the system as a whole.

In Chapters 1-10, we explained the concepts of interaction and energy and how to apply them in many situations. In particular, we have emphasized how the interaction of a system of objects can be thought of as arising from various types of "fields," (gravitational, magnetic, electric and so on). We have focused on interactions between objects as phenomena in which all of the objects participate, as part of a system in a shared, mutual way. This has involved, to a large extent, a "wholistic" or "systems" style of thinking, as described above.

In this Chapter, we have introduced the concept of "force," which requires thinking about the interacting objects individually and then two-by-two so as to identify and add up all of the forces acting on a particular object. This style of thinking is very different from much of what we have explained so far, but it is essential to understanding Newtonian theory. This theory, in turn, provides a powerful tool with which to analyze many of the systems we have already encountered, leading to additional models and increased understanding.

More specifically, we will now show how work connects the Newtonian concept of force (defined in Section 11.2) with the concept of energy (a more "systems-oriented" idea, as defined in Sections 9.2 to 9.4). We will first provide a general context for this new concept; we will then define work formally as well as with a mathematical model, explain in detail how to calculate it, and show that this calculation yields the same numerical result as do our earlier calculations of energy changes.

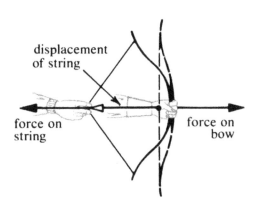

displacement
of string

force on
string

force on
bow

*Figure 11.18
Energy transfer to
the bow occurs
during displacement
of the center of the
string, where the
archer's hand exerts
a force.*

FORMAL DEFINITION
Work is the product of the magnitude of the force and the displacement component in the direction of the force.

Equation 11.5 (Work)
work (joules) $= W$
force magnitude
  (newtons) $= |\mathbf{F}|$
displacement magnitude
  in force direction
  (m) $= \Delta s_F$
  $W = |\mathbf{F}|\, \Delta s_F$

Detailed procedure for calculating work (W), illustrated in Figure 11.19 to right.

*1. Identify the force and draw the force arrow, **F**, pointing in approximately the correct direction. Find the magnitude of the force in newtons. (Don't draw the coordinate frame first; the calculation is simpler if you do that in the next step!)*
2. Now draw the coordinate frame <u>with the x-axis parallel to the force arrow</u>;
*3. Identify the direction and magnitude of the displacement (**Δs**);*
*4. Draw **Δs**, the displacement arrow, on the coordinate frame with its tail starting at the origin;*
*5. Draw the dashed line from the point of **Δs** parallel to the y-axis until it crosses the x-axis;*
*6. Draw and measure (or calculate) the length of Δs_F (the x-component of **Δs**).*
7. Calculate W from Eq. 11.5 ($W = |\mathbf{F}|\, \Delta s_F$).

We gave an informal definition of work above in Section 11.1: the energy transfer that "accompanies the displacement of an interacting object." To be more precise, work refers to the energy transfer resulting from a force acting on an object that is *displaced in the direction of the force*. For example, when the archer bends his bow, one hand interacts with the center of the string and one hand interacts with the center of the bow. Both hands exert forces, one on the string, the other on the bow. But only the center of the string is displaced; the arm holding the bow is rigid (Fig. 11.18). The energy transferred from the archer to the bow and string system is called the work done by the force his hand exerted on the string. The force exerted on the bow did no work because the center of the bow was not displaced.

Definition of work. We now present a more formal definition of work that can be used in calculations: *Work is the product of the magnitude of the force and the displacement component in the direction of the force.*

This definition is restated to the left as a mathematical model (Equation 11.5). The overall procedure for calculating work is to identify the force, find its magnitude, identify the displacement, find its component in the force direction, and multiply these two numbers. This procedure is illustrated in figure 11.19 and summarized in the left margin.

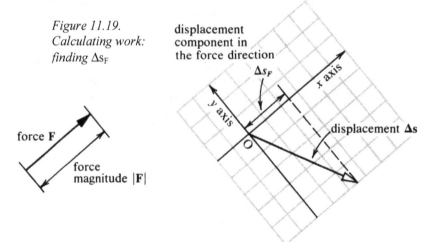

Figure 11.19. Calculating work: finding Δs_F

There are two important points to keep in mind when using the above definition to calculate work. First, *both* force and displacement have a direction in space. The work done depends on the relative direction of the force and the displacement and does not change if both arrows are reoriented without changing their relationship in space.

Second, the work concept can be associated with any force, net or partial. When several forces are acting on a body, you can use the definition to find the work done by each partial force and you can find the work done by the net force. In the example of the bent bow, for instance, the net force on the bow and string system was zero; hence, it did zero work. The partial force acting on the bow also did zero work, because the center of the bow was not displaced. Only the partial force acting on the string did work. This example shows that the work done by the net force acting on a system may *not* be equal to the sum total of the work done by all the partial forces acting on the same system.

Work is a numerical quantity measured in joules; it does *not* have a direction in space. Zero work is done when the displacement is zero. Zero work is also done when the displacement is at right angles to the force, for then the displacement component in the direction of the force is zero.

We now must verify that this definition of work really represents energy transfer in accord with at least one of the operational definitions of energy explained in Chapter 9. This is most easily accomplished for the definition based on raising an object in the gravitational field, in which energy is measured in joules (Eq. 9.2, $E = 10M_G h$, Fig. 9.8).

Example 11.1 below shows how to calculate the work done when a weight is raised or lowered and the energy stored in the gravitational field is changed. You can see that this result is in agreement with the mathematical model (Eq. 9.2) based on the operational definition. From now on we will use Eq. 11.5 to describe energy transfer whenever this occurs in the form of work, and we will not use the operational definition of energy in joules except for illustrative purposes.

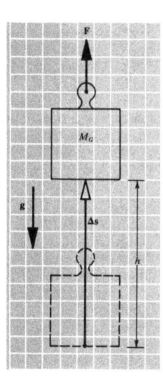

EXAMPLE 11.1. How much work is done when a weight is raised? Is the work equal to the energy transfer described by Eq. 9.2?

Solution: We must find the force and the displacement to be able to use the definition of work, Eq. 11.5. The important step is to think of raising the weight so slowly that it is always very close to mechanical equilibrium and does not acquire kinetic energy. Then the force acting on the weight is equal in magnitude, opposite in direction, to the force of gravity (Eq. 11.4),

$$|\mathbf{F}| = |\mathbf{F}_G| = |\mathbf{g}|\, M_G = 10\, M_G$$

The displacement (see diagram to left) has the magnitude h and is directed along the force. Hence the displacement component in the force direction is equal to h,

$$\Delta s_F = h$$

The work done is $W = |\mathbf{F}|\Delta s_F = |\mathbf{g}|\, M_G h = 10\, M_G h$ in agreement with Eq. 9.2.

Gravitational field energy. The fireman sliding down the pole, the ski tow pulling the skier, and the child hopping on the pogo stick are all phenomena in which energy stored in the gravitational field is either increased or decreased as the height of an object is increased or decreased.

One point that is sometimes a source of difficulty in dealing with gravitational field energy is the absence of a natural reference state in which the energy stored in the gravitational field is zero. It is always possible to increase the energy in the gravitational field by raising an object near the earth, and it is possible to decrease this energy by lowering the object.

Because there is no natural reference state, you may choose the most convenient state as reference state. Usually you should choose the state with the lowest gravitational field energy that occurs in the phenomenon, as when the fireman is on the ground floor or the skier is in the valley (Fig. 11.20). Then you can ascribe positive energy values to the states of greater gravitational field energy than the reference state, as when the fireman is

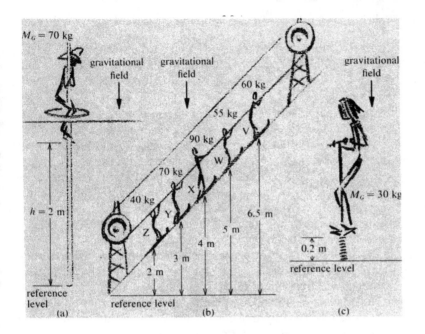

Figure 11.20 The gravitational field energy (see Eq. 11.6) is calculated relative to the specific reference levels shown in the diagram.

(a) Fireman: $E_G = |g|M_G h$
$= 10 \times 70 \times 2$
$= 1400 \ joules$

(b) Skier Y: $E_G = |g|M_G h$
$= 10 \times 70 \times 3$
$= 2100 \ joules$

Skier V: $E_G = |g|M_G h$
$=10 \times 60 \times 6.5$
$= 3900 \ joules$

(c) Child: $E_G = |g|M_G h$
$= 10 \times 30 \times 0.2$
$= 60 \ joules$

Equation 11.6

gravitational field energy
(joules) $= E_G$
gravitational mass (kg) $= M_G$
height (m) $= h$
gravitational intensity
(newtons/kg) $= g$

$$E_G = |g|M_G h \qquad (a)$$

at the earth's surface:
$$E_G = 10 \ M_G \ h \qquad (b)$$

upstairs or the skier is on his way up the hill. States that have less gravitational field energy than the reference state are described by negative energy values, as when the fireman is in the cellar of the firehouse. The height of the object in the system is measured from the level of the reference state (reference level).

Since the definition of the mechanical energy scale made use of gravitational field energy, you can obtain a mathematical model for the field energy directly from Eq. 9.2. The result is stated in Eq. 11.6. The height that occurs in this formula is measured from the chosen reference level as described above. The mathematical model works well for objects near the surface of the earth, as illustrated in Fig. 11.20. However, when an object is displaced far above or below the surface of the earth (Fig. 11.21), you

Figure 11.21 Cross-sectional view of the earth. The model for gravitational field energy (Eq. 11.6) breaks down when the object is displaced to locations where the gravitational intensity is different from its value at the earth's surface. The model is applicable in the shaded region (body B). It is not applicable at the location of bodies A or C.

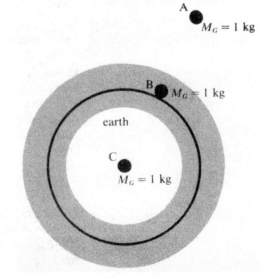

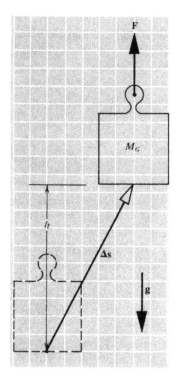

Figure 11.21 Raising a weight obliquely (along a diagonal). The work done is 10 M_Gh, which depends on the vertical height (h = Δs_F) and not the diagonal displacement (Δs).

Charles Augustin de Coulomb (1736-1806) was a French army officer of engineers who spent several years in the West Indies until failing health forced his retirement to France. After his return, he won a prize from the French Academy for a paper, Theorie des Machines Simples, *in which the law of friction was announced. Coulomb's most important work, however, was his measurement of the forces of electrical attraction and repulsion, and his formulation of the mathematical model that describes them.*

should expect the model to break down because the model in Eq. 11.4 for the force of gravity breaks down under those conditions.

Is the work done on an object near the surface of the earth always compatible with Eq. 11.6? In Example 11.1 we examined the work done on an object that was raised vertically and found agreement. Is agreement also obtained when the object is displaced to the side or obliquely?

Since the force of gravity is in the vertical direction, only the vertical component of the displacement is used to calculate the work done by the force of gravity (Example 11.2). The horizontal component of the displacement does not contribute to this work. Consequently, the horizontal component of the displacement does not change the gravitational field energy. This result is in accord with the commonsense expectation that only raising or lowering of objects affects the gravitational field energy and that sliding an object sideways (for example, a skier walking on level ground) does not affect the gravitational field energy.

EXAMPLE 11.2. How much work is done when an object is raised obliquely (that is, along a diagonal), rather than straight up, as in Example 11.1?

Solution: This problem is very similar to Example 11.1. Again, we move the weight so slowly that it is always very close to mechanical equilibrium and does not acquire kinetic energy. Then the force acting on it is equal in magnitude, and opposite in direction, to the force of gravity (Eq. 11.4),

$$|\mathbf{F}| = |\mathbf{F}_G| = |\mathbf{g}|\, M_G = 10\, M_G$$

The displacement (see Fig 11.21 to left) is not directed along the vertically upward force. We can, however, find the rectangular components of the displacement (Fig. 11.21). The vertical component is equal to h,

$$\Delta s_F = h$$

The work done is

$$W = |\mathbf{F}|\, \Delta s_F = |\mathbf{g}|\, M_G\, h = 10 M_G\, h.$$

This is the same result as found in Example 11.1.

11.5 Electrical force and energy

The interaction of electric charges was described in Section 3.5 and was applied in a crucial way in the construction of models for atoms in Chapter 8. We will now describe the magnitude of the force between electrically charged objects. Charles Augustin de Coulomb investigated this definitively with a very delicate spring balance.

By carrying out many experiments with the electrically charged objects closer together and farther apart. Coulomb constructed a mathematical model in which the electrical force of interaction varies inversely as the second power of the distance R between the charged objects (Fig. 11.22). He also found, by testing objects with various quantities of electric charge, that the mathematical model must include a dependence of the force on the

Equation 11.7
(Coulomb's Law)

electrical force
(newtons) $= \mathbf{F}$
electrical charge
(faradays) $= q_1 , q_2$
distance between
bodies (m) $= R$

$$|\mathbf{F}| = 8.4 \times 10^{19} \frac{q_1 q_2}{R^2}$$

"... the mutual attraction of the electric fluid which is called positive on the electric fluid which is ordinarily called negative is in the inverse ratio of the [second power] of the distances ..."

Charles Augustin de Coulomb
Memoires de l'Academie Royale des Sciences, 1785

two charges: the force is proportional to the product of the two charges q_1 and q_2. When the dependence on the distance and charge are combined, we obtain the model for the magnitude of the force described in Eq. 11.7, called *Coulomb's law*. The direction of the electrical force is along the line from one charged body to the other. It is attractive or repulsive, depending on whether the charges are opposite charges or like charges (Section 3.5).

The very large factor (8.4×10^{19}) in Coulomb's law means that 1 faraday is a charge that exerts enormous forces on charged objects nearby. Laboratory experiments in which objects become electrically charged, therefore, produce objects with very small charges (as measured in faradays).

The energy stored in electric fields may be calculated from the work that electrical forces can do when a charged object is displaced. Unfortunately, the electrical force changes so rapidly with displacement of the object (Eq. 11.7) that a simple average value to use in calculating the energy is not evident. We will therefore not make a mathematical model for the electrical energy, but you may like to consult a more advanced text where this is done.

Because the electrical forces are so large, electric charges are rarely and not conveniently separated from a macro-domain quantity of matter. Macro-domain electric fields are not easy to produce and do not furnish a technologically useful type of energy storage. In the micro domain, however, electric fields provide an extremely significant type of energy storage, because of their role in the many-interacting-particles model for matter (Sections 4.5, 8.1, and 11.7).

11.6 Elastic energy and elastic force

Elasticity. We have used the archer's bow as an example of a system that can store energy. The spring in the standard spring scale also can store energy when it is deformed from its equilibrium configuration by interaction with a weight or some other object. When released, it will spring back to its free equilibrium configuration. Such a system is called an *elastic system*. The energy stored by an elastic system when it is in its deformed configuration is called *elastic energy*. The force it exerts is called an *elastic force*. When an elastic system is deformed,

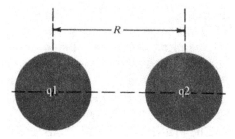

Figure 11.22 Conditions of Coulomb's experiments. Two bodies with electric charges q_1 and q_2 are separated by a distance R. The force is directed along the line joining the two bodies.

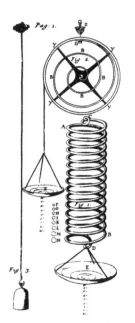

Hooke's apparatus, from his Lectures Cutlerina, 1674–1679

Equation 11·8

elastic force magnitude
  (newtons) $|\mathbf{F}|$
elastic deformation (m) Δs
force constant
  (newtons/m) κ

$$|\mathbf{F}| = \kappa \, \Delta s$$

the deforming force does work and transfers energy to the elastic system where it is stored as elastic energy. When the elastic system springs back, the elastic force does work and the energy is returned to the objects in the environment with which the elastic system is interacting. Bow and arrow, a toy car with a spring motor, or a watch with a spring are good examples of systems where energy is stored in an elastic subsystem.

Many solid objects exhibit *elasticity*, that is, a tendency to spring back after they are deformed. The deformation may occur in any one of many different ways. A coil spring, for example, may be coiled up more tightly. A fishing rod or a bow may be bent away from their original straight shapes. A piano wire may be stretched as the instrument is tuned. A paper clip may be opened slightly when it is slipped over a stack of papers. Branches of a tree may sway in the wind—even skyscrapers and bridges may sway in the wind.

Elastic limit. By contrast, there are many solid materials, such as clay and putty, which are inelastic. If they are deformed into a new shape, they retain that shape; they do not spring back. You know from everyday experience, however, that even elastic systems cannot be deformed indefinitely. If the paper clip is opened too far, it will remain in its new configuration. If the fishing rod is bent too far, it snaps. The smallest force that produces a permanent alteration of shape is called the *elastic limit* of the material.

Hooke's Law. You may remember that the spring scale in Section 11.1 had the marks for equal force increments at almost equal intervals (Fig. 11.9). This experimental result suggests that the spring can be described by the mathematical model of Eq. 11.8, in which the force applied to produce a deformation of distance Δs from the equilibrium configuration is proportional to the distance as shown in Fig. 11.23. Robert Hooke observed this property of many elastic bodies in the seventeenth century and proposed Eq. 11.8 for their description. The formula is therefore called Hooke's Law. The constant κ (Greek "kappa") in the formula is called the *force constant* and depends on the material and shape of the elastic body.

Figure 11.23 Definition of Δs, the deformation distance of the movable end of a spring, for three different spring shapes.

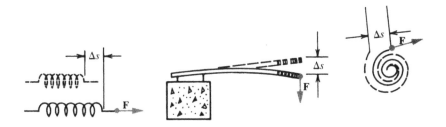

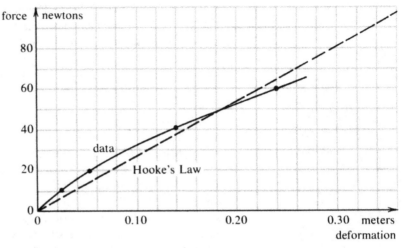

Figure 11.24 *Graph of the data for the elastic force of a bow (Table 11.2, below). The dots connected by the solid, curved line represent the experimental data. The straight (dashed) line is calculated from Hooke's law ($|\mathbf{F}| = 270\Delta s$). The relatively small discrepancy between the solid and dashed lines demonstrates that Hooke's law describes the performance of this bow quite well.*

Robert Hooke (1635-1703) became Robert Boyle's assistant after study at Oxford. In 1662, he became curator of experiments to The Royal Society and professor of geometry at Gresham College. Like many men of great ability, Hooke tended to become involved with too many things and thus found it difficult to finish anything. Nevertheless, he made a number of substantial contributions to science. He was an early exponent of the wave theory of light and was recognized by Newton as having been among the first to suggest the law of gravitation. In making observations of the structure of cork with microscopes he built, Hooke was the first to use the word "cell." Hooke is, however, best remembered for his law governing the compression and extension of elastic systems.

According to Hooke's law, the deformation returns to zero when the force on the elastic body is removed; that is, the body springs back completely. This model, therefore, clearly does not apply to bodies that are stressed beyond the elastic limit, for these do not spring back completely. The model also does not apply to rubber bands, as you can verify by trying to provide a rubber band with a calibrated scale in the way we did for the standard spring scale. The scale marks do not fall at equal intervals. In spite of its limitations, Hooke's law is extremely useful because it applies to many examples and is so simple (Table 11.2 and Fig. 11.24).

Elastic energy. Hooke's law enables us to derive a mathematical model for the elastic energy stored in a deformed body. The natural reference state (zero elastic energy) is the undeformed state. We need only calculate the work that is done by the force that deforms the elastic

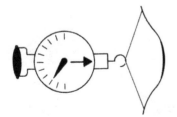

TABLE 11.2 ELASTIC FORCE OF A BOW

Displacement (Δs, meters)	Force* ($|\mathbf{F}|$, newtons)	Force** ($|\mathbf{F}|$, newtons)
0.000	0	0
0.025	10	7
0.055	20	15
0.14	40	38
0.24	60	65

* Experimental data (using apparatus shown in margin to left).
** Calculated from Hooke's law: $|\mathbf{F}| = 270\,\Delta s$ (with $\kappa = 270$ newtons per meter).

body to a distance Δs, because this work is equal to the stored elastic energy (Eq. 11.9). We therefore would like to combine the mathematical model for work (Eq. 11.5) with Hooke's law (Eq. 11.8). Unfortunately, these two cannot be combined readily because the elastic force varies proportionally to the deformation, but the problem can be solved approximately (Example 11.3). The elastic energy stored in the deformed body varies as the second power of the displacement: doubling the displacement stores four times the energy, tripling the displacement stores nine times the energy, and so on (Example 11.4).

EXAMPLE 11.3. Construct a mathematical model for the elastic energy of a system described by Hooke's law. Use Eqs. 11.5, 11.8, and 11.9.

Solution: Hooke's law, Eq. 11.8 gives the magnitude of the force that deforms the elastic body. This force varies from zero to a maximum value of $\kappa\Delta s$ when the body is deformed. As an approximation, describe the force by a constant "average" value, halfway between the minimum and the maximum,

$$|\mathbf{F}| \approx \tfrac{1}{2}\,\kappa\,\Delta s \tag{1}$$

The displacement of the end of the spring on which the force acts is parallel to the force (Fig. 11.23). Hence the displacement component is equal to the deformation distance,

$$\Delta s_F = \Delta s \tag{2}$$

The mathematical model for the elastic energy is found by using these results in Eq. 11.9:

$$E = W = |\mathbf{F}|\,\Delta s_F \approx (\tfrac{1}{2}\,\kappa\,\Delta s)\Delta s = \tfrac{1}{2}\,\kappa\,(\Delta s)^2 \tag{3}$$

EXAMPLE 11.4. Calculation of the elastic energy of a bow using the value of κ in Table 11.2 (270 newtons per meter) and the formula derived in Example 11.3.

$$E = \tfrac{1}{2}\,\kappa\,(\Delta s)^2 = \tfrac{1}{2} \times 270(\Delta s)^2 = 135\,(\Delta s)^2$$

Displacement (m)	Energy (joules)
0.00	0.0
0.05	0.34
0.10	1.4
0.15	3.0

11.7 Elastic energy and the MIP model for matter

Let us now return to consider elastic energy and ask the question, what happens to the energy when a system is deformed beyond the elastic limit? Clearly, the energy that was transferred to the system will

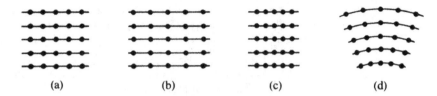

Figure 11.25 MIP model for an elastic specimen of 25 particles, (a) The specimen is unstressed, (b) The specimen is extended, (c) The specimen is compressed, (d) The specimen is bent.

not be simply returned to the environment because the system does not spring back completely. You can easily do an experiment with a paper clip (especially a 2-inch-long one), bending it rapidly back and forth (beyond the elastic limit) and touching the bent portion with your finger or above your upper lip to observe its temperature. It gets quite warm! You conclude that at least some of the energy becomes thermal energy. The same thing happens when the sidewall of an automobile tire, particularly an under-inflated one, is repeatedly flexed as the tire turns. It gets quite hot!

MIP model for an elastic system. To obtain an understanding of how elastic energy is transformed into thermal energy, we make an MIP model for elastic systems. We know that particles in solid materials are in a more or less regular arrangement that depends on the crystalline structure of the material. The electrons and nuclei of adjacent particles interact with one another via an electric field in such a way as to maintain the shape of the entire piece of material. The particles are in an equilibrium arrangement relative to one another, and are spaced at certain equilibrium distances from one another.

Now, if we bend, stretch or compress the material and thus change its shape, we are also changing the inter-particle distances and the micro-domain electric field. Figure 11.25 shows four arrangements of a system of 25 particles schematically. Elastic energy is stored in the electric field of the deformed arrangement. This energy is released by the system when the specimen is permitted to spring back to equilibrium. In

Figure 11.26 An elastic specimen is bent beyond the elastic limit.
(a) Particle arrangement before elastic energy is lost,
(b) Particle arrangement after elastic energy is lost.

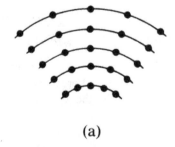

(a)

(b)

other words, what we call elastic energy in the macro domain is, according to this model, stored in micro-domain electric fields.

Elastic energy loss. Let us now consider what is likely to happen when the specimen is bent so far that it remains deformed (Fig. 11.26). Some particles are so very far apart that others from adjacent rows are attracted into the intervening spaces. This means that the particles can move toward a new equilibrium arrangement (Fig. 11.26b), which is compatible with a bent shape. As the particles move toward their new equilibrium positions, they gain kinetic energy at the expense of the electric field energy. Through collisions among the particles, the kinetic energy is shared among many of them, so that the specimen comes to a higher temperature in a macro-domain description (Section 10.5). This is a theory to explain why the paper clip gets hot.

Elastic versus chemical and phase energies. The situation we have here, where elastic energy is stored in micro-domain electric fields, is similar to the earlier example of chemical and phase changes, where energy was also gained by or lost from micro-domain electric fields. The major difference is that the particle displacements in an elastic deformation are so correlated among all the particles that the macro-domain shape of the entire system is altered: it is stretched, compressed, or bent. Therefore the elastic energy can be transferred by the elastic forces doing work in the macro domain, as when the elastic energy stored in the longbow is transferred to kinetic energy of the arrow. During a phase change, however, the displacements of the various particles are not correlated so as to produce a net displacement or work in the macro domain. The energy released during a phase change is therefore transferred to (or from) micro-domain kinetic energy, that is, macro-domain thermal energy.

The MIP model for "rigid" bodies. So far we have described elastic systems, such as springs, longbows, and fishing rods, which can be given large, easily visible deformations. The MIP model, however, leads you to expect that all solids would exhibit elastic behavior, even though the magnitude of the deformation may be very small if the interactions holding the particles in their equilibrium arrangement are very strong. Fortunately, indirect evidence of elasticity is furnished by many observations, for example, a glass marble bouncing on a steel plate (Fig. 11.27) or a drinking glass bouncing on a wood floor.

The bouncing of the glass marble and of other "rigid" bodies will make you aware that all rigid bodies are really elastic, can suffer deformation, and can store elastic energy. Thus, when the micro-domain particles are slightly displaced from their equilibrium arrangement, work is done and energy is stored; this energy is released when the particles spring back after the external stresses are removed. This fact is usually not appreciated because the elastic displacements under ordinary conditions are too small to be detected. All evidence, however,

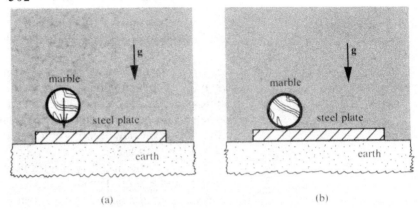

(a) (b)

*Figure 11.27 A glass marble is dropped and hits a steel plate,
(a) The marble just before impact. The gravitational field energy of the
elevated marble-earth system has been transferred to kinetic energy,
(b) The marble at impact. The gravitational field energy is zero, and the
kinetic energy is zero. Where is the energy of the system?*

indicates that the elastic deformation and the elastic force of so-called
rigid bodies are related by Hooke's law (Eq. 11.8), just as they are for
springs, diving boards, and other objects that suffer visible elastic deformations. The MIP model for rigid bodies and their elasticity also
helps to explain how it is possible for sound waves to propagate
through such bodies (Section 7.1).

11.8 Frictional force

We have already had occasion to refer to frictional interactions and
the force of friction on several occasions. The interaction of the fireman
with the brass pole was the most recent example. Others are the

*Figure 11.28 Friction
between the hand and the
brass pole.
(a) The hand slides
downward relative to the
pole. The force of friction
on the hand is directed
upward and thereby
opposes this motion,
(b) The pole slides upward
relative to the hand. The
force of friction on the pole
is directed downward and
thereby opposes this
motion.*

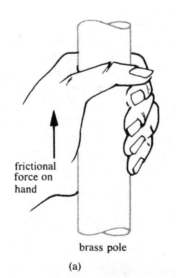

frictional
force on
hand

brass pole

(a)

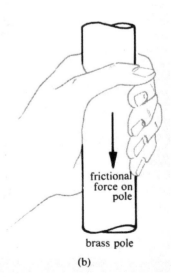

frictional
force on
pole

brass pole

(b)

interaction of your shoes with the floor while you walk, the interaction of an eraser with the paper on which it is used, and even the interaction of the archer's fingers with the arrow when he is ready to shoot.

Properties of friction. Friction occurs when two surfaces are in contact. The frictional force *opposes* relative motion of the two surfaces. When two pieces of sandpaper are rubbed on one another, the interlocking of irregularities on the two surfaces clearly causes friction. Why do even very smooth surfaces give rise to friction when the two surfaces are pressed together? Later we will describe a theory of friction that answers this question.

Friction between the fireman's hand and the brass pole, like other interactions, is described by two forces, one exerted on the hand, the other on the pole (Fig. 11.28). The direction of each frictional force is such as to drag the other surface along with it and to reduce relative motion of the two surfaces. Its magnitude is determined by the condition of the surfaces and the tightness with which they are pressed together. In one simple and fairly successful mathematical model for friction, the magnitudes of the frictional forces are directly proportional to the force pressing the two surfaces together.

Whether or not the two surfaces actually are displaced relative to one another depends on the magnitude of the frictional force in relation to the other forces that are acting. If the frictional forces on the hands of the fireman are equal to the force of gravity acting on him, he will be in mechanical equilibrium and will not slide down. If he slightly relaxes his hold and thereby decreases the frictional force, he will slide. The same is true of an eraser interacting with a piece of paper. Pressed lightly, the frictional force is small and the eraser glides over the paper easily. Pressed very heavily, the paper may tear and move along with the eraser because the frictional force is larger than the forces holding the paper fibers together.

Energy transfer by friction. If the two interacting surfaces are not displaced relative to one another, the frictional forces do no work. When there is some sliding, however, the frictional force acts on a body that is being displaced and work is done. According to the definition of work in Eq. 11.5, the work done is equal to the force of friction times the relative displacement of the two surfaces.

What happens to the energy? From your experience of rubbing your hands together on a cold day, you know that thermal energy is produced in the interacting objects, which may be your two hands, the fireman's hands and the brass pole, the eraser and the piece of paper, and so on. In the example of the eraser and the piece of paper, eraser crumbs are produced and the paper tends to be worn thin; here the shape of the interacting objects is also being changed, which requires some energy. The same is true when sandpaper is rubbed on wood or a grindstone is used to sharpen a knife.

A theory of friction. As a matter of fact, energy transfer during friction can be explained by the use of the same MIP model that we

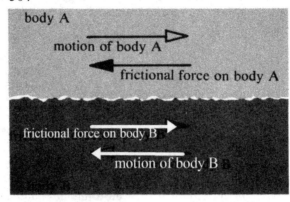

Figure 11.29 Micro-domain irregularities interlock and thereby bring about the forces of friction opposing relative motion.

introduced to explain elasticity. In the micro domain all surfaces have irregularities. When two surfaces slide over one another, very strong interactions occur at the interlocking irregularities (Fig. 11.29). The irregularities are stressed beyond their elastic limit and are deformed, with the production of thermal energy as explained in Section 11.7.

Applications. The effects of friction are often undesirable. Friction increases the energy necessary to operate machinery, and it causes wear of the moving parts. To reduce this waste, we use wheels and ball bearings and/or lubricants to minimize friction. Wheels and ball bearings roll rather than slide, whereas lubricants form a film between the interacting surfaces and prevent their irregularities from interlocking.

In many circumstances, however, friction is very desirable. When walking, driving a car (not on ice!), tying shoelaces, and holding a drinking glass, friction is indispensable in that it prevents relative motion of the interacting surfaces. The same is true of thread in fabric and nails that hold boards together. Under other conditions, frictional forces are used to do work to achieve energy transfer. For example, when you press on the brake pedal of a car, the brake shoe presses against the brake drum (or disc), and the kinetic energy of an automobile is transferred to the thermal energy of the brake linings, or when an orbiting spacecraft must be slowed down to return to earth, friction with the air converts the kinetic energy of the craft to thermal and phase energy of the heat shield.

Friction in liquids and gases. So far we have described friction between solid surfaces. Liquids and gases also exhibit friction in that relative motion of their parts is opposed by interactions among those parts. When water in a bowl is stirred near the edge, soon all the water in the bowl is rotating. Honey, though a liquid, is almost impossible to stir. Friction in liquids is called *viscosity*. Honey has high viscosity, water has low viscosity. Lubricating oils must have enough viscosity so they are not squeezed out completely from between the surfaces they are to lubricate.

Gases, too, have viscosity, although much less than liquids. Gases, therefore, make excellent lubricants, but they must be continually supplied to the lubricated surfaces because they escape rapidly. Even though air has very low viscosity, its motion past the surface of a plane flying at supersonic speeds or of a reentering spacecraft creates a great deal of frictional drag. Special materials must be used to withstand the high temperatures produced by the high rate of production (through friction) of thermal energy at an airplane's surface.

Summary

The central concept around which this chapter revolved is the Newtonian force. Instead of stressing the mutuality of the interaction among the objects in a system, Newton selected the objects' motions for detailed study one at a time. In Newton's theory, a net force acts on every body whose motion is changing. For Newton, the net force was the cause of the change in motion. A body in steady motion in a straight line is subject to a zero net force; such a body is said to be in mechanical equilibrium.

The net force acting on an object in a complex system is the sum of partial forces exerted on it by each of the other objects in the system. Each of these partial forces is assumed to be affected neither by the presence of the other objects nor by the other partial forces. A body in mechanical equilibrium may be subject to no interaction at all, or (more likely) it is subject to a set of partial forces that compensate for one another. No body ever exerts a force on itself.

Four kinds of forces were explained in some detail: the gravitational force, the electric force, the elastic force, and the frictional force. A numerical measure of force, which is described by a magnitude and a direction in space, is obtained with a spring scale calibrated in newtons. The gravitational force on a body is of special importance for all dwellers on the earth. Near the earth's surface it is described by the mathematical model in Eq. 11.4. Its direction is vertical and its magnitude is very close to 10 newtons per kilogram of gravitational mass of the body, regardless of the body's other properties.

Eq. 11.5 defines the work done by a force acting on a moving body. It is equal to the magnitude of the force times the moving body's displacement component along the force direction. The work is the energy transferred by the action of the force. When a body falls freely, for instance, the work done by the gravitational force transfers gravitational field energy to kinetic energy. Work done in the elastic deformation of a solid body is stored as elastic energy. Work done by the frictional force produces thermal energy at the surface where friction occurs.

Equation 11.4 (force of gravity)

force of gravity (newtons)
$$= F_G$$
gravitational mass (kg)
$$= M_G$$
gravitational intensity (newtons/kg) $= g$

$$F_G = g\, M_G$$

Near surface of earth:
$g = 10$ *newtons/kg, and*
$F_G = 10 M_G$ *(downward).*

Equation 11.5 (Work)

work (joules) $= W$
force magnitude (newtons) $= |F|$
displacement component in force direction (m) $= \Delta s_F$

$$W = |F|\, \Delta s_F$$

List of new terms

force	gravitational intensity	elastic limit
net force	work	force constant
spring balance	elasticity	friction
newton		

List of symbols

F	force	R	distance (between charges)
$\mathbf{F_G}$	force of gravity	κ	force constant
g	gravitational intensity	M_G	gravitational mass
W	work	h	height above reference
Δs_F	displacement component		level
	in the force direction	E_G	gravitational energy
q	electric charge		

Problems

1. Identify four examples of forces (elastic, gravitational, pressure, or frictional) in Fig. 11.1 in addition to the ones listed in the caption.

2. (a) Give two examples from everyday life of bodies that are in mechanical equilibrium relative to one reference frame and are being accelerated relative to another.
   (b) Describe the partial forces acting on the bodies in your examples.
   (c) Relate the partial forces (qualitatively) to the net force in each reference frame.

3. Comment on the assumption that the interaction between two bodies is unaffected by the nearby presence of a third, according to Newton's theory, as follows.
   (a) Do you find this assumption reasonable on the basis of your experience with inanimate objects?
   (b) Do you find this assumption applicable to human social interactions?

4. Enumerate the partial forces acting on each of these objects in Fig. 11.1:
   (a) the fisherman's left boot;
   (b) falling apple;
   (c) tip of the fishing rod.

5. Draw an approximate force diagram for all the partial forces acting on one of the objects in Problem 4.

6. Measure the spring extensions produced by various numbers of weights in Fig. 11.9 and draw a graph relating these two variable factors (extension and number of weights).

7. (a) Use a rubber band to make a "spring" scale (not necessarily calibrated in newtons).
   (b) Measure the rubber band extensions produced by various numbers of weights as you calibrate the rubber band scale.
   (c) Draw a graph relating the two variable factors in (b).
   (d) Is the rubber band described by Hooke's law?

8. Describe how the operation of a spring scale may be affected by the force of gravity acting on the elastic system in the spring scale itself.

9. Imagine the spring scale in Fig. 11.10 to be connected to the mouse's harness instead of to the wagon. What reading would you expect it to show? Explain your answer.

10. (a) At what stage of the child's swing would you expect the spring scale reading in Fig. 11.11b to be largest? Explain.
(b) At what stage would you expect the spring scale reading in Fig. 11.11b to be smallest? Explain.

11. Make an analogy between the spring scale and a thermometer. Discuss this analogy, using gravitational field energy as the analogue of thermal energy. What are the analogues of temperature and specific heat? Discuss this analogy critically.

12. Use a spring scale (bathroom scale with 1 pound equal to 4.5 newtons) to estimate the net force on yourself under the conditions in your bathroom, and then in an elevator while it is
(a) starting upward;
(b) coming to a stop on its way up; and
(c) moving uniformly. Describe and discuss your observations.

13. Calculate the reading on your bathroom scale if you were to weigh yourself (a) on the surface of the sun and (b) on the surface of the moon. (See Table 11.1.)

14. Propose one or more operational definitions of "work."

15. Calculate the work that is done when a 90 kilogram fireman slides 3 meters down a brass pole.

16. Calculate the gravitational field energies of skiers X and W in Fig. 11.20.

17. Calculate the gravitational field energy of the fireman-earth system (Fig. 11.20) when the fireman is in the firehouse cellar, 1.8 meters below ground floor level.

18. Describe how you might expect the mathematical model in Eq. 11.6 to fail for objects A and C in Fig. 11.21.

19. Your hands do work when they knead a piece of clay or dough. What happens to the energy transferred in this process?

20. Test a diving board to determine whether Hooke's Law describes its deformation. Use people as weights in this experiment. Find the force constant if the diving board satisfies Hooke's law.

21. Find an elastic system that is described by Hooke's law.
    (a) Measure the deformation displacement and the elastic force, plot a graph of these two variables, and determine the force constant.
    (b) Calculate the elastic energy and plot a graph of its relation to the deformation displacement.

22. Carry out the experiment with the paper clip described in Section 11.7. Describe and compare the effect you observe with paper clips of various sizes, shapes and materials.

23. Describe four everyday observations that give evidence that "rigid" bodies are really elastic systems.

24. Use your rubber band scale (Problem 7) to estimate the force of friction between a piece of paper and a smooth surface when various loads are placed on the paper (Fig. 11.30).
(a) Plot a graph to show the relationship between the force of friction and the force of gravity acting on the load, for two different surfaces.
(b) Make approximate mathematical models in algebraic form for the graphs in (a), if possible.

25. The Niagara Falls have a height of 50 meters. Estimate the maximum temperature rise in the water as gravitational field energy is converted to thermal energy at the Falls.

26. Interview four or more children (ages 8-12) to find their explanations for the temperature rise produced by rubbing your hands together.

Figure 11.30 Rubber band "spring" scale is used to measure the frictional force exerted by the table surface on the piece of paper (Problem 24).

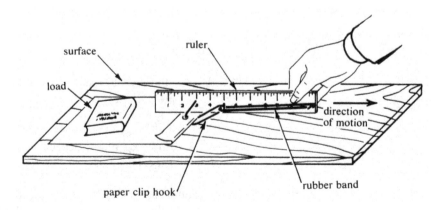

27. Find the partial force exerted by the rope tow on each of the skiers in Fig. 11.20. Assume that the frictional force acting on the skis is negligible. (Hint: The snow exerts a force at right angles to the skis if there is no friction.)

28. Identify one or more explanations or discussions in this chapter that you find inadequate. Describe the general reasons for your judgment (conclusions contradict your ideas, steps in the reasoning have been omitted, words or phrases are meaningless, equations are hard to follow, . . .), and make your criticism as specific as you can.

Bibliography

M. B. Hesse, *Forces and Fields*, Littlefield, Totowa, New Jersey, 1961.

M. Jammer, *Concepts of Force: A Study in the Foundations of Dynamics*, Harper and Row, New York, 1957.

P. Lenard, *The Great Men of Science*, Bell, London, 1950. See the section on Da Vinci.

*Electromotive action
is manifested by two sorts
of effects . . . I shall call
the first electric tension,
the second electric current.*

Ａndré Marie Amperè
Annales de Chimie et de Physique,
1820

Electric current and energy transfer

12

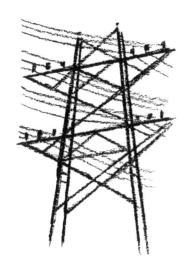

Kilowatt-hours, volts, amperes—these are terms you frequently hear when household appliances are described or the electric bill is to be paid. What do they mean? Or you may sometimes have wondered how a light switch at the top of a flight of stairs could turn off the lamp that you turned on by a switch at the bottom.

Probably you associate all these matters with the mysteries of electricity, and you may even dismiss them as incomprehensible for this reason. Electric interactions appear mysterious because macro-domain effects such as the glowing of a light bulb are produced without detectable macro-domain causes. The situation is different from that of a fire heating water or a bow shooting an arrow, where both the energy source and the energy receiver could be easily identified.

As a matter of fact, electric interactions make possible the transfer of energy over very large distances. The light bulbs and appliances in your home receive energy from a power plant that may tap the chemical energy of coal, oil, or natural gas; the nuclear energy of uranium; or the gravitational field energy associated with the water stored behind a dam. The transfer of energy occurs along transmission lines—wires—that link the power plant to your home. Remarkably, the transmission line itself gives no visible evidence whether energy is at any moment being transferred or not.

In this chapter we will describe some of the conditions under which electric interactions transfer energy in technologically useful ways. To help you with visual images of the process, we will construct an analogue model and a micro-domain model for electrical phenomena. To help you make mathematical models for the process, we will introduce two variable factors describing it: the magnitude of the electric current, and the voltage of the energy source. These two variables can be explained qualitatively by applying them to an electric spark. The brightness of the spark is related to the magnitude of the current, while the distance the spark jumps is related to the voltage of the energy source.

12.1 Electric current

Already in the eighteenth century Benjamin Franklin and others knew that electric charges were able to move along certain materials, especially metals, which were therefore called *electric conductors*. Early in the nineteenth century, Alessandro Volta invented the chemical *battery* made of two dissimilar metals and a conducting liquid (Fig. 12.1). Volta found that his battery gave only very short sparks compared to those obtainable by rubbing amber or glass, but the battery had an enormous capacity for sustained action in maintaining the spark; it could also cause frogs' legs to twitch extensively and could give repeated electric shocks to humans.

Operation of the battery required a continuous electrically conducting path between the two metals (called *electrodes*) of the battery (Fig. 12.2). Volta thought of his observations in terms of a circulating electric

Alessandro Volta (1745-1827) was born in the northern Italian town of Como. As professor of physics at Como and at the University of Pavia, he distinguished himself in research. His reputation among his contemporaries was enormous, and during a visit to Paris he was the personal guest of Napoleon. Volta's most important work was his study of the electric current flow resulting from contact between different metals. He was thus led to the invention of the voltaic pile and the voltaic battery (1800).

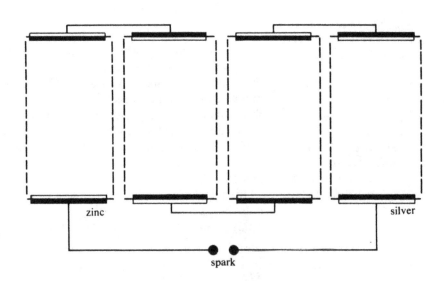

Figure 12.1 Chemical battery of zinc, silver, and brine-soaked cardboard disks constructed by Volta. This was called a voltaic pile. The zinc becomes electrically negative, the silver electrically positive. The zinc gradually dissolves.

silver disks

zinc disks

brine-soaked cardboard disks

⊕

⊖

Humphrey Davy (1778-1829), found in 1807 that electricity passing through a solution of sodium chloride (table salt) could separate it into the elements sodium (a metal) and chlorine (a gas); separating a compound into its elements with electricity in this way is known as electrolysis.

Michael Faraday (1791-1867), Davy's assistant and successor, made extensive studies of electrolysis, including the way in which electric charges pass through a conducting solution and the consequent chemical changes. By 1834, Faraday formulated what are now known as Faraday's laws of electrolysis.

fluid, which was later called an *electric current*. This is an extremely successful model, which helped to explain many other phenomena observed near wires connected to the battery electrodes. Because of this model's extraordinary success in explaining the need for a continuous conducting path, we tend today to think of electric currents as physical entities. Historic investigations that have further substantiated and elaborated the electric current model include Oersted's discovery of the interaction between magnets and electric currents, Davy and Faraday's investigations of the chemical effects of electric currents, and Joule's measurement of the heat produced during the passage of an electric current in certain conductors. Finally, the development of the electric light bulb by Edison resulted in giving the electric current concept household familiarity. For almost a hundred years before that, however, the effects of electric currents had been, as we have seen, investigated magnetically, chemically, and thermally, while light bulbs were unknown.

You may suspect that the electric current in Volta's apparatus is

Figure 12.2 Four voltaic piles connected togehter cause an electric spark when the two conductors are almost touched together in air.The solid black lines represent metal wires, which provide a continuous conducting path along which the elctric charges can travel.

zinc

silver

spark

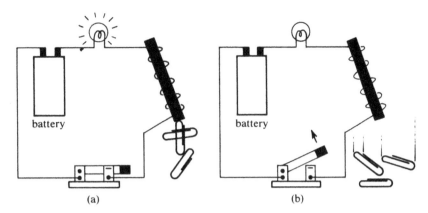

Figure 12.3 Evidence of the flow of an electric current, (a) When the switch (and thus the circuit) is closed, the bulb is lit, and the electromagnet picks up paper clips. (b) When the switch (and circuit) is open, the bulb is not lit, and the electromagnet does not hold paper clips.

related to the electric charges produced by rubbing and which are responsible for the phenomena of static (as opposed to current) electricity. The fact that the same metals and electric conductors participate in both phenomena convinced Volta that such a relation existed. This relation was implicit in Maxwell's incorporation of electric charge conservation into his electromagnetic wave theory: electric charges that disappeared from one place had to reappear somewhere else by passing through the intervening space as an electric current. Henry Rowland (1848-1901), finally, showed experimentally that electric charges produced on glass by rubbing do interact with a magnet when they move at a high speed relative to it. This observation showed conclusively that an electric current consists of electric charges in motion.

Electric circuits. An *electric circuit* is a system of conductors through which an electric current can flow. The most common conductors are metallic wires, but salt water or many other liquids can also serve, and even gases, such as the neon in a "neon light," are conductors under suitable conditions. Special objects, such as light bulbs, or motors, or electric heaters, which can operate in an electric circuit, are called *circuit elements*. A battery, an electric generator, or another energy source in a circuit is called an *electric power supply*.

A circuit element will operate only when there is a complete or *closed circuit* of electrical conductors linking it to an electric power supply. When the path is interrupted by a gap of air or some other non-conducting material, such as rubber or plastic, the system is called an *open circuit*. The lighting of a bulb, the operation of a motor, or the existence of a magnetic field near an electrical conductor is evidence that the circuit is closed. Conversely, the failure of the circuit elements to operate is evidence that the circuit is open. A switch is the most common device for opening and closing a circuit (Fig. 12.3).

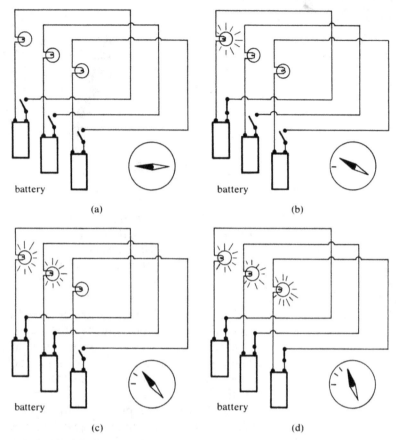

Figure 12.4 Diagrams suggesting how a compass dial may be calibrated to measure electric current. Three identical circuits are placed side by side, and the compass is held near the parallel wires,
(a) All circuits are open; the dial is marked for zero current,
(b) One circuit is closed; the dial is marked for one unit of current,
(c) Two circuits are closed; the dial is marked for two units of current,
(d) Three circuits are closed; the dial is marked for three units of current.

A closed circuit makes possible the transfer of energy from an electric power supply to one or more circuit elements. This energy transfer can take place over very large distances, as from Hoover Dam to Los Angeles, or from Niagara to New York City, or it can take place over short distances, as in a flashlight (from the battery to the bulb). Energy transfer occurs while the circuit is closed and an electric current is flowing; energy transfer ceases when the circuit is opened and the current ceases to flow. As we pointed out at the beginning of this chapter, the principal evidence of electric current flow and energy transfer is the operation of a circuit element; the connecting wires in the circuit themselves ordinarily give no such evidence. How can we unravel the mystery of what happens in the wires?

*André-Marie Ampère
(1775-1836) was
educated in Lyons,
France. Though his father
was a political victim of
the French Revolution,
this apparently had no
effect on Ampère's
academic career and he
became a member of the
prestigious French
Institute in Paris.
Oersted's discovery in
1820 that an electric
current-carrying wire
affected a magnetic
compass led to Ampère's
equally important
discovery of the magnetic
forces exerted by currents
on currents. From this
discovery, Ampère
developed the
mathematical theory that
describes these forces and
that we know as Ampère's
law.*

We will take three approaches to solving this problem. First, we will propose an operational definition of electric current that can be used to determine the current's magnitude. Second, we will construct analogue models in which water circulating through pipes corresponds to the current in the conductors. Third, we will use the models for atoms described in Chapter 8 to develop a micro-domain model for electric current.

Operational definition of electric current. Oersted discovered that there is a way of telling whether or not a current is flowing in a wire, for he found that a compass needle near an electrical conductor in a circuit is deflected when the circuit is closed; the needle returns to its original direction when the circuit is opened. The magnitude of the deflection can be used to measure the strength of the electric current. You can imagine placing a dial behind the magnetized compass needle and making marks for the deflections that indicate 1 unit of current, 2 units, and so on (Fig. 12.4). The commonly used unit of electric current, which has been rather arbitrarily chosen, is called the *ampere* after André-Marie Ampère, and the current measuring device (more sensitive and reliable than the one shown in Fig. 12.4) is called an *ammeter*. By inserting one or more ammeters into an electric circuit, you can measure and compare the electric current flowing in various parts of a closed electric circuit (Fig. 12.5).

Relation of electric charge and current. In Section 8.1 we described the faraday, a unit of electric charge. We defined one faraday as the

*OPERATIONAL
DEFINITION
Electric current is
measured by the dial
reading of a standard
ammeter (shown in Fig.
12.4) Electric current is
measured in units of
amperes (amp).*

Figure 12.5 The electric current in various parts of a closed electric circuit is indicated by ammeter readings.

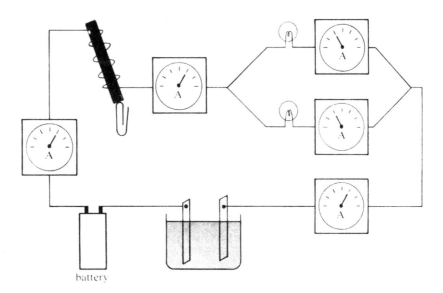

battery

Equation 12.1

electric charge
delivered (faradays) $= \Delta q$
electric current (amp) $= \mathcal{I}$
time interval (sec) $= \Delta t$

$$\Delta q = 1.0 \times 10^{-5}\ \mathcal{I}\ \Delta t$$

combinations of amperes
and seconds, which give
1 faraday:

$\mathcal{I} = 1\ amp,\ \Delta t = 10^5\ sec$

$\mathcal{I} = 1.7 \times 10^3\ amp,$
$\qquad \Delta t = 60\ sec$

$\mathcal{I} = 10^5\ amp,\ \Delta t = 1\ sec$

The modern definition of the
ampere is based on the
electrolytic effect of an electric
current on a standard solution
of silver nitrate. One ampere is
the electric current that
deposits a specified amount of
silver per second. This
definition is used to check the
accuracy of ammeters.

We can also express Equation
12.1 in another form, to show
that electric current is
essentially the same as the
transfer, or flow, of charge per
unit of time:

$$\mathcal{I} = (10^5)\frac{\Delta q\ (faradays)}{\Delta t\ (sec)}$$

One proposal for a unit of
electric current that was not
adopted more generally:
"I have expressed my quantities
of electricity on the basis of
Faraday's great discovery of
definite electrolysis; and I
venture to suggest that that
quantity of current electricity
which is able to electrolyse a
chemical equivalent expressed
in grains in one hour of time, be
called a degree."
James Prescott Joule
Philosophical Magazine, 1841

charge transferred when 1 gram of hydrogen is produced in the electrolysis of water. In this section we have defined electric current independently of electric charge, by using an operational definition in terms of an ammeter calibrated by the magnetic effect of the current on the direction of a compass needle (as in Figure 12.4). Electrolysis obviously involves *both* transfer of charge *and* an electric current; thus you may well ask exactly how the current (in amperes), is related to the charge (in faradays). Investigation shows that the amount of hydrogen produced in electrolysis of water, and therefore the electric charge delivered, is directly proportional both to the time of electrolysis and to the current strength.. You can construct a mathematical model (Equation 12.1 to the left) for this relationship once you determine what combination of amperes and seconds deliver a charge of 1 faraday, that is, produce 1 gram of hydrogen gas in electrolysis.

Water pipe analogue model for electric circuits. With the help of an ammeter, you can obtain visual evidence of an electric current in a conductor. The diagram in Fig. 12.5 illustrates the use of several ammeters to investigate the pattern of electric current in an electric circuit. Analogue models for electric circuits rely on the movement of macro-domain objects in circulating patterns to represent the electric current in wires. The flow of water through pipes is a very successful analogue model of this kind. The one-to-one correspondence between aspects of electric circuits and of the model is shown in Table 12.1.

Curiously, the flow of water even in transparent pipes or tubes shows some of the "mystery" of current flow in conductors, because it is visually undetectable. We have already pointed out that electric current flow in a wire cannot be seen without a tool such as a compass or an instrument like an ammeter. The same holds true for water. Since water is uniform, one bit of water looks like another and does not provide any visual reference point whose motion relative to the observer can be identified. To render the flow "visible," you have to mix in reference objects, such as bubbles or sand grains, which are carried along with the water and whose motion can be seen.

Water flow analogues for various circuit elements are illustrated in Fig. 12.6. Their operation can serve as evidence of water flow just as

TABLE 12.1 WATER FLOW ANALOGUE MODEL FOR ELECTRIC CIRCUITS

Electric circuit	Water flow
conductor	pipe
electric current	flowing water
electric charge	stationary water
electric power supply	pump
electric motor	water wheel
switch	faucet or valve
magnetic interaction	(no analogy)
ammeter	flowmeter
heating element	capillary tube

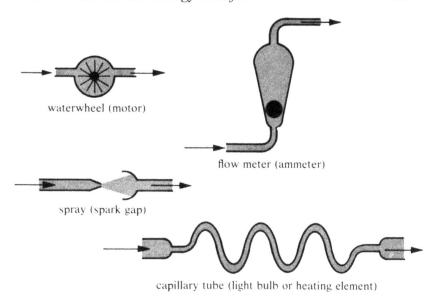

waterwheel (motor)

flow meter (ammeter)

spray (spark gap)

capillary tube (light bulb or heating element)

Figure 12.6 Water-operated devices analogous to certain electric circuit elements. In the flowmeter the water lifts the metal ball in the tapered chamber until there is enough space for the water to flow past. The faster the flow, the higher the ball. Thus the height of the ball can be used as a measure of the rate of flow.

the operation of electric circuit elements is evidence of electric current flow. "Seeing" the water flow directly, however, gives you an intuitive comprehension of water flow phenomena that is difficult to gain for electric circuit phenomena (Fig. 12.7).

Micro-domain model for electric current. A preview of some of the electrical phenomena described in this section was included in Chapter 8, where we presented evidence of the electrical nature of matter. According to the micro-domain models proposed there, matter consists of electrically charged constituents such as electrons (negative charge) and atomic nuclei (positive charge). These two constituents interact with one another and usually occur in combination in systems that may be electrically neutral (atoms or molecules) or electrically charged (ions). The charge of ions is positive or negative depending on whether the number of electrons is less than or greater than the total positive charge of all the nuclei in the ion.

"Electric fluids." With so many kinds of electrically charged constituents, it is easy to make micro-domain models for electric current. Current flows when there is motion of electric charges relative to the observer. In other words, you can think of electrons and ions as making up "electric fluids." When they accumulate on one body, it has a net electric charge; when they move through conductors, an electric current flows. These modern models, therefore, are similar to the models proposed by Franklin and others in the eighteenth century (Section 3.5). It is now recognized, however, that there are many different kinds of "electric fluids" because there are many kinds of ions.

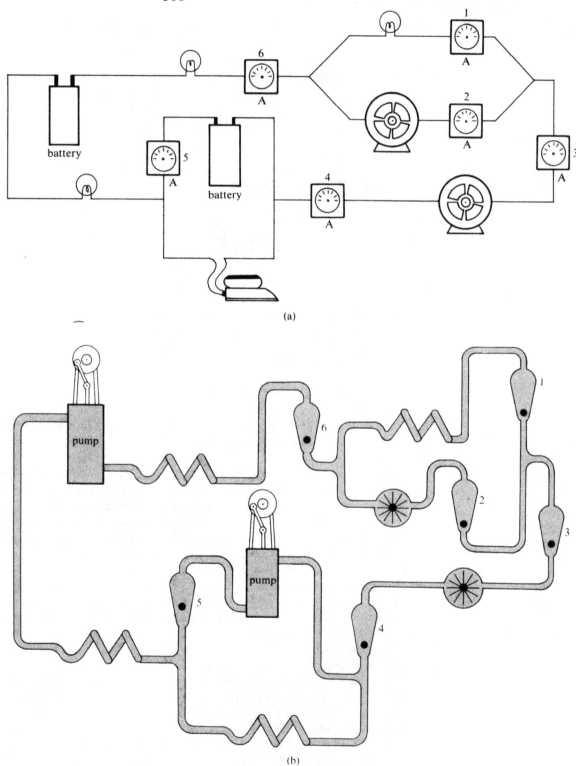

Figure 12.7 Analogue model for an electric circuit. (a) Electric circuit. (b) Water pipe analogue model.

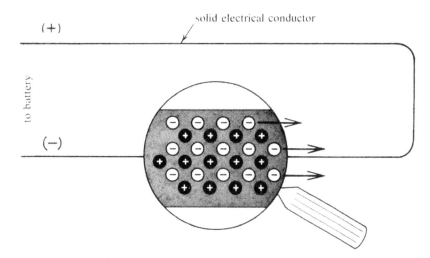

Figure 12.8 Micro-domain model for solid electrical conductors. Solid conductors contain stationary positive ions and a mobile negative "electric fluid" of electrons. The two kinds of charges are present in equal amounts, so that the conductor is electrically neutral.

There is a qualitative difference between the models for electric current in solid conductors (copper wire) and liquid or gaseous conductors (salt water, melted sodium chloride, gases at very high temperatures). In solid conductors, only electrons are in motion (Fig. 12.8). The shape of the solid material is maintained by the atoms and/or ions, which are bound together in a structure in mechanical equilibrium (Sections 4.5 and 11.7). In liquid and gaseous conductors, however, both ions (positive and negative) and/or electrons may be in motion (Fig. 12.9).

Forces on a steady current. According to these models, a *steady current* in an electrical conductor is the steady motion of electrons and/or ions, which we will call the "electric fluid." If you apply the Newtonian theory to this motion, you conclude that the electrons and ions must be

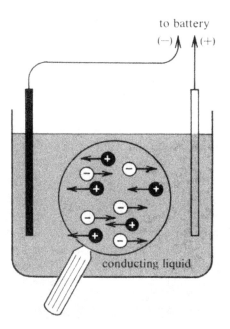

Figure 12.9 Micro-domain model for liquid or gaseous conductors. Liquid or gaseous conductors contain positive and negative mobile "electric fluids" of ions and/or electrons. The two kinds of charges are present in equal amounts so that the conductor is electrically neutral.

in mechanical equilibrium, because they are in steady motion (Section 11.2). Only when the electric current starts or stops is there a change of motion, which requires the action of a net force on the fluid.

Two partial forces play a particularly important role. One of these arises from the interaction-at-a-distance between the power supply and the electric fluid. The intermediary is an electric field set up by the power supply throughout the electric circuit. The other partial force arises from collisions between the moving electrons or ions and the stationary material of the electrical conductor in which the current flows. The combined effect of many collisions is to oppose the motion of the electric fluid relative to the conductor. This interaction is therefore analogous to the frictional interaction between two bodies sliding over one another (Section 11.8), but it occurs throughout the interior of the conductor and is not a surface effect.

When the electric current is steady, the two partial forces just described are equal in magnitude and opposite in direction. The net force acting on the electric fluid is zero. During an extremely short period of time (about 10^{-18} seconds) just after an electric circuit is closed, however, the force transmitted by the electric field is larger than the opposing force arising from collisions. The net force sets the electric fluid in motion and accelerates it until the opposing force, which increases in proportion to the speed of the electric fluid, is large enough to reduce the net force to zero. Similarly, during an extremely short time interval after a closed circuit is opened, the force transmitted by the electric field is smaller than the opposing force arising from collisions. Then the net force slows down the motion of the electric fluid and brings it to a halt.

Insulators and conductors. In this micro-domain model, *electrical conductors* are materials that contain a mobile electric fluid. Nonconductors or *insulators* are materials with no mobile electric fluid. Air, glass, and pure water are good examples of insulators. In insulators, all the electrons and ions are securely bound to one another and are not free to move even when a moderate electric field is applied. When an extremely large electric field is applied, however, an insulator may become a conductor, as when a spark jumps across an air gap in a circuit. Then the electrically neutral atoms gain energy from the electric field and are split into electrons and ions that can move freely. When the electrons and ions combine to form an atom once more, the energy of the system is radiated in the form of light that is seen as a spark.

12.2 Voltage

In the introduction to this chapter, we referred to the two variable factors needed to describe an electric spark: the current, which is related to the brightness or intensity of the spark; and the voltage, which is related to the length of the spark. In the water analogue model, a water jet emerging from a hose represents a spark by a water jet (Fig. 12.10). You can recognize the two variable factors (current and voltage) in this analogue. The flow rate of water corresponds to the current, while the pressure by which it is ejected (and thus its length) corresponds to the

Henry Cavendish (1731-1810) had identified these two key variable factors (current and voltage) long before Volta's, Oersted's, and Ampère's discoveries. In 1757 he wrote, "The strength of the shock depends rather more on the quantity of fluid [the current] which passes through our body than on the force with which it is impelled."

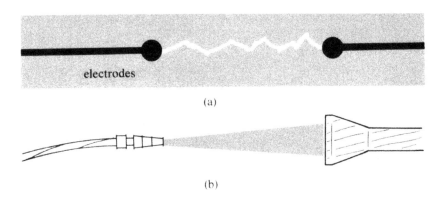

electrodes

(a)

(b)

Figure 12.10 Water-jet analogue to an electric spark,
(a) The length of the spark is determined by the voltage,
(b) The length of the water jet is determined by the water pressure.

Equation 12.2

energy transfer
   (joules) ΔE
voltage (volts) $\mathcal{V}$
electric current
   (amp) $\mathcal{J}$
time interval (sec) Δt

$$\Delta E = \mathcal{V}\,\mathcal{J}\,\Delta t$$

FORMAL DEFINITION
Voltage is equal to the
ratio of the energy transfer
to the product of electric
current and time interval.

**Equation 12.3
(Definition of voltage)**
$$\mathcal{V} = \frac{\Delta E}{\mathcal{J}\,\Delta t}$$

EXAMPLE
Energy transfer in a small
flashlight in 1 minute:

battery voltage $\mathcal{V}$ =1.5 volt
current $\mathcal{J}$=0.5 amp
time Δt=60 sec

energy transfer:
  ΔE = 1.5 volt×0.5 amp
       ×60 sec
     = 45 joules

voltage. Voltage, therefore, is a sort of "electrical pressure" built up by an electric power supply and dissipated in the wires and circuit elements, just as a water pump creates pressure that drives the water through pipes.

Formal definition of voltage. While this discussion may clarify the meaning of "voltage," it does not suggest how an electrical "pressure gauge" or *voltmeter* may be constructed. To exploit the analogy further, we will relate the water pressure to energy transfer, which is a concept that *does* apply directly to electric circuits. Though we will not describe the details, it is possible to reason from water pressure to the force exerted by the pump, then to the work done by the pump, and ultimately to the energy transfer. The outcome of this procedure is to show that the energy transfer is proportional to the pressure times the flow rate of the water times the time of operation. The voltage of an electric power supply is therefore defined to make the energy transfer from the power supply equal to the voltage (in the unit called the *volt*) times the electric current times the time of operation (Eq. 12.2). The voltage concept is applicable to circuit elements (energy receivers) if the energy transfer to the circuit element is used. This desired relation of voltage and energy transfer leads to a formal definition of voltage (Eq. 12.3) that is applicable to power supplies (energy sources) and to circuit elements (energy receivers). An energy of 1 joule is transferred by a 1-volt power supply delivering 1 ampere for 1 second. A heating element releasing 1 joule per second with a current of 1 ampere operates at 1 volt. To make the voltage concept more concrete, we will explain how to measure the voltage of a battery.

Voltage of batteries. In a battery, a chemical reaction takes place in such a way that the chemical energy of the ingredients is transferred, by means of an electric circuit, to an energy receiver such as a light bulb, a motor, or a piece of wire that gets hot. Joule compared the thermal energy released during a chemical reaction when it took place with the ingredients together in an ordinary container and when the same

*(a) Silver and zinc are in
direct contact. Chemical
energy is converted to
thermal energy as the zinc
dissolves and the solution
becomes warm.*

*(b) Silver and zinc are
connected by an electrical
conductor. Chemical
energy is converted to
electric field energy and
then to thermal energy by
the heating element in the
separate container of
water.*

OPERATIONAL
DEFINITION
*Voltage of a battery can be
measured by using the
battery to transfer energy
to a hot wire in cold water.
The voltage is equal to the
thermal energy transferred
(joules) divided by the
product of the electrical
current (amperes) and the
time elapsed (seconds). See
Example 12.1.*

*"In 1843 I showed that the
heat evolved by magneto-
electricity is proportional
to the [energy] absorbed,
and that the [energy] of the
electro-magnetic engine is
derived from the [energy]
of chemical affinity in the
battery, [an energy] which
otherwise would be evolved
in the form of heat."*
James Prescott Joule
Philosophical
Transactions, 1850

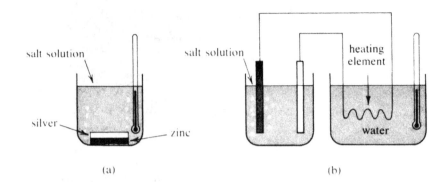

(a)   (b)

reaction took place in a battery and the electric current heated a wire
(Fig. 12.11). He found that the energy values in the two experiments
were closely equal and concluded that the electric circuit served as a
passive coupling element that transferred the energy released in the
chemical reaction but did not receive or contribute any additional energy.

By Joule's procedure, it is possible to measure the voltage of a battery.
You only have to measure the current in the circuit to the hot wire, the
time it operates, and the thermal energy transferred by the hot wire to
cold water (Example 12.1). By doing many such measurements, you

---

EXAMPLE 12.1 A battery delivers 0.5 ampere to a heating element
for 1 hour (3600 seconds) and thereby raises the temperature of 0.10
kilogram of water by 13° Celsius. What is the battery voltage?

(a) Find the energy transfer in joules (Eq.10.2 and Eq 9.3).

$$\Delta E = CM_G \Delta T = 1\,Cal/^{\circ}\ C/kg \times 0.10\ kg \times 13^{\circ}C = 1.3\ Cal$$
$$\approx 5200\ joules$$

(b) Find the voltage from Eq.12.3.

$$\mathcal{V} = \frac{\Delta E}{\mathcal{J}\Delta t} = \frac{5200\ joules}{0.5\ amp \times 3600\ sec} = \frac{5200}{1800} \approx 3.0\ volts$$

---

find that the voltage of a battery with one pair of electrodes (Fig.
12.11b) is determined by the chemical reaction occurring in it and is
not significantly affected by its size, shape, and even previous operation over short periods of time. Such a battery is called *one cell*, to distinguish it from a battery power supply in which many cells are connected in a chain, such as Volta used (Fig. 12.1). The voltage of a chain

Equation 12.4 (electrical cells connected in a chain, in "series")

battery voltage $= \mathcal{V}$
voltage of one cell $= \mathcal{V}_c$
number of cells $= n$

$$\mathcal{V} = n\mathcal{V}_c$$

Equation 12.5 (Ohm's Law, in terms of conductance)

electrical
    conductance $= \mathcal{C}$
voltage (volts) $= \mathcal{V}$
current (amps) $= \mathcal{J}$

$$\mathcal{J} = \mathcal{C}\,\mathcal{V}$$

units for
    conductance: amp/volt

George Simon Ohm (1789-1854). Ohm's father was interested in mathematics and philosophy and prepared George for the University of Erlangen. After obtaining his degree, Ohm turned his attention to the voltaic battery and published several papers on electric conduction in a circuit. His work was rejected by the Berlin Academy, and Ohm retired from scientific life in discouragement. Posterity, however, has been kinder: one of his experimental results has come down to us as Ohm's law.

of cells equals the voltage of one cell times the number of cells (Eq. 12.4 and Fig. 12.12). A large battery consisting of one cell can release more energy than a small one, but it does so by delivering current for a longer time and not at a higher voltage.

Ohm's law and electrical conductance. Voltage and current in an electric circuit correspond to water pressure and flow rate in the analogue model. Now, you would expect that the water flow rate through a particular pipe would depend on the pressure applied by the pump: the greater the pressure, the greater the flow rate, and vice versa. You might therefore expect that electrical conductors would exhibit such a relation also. Indeed, George Ohm investigated the relationship and found that the electric current was directly proportional to the voltage of the power supply. The ratio of current to voltage, which is called the conductance of the conductor, depends on the conducting material, its shape, and its temperature. For example, the larger the conductor's cross-section, the greater its conductance; and the longer the conductor, the smaller its conductance. The mathematical model, that the current in a conductor is equal to the conductance times the voltage applied to the conductor, is called *Ohm's law* (Eq. 12.5 and Fig. 12.13a). Many conductors are well described by this model but there are also many exceptions. For

Figure 12.12 Electrical cells connected in a chain (in "series") to increase the voltage of the power supply,
(a) Three flashlight "D cells" each rated at 1.5 volts, combined to deliver 4.5 volts.
(b) Common rectangular battery, consisting of six 1.5-volt cells, rated at 9 volts.
(c) Automobile storage battery (consisting of six 2-volt lead-acid "wet" cells) rated at 12 volts.

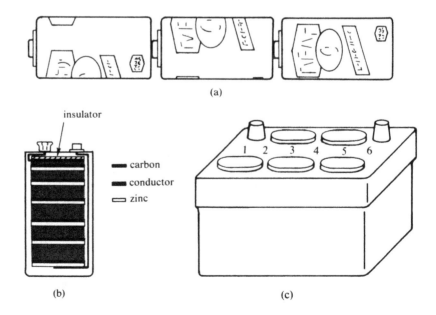

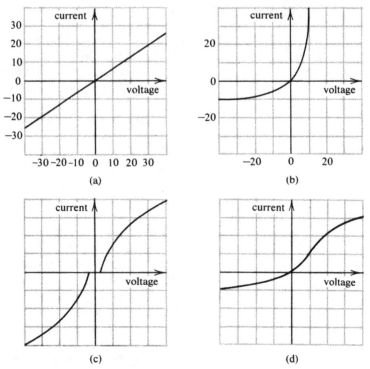

Figure 12.13 Relationship of voltage and current for various circuit elements.
(a) A circuit element described by Ohm's law (slope = conductance = 0.6 amperes per volt).
(b) A diode, which has a high conductance for positive voltage and a low conductance for negative voltage. (See Fig. 12.15)
(c) A solution of sodium chloride. (What occurs when current flows?)
(d) A radio tube.

example, Fig. 12.13 includes graphs of the relationship between current and voltage for four different systems; Ohm's law applies to only one of the graphs.

Ordinary materials show an enormous range of conductance values. It is customary to classify them into three groups: *conductors* (high conductance), *insulators* (extremely low conductance), and *semiconductors* with an intermediate conductance, but closer to that of conductors than insulators, (Table 12.2). The conductance of semiconductor materials is especially sensitive to temperature and to the presence of small amounts of impurities in the material.

Ohm's law and electrical resistance. When we described Ohm's law, we took the viewpoint that the applied voltage forced a current through the conductor. The conductance gave a measure of how much current was forced through. You can also take the viewpoint that the conductor resists the current flow by virtue of the collisions of the electrons and ions with the stationary material of the conductor. According to this viewpoint, the flow of an electric current in a conductor requires the application of a voltage that is proportional to the current. The ratio of voltage to current is called the *resistance* of the conductor. Ohm's law can be written in a form expressing this view (Eq. 12.6). Since the two

diameter of wire (cm)

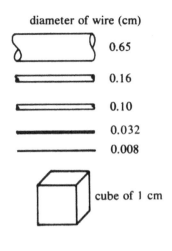

0.65

0.16

0.10

0.032

0.008

cube of 1 cm

TABLE 12·2 CONDUCTANCE AND RESISTANCE

A. Wire, 1 meter long, circular cross section

Material	Diameter (cm)	Conductance (amp/volt)	Resistance (volts/amp)
copper	0.65	2000	5×10^{-4}
	0.16	120	8×10^{-3}
	0.10	48	2.1×10^{-2}
	0.008	0.30	3.3
aluminum	0.16	73	1.4×10^{-2}
iron	0.16	21	4.8×10^{-2}
tungsten			
(20° Celsius)	0.008	0.09	11
(3000° Celsius)	0.008	4.5×10^{-3}	220
nichrome	0.16	2.1	0.48
	0.032	0.083	12.1

B. Cube-shaped piece of material, 1-centimeter-long edges

Material	Conductance (amp/volt)	Resistance (volts/amp)
copper	6×10^5	1.7×10^{-6}
iron	1×10^5	1×10^{-5}
nichrome	1×10^4	1×10^{-4}
semiconductor	$10^{-3} - 10^3$	$10^3 - 10^{-3}$
insulator (glass)	$10^{-14} - 10^{-10}$	$10^{14} - 10^{10}$

Equation 12.6 (Ohm's Law, in terms of resistance)

electrical resistance

(volt/amp) $= \mathcal{R}$

voltage (volts) $= \mathcal{C}$

current (amps) $= \mathcal{J}$

$$\mathcal{V} = \mathcal{R}\mathcal{J}$$

Equation 12.7

resistance (amp/volt) $= \mathcal{R}$

conductance

(volt/amp) $= \mathcal{C}$

$$\mathcal{R} = 1/\mathcal{C}$$

$$\mathcal{C} = 1/\mathcal{R}$$

equations state the same physical law (direct proportionality of current and voltage) in different forms, the conductance and the resistance of a conductor represent the same physical properties and can be directly calculated from one another (Eq. 12.7, Table 12.2, and Example 12.2). You might ask why we define two separate quantities that are so closely related. The reason is that they are each directly applicable to analyzing different types of electric circuits, which we will explain below. A circuit element described by Ohm's law is often called a *resistor*.

Voltmeters. We have so far not described any instruments for measuring voltage. When we discussed battery voltage, we did outline a procedure for measuring voltage by first determining thermal energy release (Example 12.1), but it is cumbersome and not useful for routine applications. After the discovery of Ohm's law, it became possible to calculate the voltage applied to a conductor from a measurement of the electric current with an ammeter (Fig. 12.14). The combination of an ammeter and a resistor with known resistance, therefore, serves as a voltmeter. The voltmeter dial must be calibrated once by using Joule's

326

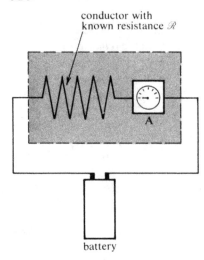

conductor with
known resistance $\mathscr{R}$

A

battery

Figure 12.14 A conductor with a known large resistance is combined with an ammeter to serve as a voltmeter.

EXAMPLE: The resistance is 10^4 volts per ampere. The ammeter dial reads 4.5×10^{-4} ampere. What is the voltage? 10^4 volts/amp x 4.5×10^{-4} amp $= 4.5$ volts

procedure (Example 12.1), but thereafter it can be used simply and efficiently for voltage measurements.

EXAMPLE 12.2

(a) Find the resistance of the circuit element described in Fig. 12.13a. For a current of 10 amperes, the voltage is 17 volts.

$\mathscr{I} = 10$ amp, $\mathscr{V} = 17$ volts

$$\mathscr{R} = \frac{\mathscr{V}}{\mathscr{I}} = \frac{17 \, volts}{10 \, amp} = 1.7 \frac{volts}{amp}; \text{ (or } \mathscr{C} = \frac{1}{1.7} = 0.6 \text{ amp/volt, as in Fig 12.13a)}$$

(b) A circuit element with a resistance of 5 volts per ampere is designed to draw an electric current of 0.6 ampere. What can you conclude? You can find the conductance (Eq. 12.7), voltage (Eq. 12.6), and energy transfer (Eq. 12.2).

Circuit elements that do not follow Ohm's Law (non-Ohmic elements). Modern electronic devices, such as vacuum tubes, diodes, transistors, and television picture tubes, are complex electric circuit elements that are not described by Ohm's law. The relationships of current and voltage are much more complicated. Diodes, for instance, are circuit elements that permit a large current to flow in one direction, but only a very small current in the opposite direction (Fig. 12.13b). In other words, they act somewhat as a one-way valve (Fig. 12.15). Vacuum tubes and transistors have more than two connections with an electric circuit. These devices can serve as "amplifiers" in that a small current flowing between two of the connections triggers a large current between two other connections.

Superconductors. Another class of circuit elements that are not described by Ohm's law is formed when certain metallic conductors are cooled to extremely low temperatures (-270° Celsius). They appear to

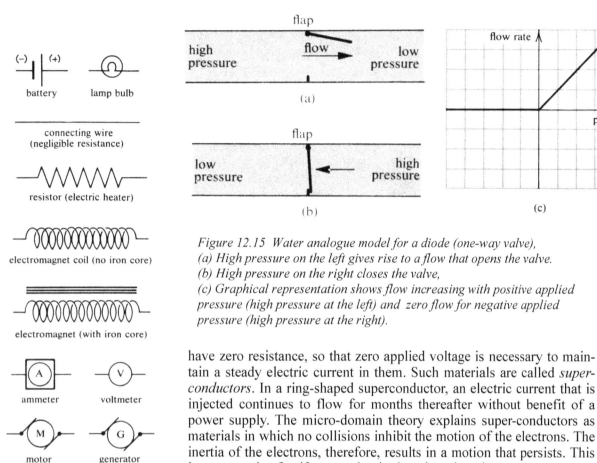

battery   lamp bulb

connecting wire
(negligible resistance)

resistor (electric heater)

electromagnet coil (no iron core)

electromagnet (with iron core)

ammeter   voltmeter

motor   generator

junction point   skip over
(wires joined) (wires not in contact)

direct current   alternating current
power supply   power supply
(Section 12·4)   (Section 12·4)

Symbols for electric circuit elements, power supplies, and connections. The connecting wires are usually drawn in rectangular shapes, as much as possible along vertical and horizontal lines.

Figure 12.15 *Water analogue model for a diode (one-way valve),*
(a) High pressure on the left gives rise to a flow that opens the valve.
(b) High pressure on the right closes the valve,
(c) Graphical representation shows flow increasing with positive applied pressure (high pressure at the left) and zero flow for negative applied pressure (high pressure at the right).

have zero resistance, so that zero applied voltage is necessary to maintain a steady electric current in them. Such materials are called *superconductors*. In a ring-shaped superconductor, an electric current that is injected continues to flow for months thereafter without benefit of a power supply. The micro-domain theory explains super-conductors as materials in which no collisions inhibit the motion of the electrons. The inertia of the electrons, therefore, results in a motion that persists. This is an example of uniform motion in the micro domain.

12.3 Analysis of simple circuits

The voltage and current concepts can be used to analyze electric circuits that include many circuit elements. To enable you to make diagrams of circuits, we have prepared a list of the generally accepted symbols that represent circuit elements.

Series and parallel connections. Two important principles of circuit analysis are derived from two especially simple types of circuit connections. In one, several circuit elements are connected end to end (Fig. 12.16). They are said to be connected *in series*. In the other one, several circuit elements are connected at two common terminals (Fig. 12.17). They are then said to be connected *in parallel*.

Series connection. The significance of the series connection is that all the circuit elements pass the same current, since there is no branch where part of the current could be diverted. The voltage applied to the entire system, however, can be divided into parts that are applied to each circuit element, in proportion to the energy transfer to each of them (Fig. 12.18). You should therefore think of a single current and partial voltages. For each resistor (circuit element that is described by

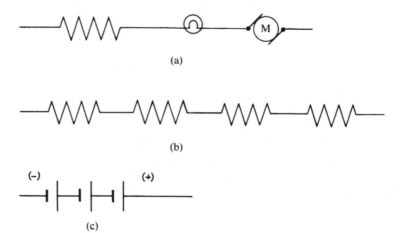

(a)

(b)

(−)   (+)

(c)

Figure 12.16 Electric circuit elements connected in series,
(a) Resistor, lamp bulb, motor, (b) Four resistors, (c) Three batteries.

Equation 12·8

current through series connection	$\mathscr{I}$
partial voltages	$\mathscr{V}_1, \mathscr{V}_2, \ldots$
resistances of resistors	$\mathscr{R}_1, \mathscr{R}_2, \ldots$

$$\mathscr{V}_1 = \mathscr{R}_1 \mathscr{I}$$
$$\mathscr{V}_2 = \mathscr{R}_2 \mathscr{I}$$
$$\cdot$$
$$\cdot$$
$$\cdot$$

Equation 12·9

voltage applied to circuit	$\mathscr{V}$
resistance of series connection	$\mathscr{R}$

$$\begin{aligned} \mathscr{V} &= \mathscr{V}_1 + \mathscr{V}_2 + \cdots \\ &= \mathscr{R}_1 \mathscr{I} + \mathscr{R}_2 \mathscr{I} + \cdots \\ &= (\mathscr{R}_1 + \mathscr{R}_2 + \cdots) \mathscr{I} \quad \text{(a)} \\ &= \mathscr{R}\mathscr{I} \end{aligned}$$

$$\mathscr{R} = \mathscr{R}_1 + \mathscr{R}_2 + \cdots \quad \text{(b)}$$

Equation 12·10

voltage applied to parallel connection	$\mathscr{V}$
partial currents	$\mathscr{I}_1, \mathscr{I}_2, \cdots$
conductances of resistors	$\mathscr{C}_1, \mathscr{C}_2, \cdots$

$$\mathscr{I}_1 = \mathscr{C}_1 \mathscr{V}$$
$$\mathscr{I}_2 = \mathscr{C}_2 \mathscr{V}$$

Ohm's law), the partial voltage is equal to the resistance times the current (Eq. 12.8). If Ohm's Law describes all the circuit elements connected in series, you can add the formulas for the partial voltages and arrive at Ohm's law for the entire circuit (Eq. 12.9a). You can see that the resistance of the entire circuit is equal to the sum of the resistances of all the circuit elements (Eq. 12.9b).

Parallel connection. The significance of the parallel connection is that the same voltage is applied to all the circuit elements, because they all are placed between the same two terminals. The current in the entire circuit can be divided into partial currents that flow through the several circuit elements (Fig. 12.19). You should therefore think of a single voltage and partial currents. For each circuit element that is described by Ohm's law, the partial current is equal to the conductance times the voltage (Eq. 12.10). If Ohm's Law describes all the circuit elements connected in parallel you can add the formulas for the partial currents

Figure 12.17 Electric circuit elements connected in parallel,
(a) Resistor, lamp bulb, motor, (b) Four resistors, (c) Three batteries.

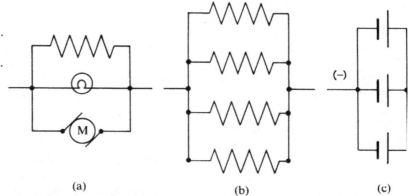

(a)   (b)   (c)

Figure 12.18 Current and voltage for a series connection.

(a) The current through the three circuit elements is measured by the ammeter connected in series. The __same current__ must flow in all elements.

(b), (c), and (d) The partial voltage applied to each circuit element is measured by the voltmeter connected in parallel with that circuit element. Thus the total voltage is equal to the sum of the three individual voltages.

If all circuit elements follow Ohm's Law, Eq. 12.9 applies, and the total resistance is equal to the sum of the three individual resistances. However, in this circuit the motor does __not__ follow Ohm's Law.

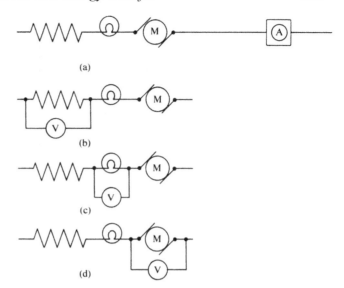

Equation 12·11

current through entire
  circuit $\mathscr{I}$
conductance of parallel
  connection $\mathscr{C}$

$$\mathscr{I} = \mathscr{I}_1 + \mathscr{I}_2 + \cdots$$
$$= \mathscr{C}_1 \mathscr{V} + \mathscr{C}_2 \mathscr{V} + \cdots$$
$$= (\mathscr{C}_1 + \mathscr{C}_2 + \cdots) \mathscr{V} \quad \text{(a)}$$
$$= \mathscr{C}\mathscr{V}$$

$$\mathscr{C} = \mathscr{C}_1 + \mathscr{C}_2 + \cdots \quad \text{(b)}$$

and arrive at Ohm's law for the entire circuit (Eq. 12.11a). You can see that for a parallel circuit, the conductance of the entire circuit is equal to the sum of the conductances of all the circuit elements (Eq. 12.11b).

Applications. Most real circuits that include a power supply and one or more circuit elements also include connecting wires whose resistance is very low compared to that of the circuit elements.

Figure 12.19 Voltage and current for a parallel connection,
(a) The voltage applied to the circuit is measured by a voltmeter in parallel with the circuit. Therefore, the voltage is the same for all circuit elements.
(b) The partial current in each circuit element is measured by an ammeter in series with that circuit element. The total current is the sum of the currents flowing through the individual circuit elements.

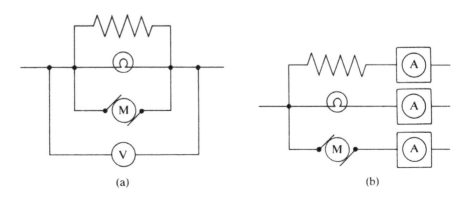

Since the wires are connected in series with the circuit element, the resistance of the circuit including the wires is only very slightly higher than that of the circuit element alone. It is customary, therefore, to make mathematical models in which the connecting wires are assigned zero resistance. When this is done, the partial voltage applied to a connecting wire and the energy transfer to it are zero (Eq. 12.8).

Since all the circuit elements in a series connection pass the same current, they all operate (or do not operate) simultaneously. Some Christmas tree light strings use a series connection. This is inconvenient in most domestic and industrial applications, however, where appliances or pieces of machinery are used individually. In these applications, therefore, all circuit elements are connected in parallel. Then each one operates under the same applied voltage and each one draws a partial current according to its requirements. The entire residence or factory, of course, draws a current that is the sum of all partial currents (Eq. 12.11a).

The series and parallel concepts can also be used to extend the entries for the conductances and resistances of pieces of wire in Table 12.2. Just think of a very long piece of wire as a certain number of 1-meter long segments in a series connection. Its resistance is therefore directly proportional to its length (Example 12.3). Similarly, a thick wire can be imagined as a bundle of thin wires (each of the same length) all connected in parallel. Thus, the conductance of a wire is directly proportional to its cross-sectional area (Example 12.4).

EXAMPLE 12.3. Using the value of resistance from Table 12.2, find the resistance of a 15-meter-long coil of 0.6 centimeter-diameter iron wire.

Solution: Think of the wire as 15 pieces, each 1 meter long, in a series connection.

$$\mathscr{R}_1 = \mathscr{R}_2 = \mathscr{R}_3 = \cdots = \mathscr{R}_{15} = 4.8 \times 10^{-2} \frac{\text{volt}}{\text{amp}}$$

By Eq. 12·9b,

$$\mathscr{R} = \mathscr{R}_1 + \mathscr{R}_2 + \cdots + \mathscr{R}_{15} = 15\mathscr{R}_1 = 15 \times 4.8 \times 10^{-2} \frac{\text{volt}}{\text{amp}}$$

$$= 0.72 \frac{\text{volt}}{\text{amp}}$$

EXAMPLE 12.4. Compare the conductance of two 1-meter-long pieces of nichrome wire that have diameters of 0.16 centimeter and 0.032 centimeter.

Solution: The diameters of the two wires are in the ratio of 0.16:0.032= 5:1. Since the area of a circle varies as the second power of the radius, the areas are in the ratio of 5^2:1 or 25:1. Thus, the thick wire

Equation 12·12

energy transfer (joules)	ΔE
voltage (volts)	$\mathcal{V}$
current (amp)	$\mathcal{I}$
power (watts)	$\mathcal{P}$
time interval (sec)	Δt

$$\mathcal{P} = \frac{\Delta E}{\Delta t} = \mathcal{V} \cdot \mathcal{I}$$

Units of power:

$$1 \ watt = 1 \ joule/sec$$
$$= 1 \ volt \times amp$$
$$1 \ kilowatt = 1000 \ watts$$
$$1 \ horsepower = 750 \ watts$$

EXAMPLE

Electric toaster: $\mathcal{V} = 110 \ volts,$

$\mathcal{I} = 8.2 \ amp$

$\mathcal{P} = \mathcal{V} \cdot \mathcal{I}$

$= 110 \ volts \times 8.2 \ amp$

$= 900 \ watts$

Equation 12·13

energy of 1 *kilowatt-hour (kWh)*

$\mathcal{P} = 1 \ kilowatt = 1000 \ watts$

$\Delta t = 1 \ hr = 3600 \ sec$

$\Delta E = \mathcal{P} \Delta t$

$= 1 \ killowatt \times 1 \ hr$

$= 1 \ kWh$

$= 1000 \ watts \times 3600 \ sec$

$= 3.6 \times 10^6 \ joules$

$\approx \dfrac{3.6 \times 10^6 \ joules}{4.0 \times 10^3 \ joules/Cal}$

$= 900 \ Cal$

EXAMPLE

Express the energy consumed by a 150-watt lamp bulb, operating for 10 hr, in various units.

$\mathcal{P} = 150 \ watts = 0.15 \ kilowatt$

$\Delta t = 10 \ hr$

$\Delta E = \mathcal{P} \Delta t$

$= 0.15 \ kilowatts \times 10 \ hr$

$= 1.5 \ kWh$

$= 5.4 \times 10^6 \ joules$

$\approx 1300 \ Cal$

contains the same amount of nichrome as a bundle of 25 pieces of the thin wire, and we can consider the thick wire as equivalent to a bundle of 25 thin wires connected in parallel. This model leads to the prediction that the thick wire has 25 times the conductance of the thin wire, $\mathcal{C} = 25 \ \mathcal{C}_t$. We can get the conductances from Table 12.2:

$$\mathcal{C} = 2.1 \ amp/volt, \ \mathcal{C}_1 = 0.083 \ amp/volt$$

$$25\mathcal{C}_1 = 25 \times 0.083 \ \frac{amp}{volt} = 2.1 \ \frac{amp}{volt}$$

12.4 Energy transfer

We have repeatedly pointed out that the electric current in a circuit acts as a passive coupling element that transfers energy from the electric power supply (battery, generator) to an energy receiver (lamp, motor). The quantity of energy transfer in a time interval is equal to the voltage times the current times the duration of the time interval (Eq. 12.2 used to define voltage). This relationship may be applied to the electric power supply to find the total energy transfer to the circuit, or it may be applied to each circuit element separately, to determine the energy transfer to that particular circuit element.

Electric power. The energy transfer in an electric circuit is proportional to the time of operation (Eq. 12.2). When an electric iron is operated for a whole hour, two times as much heat is produced as during half an hour. The significant quantity here is the rate of energy transfer ($\Delta E/\Delta t$), which is called the *electric power* (Eq. 12.12). The electric power, equal to the voltage times the current, is measured in *watts*. An appliance operating at a level of 1 watt for 1 second receives 1 joule of energy.

Most home appliances are rated in watts. Light bulbs range in power from 2.5 watts for a dim night-light to 100 watts for a good reading lamp to 500 watts for a brilliant floodlight. Electric toasters and electric irons usually are rated around 1 kilowatt. Motor-driven appliances such as washing machines and vacuum cleaners usually are rated at a few hundred watts.

Since the joule is a very small unit of energy, the much larger kilowatt-hour is used to describe the consumption of electric power. One *kilowatt-hour* is the amount of energy transferred when a 1-kilowatt appliance (such as a heater) operates for 1 hour. A kilowatt-hour is more than 3.5 million joules, or almost 1000 Calories (Eq. 12.13).

Direct and alternating current. In Section 12.1 we described electric current as a circulation of an "electric fluid" analogous to the circulation of water in pipes. The water in the analogue model is an intermediary in the transfer of energy. One way to accomplish energy transfer is shown in Fig. 12.20a, where a pump circulates the

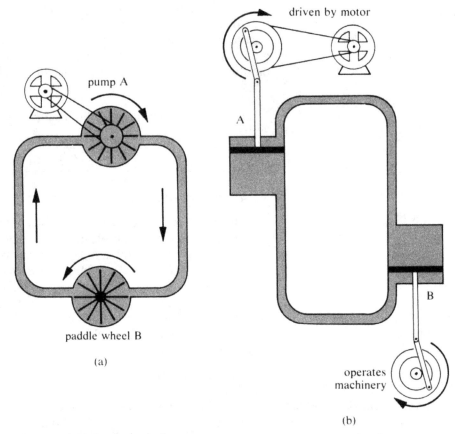

Figure 12.20 Energy transfer in a water pipe circuit,
(a) Circulating water transfers energy from pump A to paddle wheel B.
(b) Oscillating water transfers energy from piston A to piston B.

water and the water drives a paddle wheel. Another way to achieve energy transfer is shown in Fig. 12.20b, where one piston (driven by a motor) pushes the water in the pipes back and forth, and the water drives a second piston, which turns a wheel that can operate machinery. In the second system, the water does not actually circulate through the pipes, but moves alternately a short distance back and forth.

In an electric circuit, energy can be transferred in either of these two ways. In the first, the electric power supply establishes an electric field in one direction around the circuit and the current flows in that direction continuously; this is called *direct current (DC)*. Alternatively, the power supply establishes an electric field first in one direction and then in the other. The current, therefore, also alternates and is called *alternating current* (AC).

A chemical battery ordinarily gives rise to DC, because the type of metal in the battery electrodes determines the direction of the electric field. By an ingenious arrangement of switches, however, the battery connection can be reversed in a regular fashion, so as to produce AC (Fig. 12.21).

Electric generators can also produce either AC or DC, again depending on a suitable arrangement of the connections between the generator and the electric circuit. The important technological advantage of AC over DC is that the voltage and current of AC can be changed without appreciable energy loss (using a circuit element called a *transformer*). This cannot be done so simply with direct current. Electric power is most efficiently transferred over long distances (from generator to distribution) at high voltage and low current. Transformers then convert the AC to lower (and safer) voltages for residential and industrial use (usually 110 or 220 volts in the US). For this reason, alternating current is used in most industrial and residential applications. In the US, the alternating current goes through a complete cycle of flowing back and forth 60 times each second; it is therefore called 60-cycle alternating current.

Applications of alternating current. Motor-driven appliances can be designed for operation on either direct or alternating current. By referring to Fig. 12.20, you can readily see the need for matching the appliances correctly with the type of current. If the paddle wheel were to be subjected to "alternating" water flow, it would never advance significantly in either direction. Similarly, the piston would block a "direct" water flow once it had traveled to the end of its cylinder.

Appliances that produce thermal energy (lamp bulbs, heaters, electric irons, and so on) can be operated equally well on direct or alternating current. The reason for this is most easily recognized from the micro-domain model for electric current.

In Section 12.1, we pointed out that the micro-domain electric charges (electrons and ions) that comprise the "electric fluid" move at constant

Figure 12.21 A battery and a reversing switch produce alternating current,
(a) A reversing switch,
(b) The switch is closed in one direction.
(c) The switch is closed in the other direction.

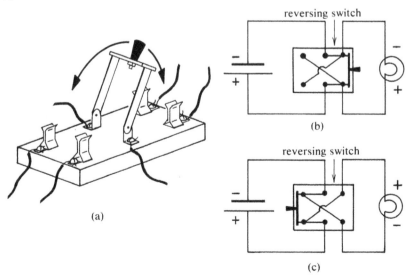

speed. They are in mechanical equilibrium subject to two interactions: the electric field established by the power supply and the effect of their collisions with the particles making up the conductor in the MIP model for matter As a result of the collisions, the conductor particles gain kinetic energy, which manifests itself as thermal energy in the macro domain (Section 10.5).

The 60-cycle alternating current switches back and forth very slowly compared to the time required for the motion of the "electric fluid" to adjust to changes in conditions. Even though the electric field alternates and the fluid flow alternates, the mechanical equilibrium description of the fluid is approximately valid. The only significant difference from direct current is that the collisions with the conductor particles take place sometimes from one side, sometimes from the other. But this variation does not affect the transfer of energy and therefore the thermal energy produced in a heating element.

In spite of the difference between alternating and direct current, the mathematical models for energy transfer (Eq. 12.2) and electric power (Eq. 12.12) can still be used. Ohm's law (Eqs. 12.5 and 12.6) also applies to many conductors of alternating current (Example 12.5).

EXAMPLE 12.5. An electric toaster is to be designed with nichrome wire heating elements. Four heating elements are needed, one for each side of two slices of bread. The toaster will operate on 110-volt alternating current. About 1 kilowatt of power can be used.

(a) Find the electric current for 1 kilowatt.

(b) Find the resistance and conductance of the toaster.

(c) Design the toaster with four equal heating elements in parallel, so that it will operate partly, even though one element is broken.

(d) For comparison, design the toaster with four elements in series.

Solution:

(a) $\mathscr{P} = 1000$ watts, $\mathscr{V} = 110$ volts

$$\mathscr{I} = \frac{1000 \text{ watts}}{110 \text{ volts}} = 9.1 \text{ amp}$$

(b) $\mathscr{I} = \mathscr{C}\mathscr{V}$

$$\mathscr{C} = \frac{\mathscr{I}}{\mathscr{V}} = \frac{9.1 \text{ amp}}{110 \text{ volts}} = 0.083 \frac{\text{amp}}{\text{volt}}$$

$$\mathscr{R} = \frac{1}{\mathscr{C}} = \frac{1}{0.083 \text{ amp/volt}} = 12 \frac{\text{volts}}{\text{amp}}$$

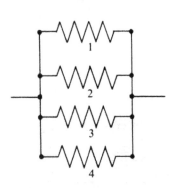

(c) Make the four elements equal, $\mathscr{C}_1 = \mathscr{C}_2 = \mathscr{C}_3 = \mathscr{C}_4$. For parallel connection, $\mathscr{C} = \mathscr{C}_1 + \mathscr{C}_2 + \mathscr{C}_3 + \mathscr{C}_4 = 4\mathscr{C}_1$.

$$\mathscr{C}_1 = \frac{1}{4}\mathscr{C} = \frac{1}{4} \times 0.083 \frac{\text{amp}}{\text{volt}} = 0.021 \frac{\text{amp}}{\text{volt}}$$

$$\mathscr{R}_1 = \frac{1}{\mathscr{C}_1} = \frac{1}{0.021\,amp/volt} = 48\,volts/amp$$

Consult Table 12.2 for properties of nichrome wire. One meter of 0.032-centimeter-diameter wire has a resistance of 12 volts per ampere. One 4-meter-long piece of wire therefore has 48 volts per ampere resistance. This is what is needed.

Each heating element is made of 4 meters of nichrome wire, 0.032 centimeter in diameter. The 4 meters of wire can be wound back and forth to radiate energy to one side of a slice of bread. The entire toaster (four elements) requires 16 meters of wire.

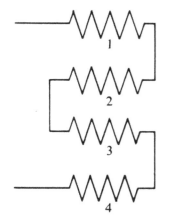

(d) Make the four elements equal, $\mathscr{R}_1 = \mathscr{R}_2 = \mathscr{R}_3 = \mathscr{R}_4$. For a series connection (shown at left), $\mathscr{R} = \mathscr{R}_1 + \mathscr{R}_2 + \mathscr{R}_3 + \mathscr{R}_4 = 4\mathscr{R}_1$.

$$\mathscr{R}_1 = \frac{1}{4}\mathscr{R} = \frac{1}{4}\times 12\frac{volts}{amp} = 3\frac{volts}{amp}$$

Consult Table 12.2 for properties of nichrome wire. One meter of 0.32-centimeter-diameter wire has a resistance of 12 volts per ampere, just what is needed for the entire toaster. Each of the four heating elements must, therefore, be made from ¼ meter or 25 centimeters (10 inches), which has a resistance of 3 volts per ampere. This is a very short piece of wire to use for radiating energy to a slice of bread. It would be better to use a much longer piece of the thicker nichrome wire with 0.16-centimeter diameter.

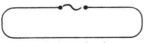

A short circuit

A "short circuit" (shown above) occurs when the power supply transfers energy solely to a conducting wire of low resistance, rather than to a circuit element. This results in a very high current, a sudden, large energy transfer, and a very hot, melted, or vaporized, wire. (See Example 12.6.) For obvious reasons, short circuits can be dangerous and should be avoided.

The "short circuit." In the description of series and parallel connections of circuit elements (Section 12.3), we pointed out that the resistance of the connecting wires is negligibly small. This statement is not true when the circuit is closed with no circuit element other than a conducting wire. Then the connecting wire is the energy receiver in the circuit and serves as a heating element. Its resistance can no longer be neglected, because it is the entire resistance of the circuit. Enormous power is delivered to the circuit under these conditions (Example 12.6). Such a circuit, which lacks a useful energy-receiving circuit element, is called a *short circuit*.

EXAMPLE 12.6. A "short circuit." A 6-foot-long extension cord is plugged into a wall outlet (110 volts) and develops a short circuit because the metal wires touch at a place of worn insulation halfway between its ends. Estimate the consequences if no fuse or circuit breaker is in the circuit. Extension cords contain a bundle of copper strands whose combined diameter is 0.10 centimeter and whose mass is 7 grams per meter.

(a) Calculate the conductance of the wire.
(b) Calculate the current in the wire.

(c) Calculate the power delivered.

(d) Calculate the thermal energy produced in the wire in 1 second and the resulting temperature rise of the wire in 1 second. (Recall Eq. 9.3, 1 Cal = 4 x 10^3 joules; Eq. 12.13, $\Delta E = \mathscr{P}\Delta t$; Equation 10.2, $\Delta E = CM_G\Delta T$; and from Table 10.4, specific heat of copper, C = 0.093 Cal/deg C/kg).

Solution:

(a)   The length of copper wire involved in the short circuit is about 6 feet or 2 meters. The resistance (Table 12.2 and Example 12.3) is

$$\mathscr{R} = 2 \times 2.1 \times 10^{-2}\,\frac{\text{volt}}{\text{amp}} = 4.2 \times 10^{-2}\,\frac{\text{volt}}{\text{amp}}$$

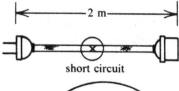

— 2 m —

short circuit

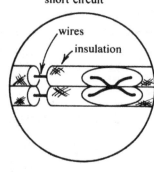

wires

insulation

The conductance is

$$\mathscr{C} = \frac{1}{4.2 \times 10^{-2}\,\text{volt/amp}} \approx 24\,\frac{\text{amp}}{\text{volt}}$$

(b) $\mathscr{I} = \mathscr{C}\mathscr{V} = 24\,\dfrac{\text{amp}}{\text{volt}} \times 110\ \text{volts} \approx 2.6 \times 10^3\ \text{amp}$

(c) $\mathscr{P} = \mathscr{V}\mathscr{I} \approx 1.1 \times 10^2\ \text{volts} \times 2.6 \times 10^3\ \text{amp} \approx 2.9 \times 10^5\ \text{watt}$

(d) Find the thermal energy produced in 1 second.

$$\Delta E = \mathscr{P}\Delta t \approx 2.9 \times 10^5\ \text{watts} \times 1\ \text{sec} = 2.9 \times 10^5\ \text{joules}$$

$$\Delta E \approx \frac{2.9 \times 10^5\ \text{joules}}{4 \times 10^3\ \text{joules/Cal}} \approx 70\ \text{Cal}$$

Find the temperature rise.

$$\Delta E = CM_G\Delta T,\ \Delta T = \frac{\Delta E}{CM_G} \approx \frac{70\ \text{Cal}}{(0.093\ \text{Cal/deg/kg}) \times (0.014\ \text{kg})}$$

$$= \frac{70}{1.3 \times 10^{-3}} = 5.4 \times 10^4\ ^\circ\ \text{Celsius}$$

The predicted temperature rise is 54,000° Celsius. This absurd result indicates that the copper will melt and possibly vaporize after a fraction of a second if no other interaction occurs. Fuses (described below) are included in circuits to prevent this.

As you can see from Example 12.6, a short circuit can be dangerous. To reduce this hazard, current-limiting circuit breakers or fuses are connected in series in electric power circuits. The current may become excessive because of a short circuit or because too many appliances are connected in parallel and operated at the same time. A fuse includes a piece of wire that melts and thus opens the circuit when its temperature increases because too large a current flows in it. A circuit breaker includes a switch that opens the circuit when the current is too large. The switch is operated by either the magnetic field or the heating effects that accompany electric currents.

Summary

The flow of an electric current in an electric circuit achieves energy transfer from an electric power supply to an energy receiver. The power supply may consume chemical energy (as in a battery or an oil- or gas-fired generating plant); it may consume gravitational field energy (as in a hydroelectric installation); it may consume nuclear energy (as in a nuclear reactor power station); or it may consume kinetic energy of air (as in a windmill-driven generator). The energy receiver may be a light bulb, a heating element, an electric motor, or an electrolysis apparatus. Even though we have referred to these systems as energy receivers, some of them are in reality only coupling elements that transfer the energy to still other systems. Thus, the light bulb produces thermal energy and radiation in the space surrounding it; the motor produces kinetic energy or does work against friction; and so on. Only in electrolysis is there really a steady accumulation of chemical energy in the reaction products, in proportion to the energy consumed at the electric power supply.

Two variable factors are particularly useful to describe the operation of an electric circuit: the current and the voltage. In an analogue model where the circuit is compared to a system of water pipes, the current corresponds to the rate of flow of the water, the voltage corresponds to the water pressure. The electric power, which is the rate of energy transfer by the circuit ($\Delta E/\Delta t$), is equal to the current times the voltage (Eq. 12.12). In many circuit elements the electric current is directly proportional to the applied voltage. The mathematical model describing this relation is called Ohm's Law, which applies to both series and parallel circuits. Ohm's Law in terms of conductance (Eq. 12.5) is most convenient for solving parallel circuits; Ohm's Law in terms of resistance (Eq. 12.6) is most convenient for solving series circuits.

Equation 12.12

$$\begin{aligned} \text{energy transfer} & \\ \text{(joules)} &= \Delta E \\ \text{voltage (volts)} &= \mathcal{V} \\ \text{current (amps)} &= \mathcal{I} \\ \text{power (watts)} &= \mathcal{P} \\ \text{time interval} & \\ \text{(sec)} &= \Delta t \end{aligned}$$

$$\begin{aligned} \mathcal{P} &= \Delta E/\Delta t \\ &= \mathcal{V}\mathcal{I} \end{aligned}$$

Equation 12.5 (Ohm's Law in terms of conductance)

$$\text{conductance (amp/volt)} = \mathcal{C}$$

$$\mathcal{I} = \mathcal{C}\mathcal{V}$$

Equation 12.6 (Ohm's Law in terms of resistance)

$$\begin{aligned} \text{electrical resistance} & \\ \text{(volt/amp)} &= \mathcal{R} \end{aligned}$$

$$\mathcal{V} = \mathcal{R}\mathcal{I}$$

Additional examples

EXAMPLE 12.7. A series circuit.

Two resistors, with resistances of 6 volts per ampere and 8 volts per ampere, are connected in series to a 10-volt power supply. Find the following items: (a) resistance and conductance of circuit; (b) current in circuit; (c) partial voltage applied to each resistor; (d) power delivered by power supply; (e) power delivered to each resistor.

Solution:

(a) $\mathcal{R} = \mathcal{R}_1 + \mathcal{R}_2 = 6\ \dfrac{\text{volts}}{\text{amp}} + 8\ \dfrac{\text{volts}}{\text{amp}} = 14\ \dfrac{\text{volts}}{\text{amp}}$

$\mathcal{C} = \dfrac{1}{\mathcal{R}} = \dfrac{1}{14\ \text{volts/amp}} \approx 0.071\ \dfrac{\text{amp}}{\text{volt}}$

(b) $\mathcal{I} = \mathcal{C}\mathcal{V} \approx 0.071\ \dfrac{\text{amp}}{\text{volt}} \times 10\ \text{volts} = 0.71\ \text{amp}$

(c) $\mathcal{V}_1 = \mathcal{R}_1\mathcal{I} \approx 6\ \dfrac{\text{volts}}{\text{amp}} \times 0.71\ \text{amp} \approx 4.3\ \text{volts}$

$\mathcal{V}_2 = \mathcal{R}_2\mathcal{I} \approx 8\ \dfrac{\text{volts}}{\text{amp}} \times 0.71\ \text{amp} \approx 5.7\ \text{volts}$

Check: $\mathcal{V} = \mathcal{V}_1 + \mathcal{V}_2 \approx 4.3\ \text{volts} + 5.7\ \text{volt} = 10\ \text{volts}$

(d) $\mathcal{P} = \mathcal{V}\mathcal{I} \approx 10\ \text{volts} \times 0.71\ \text{amp} = 7.1\ \text{watts}$

(e) $\mathcal{P}_1 = \mathcal{V}_1\mathcal{I} \approx 4.3\ \text{volts} \times 0.71\ \text{amp} \approx 3.1\ \text{watts}$

$\mathcal{P}_2 = \mathcal{V}_2\mathcal{I} \approx 5.7\ \text{volts} \times 0.71\ \text{amp} \approx 4.0\ \text{watts}$

Check: $\mathcal{P} = \mathcal{P}_1 + \mathcal{P}_2 \approx 3.1\ \text{watts} + 4.0\ \text{watts} = 7.1\ \text{watts}$

EXAMPLE 12.8. A parallel circuit.

The two resistors of Example 12.7 are connected to the same power supply in parallel. Find the same items as in Example 12.7.

Solution:

(a) In a parallel connection, we can add the conductances of the resistors (Eq. 12.11b).

$$\mathcal{C}_1 = \frac{1}{\mathcal{R}} = \frac{1}{6\ \text{volts/amp}} \approx 0.17\ \frac{\text{amp}}{\text{volt}}$$

$$\mathcal{C}_2 = \frac{1}{\mathcal{R}_2} = \frac{1}{8\ \text{volts/amp}} \approx 0.13\ \frac{\text{amp}}{\text{volt}}$$

$$\mathcal{C} = \mathcal{C}_1 + \mathcal{C}_2 \approx 0.17\ \frac{\text{amp}}{\text{volt}} + 0.13\ \frac{\text{amp}}{\text{volt}} = 0.30\ \frac{\text{amp}}{\text{volt}}$$

$$\mathcal{R} = 1/\mathcal{C} = \frac{1}{0.30\ \text{amp/volt}} \approx 3.3\ \frac{\text{volts}}{\text{amp}}$$

(b) $\mathcal{I} = \mathcal{C}\mathcal{V} \approx 0.30\ \dfrac{\text{amp}}{\text{volt}} \times 10\ \text{volts} = 3.0\ \text{amp}$

(c) $\mathcal{I}_1 = \mathcal{C}_1\mathcal{V} \approx 0.17\ \dfrac{\text{amp}}{\text{volt}} \times 10\ \text{volts} = 1.7\ \text{amp}$

$\mathcal{C}_2 = \mathcal{I}_2\mathcal{V} \approx 0.13\ \dfrac{\text{amp}}{\text{volt}} \times 10\ \text{volts} = 1.3\ \text{amp}$

Check: $\mathcal{I} = \mathcal{I}_1 + \mathcal{I}_2 \approx 1.7\ \text{amp} + 1.3\ \text{amp} = 3.0\ \text{amp}$

(d) $\mathcal{P} = \mathcal{V}\mathcal{I} \approx 10\ \text{volts} \times 3.0\ \text{amp} = 30\ \text{watts}$

(e) $\mathcal{P}_1 = \mathcal{V}\mathcal{I}_1 \approx 10\ \text{volts} \times 1.7\ \text{amp} = 17\ \text{watts}$

$\mathcal{P}_2 = \mathcal{V}\mathcal{I}_2 \approx 10\ \text{volts} \times 1.3\ \text{amp} = 13\ \text{watts}$

Check: $\mathcal{P} = \mathcal{P}_1 + \mathcal{P}_2 \approx 17\ \text{watts} + 13\ \text{watts} = 30\ \text{watts}$

List of new terms

electrical conductor	voltage	electric power
battery	volt	watt
electrode	voltmeter	kilowatt
electric current	cell	kilowatt-hour
battery	Ohm's law	direct current (DC)
electric circuit	conductance	alternating current (AC)
circuit element	semiconductor	transformer
electric power supply	resistance	heating element
closed circuit	resistor	short circuit
open circuit	superconductor	fuse
ammeter	series connection	circuit breaker
ampere	parallel connection	
insulator		

List of symbols

Δq	electric charge transfer		$\mathcal{V}$	voltage
ΔE	energy transfer		$\mathcal{R}$	electrical resistance
Δt	time interval		$\mathcal{C}$	conductance
$\mathcal{I}$	electric current		$\mathcal{P}$	power

Problems

1. (a) Determine what types of energy sources (for example, coal, oil, natural gas, nuclear, falling water, wind, solar) are used by the electric utility serving your area.
   (b) Determine from what distance electric power is imported into your area in substantial amounts.
   (c) Determine to what distances electric power is exported from your area in substantial amounts.

2. Carry out library research to investigate the early history of electric currents and batteries as studied by Luigi Galvani and Alessandro Volta.

3. (a) Touch your tongue to the two electrodes of a fresh 9 volt battery (shown in Fig 12.12b) and describe your observations. Then do the same for a "worn-out" battery.
   (b) Drop a worn-out 9-volt rectangular (shown in Fig. 12.12b) battery into a glass of table salt dissolved in water and describe your observations. Use an ionic model to explain what you see.

4. Investigate and describe the closed circuit in an operating flashlight. Identify the action of the switch.

5. Extend Table 12.1 to include energy transfer, voltage, resistance, and Ohm's law.

6. Any phenomenon that includes circulation of something that is

conserved can be used as an analogue model for an electric current flowing in a circuit. Invent such a model of your own, or use one of these ideas: (i) a conveyor belt takes coal from a stockpile to a furnace; (ii) skiers "circulate" up and down a ski slope; (iii) traffic circulates through the streets in a city. In your example, set up the correspondence as it is done for the water pipe analogue in Table 12.1. Indicate where the model is inadequate.

7. Six ammeters in Fig. 12.7 are identified by numerals. Find as many relations as you can among the currents passing through these ammeters, and explain your reasons. (Example: ammeters 3 and 4 indicate equal currents, because there is no circuit branch between them where current could be diverted or added.)

8. Explain why the "negative ion fluid" and the "positive ion fluid" in a liquid conductor move in opposite directions (Fig. 12.9).

9. Identify shortcomings of the micro-domain model for electric current presented in Section 12.1.

10. Imagine a "battery" in which hydrogen and oxygen combine to form water. Such a battery is called a fuel cell. Calculate the voltage such a fuel cell would deliver. (Hint: Use the chemical energy of the fuel system given in Table 10.3, the definition of the faraday in Section 8.1, and the relation of electric charge and current in Eq. 12.1.)

11. Use the definition of voltage in Eq. 12.3 and the quantity of chemical energy change occurring in a chain of cells connected in series to derive Eq. 12.4 for the voltage of such a chain as compared to the voltage of one cell.

12. Use the definition of voltage in Eq. 12.3 and the quantity of chemical energy change occurring in cells connected in parallel (as in the diagram to the left) to relate the voltage of this battery to the voltage of one cell.

13. Make a diagram and describe the operation of water analogue models for the systems in Problems 11 and 12.

14. The standard dry cell (flashlight battery) has a voltage of 1.5 volts. Make a diagram to show how you would make a 9-volt transistor radio battery out of single dry cells.

15. (a) Find the voltages of the batteries diagramed in Fig. 12.22 (below) in terms of the single cell voltage.

(b) Describe the operation of water analogue models for the systems in Fig. 12.22.

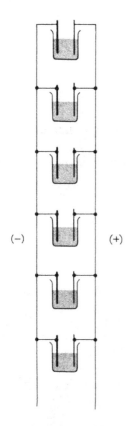

(−)　　　(+)

Diagram for problem 12.

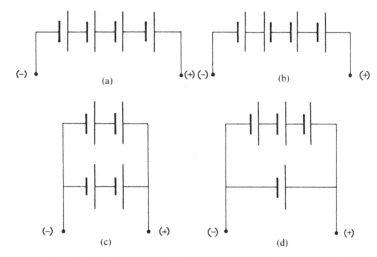

Figure 12.22 Each battery symbol represents a single standard dry cell of 1.5 volts (Problems 15, 16).

16. Find the total energy that can be delivered by each of the batteries in Fig. 12.22. Use the energy of one dry cell as the unit of energy. Explain your answers briefly.

17. A 12-volt automobile storage battery is advertised to have a capacity of 100 ampere-hours. Find the battery's energy when it is completely charged. Express the result in kilowatt-hours, joules, Calories, and ice-melting capacity (Section 9.2). The ampere-hour is a unit of electric charge; one amp-hour is the amount of charge transferred (Δq) by 1 ampere flowing for 1 hour.

18. Explain the "gap" in the current-voltage graph for a sodium chloride (table salt in water) solution (Fig. 12.13c).

19. Even though a diode is not described by Ohm's law, you can describe it approximately as having a high "forward conductance" (for positive voltage) and a low "backward conductance" (for negative voltage). Draw a graph for a circuit element that has two different conductances and estimate numerical values of these conductances for the diode in Fig. 12.13b.

20. The following data are observed for an electric circuit element.
    current: 1.8   3.0   4.6   6.2   8.6 (amp)
    voltage: 8.0   13.5   19.2   27   35 (volts)
    (a) Represent the data by a graph.
    (b) Describe the data by Ohm's law as well as you can. Find the resistance and conductance.
    (c) Evaluate the adequacy of the Ohm's law model for this circuit element.

21. (a) Explain why the resistor in a voltmeter should have a large resistance (Fig. 12.14).
(b) Define what is meant by "large resistance" in terms of the applications of the voltmeter.

22. An ammeter has a dial that is calibrated to indicate up to 5.0×10^{-3} ampere. The ammeter is to be combined with a resistor to construct a voltmeter.
(a) What is the resistance of the resistor if the dial is to indicate up to 5.0 volts?
(b) What is the resistance of the resistor if the dial is to indicate up to 150 volts?

23. Use the three water-pipe analogue models illustrated in Fig. 12.23 to carry out the following.
(a) Draw a qualitative graph of the relation of pressure and flow rate (as in Fig. 12.15) for each of models I, II and III.
(b) Describe the electric circuit element that might be represented by each water analogue, if there is one.
(c) Invent a water analogue model for a radio tube (Fig. 12.13d).

24. You are given four resistors, each with a resistance of 10 volts per ampere.
(a) Connect them in such a way (make a diagram) that the resulting circuit has the lowest possible resistance. What is the resistance?
(b) Connect them in such a way (make a diagram) that the resulting circuit has the largest possible resistance. What is the resistance?

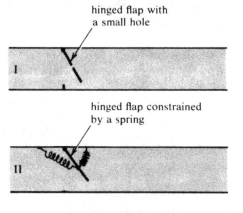

hinged flap with
a small hole

hinged flap constrained
by a spring

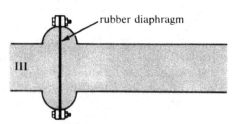

rubber diaphragm

Figure 12.23 Water pipe devices that may serve as analogue models for electric circuit elements (Problems 23, 37).

(c) Invent as many other different circuits as you can by connecting the four resistors, make a diagram for each circuit, and find its resistance.

25. Compare the conductances of the 0.16-centimeter-diameter and 0.008-centimeter-diameter copper wires (Table 12.2) in the light of the model that represents a thick wire as a bundle of thin wires with the same total cross-sectional area. (See Section 12.3, Ex. 12.4.)

26. Make a mathematical model that relates the resistance of a piece of wire to its length, cross-sectional area, and the resistance of a piece of the same material in a standard shape. (Note: the most commonly used "standard shape" is a cube with 1-centimeter-long edges, Table 12.2B.)

27. Electric toasters are usually made with nichrome ribbon heating elements. In one toaster, the ribbon was 0.085 centimeter wide and 0.010 centimeter thick. What is the diameter of a circular nichrome conductor with close to the same resistance for the same length?

Note for Problems 28-33 about light bulbs. The white-hot electric conductor in a light bulb is a piece of fine tungsten wire called the filament. (If you are curious, break open a light bulb—wrapped in a towel to confine glass splinters—and examine the filament with a magnifier. You may find that the wire is wrapped into a spiral, or helical, shape.) While the bulb operates, it is in a steady state (Section 4.6), with electric power being consumed at the same rate as radiant energy is emitted. Light bulb filaments are designed to operate near 3000° Celsius. You can assume that the electric power supply voltage in all problems is 110 volts.

28. (a) Find the electric current passing through a 60-watt bulb under normal operating conditions. Assume that 110 volts is applied to the bulb.
   (b) Find the resistance of the filament and calculate the length of tungsten wire (0.008-centimeter diameter) that must be used in the filament? (See Table 12.2 for tungsten.)
   (c) Find the electric current passing through a 60-watt light bulb when it is cold (at room temperature, just after it is turned on).
   (d) Find the power delivered to a 60-watt light bulb when it is cold.

29. Prepare a qualitative description of the process whereby a 60-watt light bulb arrives at a steady state very shortly after it is turned on.

30. About 2% of the electric power consumed by a light bulb becomes visible light. What happens to the rest?

31. To manufacture a light bulb, after the resistance and length of the tungsten filament wire for a light bulb has been determined (Problem 28, parts a and b), a suitable piece of wire must be selected and then twisted into a shape so it will attain its steady-state operating temperature near 3000° Celsius. Enumerate some of the variable factors that influence the operating temperature and indicate the direction of the influence.

32. Describe what will happen when two 60-watt light bulbs are connected in series and then connected to a 110-volt power supply. Explain.

33. Describe the relative brightness of the light bulbs in these connections: (a) a 200-watt bulb and a 60-watt bulb in parallel; (b) a 200-watt bulb and a 60-watt bulb in series; (c) four 60-watt bulbs connected as in the diagram to the left.

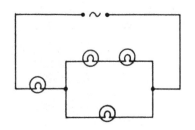

Diagram for Problem 33(a)

34. A student carries out his studies using a lamp with two 60-watt light bulbs. The lamp is turned on for 5 hours every evening. Assuming that the energy cost is 13 cents per kilowatt-hour, estimate the monthly electric bill incurred by his studiousness.

35. (a) Examine and report on the electric power rates of your local utility company.
(b) Compare the cost of the energy for maintaining one young adult for one year (10^6 Calories), when the energy is obtained from the following sources (see Table 10-6): (i) electric utility; (ii) gasoline; (iii) a diet consisting of 30% sirloin steak, 30% American cheese, 30% rye bread, and 10% sugar; (iv) your actual diet (that is, the approximate amount you spend per year on food).

36. (a) Conduct an experiment to estimate the energy used to toast two slices of bread.
(b) Estimate the energy transfer to the toast according to a model in which the only difference between toast and bread is that water has been removed by evaporation (see Table 10.5).

37. (a) Compare the operation of the water pipe device in Fig. 12.23, part III, when it is subject to "direct current" with its operation on "alternating current" (Fig. 12.20).
(b) Describe an electric circuit element for which this water pipe device is an analogue model.

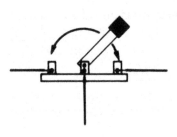

Diagram for problem of 38

38. (a) Devise a circuit whereby a staircase light can be turned off and on from two locations. (Hint: Use "double throw" switches, as in diagram to left.)
(b) Extend the procedure to the control of one light from three or more locations. If necessary, invent switches that will help solve the problem.

39. Interview four or more children (ages 8-12) to investigate their concepts of electric circuits and the functioning of electric circuits. (Suggestions: provide the children with parts from which simple circuits can be assembled—batteries, bulbs, wire, wire coils and nails for electromagnets, salt water for electrolysis. Observe their activities and ask for explanations.)

40. Identify one or more explanations or discussions in this chapter that you find inadequate. Describe the general reasons for your judgment (conclusions contradict your ideas, steps in the reasoning have been omitted, words or phrases are meaningless, equations are hard to follow, . . .), and make your criticism as specific as you can.

Bibliography

M. Born, *The Restless Universe*, Dover Publications, New York, 1957.

I. B. Cohen, *Benjamin Franklin's Experiments*, Harvard University Press, Cambridge, Massachusetts, 1941.

M. Faraday, *Experimental Researches in Electricity*, Dover Publications, New York, 1966.

A. Holden, *Conductors and Semi-Conductors*, Bell Telephone Laboratories, Murrary Hill, New Jersey, 1964.

R. A. Millikan, *Electrons (+ and -)*, University of Chicago Press, Chicago, Illinois, 1947.

S. P. Thompson, *Michael Faraday*, Cassell, London, 1901.

J. Tyndall, *Faraday as a Discoverer*, Crowell, New York, 1961.

Articles from Scientific American. Some or all of these, plus many others, can be obtained on the Internet at http://www.sciamarchive.org/.

G. de Santillana, "Alessandro Volta" (January 1965).

G. A. Hoffman, "The Electric Automobile" (October 1966).

ISAAC NEWTON
(1642–1727)

Part four: motion

*My purpose is to set forth
a very new science dealing with
a very ancient subject.
There is, in nature, perhaps
nothing older than motion,
concerning which books written
by philosophers are neither
few nor small; nevertheless
I have discovered by
experiment some properties
of it which are worth
knowing and which have not
hitherto been either observed
or demonsttrated.*

Galileo Galilei
*Dialogues Concerning Two
New Sciences,*
1638

Objects in motion

13

Galileo was the first student of nature to observe and describe carefully the motion of objects, including falling bodies. He recorded their speeds and the changes in speed that occurred when they fell (increasing speed) or encountered resistance to motion (decreasing speed). Galileo also realized that the direction of motion was of great significance. For instance, projectiles tend to curve downward, but they maintain their motion in the horizontal direction even while they first rise higher, level off, and finally arch down to the ground.

When we defined the speed of relative motion in Eq. 2.1, we referred only to the distance traveled, not to its direction. In this chapter, therefore, we will introduce a new concept, the *velocity* of a moving object, which takes into account the direction of motion as well as the speed. Like position, displacement, and force, *velocity* will be denoted by a boldface letter in the text (**s**, **Δs**, **F**, **v**) and by an arrow in a diagram. The operations of arithmetic apply to velocity in the same way as to displacements (Section 2.3).

Physical quantities, such as displacement and velocity, which must be described by both a numerical magnitude and a direction in space are examples of a class of mathematical quantities called *vectors*. To help you recognize and manipulate vectors, we are listing all the ones used in this text in Table 13.1. The table includes the algebraic and diagrammatic symbols, the algebraic symbol for the magnitude, and the text section where the quantity is defined. Note the arrows with a variety of heads to suggest close relationships among the quantities. When you are writing by hand, the boldface notation cannot be used; instead, the algebraic symbol for a vector is written with an arrow over it ($\vec{v}$) to remind the reader of the directionality. Because the word "vector" is not in common use, we will not use it further in this text, and we will refer directly to magnitude and direction whenever these are important. Other physics books generally use the vector terminology.

13.1 Velocity

The two words "speed" and "velocity" are commonly used as synonyms to describe the rate of relative motion in everyday language. However, relative motion, if you wish to specify it adequately, includes two distinct ideas: 1) the numerical rate of speed and 2) the direction of motion. Therefore, in physics, the words "speed" and "velocity" have very specific meanings: we will use the word *speed* to describe *only* the numerical rate of relative motion (Section 2.2), as might be indicated on an automobile speedometer or an anemometer (a wind speed measuring device). On the other hand, the word *velocity* includes *both* the rate *and* the direction of relative motion, for example "The wind velocity is 30 miles per hour *from the northwest*." The phrase "from the north-west" in this statement indicates the wind direction, while the phrase "30 miles per hour" indicates the wind speed. Thus velocity includes *both* the simpler numerical idea of speed *and* the direction; we can say that the velocity concept is inclusive of and more complex than the speed concept.

TABLE 13·1 PHYSICAL QUANTITIES WITH MAGNITUDE AND DIRECTION IN SPACE

Physical quantity	Algebraic symbol	Diagramatic symbol	Magnitude	Definition
position	s	▷	$|s|$	Section 2·3
displacement	Δs	▷	$|\Delta s|$	Section 2·3
force	F	►	$|F|$	Section 11·2
velocity	v	◖▷	$|v|$	Section 13·1
change of velocity	Δv	◖▷	$|\Delta v|$	Section 13·1
acceleration	a	►	$|a|$	Section 13·2
momentum	$\mathcal{M}$	◖▷	$|\mathcal{M}|$	Section 13·3

There are good reasons that the distinction between velocity and speed has not become part of everyday language. The movement of persons, animals, and vehicles on the ground is usually confined to roads or paths that are easily identified. Once the path of motion is determined, its direction is known and only the speed of motion and its sense (forward or backward) need to be communicated.

However, most objects (for example, children in a playground, boats, airplanes...) can move in a wide range of directions; therefore, *both* the speed and the direction of motion relative to the earth (or some other reference frame) are important. Therefore, to describe the motion of such objects, we must use the more complex concept of velocity rather than the simpler numerical notion of speed.

Consider, for instance, the problem faced by the navigator of a ship, who can observe the velocity (speed and direction) of the ship relative to the ocean water and who knows that the ship is in an ocean current whose velocity (speed and direction) relative to the earth's land masses has been charted. To reach his destination, he has to use the available information to compute the velocity (speed and direction) of the ship relative to the land masses. You face a similar problem when you try to

Figure 13.1 *Relation of position, displacement and average velocity.*

(*a*) *Displacement* **s** *to* **s'**, **Δs** = [5.2] *centimeters in a time* Δt = 2 sec *onds.*

(*b*) *Average velocity*:

$$\mathbf{v}_{av} = \frac{\mathbf{\Delta s}}{\Delta t} = \frac{[5, 2]\ cm}{2\ sec}$$

$$= [2.5, 1]\ cm/sec$$

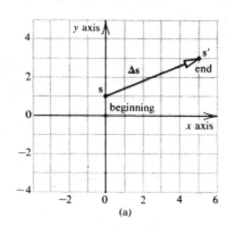

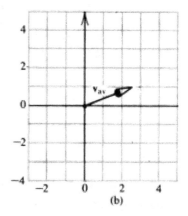

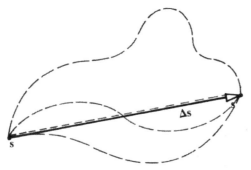

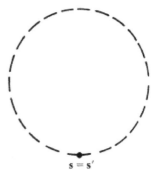

Figure 13.2 Alternate paths may result in the same displacement even though different distances are traversed. The magnitude of the displacement is equal to the actual distance traversed <u>only</u> if the actual path is along the straight line.

Figure 13.3 Displacement in a complete circle is the same as zero displacement. Δs=[0,0]. Of course, the total distance traveled is not zero.

paddle a canoe past a boulder in a swiftly moving stream. Only the velocity (speed and direction) of the canoe relative to the water is subject to your control; the velocity (speed and direction) of the water relative to the boulder is outside your control. Yet whether or not you hit the boulder depends only on the canoe's velocity (speed and direction) relative to the boulder, and this velocity is the result of adding together the other two velocities.

FORMAL DEFINITION
The average velocity is the ratio of the displacement divided by the time interval required for the displacement.

Average velocity. We will now present a formal definition of the average velocity of a moving object during a time interval. The *average velocity* ($\mathbf{v}_{av}$) is defined in terms of the duration of the time interval and the moving object's displacement during the time interval (Eq. 13.1). The definition is similar to that of speed (Eq. 2.1), but involves the displacement instead of the total distance traveled.

The average velocity describes only what is known about the motion of an object from two "snapshots" giving its position at the beginning and the end of the time interval Δt (Fig. 13.1). The snapshots give no information about what happened during the time interval: they do not reveal whether the object moved on a straight line between the two positions, or whether it made a larger detour (Fig. 13.2). An extreme case is that of a child on a merry-go-round, which may appear in the same position at the beginning and end of the interval, yet have traveled in a large circle on the merry-go-round (Fig. 13.3). According to the definition, its average velocity during the entire interval was zero because its displacement was zero. The average speed, however, was not equal to zero, since the child traveled all around a circular path of perhaps 100 feet in length.

The average velocity therefore may provide an inaccurate account of an object's motion. You can avoid this problem by using a sufficiently short time interval and defining a new quantity: the *instantaneous velocity.*

Equation 13.1

average velocity (*m/sec*)
$$= \mathbf{v}_{av}$$
displacement (*m*)
$$= \mathbf{\Delta s}$$
time interval (sec)
$$= \Delta t$$

$$\mathbf{v}_{av} = \frac{\mathbf{\Delta s}}{\Delta t}$$

Instantaneous velocity. To have a better description of the motion of the object, you must take snapshots or otherwise record its position more frequently. In fact, the position should be recorded so frequently

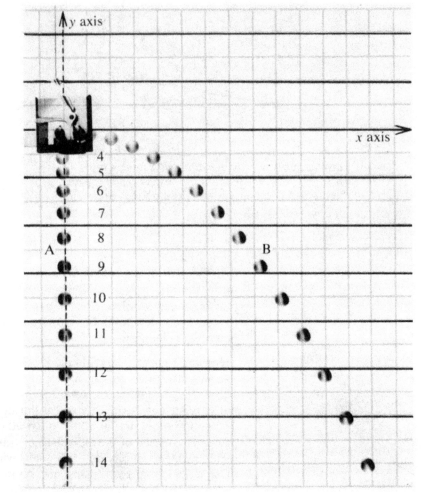

Figure 13.4 Two golf balls were released at the same time and then photographed at intervals of 1/30 second.

Scale: The black horizontal lines are 15 cm, or 0.15 m, apart. Thus each square in the grid is 0.075 m on a side and the actual distances between the positions of the balls can be found by simply estimating the position of each image in relation to the grid. For example, the bottom of ball image A9 is at about [0.0, -6.0] squares and the bottom of ball image A10 is at about [0.0, -7.4] squares. Therefore, the displacement of ball A in interval 9-10 is **Δs** *= [0.0, -1.4] squares = [0.0, -1.4 sq.](0.075 m/sq) = [0.0, -0.105] m. This agrees quite well with the vertical (y) displacement of ball B in interval 9-10 found in Table 13.2.*

that you can be sure no excursions or other irregular behavior had occurred between the recorded positions. Then the average velocity in each time interval, as defined by Eq. 13.1, would describe the actual motion of the object very accurately. When Eq. 13.1 is used with such a very short (infinitesimal) time interval, it defines the *instantaneous velocity* **v** (Eq. 13.2); the word "instantaneous," meaning over a "vanishingly small" or "infinitesimal" interval, emphasizes the distinction with the "average" velocity. Instantaneous velocity is a concept that is extremely useful for making mathematical models, but only the average

TABLE 13.2 ANALYSIS OF GOLF BALL MOTION, FIGURE 13.4

Pictures defining interval	Δt (sec)	**Δs** (m)	**v**$_{av}$ (m/sec)
4-5	0.033	[0.067, -0.048]	v_5 = [2.00, -1.4]
9-10	0.033	[0.067, -0.106]	v_{10} = [2.00, -3.2]
13-14	0.033	[0.067, -0.15]	v_{14} = [2.00, -4.5]

Equation 13.2

$$v = v_{av} = \frac{\Delta s}{\Delta t}$$

(Δt infinitesimal)

Infinitesimal means exceedingly small or vanishingly small. But how do we know when the interval is short enough? In practical terms, we always want to know the velocity to a certain accuracy set by our measuring instruments. Therefore, to decide when the time interval is short enough to be considered infinitesimal, we can continue to take position measurements at shorter time intervals until the value of the velocity calculated from the formula stays constant to within the desired accuracy.

René Descartes (1596-1650) was able to combine geometrical and arithmetical reasoning by identifying velocity components. He wrote: [the vertical component is] ... "that part which would make the ball move from above downward ...." [and the horizontal component is] "... the tendency which made it move toward the right."

velocity can be found experimentally from measurements of displacements and time intervals. As we have mentioned, you can find as close an approximation as you wish to the instantaneous velocity by measuring the average velocity over successively shorter time intervals. Using a very short time interval also means that the difference between the magnitude of the displacement and the distance traveled (Fig. 3.2) can be made as small as necessary; therefore, the magnitude of the velocity and the speed will also converge to the same value.

An example. We will analyze the familiar and important motion of a freely falling object (Fig. 13.4). We will show that the horizontal component of the velocity stays the same and only the vertical component changes, a result that we will refer to again in Chapter 14. Two techniques for determining the positions of a moving object at closely spaced time intervals were described in Section 2.2: motion pictures and multiple exposure photographs. Fig. 13.4 is a multiple exposure of two golf balls released at the same time from the mechanism at the top. Ball A was allowed to fall straight down, and ball B was shot out horizontally with an initial speed of 2.00 meters per second. The camera shutter remained open while short flashes of light illuminated the balls at intervals of 1/30 second. The black lines are horizontal strings placed 15 centimeters apart. A rectangular coordinate grid (as introduced by Descartes, Chapter 2) is shown by which you can actually measure the rectangular components of the displacement of each ball for each interval. You can then use this information to find the average velocity of each ball for each interval between pictures.

Three sample calculations of the average velocities of ball B are presented in Table 13.2. The average velocity in the fifth interval is called v_5, the average velocity in the tenth interval is v_{10}, and so on. You see that the x-component (horizontal) of the velocity does not appear to change, but that the y-component (vertical) changes greatly.

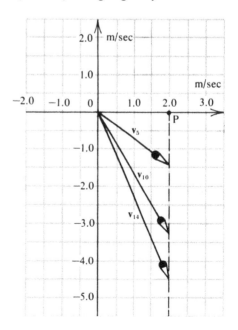

Figure 13.5 Average velocities v_5, v_{10}, and v_{14} of falling golf ball B in Fig. 13.4.

Figure 13.5 is a coordinate frame diagram to show the average velocities calculated in Table 13.2. The three arrows represent the velocities of ball B in the three intervals tabulated. This diagram gives a picture of the variation of the ball's average velocity with time. The velocity becomes progressively directed more and more strongly in the downward direction. Since the golf ball did not execute irregular motion, you may imagine that the instantaneous velocity also changed smoothly and became progressively directed downward. The instantaneous velocity at the very beginning was 2.00 meters per second to the right, as described in the conditions of the experiment. A good mathematical model for the golf ball motion is that the head of the arrow representing the instantaneous velocity begins at the point P and gradually rotates from the horizontal direction toward the vertical direction, with its tip always on the dashed line.

Change of velocity. You can carry out arithmetic operations with velocities just as you did with displacements. You must, however, remember the physical significance of these operations when they are applied as part of a mathematical model. We said above, for example, that the average velocity of the falling golf ball was changing. How much did it change between the fifth interval and the tenth interval (Fig. 13.4)? It changed by the arrow shown as $\Delta \mathbf{v}$ in Fig. 13.6. This must be added to the velocity $\mathbf{v}_5$ in the earlier interval to result in the velocity $\mathbf{v}_{10}$ in the later interval. In other words, the change of velocity $\Delta \mathbf{v}$ is the difference between the two velocities. Note that the change of velocity has a magnitude and a direction and is represented by an arrow in the diagrams. In Fig. 13.6 the change of velocity is directed vertically downward.

The relation of the arrows in Fig. 13.6 can be stated also in the form of an equation. You can think of $\mathbf{v}_{10}$ as equal to the sum of $\mathbf{v}_5$ plus $\Delta \mathbf{v}$ (Eq. 13.3), or you can think of $\Delta \mathbf{v}$ as the difference between $\mathbf{v}_{10}$ and $\mathbf{v}_5$ (Eq. 13.4). We pointed out in Chapter 3 that changes in motion are evidence of interaction of the moving object with another object. Therefore, the calculation of changes of velocity is important because it reveals the presence of interaction and helps us to identify the other objects that are the source of that interaction. For instance, the change of velocity of the falling golf ball is directed vertically downward because the force of gravity is directed vertically downward. The relation between change of velocity and force will be examined further in Chapter 14.

Finding the displacement. The relationship of average velocity, displacement, and time interval (Eq. 13.1) can be used in two different ways. So far, we have used it to find the average velocity from data of the displacement and the time interval. In addition, the relationship can be reversed to permit calculation of the displacement, which is equal to the average velocity multiplied by the time interval (Eq. 13.5).

This new relationship, a mathematical model for the displacement, is useful only while the average velocity is not changing greatly or when

Equation 13.3

$$\mathbf{v}_{10} = \mathbf{v}_5 + \Delta \mathbf{v}$$

Equation 13.4

$$\Delta \mathbf{v} = \mathbf{v}_{10} - \mathbf{v}_5$$

Equation 13.5

$$\Delta \mathbf{s} = \mathbf{v}_{av} \Delta t$$

EXAMPLE:

$\mathbf{v}_{av} = [2, 15]$ m/sec

$\Delta t = 3$ sec

$\Delta \mathbf{s} = \mathbf{v}_{av} \Delta t$

$\quad = [2, 15]$ m/sec
$\quad\quad \times 3$ sec

$\quad = [6, 45]$ m

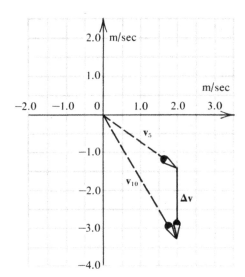

Figure 13.6 Change of the average velocity of falling golf ball B in Fig. 13.4.

Equation 2.1 (average speed)

speed (m/sec) $= v_{av}$
distance (m) $= \Delta s$
time interval (sec) $= \Delta t$

$$v_{av} = \frac{\Delta s}{\Delta t}$$

Equation 13.6 (displacement and distance)

$$|\Delta s| = \Delta s$$

Equation 13.7 (average velocity and average speed)

$$|\Delta v_{av}| = \Delta v_{av}$$

Equation 2.2 (instantaneous speed and average speed)

$$v = v_{av} = \frac{\Delta s}{\Delta t}$$

(Δt must be infinitesimal)

Equation 13.8 (instantaneous velocity and instantaneous speed)

$$|v| = v$$

the time interval is infinitesimal. It is similar to the example of the automobile trip described in Figure 1.8 except that now the direction of motion as well as the speed must be unchanging.

Velocity versus speed. At the beginning of this section we pointed out that the velocity concept describes motion more completely than does the speed concept. Speed is the rate at which distance is traversed without regard to direction. Speed and distance, therefore, are not represented by boldface symbols but by ordinary letters (Eq. 2.1). The average speed of the child on the merry-go-round (Fig. 13.3) is substantial because of the large distance traversed during one revolution, but the average velocity during one revolution is nevertheless zero. Whenever an object changes its direction of motion and moves along a curved or zigzag path, the average speed and the average velocity are *not* related simply to one another. Whenever motion is along a straight line and its direction does not change, the magnitude of the displacement is equal to the distance traversed (Eq. 13.6; see Fig. 13.2); hence, the magnitude of the average velocity is equal to the average speed (Eq. 13.7).

This same relation of speed and velocity *also* holds in the limit of infinitesimal time intervals. In this limit we speak of the instantaneous speed (Eq. 2.2) and the instantaneous velocity (Eq. 13.2). The magnitude of the instantaneous velocity is equal to the instantaneous speed (Eq. 13.8). Basically, the reason behind this relationship is that the complications arising from changes in the direction of motion do not operate during an infinitesimal time interval. The automobile speedometer, for instance, is designed to measure the instantaneous speed of a car. When you combine the speedometer reading with the direction of motion of your car (perhaps indicated on a magnetic compass mounted under the windshield), you obtain your car's instantaneous velocity (speed and direction).

Before concluding this section, we should point out an example that clearly reveals the key difference between the concepts of speed and velocity. A car travels along a road at 60 miles per hour, slows

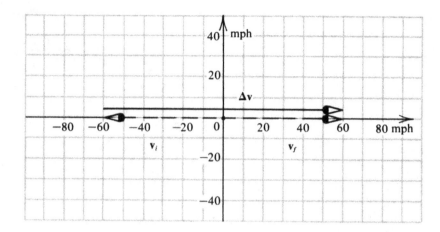

Figure 13.7 A car reverses its direction of motion from the left to the right. The change of velocity is 120 miles per hour to the right.

down and makes a U-turn, then accelerates to 60 miles per hour again. Its speed at the end is the same as at the beginning, 60 miles per hour. However, while its final velocity has the same magnitude as the initial velocity, they are pointing in opposite directions. Therefore, the *change in velocity,* illustrated in Fig. 13.7, has a magnitude of 120 miles per hour and points in the same direction as the final velocity!

13.2 Acceleration

Everyone who has been in a rapidly accelerating car or in a jet airplane has experienced the sensation of being pushed back into the seat during takeoff. When a high-speed elevator starts or stops, the passengers sometimes experience feelings of nausea. The passengers in roller coasters and other rides at the amusement park are pressed into their seats or pushed from side to side. In all these happenings, the human body experiences changes of velocity (of speed or direction of motion, or of both) that affect the internal organs and may cause "seasickness."

Acceleration is the quantity used to describe changes of velocity. The most common example is the speeding up of an automobile when the driver steps on the accelerator. Acceleration is therefore usually associated with an increase of speed. Newton's very successful theory, however, relates all changes of speed and direction of motion of objects to their interactions with other objects. As a result, we must formulate a definition of acceleration that is more general than everyday usage and that includes decrease of speed and changes in direction of motion as well as increase of speed.

Definition of acceleration. It is immediately clear that the change in velocity is by itself not directly related to the strength of interaction. Compare, for instance, the car that accelerates to 60 miles per hour

FORMAL DEFINITION
FORMAL DEFINITION
The average acceleration is the ratio of the change of instantaneous velocity to the time interval required to effect the change.

Equation 13.9 (definition of acceleration)

change in instantaneous velocity (m/sec) $= \Delta v$
time interval (sec) $= \Delta t$

$$\mathbf{a}_{av} = \frac{\Delta \mathbf{v}}{\Delta t}$$

Units of acceleration:

1 m/sec/sec = 1 meter per second per second

1 mph/sec = 1 mile per hour per second
= 0.45 meter per second per second

(Can you verify the above value, assuming that 1 mile ≈ 1600 m?)

Figure 13.8 A car drives around a curve, (a) Top view of road, with car turning from westward to southward direction at 50 miles per hour. (b) Velocity diagram, showing initial velocity $\mathbf{v}_i$, final velocity $\mathbf{v}_f$, and the velocity change $\Delta \mathbf{v}$ of 70 miles per hour to the southeast.

from a standing start in 10 seconds (with a roar of the engine and flying gravel), and the car that does the same in 1 minute with hardly any noise or notice by the passengers inside. Though the velocity change is the same (0 to 60 miles per hour), the interaction between the tires and the road (or the passengers and their seats) is much larger in the first case than in the second. The time interval during which the velocity change is accomplished appears to be significant and is included in the definition of acceleration. To pin down the time interval and the exact velocity change, the definition refers to the instantaneous and not the average velocity (Eq. 13.9).

Examples. The definition of acceleration can be applied to golf ball B in Fig. 13.4 only in approximate fashion, because instantaneous velocities have not been determined. Nevertheless, the acceleration of golf ball B will serve as an illustration in Example 13.1 below.

In Example 13.2 below we calculate the actual accelerations of the two automobiles mentioned above. The example of the car making the U-turn (Fig. 13.7) could be used to calculate an acceleration if the time interval Δt required for the U-turn were known. If the U-turn is accomplished in a short time interval, the acceleration must be large, according to Eq. 13.9. Physically, this is possible only on a road with good traction. On an icy pavement, the time interval would have to be long and the acceleration small. This shows that the maximum possible acceleration is closely related to the interaction between tires and road.

Another example of acceleration occurs when a car traveling west makes a left turn and goes south (Fig. 13.8 and Example 13.3). This situation may seem different from the ones above, because only the direction of motion changes, not the speed. Nevertheless, we can still find the change in velocity by considering the x- and y-components. The change in velocity needed is the velocity we must *add* to the initial velocity to end up with the final velocity. Therefore, the change in velocity must point both east (to cancel the initial velocity) and south (to end up with the proper final velocity). As shown in Fig. 13.8b, the change of velocity, and therefore the acceleration, are at a 45-degree angle to the initial and final directions, pointing *toward the inside* of the curve. This conclusion is generally true: *whenever an object travels in a curve, the acceleration points toward the inside of the curve.*

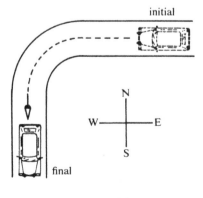

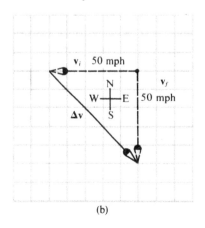

(a)   (b)

EXAMPLE 13·1. Acceleration of golf ball B, Fig. 13·4.

Data:

$$\mathbf{v}_5 = [2.0, -1.4] \text{ m/sec}; \mathbf{v}_{10} = [2.0, -3.2] \text{ m/sec}; \Delta t = 0.165 \text{ sec}$$

Solution:

$$\mathbf{\Delta v} = \mathbf{v}_{10} - \mathbf{v}_5 = [0, -1.8] \text{ m/sec}$$

$$\mathbf{a}_{av} = \frac{\mathbf{\Delta v}}{\Delta t} = \frac{[0, -1.8] \text{ m/sec}}{0.165 \text{ sec}} \approx [0, -11] \text{ m/sec/sec}$$

EXAMPLE 13·2. (a) Acceleration of an automobile from 0 miles per hour to 60 miles per hour in 10 seconds.

Data:

$$\mathbf{v}_i = [0, 0] \text{ mph}; \mathbf{v}_f = [60, 0] \text{ mph}; \Delta t = 10 \text{ sec}$$

Solution:

$$\mathbf{\Delta v} = \mathbf{v}_f - \mathbf{v}_i = [60, 0] \text{ mph}$$

$$\mathbf{a}_{av} = \frac{\mathbf{\Delta v}}{\Delta t} = \frac{[60, 0] \text{ mph}}{10 \text{ sec}} = [6, 0] \text{ mph/sec} \approx [2.7, 0] \text{ m/sec/sec}$$

(b) Acceleration of an automobile from 0 miles per hour to 60 miles per hour in 1 minute.

Data:

$$\mathbf{v}_i = [0, 0] \text{ mph}; \mathbf{v}_f = [60, 0] \text{ mph}; \Delta t = 60 \text{ sec}$$

Solution:

$$\mathbf{\Delta v} = \mathbf{v}_f - \mathbf{v}_i = [60, 0] \text{ mph}$$

$$\mathbf{a}_{av} = \frac{\mathbf{\Delta v}}{\Delta t} = \frac{[60, 0] \text{ mph}}{60 \text{ sec}} = [1, 0] \text{ mph/sec} \approx [0.45, 0] \text{ m/sec/sec}$$

EXAMPLE 13·3. Find the acceleration of the car in Fig. 13·8. It passes the curve in 3 seconds.

Data:

$$\mathbf{v}_i = [-50, 0] \text{ mph}; \mathbf{v}_f = [0, -50] \text{ mph}; \Delta t = 3 \text{ sec}$$

Solution:

$$\mathbf{\Delta v} = \mathbf{v}_f - \mathbf{v}_i = [50, -50] \text{ mph}$$

$$\mathbf{a}_{av} = \frac{\mathbf{\Delta v}}{\Delta t} = \frac{[50, -50] \text{ mph}}{3 \text{ sec}} \approx [16, -16] \text{ mph/sec}$$
$$\approx [7, -7] \text{ m/sec/sec}$$

EXAMPLE 13·4. A car slows down from 60 miles per hour to 40 miles per hour in 20 seconds. Find the acceleration.

Some useful conversions:

a) 1 mile = 1600 meters (approx.)

b) 1 mile/hr (mph) = 0.44 m/sec

c) 1 m/sec = 2.25 mph

Can you verify b) and c) by starting with a)?

Data:

$\mathbf{v}_i = [60, 0]$ mph; $\mathbf{v}_f = [40, 0]$ mph; $\Delta t = 20$ sec

Solution:

$$\Delta\mathbf{v} = \mathbf{v}_f - \mathbf{v}_i = [-20, 0] \text{ mph}$$

$$\mathbf{a}_{av} = \frac{\Delta\mathbf{v}}{\Delta t} = \frac{[-20, 0] \text{ mph}}{20 \text{ sec}} = \begin{array}{l} [-1, 0] \text{ mph/sec} \\ \approx [-0.45, 0] \text{ m/sec/sec} \end{array}$$

Finally, consider the example of a car slowing down. A car traveling initially at 60 miles per hour coasts gradually, in 20 seconds, to a speed of 40 miles per hour on a straight road (as in Example 13.4, above). What is the acceleration? You may respond that the car is not accelerated at all, that it is decelerated. This is true, in everyday language. The scientific definition of acceleration, however, refers to a velocity change and does not depend on whether the change represents an increase or a decrease in speed. The acceleration of the coasting car is calculated in Example 13.4, above, and it comes out with a negative x-component; this indicates that the acceleration is, in this case, directed in the *opposite* direction from the velocity, as expected for a decrease in speed.

It is clear from all these examples that the scientific meaning of the word "acceleration" frequently does not correspond to the intuitive meaning of the word. For interpreting changes in velocity, the definition we have given (Equation 13.9) is more useful than the everyday meaning. These examples also show how to apply the scientific definition of acceleration for all the possible ways that the velocity can change.

FORMAL DEFINITION
Momentum is the product
of inertial mass multiplied
by instantaneous velocity.

13.3 Momentum

Changes in motion brought about by interaction during a rear-end collision are a painful part of many people's experience. Imagine the cautious driver waiting at a red light, who glances into his rearview mirror and sees a trailer-truck bearing down on his car from behind! Even though the truck is advancing quite slowly, our hero may grip his steering wheel in grim anticipation. By contrast, a light motorcycle or bicycle seen in the same rearview mirror would hardly be a cause for alarm unless it was approaching at very high speed.

Equation 13.10

inertial mass $(kg) = M_I$
velocity $(m/\sec) = \mathbf{v}$
momentum $= \mathcal{M}$

$$\mathcal{M} = M_I \mathbf{v}$$

Units of momentum:
kg-m/sec = kilogram-
           meters
           per second

This example illustrates that *both* speed and mass of a moving object influence its interaction with other objects upon collision. In Section 3.4, we defined the inertial mass as a measure of a body's resistance to change in motion and the momentum as an important property of a moving body, equal to the product of the inertial mass and the speed (Eq. 3.1). We now redefine the momentum to assign it a direction in space as well as a magnitude by relating it not to the speed but to the velocity (Eq. 13.10). The direction of the momentum is defined to be exactly the same as the direction of the velocity: the momentum arrow points in exactly the same direction as the velocity arrow. However, the two arrows are a different length because the momentum and the velocity have different magnitudes. In fact, as you would expect, the magnitude of the momentum is equal to the mass times the speed (the magnitude of the velocity). The momentum concept will be used to formulate Newton's theory of moving bodies in Chapter 14.

Summary

Relative position, displacement, velocity, acceleration, and momentum are physical quantities that are described by a magnitude and a direction in space. You must distinguish between the average and the instantaneous velocity of a moving body. The average velocity, measured from observations of the path of the moving object, is equal to the displacement divided by the time interval (Eq. 13.1). The instantaneous velocity is an extrapolation of the average velocity to a time interval that is so short (infinitesimal) that the velocity does not change. The velocity must also be distinguished from the speed, which is equal to the actual distance traveled divided by the time interval.

The average acceleration is equal to the change of instantaneous velocity divided by the time interval (Eq. 13.9). The momentum of a body is equal to the velocity multiplied by the inertial mass of the body (Eq. 13.10). We will use the concepts of acceleration and momentum in the next chapter (Ch. 14) to formulate Newton's theory of moving bodies, which shows how the motion of an object is related to the forces on that object.

Equation 13.1 (average velocity)

$$\mathbf{v}_{av} = \frac{\Delta \mathbf{s}}{\Delta t}$$

Equation 13.9 (definition of acceleration)

$$\mathbf{a}_{av} = \frac{\Delta \mathbf{v}}{\Delta t}$$

Equation 13.10 (definition of momentum)

$$\mathcal{M} = M_I \mathbf{v}$$

List of new terms

average velocity
instantaneous velocity
infinitesimal time interval
average acceleration
momentum

List of symbols

$\mathbf{s}$	position	$\mathbf{v}$	instantaneous velocity
$\Delta \mathbf{s}$	displacement	v	instantaneous speed
Δs	distance	$\Delta \mathbf{v}$	change of velocity
Δt	time interval	$\mathbf{a}_{av}$	average acceleration
$\mathbf{v}_{av}$	average velocity	$\mathcal{M}$	momentum

Note: Boldface symbols, such as **s** and **v**, represent quantities with *both* magnitude and direction. Such quantities are represented as arrows with a length (magnitude) pointing in a particular direction. Such quantities cannot be represented with a single number and must be expressed mathematically with rectangular (x, y) or polar (r, θ) coordinates. Rectangular coordinates are often more convenient, especially in calculations where quantities such as displacements or velocities must be added and subtracted.

Problems

1. Formulate an operational definition for velocity. Compare your definition with the formal definition in Section 13.1. Discuss advantages and disadvantages of the formal and the operational definitions.

2. Consider the child on the merry-go-round (Fig. 13.3) and suppose it moves in a circle with a radius of 6 meters. The merry-go-round takes 12 seconds to make one revolution.
   (a) Calculate the distance the child travels during one revolution.
   (b) Calculate the average speed of the child (ratio of actual distance traveled to time interval) during one full revolution.
   (c) Calculate the average velocity of the child during one half of a revolution. Note: find the magnitude and direction (using polar or rectangular coordinates) of the velocity in a coordinate frame with origin at the child's starting position.
   (d) Calculate the average velocity during: one quarter of a revolution; one sixth of a revolution; one twelfth of a revolution. (See note in (c).)
   (e) Plot a graph of the magnitude of the average velocity versus the time of the interval Δt you used in questions (c) and (d).
   (f) Plot a graph of the direction (angle) of the average velocity versus the time of the interval Δt you used in questions (c) and (d).
   (g) Estimate the instantaneous velocity of the child by extrapolating to an infinitesimal time interval ($\Delta t = 0$) in questions (e) and (f).

3. Refer to Fig. 13.4 for this question.
   (a) Find the average velocity of golf ball A during the intervals between pictures 4 and 5, 9 and 10, and 13 and 14. Compare these average velocities with those of golf ball B recorded in Table 13.2.
   (b) Find the (approximate) average acceleration of golf ball A between the fifth and tenth intervals, and between the tenth and fourteenth intervals.
   (c) Find the (approximate) average acceleration of golf ball B between the tenth and fourteenth intervals. Compare the accelerations you found in (b), (c), and in Example 13.1.
   (d) Discuss and estimate the extent to which the instantaneous velocities of golf balls A and B in this experiment differ from the average velocities determined in part (a) and in Table 13.2.

4. (a) Describe examples (beyond those included in Section 13.2) of situations where the human body easily senses changes of velocity.
   (b) Describe the body's response (if any) to motion at high velocity that is constant (does not change). Examples of this type of motion are an airplane or train traveling at constant speed in a straight line.

5. Calculate the highest forward (speeding up) acceleration of which existing automobiles are capable. Include all the data on which you base your work. (Consult magazines, automobile dealers.)

6. Estimate the highest acceleration of which automobiles are capable

when they come to an emergency stop (with the brakes slammed on). (Note: As explained in Section 13.2, acceleration is used to denote both increases and decreases in speed; the only condition is that the velocity changes.) State your estimate numerically.

7. Estimate the acceleration of other vehicles (for example, jet aircraft on takeoff and landing, motorcycles, speedboats). Include all the data on which you base your work. State your estimates numerically.

8. The velocity of a child on a merry-go-round (Fig. 13.3) is changing continually in direction. Make a rough quantitative estimate of the acceleration involved. If you did problem 2 above, you can use the results you found there for parts (d) through (g).

9. Identify one or more explanations or discussions in this chapter that you find inadequate. Describe the general reasons for your judgment (conclusions contradict your ideas, steps in the reasoning have been omitted, words or phrases are meaningless, equations are hard to follow, . . .), and make your criticism as specific as you can.

Bibliography

L. Cooper, *An Introduction to the Meaning and Structure of Physics*, Harper and Row, New York, 1968. See the part "On the Problem of Motion."

G. Galilei, *Dialogues Concerning Two New Sciences*, Northwestern University Press, Evanston, Illinois, 1950. Look at the section called "Fourth Day."

D. A. Greenberg, *Mathematics for Science Courses*, W. A. Benjamin, New York, 1964.

E. R. Huggins, *Physics 1*, W. A. Benjamin, New York, 1968. The author has an extensive discussion of multiple exposure photographs and their use in the analysis of motion.

Newton's laws of motion

14

Isaac Newton (1642-1727) attended Trinity College, Cambridge, and obtained his degree in 1665. Although an excellent student, his early Cambridge years offered no indications of his future greatness. The plague of 1665-1666 forced Newton to return to his home in Woolsthorpe. The seclusion and lack of responsibilities over the next year and a half allowed Newton to work intensively by himself, with extraordinary results.

By 1666, at 24, he had made profound discoveries in mathematics (binomial theorem, calculus), optics (ray model), and mechanics. To this period also belongs the famous (though possibly untrue) incident of the falling apple, the realization that the force of gravity reached the "orb of the moon," and the discovery of the inverse second power law of gravitational attraction.

Returning to Cambridge, Newton continued his study of planetary motions as a problem in physics. Unfortunately, controversy with Huygens and

The subject matter of this chapter has been anticipated in several earlier sections of this text. In Section 3.4, we described inertia of motion and presented an operational definition of inertial mass. We introduced the force concept through an operational definition using a spring scale in Section 11.2. Finally, in Sections 13.2 and 13.3, the formal definitions of acceleration and momentum extended your ability to describe changes of velocity and to associate the effects of inertial mass with motion.

You are familiar with many examples of moving objects that interact with other nearby or remote objects. A few instances are the stream of water erupting from Old Faithful in Yellowstone Park, the arrow launched by the archer's bow, a sailboat racing across the wind, a stunt car rounding a curve on two wheels, and the entire earth orbiting around the sun. In Section 3.4 (as well as 11.2 and 13.2), you learned to interpret changes in motion as evidence of interaction. Both the eruption of the water of the geyser and its return to the ground are evidence of interaction between the water and something else. A body moving with a constant velocity shows no evidence of interaction by its motion and is in mechanical equilibrium. It may be completely free of interaction or it may be subject to compensating forces, like the skier who moves steadily uphill while interacting with the rope, the snow, and the earth (via the gravitational field).

14.1 Background

Galileo showed the essential similarity between a stationary object and an object moving at constant speed in a straight line, and he identified inertia as an key concept. Galileo systematically applied the concepts of velocity and acceleration to the study of moving bodies and thereby began a transformation (some would say revolution) in human thought that was completed by Newton two generations later. In the intervening years, Descartes introduced the rectangular coordinate frame so useful for a mathematical description of relative position and motion (Sections 2.1 and 13.1), and Huygens investigated the forces acting on a pendulum in circular motion. Newton, finally, formulated the science of classical mechanics, a deductive system of definitions and assumptions (called "Newton's Laws of Motion"), which replaced the Aristotelian system in use until Galileo's time. The concepts of force, mass, momentum, and acceleration were integrated into one powerful theoretical framework (summarized in Table 14.1) that is still being refined and used for the prediction of motion in macro- and cosmic-domain phenomena. Only for various forms of radiation, as we explained in Chapters 7 and 8, was Newton's theory found inadequate and replaced by a wave theory in the nineteenth and twentieth centuries. The justification for Newton's three laws of motion rests in their broad and amply-verified predictive power.

Newton's laws lead to mathematical models for the relation among force, inertial mass, acceleration, and momentum. These models will enable you to predict the motion of bodies that are subject to known forces and to make inferences about the forces from observed motion.

Hooke discouraged Newton, an extraordinarily sensitive personality, from publishing his results. Other scientists were also working on the same problems but had hit a variety of dead ends. In particular, no one had been able to demonstrate a rigorous mathematical connection between the observed elliptical shape of the planets' orbits and the dependence of the strength of gravitational attraction (the force) on the distance between the planet and the sun.

The geometrical properties of the ellipse had been well known since ancient times. As a planet moves along an elliptical path, the distance to the sun (located at one focus) changes, and the actual shape of the orbit seemed to have a close connection with the variation in the force on the planet as the distance to the sun changes. However, there were enormous mathematical difficulties involved in demonstrating this connection; the best mathematicians and "natural philosophers" (as physicists were called), despite intense effort, could not work out a convincing mathematical proof.

In 1684 Edmond Halley, an astronomer studying the motions of the planets, mentioned this problem to Newton. According to Newton's later account (as told to Demoivre), Newton immediately replied that he had solved it but couldn't find the paper and agreed to write it out again. Over the next two years, with Halley's unstinting encouragement and support, Newton wrote his monumental Principia, containing not only the promised proof but also Newton's complete theory of motion and its sweeping explanation of the existing celestial observations. The appearance of the Principia in 1687 established Newton's reputation for all time.

TABLE 14.1 DEFINITIONS AND LAWS IN NEWTON'S THEORY

Concept	As stated by Newton	As stated in this text
gravitational mass	The quantity of matter is the measure of the same, arising from its density and bulk conjunctly.	Gravitational mass of an object is measured by the number of standard units of mass that are required to balance the desired object on an equal-arm balance.
velocity	not stated.	Average velocity is the ratio of the displacement divided by the time interval required for the displacement.
momentum	The quantity of motion is the measure of the same, arising from the velocity and quantity of matter conjointly.	Momentum is the product of inertial mass multiplied by instantaneous velocity.
inertia	The inertia, or innate force of matter, is a power of resisting, by which every body, as much as in it lies, endeavours to persevere in its present state, whether it be of rest, or of moving uniformly forward in a right line.	Inertial mass is measured by the number of standard units of mass that are required to give the same rate of oscillation of an inertial balance.
force	An impressed force is an action exerted upon a body, in order to change its state, either of rest, or of moving uniformly forward in a right line	Force is measured by a standard spring scale. The scale reading indicates the magnitude of the force. The direction of the force is related to the direction of the spring scale.
Law 1	Every body perseveres in its state of rest, or of uniform motion in a right line, unless it is compelled to change that state by forces impressed thereon.	Every particle persists in its state of rest or of uniform, unaccelerated motion in a straight line unless it is compelled to change that state by the application of an external net force.

TABLE 14.1, CONTINUED

Concept	As stated by Newton	As stated in this text
Law 2	*The alteration of motion is ever proportional to the motive force impressed; and is made in the direction of the right line in which that force is impressed.*	*The change of momentum of a particle subject to a net force during a time interval is equal to the average net force times the duration of the time interval.* $\boldsymbol{F}_{av}\,\Delta t = \Delta\boldsymbol{\mathcal{M}}$
Law 3	*To every action there is always opposed an equal reaction: or the mutual actions of two bodies upon each other are always equal, and directed to contrary parts.*	*The interaction of two bodies is described by two equal and opposite forces*

The center-of-mass model. There are several ways in which Newton's theory introduces assumptions, in addition to the laws, that enabled him to reduce complicated phenomena to simpler ones. In Section 11.2 we already mentioned the assumption that the net force acting on a body is the sum of the partial forces arising within the two-body subsystems of the original system. A second assumption, used in conjunction with the first and second laws, is that a moving body can be described as one particle located at a central point called the center of mass of the body. That is, the inertial mass, position, velocity, and acceleration of the entire body are the mass, position, velocity, and acceleration of the particle located at the center of mass.

If the body is quite small (such as a drop of water or even a baseball), then the one-particle model usually permits an adequate description of its motion. If the body is large or has a complicated shape (such as a skier), then the one-particle model may be adequate for its overall motion, but a more complicated working model has to be adopted for a more complete description of its interaction, energy storage, and so on. For example, a boxer might be seen as three particles, two fists and a torso; the medium for wave propagation was represented by a model consisting of many oscillators (Section 6.1); we used the MIP model to represent an elastic body (Section 11.7); in Chapter 16, we will introduce a model of many non-interacting particles to represent a gas. Newton's third law makes possible the analysis of such systems in terms of the first and second laws. However, systems requiring a complicated description (for example, rotating rigid bodies such as bicycle wheels or gyroscopes) or working models consisting of more

"... It seems probable to me that God in the beginning form'd matter in solid, massy, hard, impenetrable, movable particles, of such sizes and figures, and with such other properties, and in such proportion to space, as most conduced to the end for which he form'd them; and that these primitive particles being solids, are incomparably harder than any porous bodies compounded of them; even so very hard, as never to wear or break in pieces; no ordinary power being able to divide what God himself made one in the first creation ..."

Isaac Newton
Opticks, 1704

Aristotle (384-322 B.C.) Aristotle's father was the physician to Philip of Macedon, and Aristotle became tutor to the son of Philip, the young Alexander the Great. Aristotle studied for many years at Plato's Academy and then founded his own school, the Lyceum, in Athens. Aristotle exerted a profound influence on the development of science. His works were so comprehensive that until the Renaissance no systematic survey comparable to his was produced in the west. Aristotle remained the authority in Western academic circles for almost 2000 years.

than one or two interacting particles will not be discussed in detail in this chapter.

Interaction mechanisms. In Newton's simplified approach to moving bodies, the various mechanisms by which they interact are not described in detail. The forces may be elastic, frictional, gravitational, electric, or magnetic. You may even find the MIP model helpful in understanding how the chair on which you sit is an elastic system that can exert a force to balance the force of gravity acting on you. Since Newton described an MIP model for matter, it is likely that he, too, used it to visualize interaction mechanisms. His theory of moving bodies, however, does not make direct reference to this model and is not logically dependent on it.

14.2 Newton's first law of motion

Newton's first law of motion, sometimes called the *law of inertia,* is as follows: *Every particle persists in its state of rest or of uniform, unaccelerated motion in a straight line unless it is compelled to change that state by the application of a net force.*

Comparison of Newtonian and Aristotelian views. The assumption expressed in this statement was contrary to that of Aristotelian scholarship, according to which force was the cause of uniform motion and only the state of rest was an equilibrium state for a non-interacting body. Newton's First Law, in contrast, asserts that motion in a straight line at constant speed is *also* an equilibrium state and, therefore, does not require a net force. As an example, consider the two opposing interpretations of a skier going downhill (Fig. 14.1 and 14.2).

The Newtonian observer would hold that the skier is subject to the force of gravity exerted by the earth and forces of support and friction by the snow. When these partial forces combine to give a zero net force, the skier remains at rest or skis at a uniform rate (Fig. 14.1). When the friction decreases (on a patch of ice, for instance), the skier accelerates (Fig. 14.1d); when the slope levels out so that gravity and support together are less than friction, the skier slows down.

The Aristotelian would insist that a force is always acting on the skier while he is moving, presumably larger or smaller depending on the skier's speed (Fig. 14.2). Only when the skier stops completely would the Aristotelian hold that there is no unbalanced force. As we have remarked in Chapter 3, some of our intuition about motion parallels the Aristotelian view because friction is so pervasive in terrestrial phenomena. We therefore tend to overlook friction and think of the partial forces that actually balance friction during steady motion as adding up to a net force.

A second and possibly even more revolutionary implication of Newton's first law is that changes in the state of motion of a body must always be ascribed to an external agency. This again is contrary to Aristotelian and commonsense views, in which motion of a person or an animal usually has an internal cause. We will discuss the skier and a

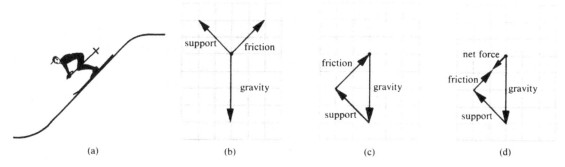

Figure 14.1 *Newtonian view of a skier sliding downhill, subject to the three partial forces of gravity, friction, and support.*
(a) Gravity is directed vertically downward, friction is along the gliding surface, and support is at right angles to the gliding surface.
(b, c) Force diagrams: three partial forces add to zero net force.
(d) Force diagram (reduced friction): three partial forces add to a nonzero net force and the skier speeds up.

man walking uphill as two examples. The modem scientist would agree that the skier accelerates downward because of the force of gravity. According to Aristotle, however, the skier moves downward because all bodies have a tendency to reach their "natural place" of repose, which is downward for solid and liquid materials.

Aristotle would ascribe motion of the man walking uphill to the man's desire to walk. The cause of motion is internal: the man's desire and his muscular energy. Where is Newton's external net force?

To answer this question, you must realize that the Newtonian theory distinguishes the energy source that is necessary for motion from the force that causes the motion. In the case of the skier going

Figure 14.2 *Aristotelian view of a skier sliding downhill.*
(a) A skier slides downhill.
(b) A force is acting to maintain motion.
(c) If the slope is steeper, or smoother, a larger force acts and causes a larger speed.

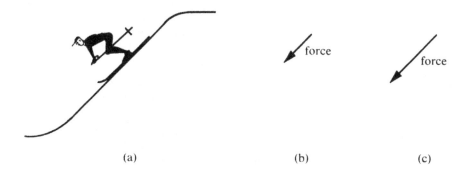

downhill, the gravitational field transmitted the force and acted as a source of the skier's kinetic energy, so the distinction was not apparent. When the man walks uphill, however, his muscles are the energy source, while now the frictional interaction between his feet and the ground supplies the force that enables him to move forward with each step. In the absence of friction (on an iceberg or on a path covered with very loose gravel), the net force on him might be inadequate for him to progress upward and he might even slide back down. Think of a skier walking uphill, for example.

Does the energy source exert a force? A few other examples will further illustrate the distinction between the energy source and the objects or systems that exert the force that sets a body in motion. When the archer shoots his arrow, the string actually exerts a net force on the arrow and is the coupling element to the bow, which originally stored the energy. When a car accelerates, the fuel in the engine is the energy source, but the force that sets the car into motion is exerted by the road through its frictional interaction with the tires. If the car is on an ice patch and the force of friction is small, then the engine turns, the wheels spin, energy is transferred, but the car does not move. When a glass marble bounces on a steel plate (Section 11.7), both the deformed glass and steel act as energy sources, but it is the upward force of the steel plate on the marble that reverses the motion of the marble. When a child rides on a horse on a merry-go-round, the motion is in a circle (not a straight line), hence there is a net force acting on the child. In this example, the energy source is the fuel in the engine, but the force on the child is exerted by the horse on which he or she rides.

Inertial reference frames. Before we conclude this section, we would like to point out one difficulty with Newton's first law. Even though it refers to the motion of a particle, it does not specify the reference frame relative to which this motion is to be observed. A reference frame in which Newton's first law applies is called an *inertial reference frame.* So far we have been using reference frames attached to the surface of the earth, and we have found reasonable agreement between theory and observation. Hence they were (nearly) inertial reference frames. In a reference frame attached to a merry-go-round, however, a stone on the ground and the rest of the earth move in a circle around the merry-go-round! To treat the merry-go-round as an inertial reference frame, you would conclude that starting the motor sets the entire earth in motion and that stopping the motor stops the earth again. Within the Newtonian theory, you then have to invent the necessary forces exerted by external objects on the massive earth to set it in motion. If you are willing to do this, you may go ahead with Newton's theory. If the invention of "fictitious" forces seems too big a price to pay, you may classify the merry-go-round as a non-inertial reference frame in which Newton's first law cannot be applied. The choice is yours, but you must stick to your decision consistently once you have made it.

Moving automobile reference frame. The same considerations apply to the reference frame attached to a moving automobile (Fig. 14.3a). As long as the speedometer reading is a constant 50 miles per hour, the

(a)

(b)

Figure 14.3 A car's brakes are applied.
(a) The reference frame is attached to the car. What object exerts the forces that cause the tree to stop and the driver to lurch forward?
(b) The reference frame is attached to the road. The force of friction exerted by the road on the tires changes the velocity of the car.

landscape moves past at a steady rate. The driver inside is at rest. Neither the landscape nor the driver, therefore, is subject to a net force, a fact compatible with your common sense. Now the driver steps on the brakes. Quickly the landscape comes to a stop, and he lurches forward in his seat. To apply Newton's theory in this reference frame, you must invent a force that causes the landscape to stop and another force that causes the driver to lurch forward. If such forces act, the stopping car is an inertial reference frame. If the invention of such forces seems too farfetched, you may classify the stopping car as a non-inertial reference frame in which Newton's first law does not apply. Again, the choice is yours. When you have ridden in an automobile, you undoubtedly have sensed a "force" that caused you to lurch forward when the car's brakes were applied. You may therefore decide to adopt this reference frame in spite of the problems with the moving landscape.

Road reference frame. Does the automobile problem become simpler in a reference frame attached to the road (Fig. 14.3b)? Such a reference frame is an inertial one. At the beginning, the landscape is at rest (zero net force), while the automobile and the driver inside are both moving steadily (zero net force). Now the driver steps on the brakes. Quickly the car comes to a stop (net frictional force exerted by the road on the car via the tires). The driver continues in motion until a net force acts to bring him to a stop; this may be exerted by the car floor on his feet or by the seat belts on his body. Relative to the car, therefore, the driver lurches forward. Relative to the road, however, he merely comes to a stop a little later than the car does, when the restraint of the

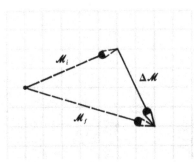

Figure 14.4 Arrows M_i and M_f represent the momenta before and after the action of the force. The change of momentum is $\Delta M = M_f - M_i$

seat belt becomes effective. In the road reference frame it is not necessary to invent interactions that provide the needed forces; the familiar interactions are sufficient. Since physicists are reluctant to invent special solutions to each problem, such as were required when the merry-go-round or the car were treated as inertial reference frames, they prefer to apply Newton's theory in the earth-fixed reference frames for which this step is unnecessary.

Selection of inertial reference frames. As a general rule, a reference frame attached to the most massive object in the system is an inertial frame. For practical purposes this means the earth for terrestrial phenomena, the sun for the solar system, and the galaxy during interstellar travel. If a system includes several objects of comparable mass, such as several stars, then an inertial frame is more difficult to find.

14.3 Newton's second law of motion

Statement of Newton's second law. We have explained that, according to the first law, a change in the state of motion of a particle is accompanied by a net force acting on the particle. The question answered by the second law is just how the unbalanced force and the change in motion are related.

In Chapter 13, velocity and momentum (inertial mass times velocity) were introduced to describe the motion of a particle. During uniform motion, each of these quantities remains constant, that is, retains its magnitude and its direction. Change of motion may be described by a change of either or both quantities. The mathematical model proposed by Newton in his second law relates the net force acting during a time interval to the change in momentum that is caused by the force, as follows.

The average net force times the time interval during which it acts is equal to the change of momentum of the particle on which the force acts (Eq. 14.1a).

The meaning of a change of momentum is illustrated in Fig. 14.4. The change of momentum is the difference between the momentum of the particle after and before the action of the net force. Note that momentum and force have direction and magnitude. The change of momentum therefore also has direction and magnitude (Examples 14.1 and 14.2).

Arrows have the following significance:

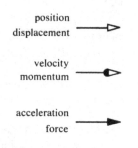

position
displacement

velocity
momentum

acceleration
force

Equation 14·1

average net force (newtons) F_{av}
time interval (sec) Δt
change of momentum (kg-m/sec) ΔM

$$F_{av}\Delta t = \Delta M \quad (a)$$

$$F_{av} = \frac{\Delta M}{\Delta t} \quad (b)$$

The momentum of the particle after the action of the net force depends on the relative direction of the change of momentum and the initial momentum, as shown in the various parts of Fig. 14.5.

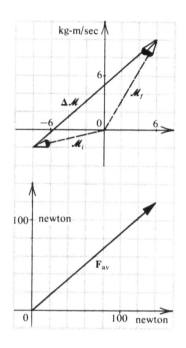

EXAMPLE 14·1. A batter changes the momentum of a baseball from $[-8, -2]$ kilogram-meters per second to $[6, 10]$ kilogram-meters per second by hitting it with his bat. Find the average force under the assumption that ball and bat were in contact for 0.10 second.

Data:

$$\mathcal{M}_i = [-8, -2] \text{ kg-m/sec}, \quad \mathcal{M}_f = [6, 10] \text{ kg-m/sec}$$
$$\Delta t = 0.10 \text{ sec}$$

Solution:

$$\Delta\mathcal{M} = \mathcal{M}_f - \mathcal{M}_i = [6, 10] \text{ kg-m/sec} - [-8, -2] \text{ kg-m/sec}$$
$$= [14, 12] \text{ kg-m/sec}$$

$$\mathbf{F}_{av} = \frac{\Delta\mathcal{M}}{\Delta t} = \frac{[14, 12] \text{ kg-m/sec}}{0.10 \text{ sec}} = [140, 120] \text{ newtons}$$

EXAMPLE 14·2. A car traveling at 60 miles per hour crashes into a tree and comes to a "stop" in 1 second. How large a force is necessary to reduce the driver's momentum so that he does not hit the windshield?

Data:

driver's mass (estimate)	$M_1 \approx 75$ kg
driver's speed	$v = 60$ mph ≈ 27 m/sec
time to stop	$\Delta t = 1$ sec

Solution:

driver's initial momentum:

$$\mathcal{M}_i = M_1\mathbf{v} \approx 75 \text{ kg} \times 27 \text{ m/sec}$$
$$\approx 2000 \text{ kg-m/sec, in the direction of the car's motion}$$

driver's final momentum:

$$\mathcal{M}_f = 0$$

momentum change:

$$\Delta\mathcal{M} = \mathcal{M}_f - \mathcal{M}_i \approx -2000 \text{ kg-m/sec in the direction of motion}$$

$$\mathbf{F}_{av} = \frac{\Delta\mathcal{M}}{\Delta t} \approx \frac{-2000 \text{ kg-m/sec}}{1 \text{ sec}} = -2000 \text{ newtons in the direction of motion}$$

The minus sign means that the force is opposite to the direction of motion. Its magnitude is 2000 newtons or about 450 pounds. This is much more than the force of friction between the driver's pants and the seat. Seat belts can provide the necessary force.

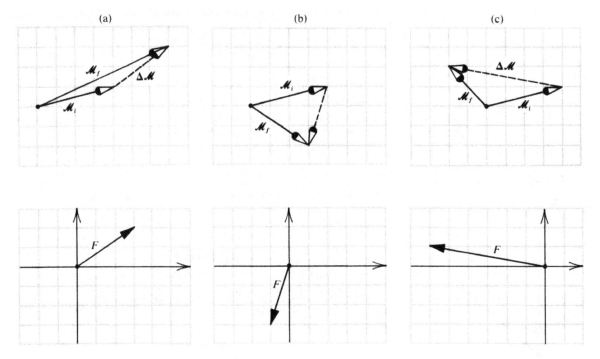

Figure 14.5 Three examples to illustrate how the final momentum of the particle after the action of the force (M_f) depends on the change of momentum (ΔM, shown as a dashed arrow), which is in the direction of the force. The same initial momentum (M_i) is used in each example. In all diagrams: $M_f = M_i + \Delta M$.

Equation 14.2
$$F_{av} = 0 \text{ implies } \Delta M = 0$$

Equation 14.3
$$\Delta t = 0 \text{ implies } \Delta M = 0$$

Immediate consequences of the second law. Several consequences follow immediately from Newton's model. First, the change of momentum is zero if the average net force is zero (Eq. 14.2). This implication is a restatement of the first law, because the change of a particle's momentum is zero only if its velocity is constant.

Second, the change of momentum is zero if the time interval Δt is zero (Eq. 14.3). A nonzero change of momentum can be achieved only by the action of a force during a nonzero time interval. In other words, there are no instantaneous jumps of momentum. Even the bouncing marble (Section 11.7) is brought to rest and reaccelerated in a finite, though extremely short, time interval. This consequence of the theory reflects the inertia of the particles on which the force acts.

Third, the *change* of momentum is in the direction of the net force. A particle initially at rest (zero momentum) acquires a momentum in the direction of the force acting on it. A particle already in motion may acquire an increased momentum, a decreased momentum, or a deflected momentum due to the action of the force (Fig. 14.5). The *final* momentum and, therefore, the final velocity after the action of the force is often *not* in the direction of the force. This is one of the most difficult consequences of the Newtonian theory to accept, because it is contrary to a deeply ingrained notion: that objects move in the same direction as the applied force. This ingrained notion, however, disregards the inertia of particles. Because particles have inertia, *both* their previous momentum *and* the force acting at any instant determine what they will do next, that is, their motion in the immediate future.

Acceleration. To use the mathematical model in Newton's second law for describing moving particles whose position and velocity are observed, we will rephrase it so that it refers directly to acceleration instead of to momentum. The definitions of acceleration and momentum given in Sections 13.2 and 13.3 are repeated here: acceleration is the change of velocity divided by the time interval (Eq. 13.9); momentum is the product of inertial mass times velocity (Eq. 13.10). You may wish to refer to Section 13.2 for an explanation of the acceleration concept in physics and how it differs from everyday usage of the word.

The momentum of a particle may change because its inertial mass changes, because its velocity changes, or because both mass and velocity change. We will pursue only the theory for particles whose velocity may change, but whose inertial mass remains constant. In this theory, the change of momentum of a particle is equal to its inertial mass multiplied by the change of velocity (Eq. 14.4). When this mathematical model for momentum change is used in Newton's second law, the latter can be put into the form that the average net force equals the inertial mass times the average acceleration (Eq. 14.5). By choosing sufficiently short time intervals, the average force and average acceleration can both be made approximately equal to the instantaneous force and acceleration as in Eqs. 2.3 and 13.2 for the speed and the velocity. The resulting form of Newton's second law, net force equals mass times acceleration (Eq. 14.6), is the one most frequently quoted.

Predictions from Newton's second law. With the help of the formulation of Newton's second law in Eq. 14.6, you can compare some of its predictions with experimental observations to determine how satisfactory the theory really is. A few consequences of the model can be inferred directly from Eq. 14.6:

1. Constant net force produces a constant acceleration.

2. Zero net force produces a zero acceleration.

3. Net force is directly proportional to acceleration (doubling the force doubles the acceleration).

4. The force and the acceleration are in the same direction.

5. Two different particles experience equal accelerations if the net forces on them are in the same direction and the more massive particle is subject to a larger force in proportion to its larger inertial mass.

Some of these predictions, such as 2, 4, and 5, can be tested experimentally by direct observation and without measurements of the magnitude of the acceleration. Predictions 1 and 3 require more extensive measurement of acceleration. How the acceleration can be calculated approximately from successive observation of the position of a moving object is explained in the next paragraphs.

Equation 13·9

$$\mathbf{a}_{av} = \frac{\Delta \mathbf{v}}{\Delta t}$$

Equation 13·10

$$\mathcal{M} = M_I \mathbf{v}$$

Equation 14·4

$$\Delta \mathcal{M} = M_I \Delta \mathbf{v}$$

Equation 14·5

$$\mathbf{F}_{av} = M_I \frac{\Delta \mathbf{v}}{\Delta t}$$

$$= M_I \mathbf{a}_{av}$$

Equation 14·6

instantaneous force   **F**
instantaneous acceleration   **a**

$$\mathbf{F} = M_I \mathbf{a}$$

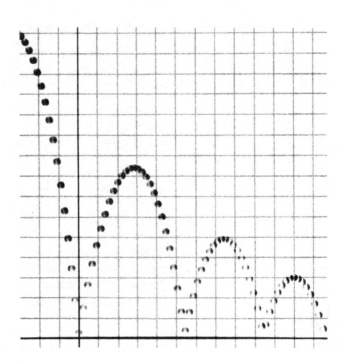

Figure 14.6 Multiflash photograph of a bouncing golf ball. The grid lines, whose spacing represents 0.11 meter, indicate the distance scale. The time intervals are 0.033 seconds.

Equation 13·1

average velocity	v_{av}
time interval between flashes	Δt
displacement between flashes	Δs

$$v_{av} = \frac{\Delta s}{\Delta t}$$

Multiflash photographs. A record of the successive positions of the moving object is made conveniently through the technique of multi-flash photography, described in Section 13.1. By flashing a light on the photographic subject repeatedly without changing the film, you obtain a multiple exposure on which the positions of the object at the times of the flashes can be seen. Eq. 13.1 relates the time interval between flashes, the average velocity, and the displacement. A multiflash photograph of a bouncing golf ball is shown in Fig. 14.6. The multiflash photograph records motion relative to the camera.

"Frictionless" motion. To conduct laboratory experiments with controlled forces, we must eliminate the effects of gravity and friction. The former can be accomplished by restricting the motion to a horizontal surface, such as a table, that exerts a balancing upward force on the moving object (imagine an MIP model for the rigid table) and thereby prevents motion in the vertical direction. Friction cannot be eliminated completely, but it can be made very slight, as we explained in Section 3.4, by maintaining a thin layer of gas between the moving object and its supporting surface. Then the net force acting on the object is equal to the sum of the horizontal partial forces acting on it.

In one technique to reduce friction, a table or track is provided with many holes through which jets of air escape. As shown in Fig. 3.4, if a smooth object is placed on such a surface, the jets lift the object so it can "float" on a thin cushion of air and thus can glide almost without friction. Another device consists of a disk or "puck" that carries a container filled with dry ice (frozen carbon dioxide) (Fig. 14.7). As the carbon dioxide vaporizes, it escapes through a small hole in the center of the bottom of the disk (Fig. 14.8) and forms a thin layer of carbon dioxide gas between the disk and the glass surface on which it rests.

Figure 14.7 A "frictionless" dry ice puck.

Figure 14.8 Carbon dioxide gas escapes through the hole in the puck base to form a thin film on which the "frictionless" puck glides.

This gas layer reduces the frictional force between the two surfaces to a very small value, so that we can study the motion of the disk without being distracted by the usual presence of friction.

Figure 14.9 shows a multiflash photograph of a dry ice puck moving from left to right alongside a meter stick. The light was flashed at the rate of 24 times in 10 seconds, or at intervals of 0.42 second. The speed of the puck can be calculated from a measurement of its displacement during each time interval (Table 14.2). Since the images are equally spaced, the displacements during equal times are equal, and the puck in this experiment approaches the ideal of constant velocity.

What would be the conclusion if the velocity were found to change? You could then conclude that there was an interaction influencing

Figure 14.9 A dry ice puck moving parallel to the meter stick was photographed at intervals of 0.42 second. Data from this experiment are analyzed in Table 14.2.

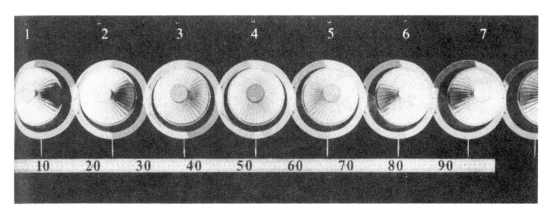

TABLE 14·2 MOTION OF THE FRICTIONLESS PUCK IN FIG. 14·9

Pictures defining interval	Time interval (sec)	Displacement (m)	Velocity (m/sec)
1–2	0.42	[0.14, 0.00]*	[0.33, 0.00]
4–5	0.42	[0.14, 0.00]	[0.33, 0.00]
6–7	0.42	[0.14, 0.00]	[0.33, 0.00]

*Displacement components are measured to the right and up on Fig. 14·9.

the puck—perhaps friction, perhaps a slight tilt in the glass surface, or perhaps an air current. In other words, this experiment does not "prove" that free bodies move with constant velocity; instead it provides evidence that the puck is a very nearly free body according to Newton's first law.

Experimental test of Newton's second law. To observe the motion of a puck subject to a constant force, we attach a string and a circular spring (Fig. 14.10), which is a convenient elastic object for monitoring the constancy of the applied force. When the spring is deformed into an ellipse of predetermined standard shape, it transmits one unit of force, which we will call an *su* (for spring unit). How this force is related to the newton we do not know.

In Fig. 14.11, two forces, each 1 su in strength, are acting on the puck in opposite directions. The net force here is zero. In Fig. 14.12, two equal magnitude forces slightly larger than 1 su are acting at an angle; the net force acts in the direction bisecting the angle between the two forces.

Figure 14.10 (below). The circular spring is attached to the dry ice puck to measure the applied force.

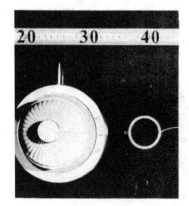

Figure 14.11 (below). Two equal-magnitude but oppositely directed forces act on the puck.
(a) The puck with deformed springs attached is in mechanical equilibrium,
(b) Force diagram, showing the combination of the two partial forces to give a zero net force.

(a)

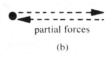

partial forces

(b)

Figure 14.12 Two equal-magnitude forces at a slight angle act on the puck.
(a) The puck with deformed springs attached is not in mechanical equilibrium,
(b) Force diagram showing the combination of the two partial forces to give a nonzero net force.

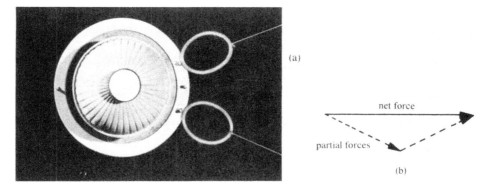

(a)

(b)

Comparison of two forces. The effects on the puck of a force of 1 su and of a force of 2 su are shown in the multiflash photographs in Fig. 14.13 and Fig. 14.14, respectively. You can see that the pucks move in a straight line (as defined by the meter stick) in the direction of the force, and that the speed is not constant but increases. According

Figure 14.13 The dry ice puck is accelerated to the right by the action of a net force of 1 su,

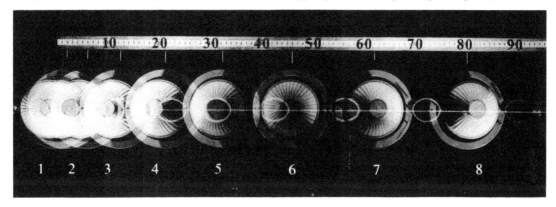

Figure 14.14 The dry ice puck is accelerated to the right by the action of a net force of 2 su.

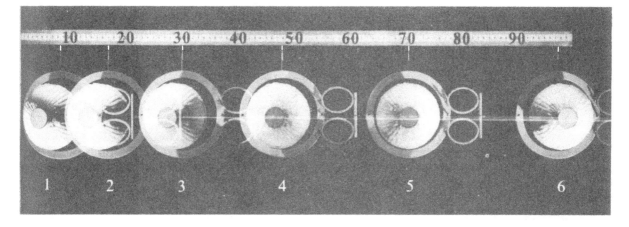

TABLE 14·3 MOTION OF PUCK IN FIG. 14·13

Pictures defining interval	Time interval (sec)	Displacement (m)	Average velocity (m/sec)	Change of average velocity (m/sec)	"Acceleration" (m/sec/sec)
1–2	*0.42*	*[0.041, 0.000]*	*[0.10, 0.00]*	*[0.05, 0.00]*	*[0.12, 0.00]*
2–3	*0.42*	*[0.063, 0.000]*	*[0.15, 0.00]*		
4–5	*0.42*	*[0.122, 0.000]*	*[0.27, 0.00]*	*[0.05, 0.00]*	*[0.12, 0.00]*
5–6	*0.42*	*[0.135, 0.000]*	*[0.32, 0.00]*		

to the Newtonian theory, the constant net force acting on the puck in each experiment should result in a constant acceleration. The force of 2 su acting in Fig. 14.14 should produce twice the acceleration as the force of 1 su in Fig. 14.13. The acceleration calculated in Tables 14.3 and 14.4 gives evidence that the Newtonian theory can be applied successfully to the motion of the dry ice puck.

Calculation of acceleration. As a matter of fact, we are unable to calculate the acceleration as defined in Eq. 13.9, because the multi-flash photographs do not enable us to determine the puck's instantaneous velocity. We have therefore computed an approximate "acceleration" in the last columns of Tables 14.3 and 14.4. You can see that the data in these two columns are in the ratio of two-to-one, just as the forces are in the ratio of two-to-one. In fact, you can see that all corresponding entries for the two pucks in Tables 14.3 and 14.4 are in this ratio. Because of this fact, you can imagine the outcome of a thought experiment in which the flashes occur at very short time intervals. The calculated average velocities will then be very close to the instantaneous velocities, the accelerations can be found from Eq. 13.9, and all numerical values will still be in the ratio two-to-one.

Other applications. With this positive outcome, we invite you to use the Newtonian theory to describe objects that are moving subject to the gravitational interaction with the earth. These investigations will lead to further evidence of the theory's usefulness. In Chapter 15, we will derive a theory of periodic motion and apply it to the planets in the solar system as well as to pendulums and other oscillators on earth.

14.4 Motion near the surface of the earth

All objects near the surface of the earth are subject to the force of gravity. They may also be subject to other forces, such as the elastic force exerted by the floor on a bouncing golf ball, the force of friction

From here on we will merely write speed, velocity, acceleration, momentum, *and* force *to refer to the instantaneous quantities. The average quantities will be designated as such so you may identify them properly (*average speed, average velocity, . . .*).*

TABLE 14·4   MOTION OF PUCK IN FIG. 14·14

Pictures defining interval	Time interval (sec)	Displacement (m)	Average velocity (m/sec)	Change of average velocity (m/sec)	"Acceleration" (m/sec/sec)
1–2	0.42	[0.083, 0.000]	[0.20, 0.00]		
				[0.10, 0.00]	[0.24, 0.00]
2–3	0.42	[0.128, 0.000]	[0.30, 0.00]		
4–5	0.42	[0.226, 0.000]	[0.54, 0.00]		
				[0.10, 0.00]	[0.24, 0.00]
5–6	0.42	[0.270, 0.000]	[0.64, 0.00]		

exerted by the road on a skidding car, or the force of friction exerted by the air on a flying airplane, but all objects near the earth are, at all times, certainly subject to the downward force of gravity.

The flat earth model. As we have described in Section 11.3, the gravitational intensity near the surface of the earth has the constant magnitude of 10 newtons per kilogram and a direction toward the center of the earth. Since the earth is a sphere that is very large compared to the macro-domain size of everyday phenomena, we will make a working model in which the earth is a flat horizontal surface and the direction of the gravitational intensity is vertical, at right angles to the horizontal plane (Fig. 14.15). The force of gravity on an object near the surface of the earth is therefore a constant force, determined by the object's gravitational mass, but independent of the object's position (Eq. 14.7). Many objects remain at rest even though they are subject to the force of gravity, because they are supported in such a way that the net force acting on them is zero.

Free fall. When all supports are removed, however, an object begins to fall and picks up speed, as shown in the multiflash photograph

Equation 14.7 (same as Eq. 11.4)

force of gravity
(newtons) $= F_G$
gravitational mass (kg) $= M_G$
gravitational intensity
(newtons/kg) $= g$

$F_G = gM_G$

(at surface of earth $g = 10$ newtons/kg, downward.

Figure 14.15 The flat earth model. The gravitational intensity is vertical everywhere and has the same magnitude everywhere.

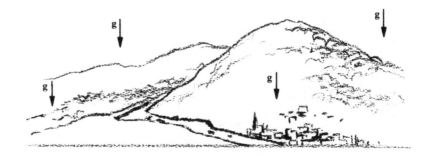

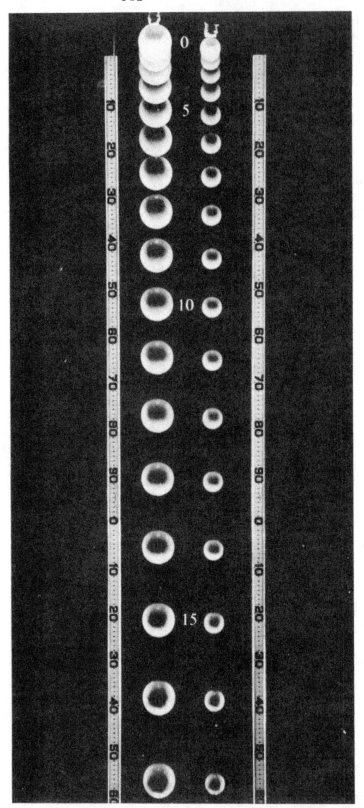

Figure 14.16 A baseball and a golf ball were released from rest at the same time. The time intervals between photographs were Δt= 0.033 second. Note the increasing displacement between successive images of each ball. In the text, we use measurements on this photo to show that the distance covered is proportional to the elapsed time to the second power.

We can also use the photo to answer a more straightforward question: does the heavier object fall faster? As you can verify in the photo, the displacements of the balls, and therefore their velocities, are equal *throughout the motion.* Conclusion: *the velocities of two objects with different masses dropped simultaneously from rest increase at the same rate in free fall. This may seem contrary to common sense, but common sense is wrong in this case! By measuring the displacements carefully on the meter sticks, can you verify Eq. 14.7 (acceleration = **g** = 10 m/sec/sec, downward)?*

Figure 14.17 (below) A particle is falling under the influence of gravity. It started at the coordinate origin with zero velocity. Its velocity and position increase as the first and second powers of the elapsed time, respectively. These math models are derived in Eqs. 14.9-14.12.

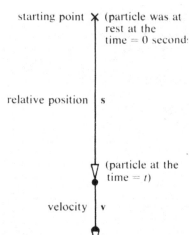

starting point ✕ (particle was at rest at the time = 0 second·

relative position │ s

(particle at the time = *t*)

velocity │ v

Equation 14.8

$$a = \frac{F_G}{M_I} = g\frac{M_G}{M_I} = a_{av}$$

Equation 14.9

actual velocity $= v$
average velocity $= v_{av}$

$$v_{av} = \frac{1}{2}v$$

Equation 14.10

elapsed time of
falling $= t$
position relative to
starting point $= s$

$$s = v_{av}t = \frac{1}{2}vt$$

Equation 14.11

$$v = a_{av}t$$

Equation 14.12

$$s = \frac{1}{2}vt = \frac{1}{2}a_{av}t^2$$

$$= \frac{1}{2}g\frac{M_G}{M_I}t^2$$

Equation 14.13

$$\frac{M_G}{M_I} = 1 \text{ or } M_G = M_I$$

of a falling baseball and a falling golf ball in Fig. 14.16. Since each ball is subject to a constant net force, the Newtonian theory (Eq. 14.6, $\mathbf{F} = M_I\mathbf{a}$, combined with Eq. 14.7, $\mathbf{F_G} = \mathbf{g}M_G$) predicts that it should fall with *constant* acceleration equal to the gravitational intensity times the ratio of gravitational to inertial masses [$\mathbf{a} = \mathbf{g}(M_G/M_I)$]. The acceleration must be constant because $\mathbf{g}$, M_G, and M_I are all constant; in addition, a constant acceleration is always equal to the average acceleration (Eq. 14.8).

Velocity and position during free fall. Since the two balls start out at rest and suffer a constant acceleration, it is relatively easy to make a mathematical model for relating the distance they fall to the elapsed time. First, for a particle that starts from rest (zero initial velocity), the change of velocity is equal to the actual velocity and the displacement is equal to the position relative to the starting point (Fig. 14.17). Second, the average velocity of such a particle, whose speed builds up steadily from zero to its final value, is equal to one half of the actual velocity (Eq. 14.9). Third, the average velocity and the elapsed time can be multiplied, according to Eq. 13.5, to give the position of the particle relative to the starting point (Eq. 14.10). Fourth, the equation defining the average acceleration (Eq. 13.9, $\mathbf{a}_{av} = \Delta\mathbf{v}/\Delta t$) can then be rewritten to give the velocity of the falling particle, which is directly proportional to the time of falling (Eq. 14.11). Finally, this formula for the velocity and Eq. 14.8 are used to relate the distance of fall to the elapsed time (Eq. 14.12). The result is that the distance is proportional to the elapsed time raised to the second power.

In Table 14.5 this mathematical model is compared with the measurements made on the falling balls in Fig. 14.16. The prediction depends, as you might have expected, on the ratio of gravitational to inertial masses of each ball. You can see, however, that the prediction agrees well with the data if the ratio of masses is equal to one (Eq. 14.13).

Comparison of gravitational and inertial masses. A close relation between the two masses of each ball can be inferred even without measurements. The photograph shows that the accelerations of the two balls are at least approximately equal because they appear to start out and remain side by side. According to Newton's theory, the net forces on them are in proportion to their inertial masses (prediction 5, page 375).

TABLE 14·5 DISTANCE OF FREE FALL FROM EXPERIMENT AND THEORY

Time (sec)	Distance – experiment (Fig. 14·16) (m)	Distance – theory (Eq. 14·12) (m)
0.000	0.00	0.00 (M_G/M_I)
0.167	0.15	0.14 (M_G/M_I)
0.333	0.56	0.55 (M_G/M_I)
0.500	1.25	1.25 (M_G/M_I)
0.567	1.59	1.60 (M_G/M_I)

But you know from Eq. 14.7 that the forces of gravity acting on them are proportional to their gravitational masses. It appears, therefore, that gravitational and inertial masses are proportional to one another! You can infer from the data in Table 14.5 that the two masses of each ball are numerically equal (Eq. 14.13).

As a matter of fact, it is found that all objects that fall freely in the earth's gravitational field experience the same acceleration of 10 meters per second per second. Even objects such as feathers, which ordinarily are affected greatly by the air, experience this acceleration in an air-free tube. This acceleration is numerically equal to the magnitude of the gravitational intensity (Eq. 11.3).

There are two physically equivalent but logically distinct views you may now take. One is that the two types of mass are *not* exactly equal and that the relative degree of equality must be established experimentally. As explained above, the data in Figure 14.16 and Table 14.13 show that $M_G = M_I$ to an accuracy of about 1% (1 part in 100). This issue is of great importance in physics, and experimenters (including Newton himself) have carried out a variety of ingenious and sensitive tests. The latest attack on this problem was by Adelberger in 1990, confirming that $M_G = M_I$ to an accuracy of 1 part in 10^{12}! A new experiment is now planned which would measure with great accuracy any differences in acceleration of two weights in "free fall" while orbiting the earth; the experimenters expect an accuracy of 1 part in 10^{18}!

The second view is to modify Newton's theory by simply adding the assumption that M_G and M_I are equal. This agrees with common sense; the modified theory then predicts that the acceleration of free fall is exactly equal to the gravitational intensity (Eq. 14.14, from Eqs. 14.8 and 14.13). Newton took this view, as you can verify in Table 14.1. He did not state a definition for inertial mass, but based his definitions of momentum and inertia on the gravitational mass concept he had defined earlier. Einstein also adopted this assumption (called the Principle of Equivalence) in his general theory of relativity. We will take the same view and assume for the rest of this text that $M_G = M_I$ (Eq. 14.13) holds exactly with no experimental uncertainties.

Acceleration of gravity. The gravitational intensity (**g**) is often called the *acceleration of gravity.* Its value in newtons per kilogram is equal to its value in meters per second per second. To achieve this equality, we based the definition of the joule (Sect. 9.2) and the definition of the newton (Sect. 11.2) on an object with mass of 0.10 kilogram. As you saw in Sect. 11.2, this choice for the newton determined the value of **g** (Eq. 11.3).

The acceleration of all objects falling freely near the surface of the earth is close to 10 meter/sec/sec in the vertical direction (Eq. 14.14). A very useful property of all uniformly accelerated objects is that their average velocity for any time interval is halfway between their initial and final velocities (Eq. 14.15). An application of this formula is worked out in Example 14.3.

You must remember that these results are based on the Newtonian one-particle theory and take into account only the force of gravity. If

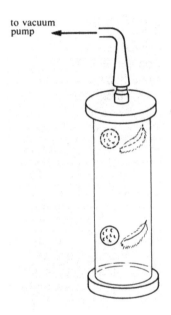

to vacuum pump

Equation 14.14

free fall near earth's surface

$$a = g = 10 \text{ m/sec/sec, downward}$$

Equation 14.15

initial velocity $= v_i$
final velocity $= v_f$
average velocity $= v_{av}$
$$v_{av} = \tfrac{1}{2}(v_i + v_f)$$

air resistance is important, if the object is partially supported by a sloping surface (like the skier on a hill), or if still other interactions must be considered, then a more complete theory within the Newtonian framework must be used.

EXAMPLE 14·3.   A stone is dropped from the Leaning Tower of Pisa. (a) Find the speed at various times after release. (b) Find the distance fallen at various times after release.

Solution:

(a) Use Eqs. 14·11, 14·13, and 14·14 or $v = 10t$.

$t = 0$ sec: $v = 10 \times 0 = 0$ m/sec
$t = 1$ sec: $v = 10 \times 1 = 10$ m/sec
(and so on for $t = 2, 3, 4, \ldots$, see table).

(b) I. Use Eqs. 14·12, 14·13, and 14·14: $s = \frac{1}{2}gt^2$ or $s = (5t^2$, downward), $|s| = 5t^2 = s$.

$t = 0$ sec: $s = 5 \times 0 = 0$ m
$t = 1$ sec: $s = 5 \times 1^2 = 5$ m
$t = 2$ sec: $s = 5 \times 2^2 = 20$ m
(and so on for $t = 3, 4, \ldots$, see table).

II (alternate). Use Eq. 14·10 and solution to (a).

$$|s| = \frac{1}{2} vt \text{ and } v = 10t$$

$t = 1$ sec, $v = 10$ m/sec:

$$|s| = \tfrac{1}{2}vt = \tfrac{1}{2} \times 10 \text{ m/sec} \times 1 \text{ sec} = 5 \text{ m}$$

Table of speeds and distances of free fall

Time (sec)	Speed (m/sec)	Average speed (m/sec)	Distance (m)
0	0	0	0
1	10	5	5
2	20	10	20
3	30	15	45
4	40	20	80*

*The leaning tower is only 54 meters high. Hence the stone strikes the ground between 3 and 4 seconds after release and the mathematical models used in the calculation are no longer valid. To calculate the time before impact, use $|s| = 5t^2$.

$|s| = 54$ m, or $t = \sqrt{|s|/5} = \sqrt{54/5} \approx \sqrt{10.8}$ sec ≈ 3.3 sec.

14.5 Newton's third law of motion

Even though the one-particle model for a moving body has been extremely valuable in leading to a successful description of the overall behavior of moving bodies, it clearly has limitations. In fact, these limitations are evident whenever you look more closely at a moving body and realize that it actually is a system of interacting parts, like the skier, who can maneuver his legs to make a turn; the stunt car, the wheels of which spin when it accelerates; and the interplanetary rocket, which leaves behind a trail of hot exhaust as it hurtles into space. For a more complete understanding of how these systems function, you must relate their motion to the operation and interaction of their various parts. The one-particle model, in which the entire system is represented by a single particle at the center of mass, is not adequate.

Newton's third law supplements the first two in such a way that they can be applied to systems represented by complicated models consisting of many particles, ultimately even atoms or molecules if necessary. The most widely known statement of Newton's third law follows. *For every action, there is an equal and opposite reaction.*

Interpretations of the third law. The words "action" and "reaction," which are no longer part of the physics vocabulary, signify forces acting during a time interval. We may, therefore, interpret the third law as stating that the interaction of two bodies is described by two forces equal in magnitude and opposite in direction. This interpretation (see Section 11.2) applies directly to the two forces (Fig. 11.2). We have another option, however. Since a force acting during a time interval produces a change of momentum, two equal but opposite forces acting during the same time interval produce equal but opposite changes of momentum. We may, therefore, also interpret the third law as stating that the interaction of two bodies results in changes of their individual momenta that are of equal magnitude but oppositely directed.

Conservation of momentum. The second of these interpretations is especially significant, because it leads to the law of *conservation of momentum.* If the interaction of two bodies results in equal and opposite changes of momentum, the sum of the momenta of the two bodies is not changed by their interaction. The momentum gained by one body is lost by the other. There is transfer of momentum from one body to the other. If three or more bodies in a system interact at once, momentum may be transferred between the members of each interacting pair, with the result that the gains and losses of momentum balance out and the sum of all the momenta does not change. This is the law of conservation of momentum: The bodies in an isolated system can exchange momentum, but the total momentum of the system is conserved (constant).

The law of conservation of momentum, which we have here derived from Newton's theory, has turned out to be much more generally valid than Newton's theory itself. For instance, the law applies to radiation and to phenomena in the micro domain, for both of which

"The mutual actions of two bodies upon each other are always equal and oppositely directed."

Isaac Newton
Principia, **I**, 1687

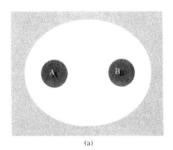

(a)

(b)

Figure 11.2
(a) Two interacting bodies
(b) The two forces of interaction. One force acts on each body.

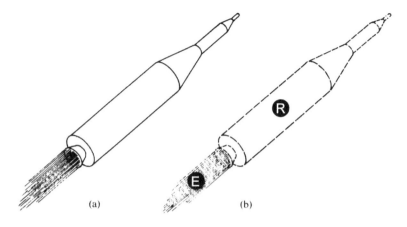

Figure 14.18 Two-particle model for a rocket,
(a) The rocket while engine is in operation,
(b) The rocket, with particle R representing the rocket-plus-fuel subsystem, and particle E representing the exhaust sub-system.

Newtonian mechanics is inadequate (see Chapters 7 and 8). This law, along with the law of conservation of energy (Section 4.1), is one of the cornerstones on which all current physical theories are built.

Rocket propulsion. As an example of how to apply Newton's theory, we present a two-particle theory of rocket propulsion (Fig. 14.18). A rocket is a system that ejects material through a nozzle at one end. Consider a particular 1-second interval while the rocket engine is burning fuel and ejecting the combustion products through nozzles at the rear of the rocket. The rocket-plus-remaining-fuel after the interval is one "particle" in the model, the material ejected during the 1-second interval is the other "particle." At the beginning of the 1-second interval, the two "particles" move at the same velocity (Fig. 14.19a). Each has a momentum determined by this velocity and its own mass. At the end of the 1-second interval, the ejected material is in the exhaust trail and no longer has the velocity it had before (Fig. 14.19b). It therefore has a

Figure 14.19 (to right). Conservation of momentum in rocket propulsion.
(a) Rocket (R) and the fuel-oxygen subsystem about to burn (E) are traveling at the same speed (v_i) relative to the earth reference frame.
(b) Fuel-oxygen subsystem, now exhaust (E), is ejected, at speed $-v_{exhaust}$, (relative to the rocket) so it loses momentum relative to the earth. This momentum is transferred to the rocket subsystem (R), which then moves faster (v_f) relative to the earth reference frame.

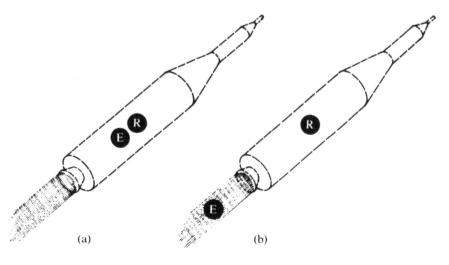

changed momentum. According to Newton's third law, the momentum is not lost from the two-particle system, but it is transferred to the rocket and the remaining fuel. The rocket "particle" therefore moves on with an increased velocity. The process is now repeated, with a new two-particle system being made out of the rocket and its fuel to describe the ejection of material during the next 1-second time interval. And so on.

The quantity of momentum transferred in 1 second is a measure of the strength of the interaction between the two "particles." It is determined by the rate of fuel consumption, temperature in the combustion chamber, and other construction details of the rocket engine. The momentum transferred in 1 second is called the *thrust* of the engine (Example 14.4) and is usually measured in newtons or pounds (1 pound of thrust is approximately 4.5 newtons). A rocket engine is sometimes called a *reaction motor* because its functioning depends on the action-reaction principle of Newton's third law.

EXAMPLE 14·4. A rocket motor burns 1 ton of fuel–oxygen mixture per second. The exhaust speed is 700 meters per second relative to the rocket. Find the thrust of the rocket. (1 ton = 1000 kilograms.)

Solution: Calculate the thrust while the rocket is strapped down for a test firing.

Data: for "exhaust particle,"
$M = 1000$ kg; $|\Delta v| = 700$ m/sec; $\Delta t = 1$ sec

$$|F| = \frac{M|\Delta v|}{\Delta t} = \frac{1000 \text{ kg} \times 700 \text{ m/sec}}{1 \text{ sec}} = 700,000 \text{ newtons}$$

thrust = 700,000 newtons or 160,000 lbs

Question: Does the thrust change while the rocket is moving? Repeat the calculation for the rocket moving at 1000 meters per second. Then the exhaust moves at 300 meters per second.

14.6 Kinetic energy

If you have a driver's license, you very likely were taught that the distance required to stop a car increases fourfold when its speed doubles. Have you ever wondered why? When a bicycle rider approaches a hill, she usually pedals as fast as she can so that she will get to the top of the hill more easily. Just how far up will her speed carry her? In both these examples, there is a transfer of energy from kinetic energy to another type: thermal energy of the brakes, or gravitational field energy of the bicycle, rider, and earth system.

As we have said in Chapter 4, kinetic energy is the energy stored in moving objects. Thus, the kinetic energy of the car determines how far it will advance as the brakes bring it to a stop. The bicyclist maximizes her kinetic energy as she approaches the hill.

When a force acts on a particle, its velocity or momentum changes, and usually its energy changes also. In this section we will derive a mathematical model for the relation of kinetic energy to speed. We will show how this relation can be used in conjunction with the law of conservation of energy to predict the motion of objects under many circumstances, such as the car coming to a stop and the bicycle moving uphill.

Equation 14·16

kinetic energy   KE
work   W
net force   F
displacement component
  along the force
  direction Δs_F

$$KE = W = |F|\Delta s_F$$

Derivation. Instead of constructing the model in the light of experimental results, we will derive it from Newton's theory. Imagine a particle at rest (zero speed, zero kinetic energy) that is acted upon by a constant net force until it is moving with the velocity v. The kinetic energy of the particle is, according to the law of conservation of energy, equal to the work done by the net force (Eq. 14.16). To find the work, we have to calculate the distance through which the particle moved while it was being accelerated by the action of the force.

This problem is very similar to the problem of free fall solved in Section 14.4. There, too, a constant force speeded up a particle that was initially at rest. The principal differences between that and the present tasks are that now the force can be any force (not only the force of gravity), and the motion can occur in any direction (not only vertically). Still, the motion and the force are in the same direction, because the particle starts from rest (Fig. 14.20).

Equation 14·17

position relative to starting
  point   s
velocity   v
elapsed time   t

$$s = \tfrac{1}{2}vt$$

The relative position of the particle is equal to one half of the velocity times the time (Eq. 14.17 from Eq. 14.10, $s = v_{av}t = (1/2)vt$). The net force also can be related to the actual velocity (equal to the change of velocity) and to the elapsed time (Eq. 14.18 from Eq. 14.5, $F_{av} = M_1a_{av}$). Since the force, the velocity, and the relative position are all in the same direction, the component of the displacement along the force direction is equal to the magnitude of the relative position (Eq. 14.19). When the formulas are combined to calculate the work and therefore the kinetic energy, we obtain a mathematical model (Eq. 14.20), which has been simplified using the fact that the magnitude of the velocity is

Equation 14·18

mass   M

$$F = M\frac{v}{t}$$

Equation 14·19

speed   v

$$\Delta s_F = |s| = \tfrac{1}{2}vt$$

Equation 14·20

$$KE = |F|\Delta s_F = M\,\frac{|v|}{t} \times \frac{1}{2}\,vt$$

$$= \frac{1}{2}\,Mv^2$$

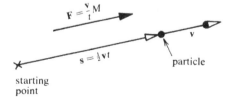

Figure 14.20 The kinetic energy of a particle is equal to the work done by a constant force that accelerates the particle from zero velocity to its actual velocity. The force required and the position relative to the starting point reached by the particle are related to the velocity by Eqs. 14.17 and 14.18.

equal to the speed (Eq. 13.8). Thus, the kinetic energy equals one half times the mass times the speed to the second power (Example 14.5).

Applications. This mathematical model for kinetic energy may be applied to moving bodies, such as automobiles, footballs, spaceships, hailstones, pendulums, and the earth in its orbit around the sun. However, this model does not apply directly to sound and light, since both of these phenomena are described better by a wave model than by

---

EXAMPLE 14·5. Find the kinetic energy of a car with mass of 1500 kilograms moving at a speed of 27 meters per second (60 miles per hour).

Solution:

$$KE = \frac{1}{2} Mv^2 \approx \frac{1}{2} \times 1500 \text{ kg} \times (27 \text{ m/sec})^2$$

$$\approx 750 \times 730 \approx 550,000 \text{ joules}$$

EXAMPLE 14·6. Calculate the braking distance of a car. The force of friction on good roads is about one half the force of gravity. What is the braking distance at various speeds?

Data:

braking distance s, car speed v.

force of friction $|\mathbf{F}| \approx \frac{1}{2} |\mathbf{g}| M$, direction opposite to motion

work done by car $W = |\mathbf{F}| \Delta s_F \approx \frac{1}{2} |\mathbf{g}| Ms$

kinetic energy $KE = \frac{1}{2} Mv^2$

Solution:

Kinetic energy is completely converted to thermal energy by friction:

$$KE = W \quad \text{or} \quad \frac{1}{2} Mv^2 \approx \frac{1}{2} |\mathbf{g}| Ms$$

$$s = \frac{v^2}{|\mathbf{g}|} = \frac{1}{10} v^2$$

Table of results

Speed		Braking distance	
(m/sec)	(mph)	(m)	(ft.)
0	0	0	0
10	23	10	33
20	45	40	130
30	68	90	295
40	90	160	525

a particle model. The mathematical model for kinetic energy (Eq. 14.20, KE = ½Mv²) *can* be used in conjunction with the MIP model for matter by applying the equation to each separate particle in the model system.

With the mathematical model for kinetic energy, you can apply the law of conservation of energy to processes involving moving particles, just as you used the mathematical model for thermal energy to predict temperature changes. As a first example, we return to the braking distance of a speeding automobile. When the brakes are applied, energy is transferred from the kinetic energy of the car to thermal energy of the brake linings by way of the frictional interaction of the brakes. Doubling the speed of the car increases its kinetic energy fourfold (KE depends on v², Eq. 14.20); therefore, the car must do four times the work, which requires traveling four times as far, so as to permit the transfer of all this energy to the brakes (Example 14.6 above). We would thus expect to find four times as much thermal energy in the brakes after the stop.

Summary

Isaac Newton formulated a theory of moving bodies that applies to all macro-domain phenomena except radiation and is still one of the foundations of physics as well as an outstanding model of a successful physical theory. The central quantitative concepts in Newton's theory are force, mass, velocity, momentum, and acceleration. To deal with instantaneous velocity and acceleration, Newton developed a new branch of mathematics now called the calculus.

Newton's approach was one of reducing complex phenomena to simple ones. Just as a house may be built of bricks, so it is possible to build theories of motion for complex systems (the entire solar system, a bicycle with many moving parts, and even fluids), out of Newton's laws of motion for one particle.

Mass, momentum, velocity, and acceleration are used to describe the motion of each particle. The influence on each particle of its interaction with all other particles is summarized in the net force concept. This influence is described in the first and second laws of motion.

Newton's first law: Every particle persists in its state of rest or of uniform, unaccelerated motion unless it is compelled to change that state by the application of an external net force.

Newton's second law: The change of momentum of a particle subject to a net force during a time interval is equal to the average net force times the duration of the time interval.

Since motion is defined only in relation to a reference frame, the applications of Newton's first and second laws require the prior selection of a suitable reference frame, which is called an inertial frame. Newton's first law may be used to select such a reference frame; those frames in which the first law is contradicted by observations are not inertial frames and must not be used. Once an inertial frame has been selected, the second law is used to make quantitative predictions of the motion of particles relative to this frame.

Equation 14.5

net force   = *F*
mass   = *M*
acceleration   = *a*

$$F = Ma$$

Equation 14.13

inertial mass   = M_I
gravitational mass = M_G

$$M_I = M_G$$

Equation 14.20

kinetic energy = *KE*
mass   = *M*
speed   = *v*

$$KE = \tfrac{1}{2}mv^2$$

The most convenient mathematical statement of Newton's second law is given in Eq. 14.5. This formula may be applied with great effectiveness to the motion of particles near the surface of the earth, where they are subject to the force of gravity described in Section 11.3. One of the outcomes of these applications is support for the assumption that the gravitational mass of a particle equals its inertial mass (Eq. 14.13). Newton's theory of particle motion can also be applied to predict the overall motion of complex systems by means of a one-particle center-of-mass model.

To build up a more complete theory of complex systems, Newton introduced his third law of motion.

Newton's third law: The interaction of two bodies is described by two equal and opposite forces; or, the interaction of two bodies results in a transfer of momentum between them.

This law is used to relate the motion of one particle, controlled by the forces acting on it, to the forces it in turn exerts on the other particles. Newton's third law also leads to the law of conservation of momentum for complex systems.

A mathematical model for the kinetic energy of a particle moving with a certain speed can be derived from Newton's laws of motion. The kinetic energy is proportional to the speed to the second power (Eq. 14.20).

Additional examples

EXAMPLE 14.7. An arrow with a mass of 0.1 kilogram is shot 50 meters straight up into the air and then falls to the ground.

(a) How much elastic energy was stored in the bow?

(b) What was the arrow's initial upward speed?

(c) With what speed will the arrow strike the ground?

(d) How long will the arrow remain in the air?

Solution: We use the following model: the arrow is one particle; its interaction with the air is negligible. The only force acting on the flying arrow is the force of gravity; therefore, the net force is equal to the force of gravity.

(a) The energy stored in the bow must equal the gravitational field energy of the arrow-earth system when the arrow is 50 meters high. Use ground level as the reference level for gravitational field energy.

gravitational field energy:

$E_G = |g|Mh = 10$ newtons/kg x 0.1 kg x 50 m = 50 joules

elastic energy of bow = 50 joules

(b) The arrow's initial kinetic energy was equal to the elastic energy stored in the bow. This determines the speed of the arrow.

$KE = \tfrac{1}{2}Mv^2$ or $50 = \tfrac{1}{2}$ x 0.1 x v^2, or $1000 = v^2$, or 32 m/sec = v.

Initial speed of arrow = 32 m/sec

(c) When the arrow returns to the ground, all the gravitational field energy has been transferred back to kinetic energy. Hence the final speed is equal to the initial speed.

final speed of arrow = 32 m/sec

(Note: Similar reasoning can be applied at any instant of the downward motion and leads to the conclusion that the downward motion is just the reverse of the upward motion.)

(d) The arrow's total time in the air is the time to rise plus the time to fall. The time to rise is determined by the distance and average speed.

distance:

$s = 50$ m, $v_{av} = \frac{1}{2}v = \frac{1}{2}$ x 32 = 16 m/sec

time rising:

$$t = \frac{s}{v_{av}} = \frac{50}{16} = 3.1 \text{ sec}$$

The distance and the average speed during downward motion are the same as those during upward motion. Hence, the time the arrow takes to fall equals the time it takes to rise.

time falling:

t= 3.1 sec

total time:

3.1 sec + 3.1 sec = 6.2 sec

EXAMPLE 14.8. Two children on roller skates face each other and push one another apart. Child A has a mass of 30 kilograms, child B 60 kilograms. Child A rolls away with a speed of 2 meters per second relative to the ground.

(a) What is the speed of child B relative to the ground?
(b) What is the speed of child A relative to child B?
(c) What is the kinetic energy of the two-child system?

Solution: Use conservation of momentum of the two-child system, under the assumption that no net external force acts on this system as a whole. The partial force exerted by the ground on each child just cancels the force of gravity on each child.

(a) total momentum before push $\mathcal{M}$ = [0, 0] kg-m/sec
total momentum after push $\mathcal{M}$ = [0, 0] kg-m/sec
velocity of child A after push $\mathbf{v}_A$ = [2, 0] m/sec
child A, momentum after push $\mathcal{M}_A = M_A\mathbf{v}_A = $ 30 kg x [2, 0] m/sec
= [60, 0] kg-m/sec
child B, momentum after push $\mathcal{M}_B = M_B\mathbf{v}_B = 60 \ \mathbf{v}_B$

$$\mathcal{M}_A + \mathcal{M}_B = \mathcal{M}$$
$$[60, 0] + 60 \mathbf{v}_B = 0$$
$$\mathbf{v}_B = [-1, 0] \text{ m/sec}$$

The speed of child B is 1 meter per second, directed opposite to that of child A.

(b) The children are moving apart from their starting point at 2 meters per second and 1 meter per second. Hence the speed of A relative to B is 3 meters per second, away from B.

(c) Kinetic energy of child A:

$$KE = \tfrac{1}{2}Mv^2 = \tfrac{1}{2} \times 30 \text{ kg} \times (2 \text{ m/sec})^2 = 60 \text{ joules}$$

Kinetic energy of child B:

$$KE = \tfrac{1}{2}Mv^2 = \tfrac{1}{2} \times 60 \text{ kg} \times (1 \text{ m/sec})^2 = 30 \text{ joules}$$

total kinetic energy: 90 joules

List of new terms

center of mass   acceleration of gravity
inertial reference frame   free fall
                                       momentum transfer

List of symbols

F	force	$\Delta\mathbf{v}$	change of velocity
$\vert F \vert$	force magnitude	$\mathbf{F}_G$	force of gravity
$\mathbf{F}_{av}$	average force	M_G	gravitational mass
$\mathcal{M}$	momentum	$\mathbf{g}$	gravitational intensity
$\Delta\mathcal{M}$	change of momentum	$\mathbf{s}$	position
Δt	time interval	t	elapsed time
$\mathbf{a}_{av}$	average acceleration	KE	kinetic energy
$\mathbf{a}$	acceleration	W	work
M_I	inertial mass	Δs_F	displacement component
$\mathbf{v}$	velocity		in force direction
$\mathbf{v}_{av}$	average velocity	$\vert\mathbf{s}\vert$	position magnitude
v	speed		

Problems

1. Identify the partial forces that act on: (a) an arrow being shot (immediately after the archer releases the string); (b) a sailboat in a race; (c) a stunt car rounding a curve on two wheels; (d) the earth; (e) a drop of water erupting from Old Faithful; (f) a raindrop.

2. Are any of the objects in Problem 1 in mechanical equilibrium? Explain your answer.

3. Compare Newton's and this text's formulations of Newton's theory (Table 14.1). Comment on the nature (operational or formal) of the definitions, undefined quantities, and hidden assumptions. (You may consult any reference you wish.)

4. Describe the causes of motion of two or three moving objects according to Newton's theory and compare them with your common-sense view.

5. Identify three or more bodies for which the center-of-mass model should be adequate and three or more for which it might be very misleading. Explain your reasons.

6. Give two examples from everyday life in which the source of kinetic energy for a moving body is *not* the system that exerts the force setting the body in motion. Explain your answer.

7. Describe two or more non-inertial reference frames. Explain why you believe they are non-inertial.

8. Describe two or more inertial reference frames. Explain why you believe they are inertial.

9. A long time exposure of the night sky, made by a camera fixed on the ground, shows star "trails" in the shape of circular arcs centered on the North Star. Use this evidence to discuss whether a reference frame attached to the earth is an inertial frame. If all the star "trails" are quarter circles, for how long was the film exposed?

10. Give two examples from everyday experience that are easier to explain in the Aristotelian theory than the Newtonian theory. Include both types of explanations for each example.

11. (a) Restate Newton's second law so it applies directly to acceleration instead of to momentum.
(b) Do you expect that this form of the law might be *more* or *less* general in its applicability? Explain.

12. Give two or more examples from everyday life in which the velocity of a moving body is *not* in the direction of the net force acting on it.

13. A 1500-kilogram car is advertised to accelerate from a standing start to 60 miles/hr in 10 seconds.
(a) What is the average horizontal force (in newtons) exerted by the road on the car? Explain why you do not need to consider the vertical force. Hint: convert mi/hr to m/sec using 1 mi = 1600 m and 1 hr = 3600 sec; thus 1 mi/hr = 1600 m/3600 sec = 0.44 m/sec.
(b) What is the average horizontal force exerted by the car on the road? (Use Newton's third law.)
(c) How far does the car travel in the 10 seconds?
(d) Calculate the work done by the force in you found in (a) acting through the distance you found in (c). Compare the work to the kinetic energy of the car (Example 14.5).

396

14. A ball (mass 0.1 kilogram) is thrown upward from ground level so it rises to a height of 30 meters (100 feet) and falls back. Find each of the following:
    (a) the time it was in the air;
    (b) the initial upward speed;
    (c) the kinetic energy it had just after it was thrown;
    (d) the gravitational field energy of the ball-earth system when the ball is at the top of its trajectory.

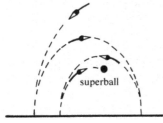

15. Obtain a "super ball" and carry out some experiments in which it bounces strangely. (See the diagram to left.) Explain your observations qualitatively in terms of Newton's theory.

16. Calculate the average acceleration of the golf ball in Fig. 14.6 from measurements made on the photograph, as follows.
    (a) Select three time intervals during which the ball was *not* bouncing on the table. Comment on your result.
    (b) Select two time intervals during which the ball *was* bouncing on the table. Comment on your result and compare with (a).

17. Suppose you did *not* know the time interval between flashes in Fig. 14.6. Devise a procedure and apply it to determine the time interval. You may assume that all time intervals are equal and that the gridlines in the figure are (as stated in the caption) 0.11 meter apart.

18. Figure 14.9 was used to verify that the dry ice puck was in mechanical equilibrium according to Newton's first law (constant speed, straight-line motion). What definitions of "speed" and "straight line" were used implicitly?

19. What is the minimum number of successive flash photographs that are needed for a determination of acceleration? Explain.

20. Explain in what way the acceleration listed in Tables 14.3 and 14.4 is an "approximate" rather than an average value. Describe conditions under which the approximation should be a good one and those when you expect it to be poor.

21. Describe one or more familiar phenomena for which the flat earth model (Section 14.4) is *not* adequate.

22. Compare the two possible points of view toward Eq. 14.13 and explain your preference.

23. Imagine the thought experiment in which a marble is dropped in a freely falling chamber.
    (a) Describe the observations made in such an experiment, using the chamber as reference frame.
    (b) Is the chamber an inertial reference frame while it is falling freely? Explain your answer.

24. You often read in the newspaper about the "weightlessness" of astronauts in an orbiting spacecraft.
(a) Discuss this description from the Newtonian point of view.
(b) Comment on the suitability of the spacecraft as an inertial reference frame before launching, during takeoff, during orbital motion, and during reentry.

25. Interview four or more children (ages 8-12) concerning their explanation of motion. (Suggestion: Select some moving objects and ask, "Why does . . . move?" or, "What makes . . . move?" or another form of the question that you find effective.) Interpret the children's answers in terms of Newtonian and Aristotelian theories.

Mother: "Johnny! Don't pull the cat's tail like that."
Johnny: "I'm only holding it, Mom. The cat is pulling."

26. Interpret the conversation in the margin to the left in the context of the Newtonian theory. Identify the partial forces that act, the net force on the cat's tail, and the aptness of Johnny's remark.

27. Give two or more examples from everyday life of the conservation of momentum. If possible, explain any apparent loss of momentum.

28. Apply Newton's third law to identify the "reaction" forces in the following examples of forces.
(a) A rocket exerts a force expelling the exhaust gases.
(b) A car's tires exert a frictional force against the road.
(c) The earth exerts a gravitational force on you.
(d) A stretched rubber band exerts an elastic force on your fingers stretching it.
(e) The sun exerts a gravitational force on the earth.

29. Because momentum is conserved, it makes sense to speak of a "momentum source" that exerts a force and thereby transfers momentum to a moving object. Identify the momentum source (which exerts force) and the energy source for the moving object in each of the following examples. In each case, the momentum source *may or may not* be the same as the energy source.
(a) A bullet is fired by a gun.
(b) A baseball is hit by a bat.
(c) A bowling pin is hit by a bowling ball.
(d) An automobile accelerates.
(e) A broad jumper leaps.

30. Two children on ice skates face each other and push with their hands against one another. Assuming that the children have different masses, describe what will happen.

31. A 75-kilogram man stands in the stern (back) of a 50-kilogram canoe while the canoe is stationary in the water. He walks toward the bow (front) with a speed of 1 meter per second relative to the canoe.
(a) Describe qualitatively what will happen.
(b) Calculate the speed of the man relative to a shore-based reference frame. (Neglect friction between the boat and the water.)

32. A small rocket motor ejects 5 kilograms of exhaust gases per second, with an exhaust speed of 1700 m/sec (relative to the rocket). The rocket is moving at 600 meters per second relative to the earth and its mass is 200 kilograms at that time.
(a) Calculate the rocket motor's thrust.
(b) Calculate the speed change of the rocket in one second.
(c) Calculate the kinetic energy change of the rocket in one second.
(d) Calculate the kinetic energy change of the fuel-oxygen mixture burned in one second and exhausted.
(e) Four and a half kilograms of exhaust are produced by the combustion of 1 kilogram kerosene fuel. Calculate the chemical energy consumed during 1 second of operation (see Table 10.6). Discuss what happens to this energy from the viewpoint of energy conservation.

33. (a) How much kinetic energy does an 80-kilogram parachutist acquire when he jumps from a plane and falls 500 meters before the chute opens? (Assume that there is no friction with the air.)
(b) What happens to the kinetic energy when the chute opens and the man-chute system slows down suddenly?
(c) Enumerate the partial force (or forces) that bring about the change of velocity after the chute opens.

34. Make a mathematical model that relates the speed of a falling object to the distance it has fallen from rest. *(Hint:* The kinetic energy of a falling object is transferred from the gravitational field energy stored in the object-earth system.)

35. Romeo would like to throw flowers to Juliet, whose balcony is 6 meters above street level. What upward speed does Romeo have to impart to the flowers? *(Hint:* Kinetic energy is transferred to gravitational field energy.)

36. A pole vaulter wishes to clear a bar 5 meters high. Approximately what speed must he attain during his running start?

37. Write a critique of the Newtonian theory from *your* point of view. Identify those features (if any) that you find useful, intellectually satisfying, difficult, confusing, or contrary to your experience.

38. Give a qualitative analysis of the following situations in terms of Newtonian theory. Identify the partial forces acting on the italicized objects, point out a nonzero net force acting on any object, and mention any relationships between forces that are consequences of Newton's laws.
(a) A stalemated tug-of-war between *two children* using a *rope*.
(b) A *car* accelerating on a level *road*.
(c) A *pole vaulter* vaulting over a *bar* with a fiberglass *pole*. (Select two stages during the motion for analysis.)

Bibliography

D. Cassidy, G. Holton, J. Rutherford, *Understanding Physics*, Springer Verlag, 2002.

S. Chandrasekhar, *Newton's Principia for the Common Reader*, Oxford University Press, 1995 (hardback), 2003 (paperback). A version in relatively down-to-earth language with commentary by an outstanding theoretical physicist.

I. B. Cohen, *The Birth of a New Physics*, Doubleday, Garden City, New York, 1960.

L. N. Cooper, *An Introduction to the Meaning and Structure of Physics*, Harper and Row, New York, 1968. See the chapter "The Newtonian World."

S. Drake, *Galileo at Work: His Scientific Biography*, University of Chicago Press, Chicago, Illinois, 1978.

A. Eddington, *The Nature of the Physical World*, University of Michigan Press, Ann Arbor, Michigan, 1958.

A. Eddington, *Space, Time, and Gravitation*, Harper and Row, New York, 1959.

G. Galilei, *Dialogues Concerning Two New Sciences*, Northwestern University Press, Evanston, Illinois, 1950. See the sections Galileo called "First Day" and "Third Day."

M. B. Hesse, *Forces and Fields*, Littlefield, Totowa, New Jersey, 1961.

M. Jammer, *Concepts of Force: A Study in the Foundations of Dynamics*, Harper and Row, New York, 1957.

F. Manuel, *A Portrait of Isaac Newton*, DaCapo Press (reissue), 1990.

I. Newton, *The Principia, Mathematical Principles of Natural Philosophy*, translated by I. B. Cohen and A Whitman, University of California Press, 1999. This is the most recent translation and appears to be in the process of becoming the standard. There are many older translations; by far the best known, and the standard for many years, was the translation by Andrew Motte (done in 1729) as revised by Cajori (published as two volumes in 1962 by the University of California Press). [Editor's Note: all quotations from the *Principia* in this textbook are from the Motte/Cajori translation.]

R. E. Peierls, *The Laws of Nature*, Scribner's, New York, 1956.

R. S. Westfall, *Never at Rest, A Biography of Isaac Newton*, Cambridge University Press, 1983.

. . . after I had by unceasing toil through a long period of time, using the observations of Brahe, discovered the true distances of the orbits, at last, at last, the true relations. . . overcame by storm the shadows of my mind, with such fullness of agreement between my seventeen years' labor on the observations of Brahe and this present study of mine that I at first believed that I was dreaming. . .

JOHANNES KEPLER
Harmony of the World,
1619

Periodic motion

15

Have you ever wondered how extensively human civilization depends on clocks? Western culture is especially time conscious, since daily schedules in the complex worlds of business, industry, and education require synchronized cooperation by many individuals. Sometimes it seems as though clocks are the masters and humanity is enslaved!

Clocks are instruments that make possible the operational definition of time intervals. Each clock has a regulating mechanism, such as a pendulum, a balance wheel, or a vibrating crystal, which moves in a repeating pattern (called *periodic motion)* and defines equal time intervals, one after another. In this chapter, we will formulate a mathematical model for periodic motion and apply it to the solar system, the pendulum, and the elastic oscillator. We will not be concerned with other aspects of clocks, such as their energy sources, the internal connections that regulate the movement, and the face or dial on which the elapsed time is indicated.

Periodic motion also occurs under natural conditions in the macro domain. The child in her swing, the bird swaying gently on a tree branch, the bicycle wheel spinning on its axle, your arm swinging by your side as you walk—all these are examples of periodic motion from the everyday world.

There are several sections in this text where we have already referred to periodic motion. In Section 1.5, the operational definition of time intervals was based on periodic motions such as the earth's revolution on its axis and the swinging of a pendulum. In Section 3.4, we introduced the inertial balance, whose periodic motion is the basis for the operational definition of inertial mass. In Section 6.1, we introduced the oscillator model for a medium that permits wave propagation, and we described the periodic motion of each oscillator (Fig. 6.4).

15.1 Properties of periodic motion

Periodic motion can occur in a system only when the objects in the system interact with one another in such a way that they remain near one another. In the absence of interaction, the objects would move apart and not return to repeat their motion once they had passed each other. Even though systems carrying out periodic motion will eventually stop moving, they do have a stable existence over many cycles of their motion, as shown by the example of the solar system.

In this chapter we will examine periodic motion from the viewpoint of the Newtonian theory. The two principal quantities that are used to describe periodic motion of a particle are the time required for one cycle, which is called the *period* ($\mathcal{T}$, seconds), and the radius or width of the orbit (R, meters). We will construct a mathematical model for periodic motion and thereby relate the period and radius to the mass of the particle and the force that maintains the periodic motion.

The qualitative nature of the relationship among period, radius, mass, and force is easy to infer. The moving particle has a velocity

401

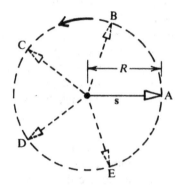

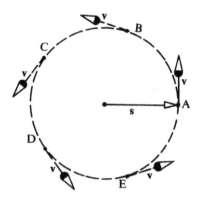

Figure 15.1 Circular motion with radius R and period T. In the time T, the arrow representing the relative position of the orbiting particle rotates once around the circle counterclockwise.

Figure 15.2 Arrows representing the instantaneous velocity at five instants (A, B, C, D, E) during one orbital revolution.

whose magnitude depends on the radius (distance traveled) and period (time required) of the orbit. Since the particle is moving back and forth, the velocity cannot be constant but must change, giving rise to an acceleration. The acceleration is related to both the net force and inertial mass according to Newton's second law (Eq. 14.6, $F_{net} = M_I a$). As explained in Section 14.4, we can assume that the inertial mass of a particle is equal to its gravitational mass. Therefore, from now on we will refer to the mass without specifying it further; thus, $a = F_{net}/M$.

15.2 A mathematical model for circular motion

Circular motion with a constant speed is a particularly simple example of periodic motion. The moon in its orbit around the earth, the tetherball whirling at the end of its string, and the child on the merry-go-round exemplify more or less closely this kind of periodic motion. The arrow representing the position of the moving particle in a diagram rotates around the circle (Fig. 15.1). The acceleration in this example arises only from changes of direction of the instantaneous velocity, since the instantaneous speed (magnitude of the velocity) does not change (Section 13.2). To apply Newton's theory, we must first find how the acceleration is related to the radius and period.

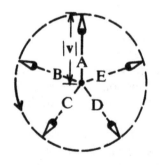

Figure 15.3 Diagram of the velocities at five instants during one revolution of the orbiting particle. Since the magnitude of the velocity is constant, the velocity arrows all extend from the origin to the dashed circle of radius |v|.

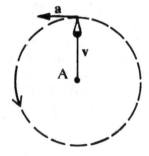

Figure 15.4 The arrow representing the acceleration is in the direction of the velocity change, which is to the left as the velocity arrow rotates around the circle counterclockwise.

Equation 15.1

*instantaneous
velocity* $= \mathbf{v}$
*instantaneous
speed* $= v$
average speed $= v_{av}$

$$|\mathbf{v}| = v = v_{av}$$

**Equation 15.2 (velocity
of particle moving in a
circle)**

distance traveled $= \Delta s$
time elapsed $= \Delta t$
orbital radius $= R$
orbital period $= \mathcal{T}$

$$|\mathbf{v}| = v_{av} = \frac{\Delta s}{\Delta t} = \frac{2\pi R}{\mathcal{T}}$$

*From here on we will
merely write "speed"
and "velocity"for the in-
stantaneous quantities.
The average quantities
will be designated as
such so you may identify
them properly.*

Equation 15.3

acceleration $= \mathbf{a}$

$$|\mathbf{a}| = \frac{2\pi|\mathbf{v}|}{\mathcal{T}}$$

**Equation 15.4 (Accel-
eration of particle mov-
ing in a circle)**

$$|\mathbf{a}| = 2\pi\left(\frac{2\pi R/\mathcal{T}}{\mathcal{T}}\right)$$

$$= \frac{4\pi^2 R}{\mathcal{T}^2}$$

Velocity of circular motion. The instantaneous velocity is directed along the circumference of the circular orbit, at right angles to the position arrow (Fig. 15.2). The magnitude of the instantaneous velocity is equal to the instantaneous speed, and this, in turn, is equal to the average speed since the latter does not vary (Eq. 15.1). The average speed is equal to the circumference of the circular orbit divided by the period (Eq. 15.2, $|\mathbf{v}| = 2\pi R/T$).

It is instructive to make a diagram in which the instantaneous velocity arrows are compared (Fig. 15.3). The tails of all the arrows are placed at the origin of the coordinate frame and the heads of the arrows fall on the circle whose radius is the velocity magnitude given by Eqs. 15.1 and 15.2. You can see that the velocity diagram (Fig. 15.3) is very similar to the position diagram (Fig. 15.1). One way of describing the circular motion is to say that both the position arrow and the velocity arrow rotate once around their circles during one period.

Acceleration in circular motion. The easiest way to infer the direction and magnitude of the acceleration is to reason from the formal similarity of the relations "position-velocity" and "velocity-acceleration." We have just pointed out that the geometry of the position arrow, which rotates counterclockwise around a circle (radius *R)* in one period (Fig. 15.1), is analogous to the geometry of the velocity arrow, which also rotates counterclockwise around a circle (with radius $|\mathbf{v}|$) in one period (Fig. 15.3). The magnitude of the acceleration is therefore given by a formula like Eq. 15.2, except using the radius of the velocity circle instead of the radius of the position circle (Eq. 15.3). In fact, the head of the velocity arrow travels around the circumference of a circle, a distance of $2\pi v$; therefore, over the full circle a = $\Delta v/\Delta t = 2\pi v/\mathcal{T}$. If you replace v with its value from Eq. 15.2 you find Eq. 15.4, in which the magnitude of the acceleration is directly proportional to the orbital radius and inversely proportional to the period raised to the second power.

The direction of the acceleration can be found by using the analogy of the position circle in Fig. 15.2 to the velocity circle in Fig. 15.3. The velocity arrow in Fig. 15.2 points at right angles to the position arrow, and to the left. The acceleration implied by Fig. 15.3 is therefore directed at right angles to the velocity, and to the left (Fig. 15.4). The position, velocity, and acceleration at instant A are summarized in Fig. 15.5. You can see that the acceleration is directed from the

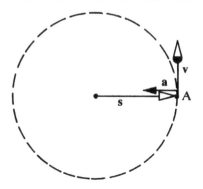

Figure 15.5 Comparison of the position, velocity, and acceleration of a particle moving at uniform speed along a circular path. The tails of the velocity and acceleration arrows are placed at the position of the particle (A). The velocity points along the <u>tangent to the path</u>; the acceleration points toward the <u>center</u> of the circle.

Equation 15.5 (Centri - petal force for motion in a circle)

force (newtons) $= \mathbf{F}$
mass (kg) $= \mathrm{M}$

$$\left|\mathbf{F}_{net}\right| = \mathrm{M}\left|\mathbf{a}\right|$$
(Newton's Second Law)

$$\left|\mathbf{F}_{net}\right| = \frac{4\pi^2 MR}{\mathcal{T}^2}$$
(from Eq. 15.4)

"A centripetal force is that by which bodies are drawn or impelled, or any way tend, towards a point as a centre.... Of this sort is gravity, by which bodies tend to the centre of the earth; magnetism, by which iron tends to the lode- stone; and that force, what- ever it is, by which the planets are continually drawn aside from the recti- linear (straight line) mo- tions, which otherwise they would pursue, and made to revolve in curvilinear or- bits."

Isaac Newton
Principia, 1687

point A toward the *center* of the circle, a result that also applies at any point on the circle (Fig. 15.3, B, C, . . .) . In his studies of circular planetary orbits, Newton therefore introduced the term *centripetal ac- celeration* for the acceleration of any object moving in a circle at con- stant speed.

Centripetal force. When you use Newton's second law ($\mathbf{F}_{net} = \mathbf{Ma}$) to calculate the force required to cause this acceleration (Eq. 15.5), you find that the magnitude of the force is constant and that it is always di- rected toward the center of the circle. The force is therefore often called a *centripetal force*. The magnitude of the force is directly proportional to the mass of the particle and the radius of the circle; it is inversely proportional to the second power of the orbital period (Eq. 15.5). This equation is a mathematical model for the centripetal force during circu- lar motion. The application of this model to a simple laboratory ex- periment is described in Fig. 15.6. In the experiment, a weight hanging from a string supplies the centripetal force, which is transmitted by the string to the orbiting weight.

15.3 The solar system and gravitation

Newton applied the theory of circular motion and its extension for el- liptical motion to the moon and the planets in the solar system. He used the observations of orbital motion to deduce the centripetal forces that were acting. He then identified these forces with the terrestrial force of gravity and thereby unified terrestrial and celestial phenomena.

Newton took the point of view that the sun exerted the force that kept the planets in their orbits. Unlike his predecessors Copernicus, Kepler, and Galileo, Newton had a mathematical model relating force to motion ($\mathbf{F}_{net}\Delta t = \Delta\mathbf{M}$, or $\mathbf{F}_{net} = \mathbf{Ma}$). This was a powerful method for testing whether the heliocentric point of view resulted in a simple explanation for the observations that had been made on the solar system. In particu- lar, Newton invented a simple mathematical model for the force exerted

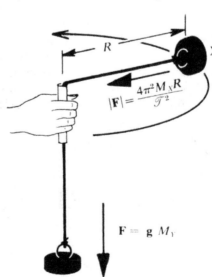

$$\left|\mathbf{F}\right| = \frac{4\pi^2 M_X R}{\mathcal{T}^2}$$

$$\mathbf{F} = \mathbf{g}\, M_y$$

Y

Figure 15.6 Weight X is twirled in a circle to hold weight Y in mechanical equilibrium. The centripetal force on weight X is equal in magnitude to the gravi- tational force $|g|M_Y$ acting on weight Y. Measurements of the radius R, the period T, and the masses of the two weights lead to a direct experimental test of Eq. 15-5.

Claudius Ptolemy (approx. 140 A.D.) was probably an Egyptian Greek who lived for a period in or near Alexandria in Egypt (127-ca. 150 A.D.). His two great works, the Almagest (on mathematical astronomy) and the Geography (on mathematical geography) remained the standard text references in their respective fields for over 14 centuries.

Nicolaus Copernicus (1473-1543), a Polish astronomer, proposed replacing the complex geocentric universe with a simpler, sun-centered (heliocentric) system. Unfortunately, to advance such views in the early 16th century was heresy, and even Martin Luther spoke of Copernicus as "the fool who would overturn the whole science of astronomy." While the heliocentric system was simpler, the existing observations did not decisively favor either system, and there was not yet a consensus among scientists. Consequently, Copernicus would not permit his book on the subject to be published until he lay on his deathbed.

by the sun on the planets. This model relates the magnitude of the force acting on each planet to the distance between the sun and the planet.

Motion in the solar system. Astronomy, the oldest science, has been important for millenia, because it has enabled us to anticipate seasonal changes and to schedule agricultural activities. The observed motion of the sun, moon and stars around the earth was first attributed to gods and goddesses. Later, Greek philosophers explained the motion by means of celestial concentric rotating spheres to which the heavenly bodies were attached. The earth was at the center of all the spheres; hence, this model is called *geocentric* (earth-centered). This model seems intuitively reasonable, and it became generally accepted.

However, the motion of the sun and the planets in the sky was more complicated than that of the stars. In the third century BC, a Greek astronomer, Aristarchus of Samos, suggested that *"the fixed stars and the sun remain unmoved, that the earth revolves about the sun on the circumference of a circle, the sun lying in the middle of the orbit"* (as reported by Archimedes in *The Sandreckoner).* We will call this model *heliocentric* (sun is *helios* in Greek, thus sun-centered). This model attracted little attention throughout ancient and medieval times.

Ptolemy's geocentric theory. The major trend of ancient thought elaborated the geocentric point of view, which offered three great advantages: first, philosophical doctrines required that the earth be stationary at the center of the universe; second, the spherical motion of the heavenly spheres was a "natural" and perfect motion; third, the data gathered by observers on earth could be used for the prediction of stellar movements in a geocentric reference frame but were not sufficiently accurate to permit their correct transformation to a heliocentric reference frame. Claudius Ptolemy developed this theory to a perfection that served mankind until it was abandoned in favor of the heliocentric theory after more than fourteen hundred years.

The Copernican heliocentric theory. One of the weaknesses of the Ptolemaic theory was the need to make it more complicated as astronomical data improved in accuracy. In fact, Ptolemy himself recognized that not all celestial spheres moved around the earth as center of rotation. In the sixteenth century, Nicolaus Copernicus revived the heliocentric picture of the universe and thereby touched off a controversy between religious dogma and science that lasted for a hundred years. Copernicus realized that the rotation of the earth on its axis could explain the motion of the fixed stars, which he placed on an immobile celestial sphere. In his theory, planets were on smaller concentric spheres, with the sun at the center of the universe. Unlike earlier proponents of the heliocentric point of view, however, Copernicus had the data with which he could find the period of motion and the orbital radius of each planet around the sun. His unit of distance was the distance between the sun and the earth, known as the astronomical unit (AU). Copernicus' values for the planets' periods and orbital radii were very close to the modern values (Table 15.1).

Copernicus succeeded in showing that heliocentric theory was as

TABLE 15.1 DATA ON THE SOLAR SYSTEM

Planet	Radius (m)	Mean radius of orbit (m)	(AU*)	Period (years)	(Radius)³ (AU*)³	(Period)² (years)²
Mercury	2.5×10^6	5.8×10^{10}	0.39	0.24	0.059	0.058
Venus	6.1×10^6	1.1×10^{11}	0.72	0.62	0.38	0.38
Earth	6.4×10^6	1.5×10^{11}	1.00	1.00	1.00	1.00
Mars	3.4×10^6	2.4×10^{11}	1.52	1.9	3.5	3.6
Jupiter	7.2×10^7	7.8×10^{11}	5.2	11.9	140.	141.
Saturn	5.8×10^7	1.4×10^{11}	9.6	30.	890.	900.
Uranus**	2.7×10^7	2.9×10^{11}	19.2	84.	7100.	7000.

*AU stands for astronomical unit, a distance measure equal to the mean orbital radius of the earth (1 AU = 1.5×10^{11} m).
**Not known in Copernicus' and Newton's times.

Johannes Kepler (1571-1630) was born in Weil der Stadt, Germany, and studied mathematics at the University of Tübingen. After obtaining his degree, Kepler became Tycho Brahe's assistant at Prague and eventually succeeded him as Court Mathematician. Kepler argued for the Copernican system in his first book The Cosmographic Mystery (1597). Through his studies of the orbit of Mars, Kepler arrived at his first two laws, and he published his findings in the New Astronomy or Celestial Physics (1609). The third of his great laws was announced in The Harmonies of the World (1619). In all of his research, Kepler benefited immensely from the huge collection of accurate, long-term astronomical measurements that Tycho bequeathed to him.

good as geocentric theory in summarizing the astronomical data. His theory, however, did not have a decisive scientific advantage; it merely appealed to a different sense of order or simplicity than did the geocentric theory. It is now recognized that Copernicus started a scientific revolution by advancing a new interpretation of data, but that others, especially Isaac Newton, really exploited the new point of view.

Kepler's laws. New and especially accurate data on the motion of Mars were collected by the Danish astronomer Tycho Brahe (1546-1601) and were left for his assistant Johannes Kepler to analyze. After painstaking work, Kepler concluded that Mars did not carry out uniform circular motion, either geocentric or heliocentric. He therefore used the observations of Brahe, as a surveyor would use his sightings, to identify the geometrical shapes of the orbits of the earth (the surveyor's "base") and of Mars relative to the sun as fixed point.

Kepler's conclusions are summarized in three laws that bear his name. The first law states that the planets' orbits have the shape of ellipses, with the sun located at one focus of the ellipse (Fig. 15.7). Actually, the ellipses of most planets are quite close to circles, in that the two diameters differ in length by only a few percent. This fact explains in part why the model of circular orbits had not been discarded earlier. Kepler's second law points out that a planet does not move with the same speed along all parts of its orbit. It moves faster than average when it is close to the sun (point A, Fig. 15.7) and slower when it is far from the sun (point B, Fig. 15.7), in such a way that the line connecting it to the sun sweeps out equal areas in equal times. Here also the deviations from uniform motion are so small that they were not detected before Brahe's accurate observations.

Equation 15.6
(Kepler's Third Law)

planetary period
(years) $= \mathcal{J}$
orbital radius (AU) $= R$

$$\mathcal{J}^2 = R^3 \qquad (a)$$

planetary period
(sec) $= \mathcal{J}$
orbital radius (m) $= R$

$$\mathcal{J}^2 = (3 \times 10^{-18})R^3 \quad (b)$$

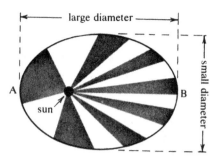

large diameter

small diameter

A   B

sun

Figure 15.7 Kepler's Laws
Kepler's First Law: *The orbit of a planet around the sun is in the shape of an ellipse with the sun at one focus.*

Kepler's Second Law: *the line from the sun to the planet sweeps out equal areas in equal times.*

Figure 15.8 (below) Two drawings made by Galileo from his telescopic observations of the moon. These drawings showed the mountains and valleys on the moon and directly contradicted existing beliefs. He published them in his "best seller" Sidereus Nuncius (The Starry Messenger, 1610). Galileo's extraordinary observations, published as a 63-page pamphlet in popular Italian rather than the traditional Latin, provided clear-cut support for the heliocentric system over the then widely-accepted geocentric theory. Unfortunately, the geocentric system had been incorporated into Catholic dogma, and at the same time as Galileo was advocating the heliocentric system, the Church was becoming more conservative as Protestantism grew. Galileo's conflict with the Church finally led to his "trial" by the Inquisition, the threat of torture, and his public "confession" to the crime of heresy! This was, and remains, an unforgettable episode, a searing reminder of the importance of the struggle required to develop civil society with an independent justice system and limits on religious authority.

Kepler's first two laws apply to the motion of each planet separately. Kepler's third law relates the motions of all the planets together. In modern terminology, the third law states that the planetary period (in years) to the second power equals the mean orbital radius (in astronomical units) to the third power (Eq. 15.6a). Note how the orbital motion of the earth furnishes the units of distance and time. Note also that Kepler's third law, though published a whole decade after his first two laws, made use basically of the data of Copernicus and did not depend on Kepler's other discoveries. For later use, we state Kepler's third law in ordinary units (meters, seconds) in Eq. 15.6b.

The nature of the sun and planets. One of the important assumptions of geocentric theory was that terrestrial matter was essentially different from the celestial. The latter was perfect, uniform, and spherical in shape and motion; the former was irregular, rough, and massive, falling toward the ground when unsupported. Copernicus had addressed himself to a description of the relative motion of the heavenly bodies without concerning himself about its causes. Kepler's search for simple mathematical patterns in the description had been motivated by his vision that a single force exerted by the sun was responsible. However, he had no quantitative techniques for analyzing this problem, but he speculated that magnetism might be the agency. In his view, the planets and the sun were magnets.

The first direct observations of heavenly bodies to give evidence of their nature were made by Galileo with a telescope that had been recently invented and that he had improved. Galileo examined the moon, the sun, the Milky Way, the planet Jupiter, and the planet Venus.

What Galileo saw supported his belief in the heliocentric Copernican

Galileo Galilei (1564-1642) was born in the year of Shakespeare's birth and Michelangelo's death. By the age of 25, he was a teaching member of the University of Pisa. Galileo's intense criticism of Aristotelian natural philosophy provoked a controversy that drove him to Padua in 1592, where he constructed his first telescope and began his battle on behalf of the heretical Copernican theory. His fame became so widespread that he was even recalled to Pisa as Court Mathematician and Philosopher. In spite of prohibitions from the Church, Galileo published his Dialogue Concerning the Two Chief Systems of the World, *which praised Copernican theory at the expense of orthodox Ptolemaic theory. Galileo was again brought before the Inquisition. Weakened by age and fearing for his safety, he publicly abjured his belief in the heliocentric system and pledged not to discuss nor write about it again. However, Galileo survived to publish his masterpiece on motion on earth*: Discourses Concerning Two New Sciences (1638).

Galileo's extraordinary example of the value of individual freedom in scientific inquiry, and the high price he paid for it, stimulated the search for a less destructive relationship between science and religion. Today, we sometimes take freedom of inquiry and freedom of religion for granted, but this was (and is) not always so. Maintaining the uneasy balance between reason and faith (or freedom and security) requires ongoing vigilance and hard work.

model of the solar system, for many reasons: 1. The moon exhibited a rough landscape with mountains and valleys much like the earth and was not perfect, spherical, and uniform (Fig. 15.8). 2. The sun's disk likewise was not uniformly bright, but showed dark "spots" that moved on its surface; this further contradicted the idea of celestial perfection and strongly suggested that the sun itself was rotating. 3. The Milky Way was not a continuous ribbon of light but was made up of many faint stars, individually invisible, that could not conceivably serve the purpose of providing light for men to see at night. 4. The planet Jupiter was surrounded by four small bright stars that, Galileo showed, were really satellites revolving around Jupiter much like the moon revolves around the earth. This was direct observation of orbital motion not centered on the earth. 5. Venus exhibited moon-like phases; thus Venus must *reflect* sunlight like the moon and was not self-luminous, again contradicting Aristotle's assertions about celestial objects. 6. Finally, Venus showed great variations in apparent size and hence in distance from the earth; therefore, Venus *cannot* move in a circular orbit around the earth, as required by the geocentric system. On the other hand, as Galileo enthusiastically explained, the Copernican system could easily explain every detail of his observations of Venus. In the last hundred years, spectroscopy has offered final chemical confirmation of Galileo's hypothesis that "terrestrial" and "celestial" matter shared the same fundamental nature and thus were subject to the same scientific laws.

Galileo publicized his findings widely, but he did not convince most proponents of the geocentric view, many of whom simply refused to accept the telescopic observations as evidence. Because his findings were believed to threaten the Church and the existing social order, Galileo was required to cease his teaching and even to deny the Copernican theory. Nevertheless, the process of free inquiry had begun to show significant results, and the effort to understand matter and motion both in the heavens and on earth had begun.

The Newtonian theory of the solar system. On the basis of his theory of motion, Newton investigated the shapes of particle orbits around an attracting center that subjected the particles to a centripetal force. The first result was that the lines from a particle to the attracting center swept out equal areas in equal times. Since this finding was in accord with Kepler's second law, it was clear that planetary motion was caused by a centripetal force attracting the planets to the sun and that there was no need for other forces propelling them along their orbits.

Newton's second result concerned the magnitude of the centripetal force. By following a line of reasoning similar to that in Section 15.1, Newton found a mathematical model for the centripetal acceleration of a particle moving in an elliptical orbit. The centripetal acceleration did not have a constant magnitude at all points of the orbit, but varied inversely as the second power of the particle's distance from the attracting center. Consequently the required centripetal force, proportional to the acceleration according to Newton's second law, also must vary inversely as the second power of the distance. Thus Kepler's first law regarding the shape of the orbit permitted Newton to infer

Equation 15.7

force on planet
(newtons) $= F_{planet}$

mass of planet (kg) $= M_{planet}$

orbital radius (m) $= R_{planet}$

orbital period (sec) $= \mathcal{T}_{planet}$

$$\left|\mathbf{F}_{planet}\right| = \frac{4\pi^2 M_{planet} R_{planet}}{\left(\mathcal{T}_{planet}\right)^2}$$

Equation 15.8

$$\left|\mathbf{F}_{planet}\right| = \frac{4\pi^2 M_{planet} R_{planet}}{3\times10^{-18}\left(R_{planet}\right)^3}$$

$$= \frac{1.3\times10^{-19} M_{planet}}{\left(R_{planet}\right)^2}$$

"Hitherto [I] have explained the phenomena of the heavens and of our sea by the power of gravity . . . and even . . . the remotest [motions] of the comets . . . But I have not been able to discover the cause of [the] properties of gravity from phenomena, and I frame no hypotheses; and hypotheses, whether metaphysical or physical, whether of occult qualities or mechanical, have no place in experimental philosophy."

Isaac Newton
Principia, 1687

Equation 15.9

acceleration of gravity
(m/sec/sec) $= \mathbf{g}$

distance from earth's
center (m) $= R$

$$\left|\mathbf{g}\right| = \frac{4.0\times10^{14}}{R^2}$$

the magnitude of the force, just as Kepler's second law had permitted Newton to infer the direction.

Newton was also able to show that Kepler's third law led to the conclusion that the force holding the planets in their orbits varied inversely as the second power of the distance from the sun to the planet. For this reasoning, the planetary orbits can be described approximately as circles. Then the centripetal force acting on the planets is given by Eq. 15.7 (from Eq. 15.5) in terms of the orbital radius and period. Now, since Kepler's third law relates the period to the radius (Eq. 15.6b), reference to the orbital period can be eliminated from Eq. 15.7 to give a mathematical model for the centripetal force that depends only on the radius (Eq. 15.8). You can see that the force varies inversely as the second power of the orbital radius and directly as the mass of the planet. Kepler's first and third laws, therefore, led Newton to the same conclusion, a result that must have increased Newton's confidence in his findings.

Newton furthermore proposed a dramatic solution to the problem of the nature of the interaction. Kepler had speculated that the force was magnetic, and Descartes had worked out a "theory of vortices" in which the interaction was transmitted by a swirling fluid. But Newton argued that the interaction was gravitational, that it was of the same type as the interaction that causes an apple on earth to fall to the ground, that it varied inversely as the second power of the distance of the particle from the attracting center, and that it varied directly as the mass of the particle. The gravitational force exerted by the sun on a planet is given in Eq. 15.8, and the force exerted by other gravitating bodies (as by the planets or by their satellites) is given by a similar mathematical model, but with a different numerical factor. To justify this proposal, Newton showed that his theory could correctly account for the orbital motion of the moon as it is described in Table 15.2.

The force of gravity exerted by the earth governs the motion of the moon around the earth. The acceleration caused by this force is equal to the acceleration of gravity; at the surface of the earth, this acceleration is 10 meters per second per second (Eq. 14.14). As the force of gravity decreases with increasing distance from any gravitating body, the acceleration of gravity decreases likewise, inversely as the second power of the distance. The mathematical model in Eq. 15.9 describes this variation and gives the correct value for the acceleration of

TABLE 15.2 DATA ON EARTH SATELLITES

Satellite	Orbital radius*		Orbital period	
	(m)	(earth radii)	(sec)	(days)
Moon	3.8×10^8	60.	2.3×10^6	27.3
Syncom	4.3×10^7	6.7	8.6×10^4	1.0
Explorer 1	6.6×10^6	1.03	5.1×10^3	0.062

*The radius of the earth is a convenient unit to use for describing distances to earth satellites.

gravity at the earth's surface, at a distance of one earth radius from the earth's center (Example 15.1). The acceleration of gravity predicted by this theory at the position of the moon is much smaller, as calculated also in Example 15.1.

What is the observed centripetal acceleration of the moon? This can be calculated from the moon's orbital data and the mathematical model in Eq. 15.4. It is found to agree closely with the prediction (Example 15.2), thus lending further support to Newton's theory of gravitation as the binding force of the solar system. This was the final blow to the ancient view, according to which terrestrial and celestial phenomena were qualitatively different.

This extraordinary conceptual transformation (some would say revolution) came about as the result of two key developments: first, new mathematical techniques, including the rectangular coordinates of Descartes and methods of dealing with infinitesimal quantities invented by Newton and Leibniz (the calculus), and second, Galileo, Brahe, Kepler, Hooke, Huygens and others' use of experiments as a way to gather data and thus understand the details of real world motions. This is, indeed, an example of how a scientific breakthrough really rests squarely on the contributions of many other individuals.

EXAMPLE 15·1. Use Eq. 15·9 to find the acceleration of gravity caused by the earth at the surface of the earth and at the position of the moon.

(a) At the surface of the earth (Table 15·1), $R = 6.4 \times 10^6$ m

$$|\mathbf{g}| = \frac{4.0 \times 10^{14}}{R^2} = \frac{4.0 \times 10^{14}}{(6.4 \times 10^6 \text{ m})^2} = \frac{4.0 \times 10^{14}}{4.0 \times 10^{13}}$$

$$= 10 \text{ m/sec/sec} = 10 \text{ newtons/kg}$$

Henry Cavendish (1731-1810) inherited a fortune through the death of an uncle and withdrew from society to devote himself to scientific pursuits. Unfortunately, the same shyness that produced withdrawal from society also made him reluctant to publish his manuscripts. Although Cavendish was known as a chemist, he was the first to measure gravitational forces directly. His unpublished experiments were later found (by Maxwell in 1879) to have anticipated some of the electrical discoveries of Coulomb and Faraday.

(b) At the position of the moon (Table 15·2), $R = 3.8 \times 10^8$ m

$$|\mathbf{g}| = \frac{4.0 \times 10^{14}}{R^2} = \frac{4.0 \times 10^{14}}{(3.8 \times 10^8 \text{ m})^2} = \frac{4.0 \times 10^{14}}{14 \times 10^{16}}$$

$$= 2.8 \times 10^{-3} \text{ newton/kg} = 2.8 \times 10^{-3} \text{ m/sec/sec}$$

EXAMPLE 15·2. Use Eq. 15·4 to find the centripetal acceleration of the moon.

orbital radius $R = 3.8 \times 10^8$ m, orbital period $\mathcal{T} = 2.3 \times 10^6$ sec

$$|\mathbf{a}| = \frac{4\pi^2 R}{\mathcal{T}^2} = \frac{4 \times 10 \times 3.8 \times 10^8 \text{ m}}{(2.3 \times 10^6 \text{ sec})^2} = \frac{1.5 \times 10^{10}}{5.3 \times 10^{12}}$$

$$= 2.8 \times 10^{-3} \text{ m/sec/sec}$$

Law of universal gravitation. Laboratory experiments to test Newton's mathematical model for the gravitational force had to await the construction of the delicate apparatus that was necessary. About the

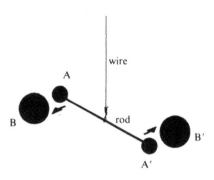

Figure 15.9 Cavendish's experiment to measure the force of gravity between lead spheres A-B and A'-B'. The top end of the wire is clamped firmly. As the rod rotates around the wire due to the gravitational attraction between the lead spheres, the bottom end of the wire is twisted elastically until it comes to mechanical equilibrium subject to the elastic and gravitational forces. The position of the rod is recorded accurately, and the spheres B and B' are then moved to the other side of spheres A and A'. The rod then rotates and comes to equilibrium at a slightly different position. The small change in the angular position of the rod is measured by using a light beam reflected from a mirror mounted on the rod.

year 1800, Henry Cavendish succeeded by a method described theoretically by Newton and illustrated in Fig. 15.9. Cavendish found what Newton had surmised, that the gravitational force of interaction is directly proportional to the product of the masses of the two interacting bodies and varies inversely as the second power of the distance between them (Eq. 15.10). The gravitational constant has a small numerical value in the units of newtons, kilograms, and meters because the force of gravity is extremely weak unless one of the two interacting bodies has a very large mass. Equation 15.10 is called the *law of universal gravitation* because it describes the ability of all objects, terrestrial and celestial, to participate in gravitational interaction. Applications of the law of universal gravitation are described in Examples 15.4 and 15.9 at the end of this chapter.

**Equation 15.10
(Law of Universal
Gravitation)**

force of gravity $= \mathbf{F}_G$
masses of
interacting bodies $= M_1, M_2$
distance between
 centers of bodies $= R$
gravitational
 constant $= G$

$$\left|\mathbf{F}_G\right| = G\frac{M_1 M_2}{R^2} \qquad (a)$$

$$G = 6.7 \times 10^{-11}\,\frac{newton \cdot m^2}{kg^2} \qquad (b)$$

15.4 The pendulum

The simple pendulum model. A pendulum is a system that carries out periodic swinging motion in gravitational interaction with the earth (Fig. 15.10). At equilibrium, the system hangs in the vertical direction. When the system is released after being displaced from its equilibrium position, it swings back through the equilibrium position

Figure 15.10 A pendulum consists of a massive object that is supported, but is free to swing in the gravitational field.

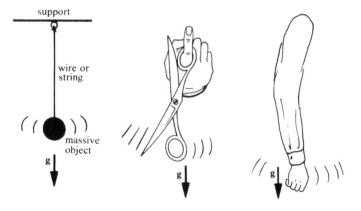

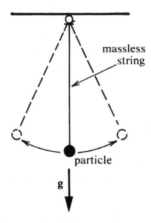

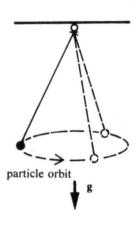

Figure 15.11 The simple pendulum is a working model for a real pendulum.

Figure 15.12 The simple pendulum particle can also swing in an orbit around the equilibrium position.

to the other side. A *simple pendulum* is a working model for a pendulum; it consists of a particle supported by a massless string (Fig. 15.11). Clearly, the simple pendulum is a better model for the first pendulum illustrated in Fig. 15.10 than for the other two. A simple pendulum can swing back and forth; it can also be given a sideways push and let swing in an orbit around the equilibrium position (Fig. 15.12). Not all real pendulums can carry out such motion, however (refer to Fig. 15.10).

Clearly, some energy is stored in the swinging pendulum. As it gradually transfers energy to the air through which it moves and to the support, where there is the inevitable friction, the pendulum's swings become smaller and smaller and finally stop altogether. With some care, it is possible to make a pendulum that swings very many times—50 or 100 times—before coming to rest. The loss of energy during one swing, therefore, is small. It is possible to describe the motion as being approximately periodic (each swing almost repeats the motion during the prior one). The simple pendulum model does not lose energy to other systems, and executes genuine periodic motion. In the following discussion we will first make a mathematical model for the motion of a simple pendulum that executes a circular orbit (Fig. 15.12) and then for one that oscillates through a small angle (Fig. 15.11).

Simple pendulum in a circular orbit. Huygens already had studied this system extensively when he was developing pendulum clocks in the seventeenth century. The particle in the simple pendulum is subject to interaction with the earth and with the string. The net force is obtained by adding the two forces. The force of gravity is given in Eq. 15.11. The force exerted by the string is of unknown magnitude, but we do know that it is directed along the string. The net force is a centripetal force directed horizontally. These facts are illustrated in Fig. 15.13 and are sufficient to permit a calculation of the magnitude of the net

Christian Huygens (1629-1695), the great Dutch contemporary of Newton, made thorough studies of centripetal acceleration while investigating pendulum motion. To Newton's chagrin, Huygens published his studies first in 1673.

Equation 15.11

force of gravity $= \mathbf{F}_G$
particle mass $= M$
gravitational
 intensity $= \mathbf{g}$

$$\mathbf{F}_G = M\mathbf{g}$$

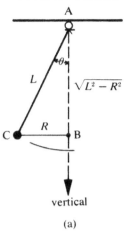

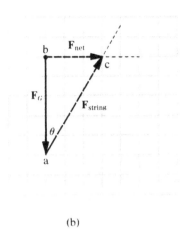

(a)

(b)

Figure 15.13 Theory of the simple pendulum.
(a) Diagram of the pendulum, showing the string length L, the angle θ, and the radius of the circular orbit R. The triangle ABC is a right triangle with acute angle θ.
(b) Force diagram for the simple pendulum. The dotted lines indicate the directions of the two unknown forces. Their magnitudes are found from the force addition formula, $F_{(net)} = F_G + F_{(string)}$. Triangle abc is similar to triangle ABC.

Equation 15.12

$$\frac{|\mathbf{F}_{net}|}{|\mathbf{F}_G|} = \frac{R}{\sqrt{L^2 - R^2}} \quad (a)$$

$$|\mathbf{F}_{net}| = \frac{R}{\sqrt{L^2 - R^2}} |\mathbf{g}| M \quad (b)$$

Equation 15.13

$$|\mathbf{F}_{net}| = \frac{R}{\sqrt{L^2 - R^2}} |\mathbf{g}| M$$

$$= \frac{4\pi^2 MR}{\mathcal{T}^2}$$

force by the procedure illustrated in Fig. 11.12 to Fig. 11.14. Since the force triangle (Fig. 15.13b) and the right triangle formed by the string and the vertical (Fig. 15.13a) are similar, the net force magnitude is to the gravitational force magnitude as the radius is to the vertical side (Eq. 15.12). The net force could be measured with a spring scale that is used to hold the pendulum in a deflected position (Fig. 15.14).

The net force given in Eq. 15.12b is then, according to Newton's second law, equal to the centripetal force required for circular motion (Eq. 15.5), resulting in Fig. 15.13. We find, therefore, that the radius of the orbit and the particle mass cancel out, leaving a relations between the length of the pendulum, its period, and the gravitational intensity (Eq.

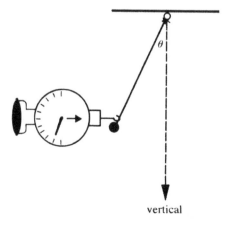

Figure 15.14 Use of spring scale to measure the net force on an orbiting pendulum.

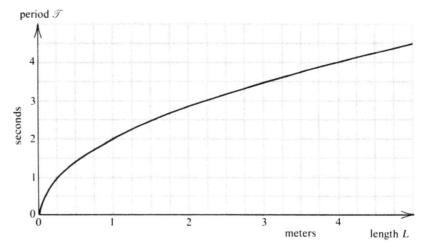

Figure 15.15 Graphical representation of the law of the pendulum at the surface of the earth. The algebraic form of this law is $F = 2.0\sqrt{L}$, where the period T is measured in seconds and the length L in meters.

"As to the times of vibration of bodies suspended by threads of different lengths, they bear to each other the same proportion as the square roots of the lengths of the thread; or one might say the lengths are to each other as the squares of the times; so that if one wishes to make the vibration-time of one pendulum twice that of another, he must make its suspension four times as long."

Galileo Galilei
Dialoghi delle Due
Nuove Scienze,
1638

Equation 15.14

$$\frac{|g|}{\sqrt{L^2 - R^2}} = \frac{4\pi^2}{\mathcal{T}^2}$$

Equation 15.15

$$\sqrt{L^2 - R^2} \approx L \qquad \text{(a)}$$

$$\frac{|g|}{L} = \frac{4\pi^2}{\mathcal{T}^2} \qquad \text{(b)}$$

Equation 15.16

$$L \approx \frac{|g|\mathcal{T}^2}{4\pi^2} \qquad \text{(a)}$$

$$\mathcal{T} \approx 2\pi\sqrt{\frac{L}{|g|}} \qquad \text{(b)}$$

15.14). For a very small deflection angle (θ), the radius (R) is small compared to the length of the string (L), so that the square root in Eq. 15.14 can be approximated by the length of the string (Eq. 15.15). You can rearrange this formula to express the length in terms of the period (Eq. 15.16a) or the period in terms of the length (Eq. 15.16b). A graph of this relation, called *the law of the pendulum* (discovered by Galileo), is shown in Fig. 15.15. Since the mass canceled out, the relation holds for every simple pendulum at the surface of the earth with a small angle of deflection (θ). Equation 15.16b shows that the period is completely independent of the mass of the particle and, for small angles of swing, the period is also independent of the angle. Thus if a pendulum is started at a small angle, the period will be constant as the pendulum gradually comes to rest. Galileo describes how he first discovered this aspect of the law of the pendulum by using his pulse to time the swings of a hanging church lamp!

The oscillating pendulum. The oscillating motion of a pendulum is observed when the particle is displaced to the side and then released with zero initial speed (Fig. 15.10). The oscillating motion is more difficult to describe mathematically than the circular motion of the orbiting pendulum because the magnitude of the velocity and acceleration are changing all the time. We will therefore not analyze it directly. As long as the angle of deflection (θ) is small, however, the same mathematical models and the same law of the pendulum (Eq. 15.16 and Fig. 15.15) applies to the oscillating pendulum as to the orbiting pendulum.

Applications. Many familiar applications of the simple pendulum make use of the regularity of the orbital or oscillating motion. Both a pendulum clock and a child's swing are characterized by a rhythm that satisfies the law of the pendulum. The period does not change with the width of the swing or the mass of the particle (child). Only

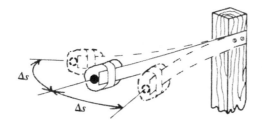

Figure 15.16 One-particle model for an elastic oscillator, The particle of mass M is displaced to a distance Δs from its equilibrium position.

changing the length of the pendulum's support or the gravitational intensity changes the period.

15.5 Elastic oscillators

A weight executes periodic motion when it bounces up and down at the end of a spring. Other examples of periodic motion caused by elastic objects were mentioned in the introduction to this chapter: the inertial balance we used in Section 3.4 to define inertial mass (Fig. 15.16), the bird swaying on a branch of a tree, and the oscillators in an elastic medium, which make wave propagation possible.

We will investigate the motion of a model elastic oscillator in which a massive particle is moving subject to interaction with a "massless" elastic object (a spring, a tree branch) whose behavior is described by Hooke's law (see Section 11.6 and Equation 11.8, in left margin). The one-particle Newtonian approach is adequate for such a working model.

The massive particle has an equilibrium position from which it may be displaced (Fig. 15.16). When it is displaced, the elastic object exerts a restoring force that pulls the particle back to its equilibrium position. For elastic objects described by Hooke's law (Section 11.6), the net force is proportional to the displacement (Eq. 11.8). This situation is similar to the one you encountered with the pendulum (Eq. 15.12b), where the net force was proportional to the radius of the orbit. We therefore select the simple pendulum as an analogue model for the elastic oscillator (Table 15.3).

From the simple pendulum analogue you obtain the mathematical model in Eq. 15.17, according to which the period does not depend on the displacement, but does depend on the mass of the particle and on the strength of the spring. The period increases with greater mass but decreases for stronger springs. This result is consistent with what we found qualitatively in Section 3.4 for the effects of inertia (mass) on the period of the inertial balance.

The most serious limitation of the model is caused by the neglect of the mass of the spring. When the mass of the "massive" particle is equal to zero (no particle is placed on the spring), the model predicts a period of zero seconds. Even an unloaded spring or branch has inertia and therefore an oscillation period unequal to zero, however.

Nevertheless, the model is extremely useful. It can be applied not only to a leaf spring as in the inertial balance, but also to helical springs and coil springs (Fig. 11.23). As a matter of fact, Eq. 15.17 is used to

Equation 11.8 (Hooke's Law)

force $= \mathbf{F}$
distance displaced $= \Delta s$
force constant $= \kappa$

$$|\mathbf{F}| = \kappa \Delta s$$

Equation 15.17 (Elastic oscillator)

Inertial mass (kg) $= M$
force constant
(newtons/m) $= \kappa$
period (sec) $= \mathfrak{I}$

$$\mathfrak{I} = 2\pi \sqrt{\frac{M}{\kappa}}$$

TABLE 15·3 PENDULUM ANALOGUE FOR ELASTIC OSCILLATOR

Pendulum		*Elastic oscillator*	
particle mass	*M*	*particle mass*	*M*
period	$\mathcal{T}$	*period*	$\mathcal{T}$
radius	*R*	*distance*	Δs
orbital motion		*oscillation*	
net force $|\mathbf{F}| = \dfrac{|\mathbf{g}|MR}{L}$		*elastic force* $|\mathbf{F}| = \kappa\Delta s$	
law of the pendulum		*law of the oscillator*	
$\mathcal{T} = 2\pi\sqrt{\dfrac{MR}{|\mathbf{F}|}} = 2\pi\sqrt{\dfrac{L}{|\mathbf{g}|}}$		$\mathcal{T} = 2\pi\sqrt{\dfrac{M\Delta s}{|\mathbf{F}|}} = 2\pi\sqrt{\dfrac{M}{\kappa}}$	

Equation 15.18 (from Eq.15.5) (centripetal force)

centripetal force (newtons)
$\qquad\qquad\qquad = |\mathbf{F}|$
radius of circular
$\qquad$ motion (m) $\quad = R$
period of circular
$\qquad$ motion (sec) $\quad = \mathcal{T}$

$$|\mathbf{F}| = \frac{4\pi^2 MR}{\mathcal{T}^2}$$

Equation 15.16b
(Law of the pendulum)

$$\mathcal{T} \approx 2\pi\sqrt{\frac{L}{|\mathbf{g}|}}$$

Equation 15.17
(elastic oscillator)

inertial mass (kg) $\qquad = M$

strength of interaction (force
constant, newtons/m) $\quad = \kappa$

$$\mathcal{T} = 2\pi\sqrt{\frac{M}{\kappa}}$$

determine the force constant of very delicate springs in some spring scales through a measurement of their oscillation periods. Once the force constant is known, the spring scale can be calibrated without resort to the complicated procedure described in Section 11.2. Cavendish used this technique when he measured the force of gravity between lead spheres in his laboratory (Fig. 15.9), and Coulomb used it when he measured the electrical force between charged spheres (Section 11.5).

Summary

The periodic motion of interacting objects played an important part in the history of science. The periodic motion of the planets in the solar system stimulated Newton's development of his theory of moving bodies and the law of gravitation (Eq. 15.10). The mathematical model for centripetal force (Eq. 15.18) was an important intermediate step that enabled Newton to use Kepler's laws of planetary motion in his investigations. The simple pendulum and the elastic oscillator are systems that are used extensively in the regulation of clocks and in scientific studies. The mathematical models governing their motion are given in Eqs. 15.16b and 15.17, respectively.

Additional examples

EXAMPLE 15.3. Calculate the orbital period of a low-altitude satellite (Table 15.2). (Assume that the satellite is sufficiently close to the earth that the radius of its orbit can be assumed to be the same as the radius of the earth.)

Data:

$|\mathbf{g}| = 10$ newtons/kg; $R = 6.6 \times 10^6$ m.

Solution: Near the surface of the earth, the gravitational intensity is 10 newtons per kilogram. This is equal to the satellite's centripetal acceleration.

From Eq. 15·4

$$|\mathbf{g}| = \frac{4\pi^2 R}{\mathscr{T}^2}$$

$$\mathscr{T} = 2\pi\sqrt{\frac{R}{|\mathbf{g}|}} = 2 \times 3.14 \times \sqrt{\frac{6.6 \times 10^6}{10}} \approx 6.3 \times \sqrt{66 \times 10^4}$$

$$= 6.3 \times 8.1 \times 10^2 = 5.1 \times 10^3 \text{ sec} \approx 85 \text{ minutes}$$

EXAMPLE 15·4. Find the mass of the earth.

Data:

$$R = 6.4 \times 10^6 \text{ m}; \ |\mathbf{g}| = 10 \text{ newtons/kg}$$

Solution: Use the law of universal gravitation (Eq. 15·10) and the gravitational intensity of the earth (Eq. 15·9). Call the mass of the earth M_E. The force on a body of mass M at the earth's surface is

$$|\mathbf{F}_G| = |\mathbf{g}|M = \left(G\frac{M_E}{R^2}\right) M$$

$$M_E = \frac{R^2|\mathbf{g}|}{G} = \frac{(6.4 \times 10^6)^2 \times 10}{6.7 \times 10^{-11}} \approx 6 \times 10^{24} \text{ kg}$$

EXAMPLE 15·5. What mathematical model for the centripetal force would you infer if Kepler's law had been $\mathscr{T}^2 = KR$?

Solution: Use Eq. 15·18 and insert the hypothesis for $\mathscr{T}^2$

$$|\mathbf{F}| = \frac{4\pi^2 RM}{\mathscr{T}^2} = \frac{4\pi^2 RM}{KR} = \frac{4\pi^2 M}{K}$$

The centripetal force is independent of radius.

EXAMPLE 15·6. What law relating period and radius of circular motion would you predict if the force decreased inversely as the third power of the radius, $|\mathbf{F}| = k/R^3$?

Solution: Use Eq. 15·18 and solve it for the period to the second power, $\mathscr{T}^2$.

$$|\mathbf{F}| = \frac{4\pi^2 MR}{\mathscr{T}^2}, \qquad \mathscr{T}^2 = \frac{4\pi^2 MR}{|\mathbf{F}|}$$

Substitute the force formula

$$\mathscr{T}^2 = \frac{4\pi^2 MR}{k/R^3}$$

$$= \frac{4\pi^2 MR^4}{k}$$

The second power of the period varies as the fourth power of the radius.

EXAMPLE 15·7. A simple pendulum is raised 0.40 meter above its equilibrium position. With what speed will it pass through the equilibrium position?

Data:

$h = 0.40$ m; $|\mathbf{g}| = 10$ m/sec/sec

Solution: Conservation of energy can be used to solve the problem. The mass and length of the pendulum need not be known.

kinetic energy = gravitational energy

$\frac{1}{2} M v^2 = M |\mathbf{g}| h$

$$v = \sqrt{2|\mathbf{g}|h} = \sqrt{2 \times 10 \text{ m/sec/sec} \times 0.4 \text{ m}}$$

$$= \sqrt{8.0} = 2.8 \text{ m/sec}$$

EXAMPLE 15·8. In one of Cavendish's experiments (Fig. 15·9), the rod suspended by the thin wire had a vibration period of 15 minutes when the two lead spheres A and A' at the ends each had a mass of 0.75 kilogram. What is the force constant of Cavendish's thin wire?

Data:

total mass of oscillator $M = 2 \times 0.75$ kg $= 1.5$ kg;
period $\mathcal{T} = 15$ minutes $= 900$ sec

Solution: Use Eq. 15·17.

$$\mathcal{T} = 2\pi \sqrt{\frac{M}{\kappa}}$$

Solving Eq. 15·17 for κ, we get

$$\kappa = 4\pi^2 \frac{M}{\mathcal{T}^2} = 40 \times \frac{1.5 \text{ kg}}{(900 \text{ sec})^2} \approx \frac{60}{8 \times 10^5} = 7.5 \times 10^{-5} \text{ newton/m}$$

(See also Example 15·9.)

EXAMPLE 15·9. What is the deflection of the Cavendish spring scale in Example 15·8 if the large lead spheres B and B' have a mass of 15 kilograms each and the distance between centers of spheres A and B is 0.13 meter?

Data:

Mass $M_1 = 0.75$ kg; mass $M_2 = 15$ kg; $R = 0.13$ m;
$\kappa = 7.5 \times 10^{-5}$ newton/m; $G = 6.7 \times 10^{-11}$ newton-m²/kg²

Solution: Use Eqs. 11·8 and 15·10.

Note that the gravitational force is that due to *two* pairs of interacting spheres.

$$|\mathbf{F}| = \kappa \Delta s = 2G \frac{M_1 M_2}{R^2}$$

$$= 2 \times 6.7 \times 10^{-11} \frac{\text{newton-m}^2}{(\text{kg})^2} \times \frac{0.75 \text{ kg} \times 15 \text{ kg}}{(0.13 \text{ m})^2}$$

$$7.5 \times 10^{-5} \Delta s \approx \frac{2 \times 6.7 \times 11 \times 10^{-11}}{1.7 \times 10^{-2}} \approx 8.5 \times 10^{-8}$$

$$\Delta s \approx \frac{8.5 \times 10^{-8}}{7.5 \times 10^{-5}} \approx 1.1 \times 10^{-3} \text{ m} = 1.1 \text{ mm}$$

The deflection on the sensitive scale is 1.1 millimeters.

EXAMPLE 15·10. (a) Find the average velocity of a particle in uniform circular motion (radius R) during a time interval $\tfrac{1}{4}\mathcal{T}$.

Solution: In a time interval $\Delta t = \tfrac{1}{4}\mathcal{T}$, the particle makes one fourth of a revolution (see diagram in margin). The positions of the particle are

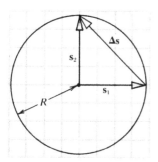

$$\mathbf{s}_1 = [R, 0]$$

$$\mathbf{s}_2 = [0, R]$$

$$\Delta \mathbf{s} = \mathbf{s}_2 - \mathbf{s}_1 = [0, R] - [R, 0] = [-R, R]$$

$$\mathbf{v}_{\text{av}} = \frac{\Delta \mathbf{s}}{\Delta t} = \frac{[-R, R]}{1/4\mathcal{T}} = \left[-\frac{4R}{\mathcal{T}}, \frac{4R}{\mathcal{T}} \right]$$

(b) Find the magnitude of the average velocity in part (a) and compare it with the instantaneous velocity in Eq. 15·2.

Solution:

$$|\mathbf{v}_{\text{av}}| = \sqrt{\left(-\frac{4R}{\mathcal{T}}\right)^2 + \left(\frac{4R}{\mathcal{T}}\right)^2} = \sqrt{\frac{16R^2}{\mathcal{T}^2} + \frac{16R^2}{\mathcal{T}^2}} = \sqrt{\frac{32R^2}{\mathcal{T}^2}}$$

$$= 4\sqrt{2}\,\frac{R}{\mathcal{T}} \approx 5.6\,\frac{R}{\mathcal{T}}$$

From Eq. 15·2,

$$|\mathbf{v}| = \frac{2\pi R}{\mathcal{T}} = 6.28\,\frac{R}{\mathcal{T}}$$

which is greater than

$$|\mathbf{v}_{\text{av}}| = 5.6\,\frac{R}{\mathcal{T}}$$

List of new terms

centripetal acceleration	heliocentric	simple pendulum
centripetal force	law of universal	law of the pendulum
geocentric	gravitation	elastic oscillator

List of symbols

a	acceleration	π	3.1415...
|**a**|	acceleration magnitude	**F**	force
v	velocity	|**F**|	force magnitude
|**v**|	velocity magnitude	**F**$_G$	force of gravity
v	speed	M	mass
v$_{av}$	average speed	G	gravitational constant
s	position	**g**	gravitational intensity (acceleration of gravity)
Δs	displacement		
Δs	distance		
Δt	time interval	L	length of simple pendulum
$\mathcal{I}$	period		
R	radius of circular orbit	θ	deflection angle
		κ	force constant

Problems

1. Propose two operational definitions of *time interval* that could have been used before the invention of clocks (for example, in ancient times).

2. Undertake library research to determine the history of clocks, especially the gradual improvement of clock accuracy. Point out some of the important uses of clocks at various stages of the development.

3. Calculate the average velocity of a particle moving in a circular orbit of 1 meter radius at a constant speed of 6.28 meters per second. Use the following time intervals: (a) 1 second; (b) 5/6 second; (c) 2/3 second; (d) 1/2 second; (e) 1/3 second; (f) 1/6 second; (g) 1/12 second.

4. (a) Make a graph of the magnitudes of the average velocities in Problem 3 to show their dependence on the time interval.
(b) Extrapolate to zero time interval on your graph to find the magnitude of the instantaneous velocity.
(c) Compare the result with that calculated according to Eq. 15.2.

5. Calculate the average acceleration of a particle moving in a circular orbit of 1 meter radius with a constant speed of 6.28 meters per second. Use the following time intervals: (a) 1 second; (b) 5/6 second; (c) 2/3 second; (d) 1/2 second; (e) 1/3 second; (f) 1/6 second; (g) 1/12 second.

6. (a) Make a graph of the magnitudes of the average accelerations vs. the time interval in Problem 5 to show their dependence on the time interval.

(b) Extrapolate to zero time interval on your graph to find the magnitude of the instantaneous acceleration.

(c) Compare the result of the extrapolation with that calculated from Eq. 15.4.

7. The orbital data for the four moons of Jupiter (Io, Europa, Ganymede and Callisto) discovered by Galileo are given below. Determine whether they satisfy Kepler's third law. The units of distance and time are the orbital radius and period of the innermost moon, which are 4×10^8 meters and 42.5 hours, respectively

	Io	Europa	Ganymede	Callisto
Orbital radius	1.0	1.6	2.5	4.5
Orbital period	1.0	2.0	4.0	9.5

Optional: Find the mass of Jupiter.

8. (a) Find the centripetal acceleration of a 40-kilogram child on a merry-go-round. The child is at a distance of 5 meters from the center of the merry-go-round, which turns at a rate of six revolutions per minute.
(b) Find the centripetal force on this child.
(c) What object exerts this centripetal force?

9. (a) Obtain or construct an apparatus like that shown in Fig. 15.6 and conduct experiments to test the mathematical model for circular motion (Eq. 15.5).
(b) Discuss some of the sources of experimental error in this experiment.

10. Consult references to determine Copernicus' reasons for preferring a heliocentric over a geocentric model for the solar system. Report and discuss your conclusions.

11. Consult references to find the basis on which Kepler tried to explain planetary motion by forces. Report and discuss his approach.

12. Find the mass of the sun by using data on planetary motion.

13. Explain why the law of gravitation and the moon's orbit around the earth do not allow you to calculate the moon's mass.

14. Verify that the data on the Syncom satellite (Table 15.2) are compatible with the law of gravitation. Use the same method that Newton used, as explained in Section 15.3.

15. Suppose the force of attraction between objects varies inversely as the radius, $|F| = k/R$. What is the form of Kepler's third law appropriate to orbital motion of particles subject to this force?

16. Give four examples from everyday life of systems to which the simple pendulum model should apply.

17. Test one of the systems mentioned in your answer to Problem 16 to determine whether the law of the pendulum applies to it.

18. A large pendulum in a science museum is found to have a period of 8 seconds. How long is the wire supporting it?

19. Find the period of each of the following systems allowed to swing naturally: (a) your right arm held stiffly; (b) your left leg held stiffly; (c) your right forearm (hold the upper arm stationary). Compare and discuss your findings and relate them to your speed of walking. *(Hint:* Measure the time for 20 swings and divide by 20 to find the period.)

20. Give four examples of elastic oscillators from everyday life.

21. Experiment with one of the systems you identified in Problem 20 to find its force constant by: (a) attaching weights or a spring scale and observing the elastic displacement (direct measurement); and (b) making an elastic oscillator and measuring the period (indirect method).

22. Write a critique concerning the application of Eq. 15.17 to two of the systems you selected in answer to Problem 20.

23. (a) Explain under what circumstances you would or would not expect an object hanging from a rubber band to have a period described by the mathematical model in Eq. 15.17.
    (b) Test your ideas by experimenting with such an oscillator.

24. Calculate the centripetal acceleration of an object attached to the earth at the equator (due to the rotation of the earth). Compare your result to the acceleration of gravity.

25. Identify one or more explanations or discussions in this chapter that you find inadequate. Describe the general reasons for your judgment (conclusions contradict your ideas, steps in the reasoning have been omitted, words or phrases are meaningless, equations are hard to follow, . . .), and make your criticism as specific as you can.

Bibliography

A. Armitage, *The World of Copernicus,* New American Library, New York, 1951.

B. Brecht, *Galileo,* Grove Press, New York, 1966.

D. Cassidy, G. Holton, J. Rutherford, *Understanding Physics*, Springer Verlag, 2002.

I. B. Cohen, *The Birth of a New Physics,* Norton, New York, 1985.

G. De Santillana, *The Crime of Galileo,* University of Chicago Press, Chicago, Illinois, 1955.

S. Drake, *Galileo at Work: His Scientific Biography*, University of Chicago Press, Chicago, Illinois, 1978.

A. Eddington, *Space, Time, and Gravitation,* Harper and Row, New York, 1959.

G. Galilei, *Dialogue on the Great World Systems,* University of California Press, 1967.

A. R. Hall, *The Scientific Revolution 1500-1800,* Beacon Press, Boston, Massachusetts, 1954.

G. Hawkins and J. B. White, *Stonehenge Decoded,* Doubleday, Garden City, New York, 1965.

F. Hoyle, *Frontiers of Astronomy,* Harper and Row, New York, 1955.

F. Hoyle, *The Nature of the Universe,* Harper and Row, New York, 1960.

R. Jastrow, *Red Giants and White Dwarfs, The Evolution of Stars, Planets, and Life,* Harper and Row, New York, 1967.

A. Koestler, *The Watershed: A Biography of Johannes Kepler,* University Press of America, 1985.

A. Koyré, *From the Closed World to the Infinite Universe,* Johns Hopkins Press, Baltimore, Maryland, 1957.

T. Kuhn, *The Copernican Revolution,* Harvard University Press, Cambridge, Massachusetts, 1990.

P. Moore, *Basic Astronomy,* Oliver and Boyd, Edinburgh and London, 1967.

D. Sobel, *Galileo's Daughter*, Walker and Co., New York, 1999.

D. Sobel, *Longitude*, Walker and Co., New York, 1995. The story of John Harrison, the extraordinary instrument-maker who invented and built a clock that ran sufficiently accurately at sea to permit the determination of longitude and thus opened the door to reliable sea navigation. Newton and others had attempted and failed to develop a reliable astronomical method of determining longitude at sea. The unpredictable motion of a ship at sea posed a difficult challenge both to keeping time as well as to measuring the positions of objects in the sky.

F. L. Whipple, *Earth, Moon, and Planets,* Harvard University Press, Cambridge, Massachusetts, 1968.

Article from Scientific American. Articles can be obtained on the Internet at http://www.sciamarchive.org/.

J. E. Ravetz, 'The Origin of the Copernican Revolution" (October 1966).

So many of the properties of matter, especially when in the gaseous form, can be deduced from the hypothesis that these minute parts are in rapid motion, the velocity increasing with the temperature, that the precise nature of this motion becomes a subject of rational curiosity.

JAMES CLERK MAXWELL
Philosophical Magazine,
1860

Heat and motion

16

The gasoline engine in automobiles, motorcycles, lawnmowers, and other similar appliances is one of the major conveniences of modern technology, but it is also, because of the air pollution it causes, one of the major threats to civilization. The gasoline engine, diesel engine, and gas turbine jet engine are modern counterparts of the coal-burning steam engine, whose development by James Watt (1736-1819) led to the industrial revolution in the nineteenth century. All of them transform heat released by burning fuel into useful work for manufacturing, construction, and transportation; they are special types of what are called *heat engines*.

Before the industrial revolution, the major sources of energy transfer in the form of work were human beings and animals, which served as coupling elements to convert chemical energy of food-oxygen systems into kinetic energy (transportation), gravitational field energy (construction of buildings), and so on. The energy of fuel-oxygen systems was released as heat for cooking and warmth, but was not otherwise applied. Since the industrial revolution, the heat engines we enumerated above are used as coupling elements to convert the chemical energy of fuel-oxygen systems to useful work. In industrial nations, in fact, human and animal labor are fast becoming surplus commodities.

The conversion of heat to work in a heat engine is the reverse of the conversion of work to heat by friction. A very simple device to accomplish this was Hero's steam engine, invented in ancient times. Hero's engine operates like a whirling lawn sprinkler in that jets of steam emerge from a boiler through bent tubes and force the boiler to spin in the opposite direction. In modern engines, the steam or another hot gas at high pressure pushes against a piston in a cylinder or against the fan-like blades of a turbine (Fig. 16.1). In all heat engines, a gas is the essential coupling element that accomplishes the conversion of heat to work.

The engineering development of heat engines in the last 200 years illustrates the continued interplay of science and technology. The invention of the steam engine and the need for its further improvement focused attention on the nature of heat and properties of gases. As a result, the caloric theory of heat was replaced by the kinetic theory (Section 10.5), and the many-interacting-particles (MIP) model for matter became firmly established. This scientific progress, in turn, led to the invention of gasoline and turbine engines, which have now replaced the steam engine wherever high power and low mass are important.

Since gases play a central role in the operation of heat engines, the physics of the industrial revolution is linked closely to the physics of gases. We therefore begin this chapter with a description of the properties of gases, such as air, oxygen, nitrogen, and water vapor (gaseous water). The concept of gas pressure is related to the force exerted by a gas on its container. After an examination of the energy transfer to and from a gas sample, we explain the operation of a working model for a heat engine. Finally, we trace the introduction of the kinetic theory of gases and outline its successes.

425

Figure 16.1 This turbo fan aircraft engine is capable of producing a thrust of more than 200,000 newtons and has a mass of 4 tons.

16.1 Properties of gases

Air is such a tenuous material that most people take it for granted and are hardly aware of its existence. Nevertheless, the atmosphere of air surrounding the planet earth fulfills many functions that are essential for the existence of life: it absorbs ultraviolet radiation from the sun; it absorbs cosmic radiation; it stores thermal energy from the sun; and it contains oxygen, carbon dioxide, and gaseous water, which participate in metabolic processes of animals and plants. In this section, however, we would like you to ignore these functions temporarily and to think of air as an example of the kind of material called gases. Air possesses the physical properties that most gases exhibit: fluidity, low density, ability to fill its container completely and to exert pressure on the container walls, compressibility, and a large thermal expansion compared to solid and liquid materials.

We briefly discussed gases in Section 4.5, where we used the MIP model for matter to explain some of the properties that distinguish the gas phase from the solid and liquid phases. First, gases have much lower density than solids and liquids. Whereas 1 liter of water has a mass of 1 kilogram, 1 liter of atmospheric air has a mass of only about 0.0012 kilogram (Fig. 16.2a). Conversely, 1 kilogram of water occupies a space of 1 liter, but 1 kilogram of atmospheric air fills more than 800 liters (Fig. 16.2b). Other gases are generally similar to air in that their density is very low compared to that of solids or liquids.

A second distinguishing characteristic of the gas phase is its compressibility. Even with the application of only moderate forces, the volume filled by a sample of air can be greatly reduced or increased compared to the volume it filled as part of the atmosphere (Fig. 16.3). A good example of this is when you pump up a bicycle tire. As the volume is reduced, however, larger forces are required to reduce it still further. Solid and liquid materials hardly change in volume unless extremely large forces are applied.

A third special aspect of gases is their thermal expansion. Gases expand or contract much more for a certain temperature change than do solid or liquid materials. In Section 10.1 we referred to Galileo's thermoscope, in which the expansion and contraction of a sample of air in a glass bulb served as a temperature indicator. We nevertheless

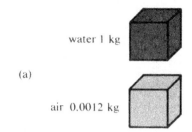

water 1 kg

air 0.0012 kg

(a)

Figure 16·2   Comparison of the density of water and air (0.1 meter = 1 meter). (a) Equal volumes of water and of air. (b) Equal masses of water and of air.

water 1 kg

(b)

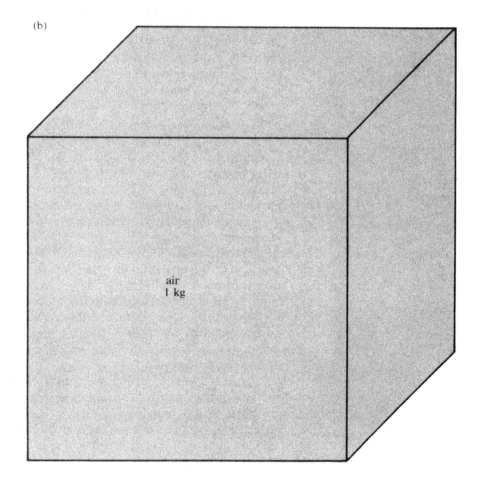

air
1 kg

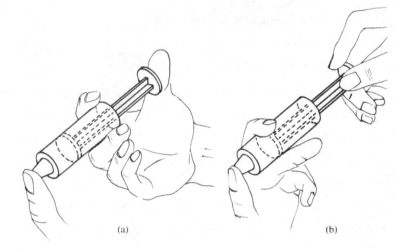

Figure 16.3 Air in a syringe with a finger over the open end.
(a) The air volume is compressed when the plunger is pushed in.
(b) The air volume is expanded when the plunger is withdrawn.

(a)   (b)

selected the mercury thermometer instead of an air thermometer for the operational definition of temperature. As pointed out in Section 10.1, the mercury must be confined to a narrow glass tube so as to make the small temperature-related volume changes visible (Fig. 10.3).

16.2 Gas pressure

Gases exert pressure on the walls of the container in which they are confined. Tennis balls, automobile tires, and children's balloons all contain compressed air, which makes these objects relatively stiff. If the container walls are too weak to withstand the pressure, they burst. In the experiment shown to the left, air is pumped out of a sealed container, which reduces the pressure inside, as a result, the pressure of the outside atmosphere crushes the container.

to pump

Force and pressure. Pressure (rather than force) is the measure of interaction strength that we use when a confined gas interacts with its container wall. The reason the force concept is not directly useful for gases is that a gas cannot be described by a single-particle center-of-mass model (Section 14.1). In the center-of-mass model, the force of interaction is concentrated on one particle, but the interaction between a gas and its container is distributed over the entire surface of the container wall (Fig. 16.4).

Pressure can nevertheless be related to force with the help of the following working model for the gas. Imagine the outside layer of gas represented by a layer of "gasbags" (Fig. 16.5) that press on the container wall. You can represent the interaction between each bag and the wall with two forces, one exerted by the gasbag on the container, and the other exerted by the container on the gasbag. According to Newton's third law, these two sets of forces are equal in magnitude and oppositely directed.

You can see from Fig. 16.5 that the forces produced by the gas pressure are directed at right angles to the boundary surface. You can also recognize that the magnitude of the forces depends on the size of the contact area between one gasbag and the boundary surface. If, for example, you were to divide one bag in Fig. 16.5 into smaller ones, then each smaller bag would exert a smaller partial force in proportion to its area (Fig. 16.6).

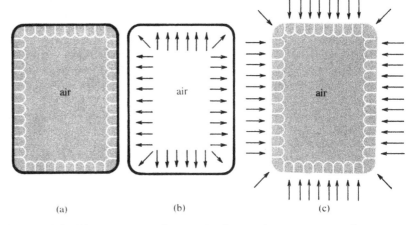

Figure 16·4 The interaction of a sample of air and its container is distributed all over the boundary surface between the two materials.

Figure 16·5 The interaction of a sample of gas with its container wall.
(a) "Gasbag" layer model for the gas.
(b) Forces exerted on the container by the gasbags.
(c) Gasbags and forces exerted on them by the container.

Gas pressure gauges. Gauges for measuring gas pressure can be understood with the help of the "gasbag" working model pictured in Fig. 16.5. In effect, a gauge replaces the section of boundary surface with a device that registers the force exerted by the gasbag. You are undoubtedly familiar with several different types of pressure gauges and pressure readings. Air pressure in automobile and bicycle tires is measured in "pounds" (more accurately "pounds per square inch"), and usually ranges from 20 pounds (automobiles) to more than 100 pounds (racing bicycles). Atmospheric pressure as reported by the weather bureau is measured in "inches" of mercury and is usually about 30 inches. Other commonly used units of pressure are the "millimeter" of mercury and the "atmosphere." Since gas pressure results in the action of forces, it is also possible to relate the pressure to the newton.

Discovery of atmospheric pressure and the barometer. Evangelista Torricelli carried out a series of key experiments that showed that the atmospheric air could exert pressure and thus support a column of mercury about 0.76 meter (30 inches) high. He filled a long glass tube closed at one end with mercury and then inverted the tube, as shown in

Figure 16.6 Interaction of a gas and its container (detail of Fig. 16-5).
(a) One gasbag is divided into three equal smaller bags that press individually on container wall section A.
(b) Forces exerted on container sections A and B.
(c) Addition of the partial forces acting on container section A gives a force equal to that acting on section B.

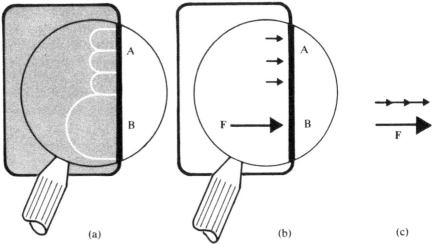

Fig. 16.7, with the other end of tube open to the atmosphere. Some of the mercury ran out and left an empty space at the top of the tube; the difference in height between the higher end of the mercury column and the lower end was always about 30 inches (0.76 m). Torricelli and his successors reasoned that the mercury remaining in the tube was in mechanical equilibrium (zero net force) and thus subject to two equal and opposed forces: the force of gravity (downward) and the force exerted by the atmosphere on the lower end of the mercury column. Torricelli reasoned that the downward force exerted by the atmosphere was transmitted from the open end of the tube by the mercury in the curved section (Fig. 16.7) and thus applied an upward force on the vertical column just large enough to support the observed height of 0.76 m.

Torricelli encountered substantial resistance from traditionalists, who refused to accept the idea that an empty space (or vacuum) existed at the top of the tube and who also found the idea that the weight of the atmosphere pushed the mercury *up* to be contrary to common sense.

Torricelli responded to such objections in the manner of a true scientist: with an experimental test. Torricelli reasoned that the atmospheric pressure should vary with height above (or below) sea level, and he tested this idea by carrying his experimental apparatus up a mountain and measuring the length of the mercury column at various points on the way up to the summit. The experimental results confirmed Torricelli's prediction. These and other experiments convinced most scientists of the validity of the concept of atmospheric pressure.

Torricelli's inverted mercury-filled tube is now called a *mercury barometer* and used as a *pressure gauge*. The length of the mercury column is an excellent measure of (and operational definition for) the atmospheric pressure and is measured in millimeters, inches, or any other unit of length. This is the barometric pressure scale used in weather reports. Water or oil is sometimes used as the barometric liquid instead of mercury. Can you identify the advantages and disadvantages of using water or oil rather than mercury?

Spring scale gauge. A second type of pressure gauge makes use of a suitably designed spring scale that can be attached to the gas container (Fig. 16.8). At a low gas pressure, the force reading is small; at a high pressure, it is large. Tire pressure is usually measured by this type of pressure gauge; hence the "pound" (1 pound equals 4.5 newtons) is used to describe tire pressure. The *aneroid barometer* is a meteorological instrument that incorporates a spring scale for measuring atmospheric pressure (Fig. 16.9); its dial, however, is usually calibrated in inches or millimeters of mercury, because those are the pressure units commonly used in weather reports.

Formal definition of pressure. Because we will be concerned with energy transfer from an expanding gas, and therefore with the force it exerts, we will introduce a formal definition for pressure as the force per unit area (Eq. 16.1). Whenever the force exerted by a gas is proportional to the area (and this is generally the case), the numerical value of the pressure does not depend on the size of the area chosen.

Before Torricelli, the theory that "nature abhors a vacuum" was used to explain why liquids could not be poured out of containers with only one small hole.

FORMAL DEFINITION
A gas exerts a force on a surface of area A. Then the "pressure" of the gas is equal to the magnitude of the force divided by the area.

Equation 16·1

force	**F**
magnitude of the force	**|F|**
area	*A*
pressure	*P*

$$P = \frac{|\mathbf{F}|}{A}$$

unit of pressure:
newtons per square meter
= newtons/m²

EXAMPLE

$$|\mathbf{F}| = 1000 \text{ newtons}$$
$$= 10^3 \text{ newtons}$$
$$A = .01 \ m^2 = 10^{-2} \ m^2$$

$$P = \frac{|\mathbf{F}|}{A}$$

$$= \frac{10^3 \text{ newtons}}{10^{-2} \ m^2}$$

$$= 10^5 \text{ newtons/m}^2$$

Evangelista Torricelli (1608-1647) was an Italian noble who came to Rome in 1628 to study mathematics. Upon becoming acquainted with Galileo's work on motion, Torricelli was overcome with admiration and wrote Galileo on the subject. Galileo was impressed and invited Torricelli to become his assistant. After his master's death, Torricelli became his successor at the Academy of Florence. He is best known as the inventor of the barometer, which created a sensation because Torricelli and his colleagues argued that the space above the inverted mercury column was essentially empty. This was a realization of tremendous significance, because Aristotle had held that it was impossible to create a vacuum. Like his mentor Galileo, Torricelli criticized Aristotelian physics and helped lay the groundwork of the new physics.

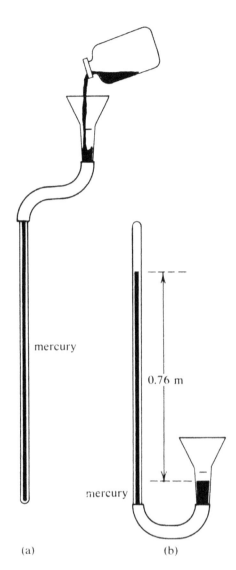

(a) (b)

Figure 16.7 (to left) Torricelli's barometer and his experiment revealing the pressure of the atmosphere.
(a) A long glass tube (more than 30 inches or 0.76 m) is filled with mercury.
(b) The glass tube is turned upside-down. A gap appears at the top of the tube above the mercury column. Torricelli argued that the gap was empty, or a vacuum. He showed that the length of the gap (and the height of the mercury column) depends upon the atmospheric pressure. Torricelli interpreted this to mean that the weight of the 0.76 m mercury column was being supported by the weight, or pressure, of the atmosphere on the mercury surface in the open funnel.

Figure 16.8 (below) Spring scale pressure gauge is connected to a gas sample. Note the enclosed plunger, behind which there is a vacuum.

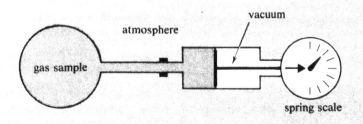

Figure 16.9 (to left) An aneroid barometer contains a thin metal membrane on a sealed, evacuated cylinder (at the back). The membrane flexes in and out in response to the outside air pressure. The long needle is connected by a sensitive gear to the membrane and thus registers the air pressure on a scale (not shown).

TABLE 16.1 RELATION OF UNITS FOR PRESSURE MEASUREMENT

Unit	Equivalent		Use of unit
	(newtons/m^2)	(atm)	
1 newton per square meter	1.0	1.0 x 10^{-5}	physics
1 millimeter of mercury	130.	$\frac{1}{760} = 1.3\times10^{-3}$	meteorology, physics
1 inch of mercury	3.4 x 10^3	$\frac{1}{30} = 3.4\times10^{-2}$	meteorology
1 pound per square inch	6.8 x 10^3	$\frac{1}{14.7} = 6.8\times10^{-2}$	technology
1 atmosphere	1.0 x 10^5	1.0	high-pressure measurements

A pressure gauge based on the formal definition of pressure can be made from the standard spring scale by suitably marking its dial in relation to the area on which the gas exerts pressure (Example 16.1). We will use such a pressure gauge for illustrative purposes when describing experiments.

Units for measuring pressure. We have mentioned a large number of different standard units of pressure in this section (Table 16.1). Some of them are used only under special conditions, as indicated. In the remainder of this chapter we will use only two of these units. One of them is the *newton per square meter*, which is indicated by the standard pressure gauge (Example 16.1). The other unit is the *atmosphere*. One atmosphere is the average pressure of the air in the earth's atmosphere

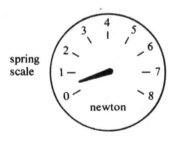

spring scale

newton

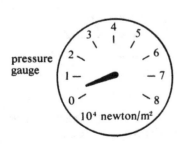

pressure gauge

10^4 newton/m^2

EXAMPLE 16.1. We wish to calibrate a spring scale pressure gauge with a plunger area of 1 square centimeter. The spring scale is to be calibrated in newtons.

$A = 1 \text{ cm}^2 = (0.01 \text{ m})^2 = 10^{-4} \text{ m}^2$

(a) For a dial reading $|\mathbf{F}| = 1$ newton.

$$\text{Pressure } P = \frac{|\mathbf{F}|}{A} = \frac{1 \text{ newton}}{10^{-4} \text{ m}^2} = 1\times10^4 \text{ newtons/m}^2$$

(b) For dial reading $|\mathbf{F}| = 2$ newtons.

$$\text{Pressure } P = \frac{|\mathbf{F}|}{A} = \frac{2 \text{ newtons}}{10^{-4} \text{ m}^2} = 2\times10^4 \text{ newtons/m}^2$$

Therefore, the spring scale dial can be used as a pressure gauge dial if each newton is interpreted as 10^4 newtons per square meter.

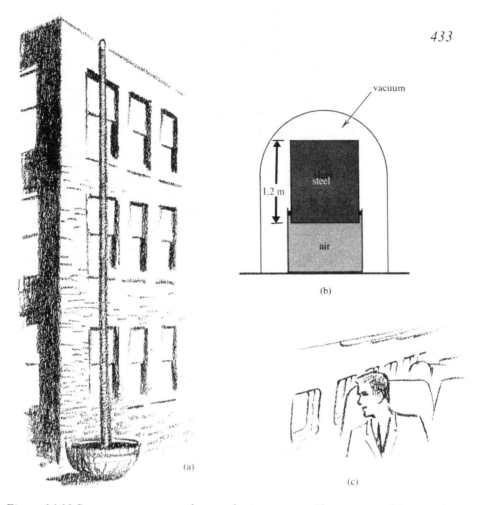

Figure 16.10 Some consequences of atmospheric pressure. The pressure of 1 atmosphere may be described in the following terms: 10 meters of water, 1.20 meters of steel, 0.76 meter (30 inches) of mercury, 10^5 newtons per square meter, and 14.7 pounds per square inch.

(a) The atmosphere pressing on the water surface supports a 10-meter-high column of water. On this basis, Galileo explained why, in his day, water could not be pumped from wells deeper than 10 meters.

(b) The trapped air at a pressure of 1 atmosphere supports a 10-ton steel block.

(c) Have you wondered why jet plane windows are so small? The net pressure force on an 8-by-8-inch window can be 90 pounds (8 in. x 8 in x 14.7 lbs/in^2). Is this net force directed inward or out-ward?

at sea level. This is very slightly more than 100,000 newtons per square meter. The atmosphere (unit) is therefore a much larger unit of pressure than the newton/m^2. Table 16.1 shows the relation of the newton per square meter and of the atmosphere to the other units that are in common use.

Because the earth's atmosphere is so important, we illustrate some aspects of atmospheric pressure in Fig. 16.10 and implications for the weather in Example 16.2. Since Torricelli's day, atmospheric pressure has been ascribed to the fact that the air near the ground supports the layers of air above. As you proceed upward through the atmosphere, the remaining load of air on you becomes less and less, with the result that the pressure of the supporting air also is gradually reduced. At the peak of Mount Everest the air pressure is only one third of its magnitude at sea level.

EXAMPLE 16.2. Air pressure plays an important role in weather forecasting, because air pressure differences give rise to net forces that set air masses into motion. Air masses move from high-pressure regions to low-pressure regions and carry thermal energy and moisture with them. Estimate the wind speed created by a pressure difference of 10^3 newtons per square meter (0.3 inch of mercury) between a high-pressure region and a low-pressure region.

Solution: The speed can be found either from the net pressure force and the time over which it acts (Newton's second law) or from the kinetic energy acquired by the air as work is done on it by the net pressure force. We will use the kinetic energy method.

Take an air sample of 1 cubic meter, with a mass of 1.2 kilograms, that is blown for a distance Δs meters from the high-pressure region to the low-pressure region. Since the total pressure difference of 10^3 newtons per square meter is spread out over a distance of Δs meters, the magnitude of the net force acting on the air sample is $|\mathbf{F}| = (10^3/\Delta s)$ newtons.

The work done:

$$W = |F| \Delta s = [(10^3 / \Delta s) \text{ newtons}] \times \Delta s \text{ meters} = 10^3 \text{ joules}.$$

Note that this result does not depend on the distance Δs.

The kinetic energy:

$$KE = \tfrac{1}{2} Mv^2 = \tfrac{1}{2} \times 1.2 \times v^2 = 0.6v^2$$

The kinetic energy equals the work:

$$0.6v^2 = 10^3 \text{ joules}$$

$$v^2 = (1.0/0.6) \times 10^3 = 17 \times 10^2$$

$$v = \sqrt{17 \times 10^2} = 40 \text{ m/sec}$$

Answer: 40 meters per second or $(40 \text{ m/s})[(2.25 \text{ mi/hr})/(\text{m/s})] = 90$ mi/hr.

16.3 Energy storage by gases

A cylinder with a movable piston, such as in an automobile engine, is a convenient container for studying energy transfer to or from a gas sample. The gas may be heated as it interacts with a system at higher temperature than itself (for example, a flame), or the gas may be compressed by a force acting on the piston and doing work on the gas in

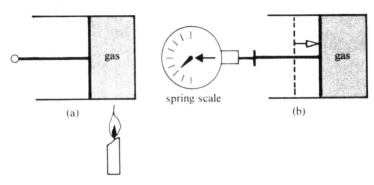

Figure 16.11 Energy transfer to a gas sample,
(a) Heat is absorbed by the gas.
(b) Work is done on the gas.

Robert Boyle (1627-1691)
was born in Ireland.
After education at Eton
and on the continent,
Boyle returned home in
1644 when his father
died. Following 10
years of seclusion in
Ireland, Boyle moved to
Oxford and began to
take active part in
English scientific life. It
was Boyle's celebrated
experiments with gases,
described in his New
Experiments
Physicomechanical,
Touching the Spring of
the Air *(1662), that*
showed how changes in
the volume of a gas
were related to the gas
pressure. Not the least
of Boyle's achievements
was the important part
he played in
establishing The Royal
Society in 1662.

displacing the piston (Figure 16.11). Similarly, there are two ways in which the gas can release energy: it may be cooled by interacting with a system at lower temperature than itself or it may expand and push the piston out, thereby doing work. The heating and cooling processes, which we described in their applications to liquid and solid materials in Chapter 10, lead to the concept of specific heat and thermal energy storage for gases. The compression and expansion processes, which we described in their application to elastic systems in Section 11.6, lead to the concept of "elastic" energy storage by gases. As you will see in this section, experiments show that these two forms of energy storage by a gas are almost identical. Gases store thermal energy, which may be released either as heat (when the gas cools down at constant volume) or as work (when the gas expands against a piston).

"Elasticity" of gases. Everyone is familiar with the "spring" of air in an automobile tire. Air in a pump or syringe (with stopped-up opening) can be squeezed into a smaller volume by a plunger, but the plunger is forced back out when the hand is taken off (as illustrated in Fig. 16.3).

Boyle's law. In 1662, Robert Boyle published a description of "Two New Experiments Touching the Measure of the Force of the Spring of Air Compressed and Dilated." Boyle concluded from his observations that the volume of a gas sample is inversely proportional to its pressure, a relationship known as Boyle's law (Eq. 16.2). A sequence of Boyle's experiments is illustrated schematically in Fig. 16.12. You may recognize that gases are analogous to springs even though the mathematical models of Boyle's and Hooke's laws are quite different (Table 16.2). In fact, Hooke was a contemporary and occasional collaborator of Robert Boyle.

Charles' law. About 100 years after Boyle's publication, Jacques-Alexandre Charles (1746-1823) and Joseph Gay-Lussac (1778-1850) independently discovered a relationship between the volume and the

**Equation 16.2
(Boyle's Law)**

gas volume (m^3) = V
gas pressure
  $(newtons/m^2 = P$
proportionality
  sign $\propto$
$V \propto \dfrac{1}{P}$,
or VP = *constant*
at fixed temperature

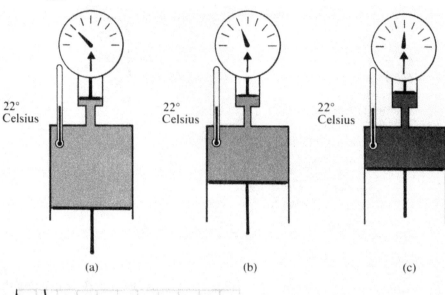

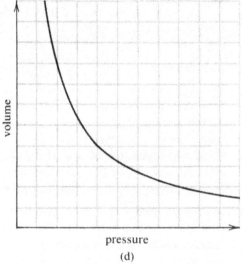

Figure 16·12 Boyle's law.
(a) A gas sample in a certain
volume has pressure.
(b), (c) The same gas sample
in a smaller volume has
proportionately larger pressure.
(d) Graphical representation of
the inverse proportion between
pressure and volume.

TABLE 16.2 ANALOGY BETWEEN GASES AND ELASTIC SYSTEMS

elastic system (such as a spring)	Gas
elastic force	pressure
elastic deformation	volume change
Hooke's law	Boyle's law
elastic energy	"elastic" energy
isolated spring in mechanical equilibrium	(isolated gas in mechanical equilibrium)*

*The analogy breaks down because an isolated gas, not interacting
with any other system, spreads throughout space and is not in mechanical equilibrium. Therefore, a gas has no natural equilibrium state in
which its "elastic" energy is zero.

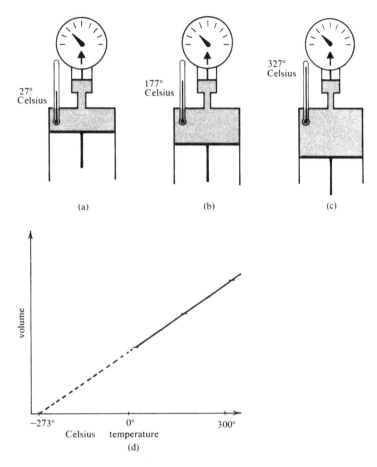

Figure 16.13 Charles' law.
(a) A gas sample at a certain temperature and pressure occupies a certain volume.
(b), (c) The same gas sample at higher temperatures but the same pressure.
(d) Graphical representation of the proportional increases in temperature and volume.

Equation 16.3
(Charles' Law)

temperature
  (deg Celsius)
                 $= T$

$V \propto T + 273$

temperature of gas samples confined in a container at constant pressure (Fig. 16.13). The relationship that the volume is proportional to the "temperature plus 273 degrees" is known as Charles' law (Eq. 16.3). Basically, Charles' law provides a quantitative basis for the operation of Galileo's thermoscope (Fig. 10.1). The one condition, not recognized by Galileo, that must be kept constant is the pressure of the gas sample; otherwise there are volume changes derived from a gas's compressibility (Boyle's law) in addition to those associated with the temperature change.

Ideal gas model. You will now realize that the quantitative description of gas samples is made by means of four variable factors: the mass, the pressure, the temperature, and the volume. These four factors cannot be changed arbitrarily and independently, but are related by Boyle's law

TABLE 16.3 IDEAL GAS MODELS

Gas	Mathematical model	
carbon dioxide	$PV =$	$190M(T + 273)$
helium	$PV =$	$2100M(T + 273)$
hydrogen	$PV =$	$4100M(T + 273)$
nitrogen	$PV =$	$290M(T + 273)$
oxygen	$PV =$	$260M(T + 273)$

Equation 16.4 (Ideal gas law for air)

pressure (newtons/m^2) $= P$
volume (m^3) $= V$
mass of gas (kg) $= M$
temperature
   (deg Celsius) $= T$

$PV = 280M(T + 273)$
(for air)

and Charles' law. A mathematical model that quite accurately describes the interdependence of the four variable factors for air is given in Eq. 16.4. This model is called an *ideal gas model.* You can see that it includes Boyle's law (when the mass and temperature are kept constant) and Charles' law (when the mass and pressure are kept constant). According to the model, you can arbitrarily choose values for any three of the variable factors listed above; the value of the fourth one is then predicted accurately by the model and can be calculated from Eq. 16.4.

An investigation of many different gases reveals that each is described very closely by a similar ideal gas model (Table 16.3). The only difference between the models for various gases is in the specific value of the numerical factor. All these gases and air, therefore, behave generally like the gas helium examined in Example 16.3, below.

Gases increase in volume without limit when their pressure is reduced. In this respect they differ greatly from solid and liquid materials, whose volume hardly changes when the pressure is reduced except that they might evaporate. If the pressure of a gas is increased and its volume reduced, the ideal gas model eventually ceases to be applicable, especially when conditions for liquefaction of the gas are approached.

Absolute zero. The factor $(T + 273)$ in the ideal gas model is particularly curious. According to this factor, the pressure or the volume of every gas becomes zero if the gas temperature is brought to -273° Celsius (Fig. 16.13d). This prediction applies to all the gases regardless of the initial state from which the gas is cooled to -273° Celsius. At this temperature the gas has lost all its "spring" and does not exert any pressure on its container. This does not happen in reality, of course, because the ideal gas model is mainly accurate near and above room temperature. It fails at lower temperatures, as it does at very high pressure, when the conditions for liquefaction are approached. Nevertheless, the temperature of -273° Celsius, which is called the *absolute zero* of temperature, has come to play a very important role in the theory of thermal phenomena.

Work done on or by a gas. The analogy between gases and elastic systems (Table 16.2) can be pursued to calculate the work done on a

**Equation 11.5
(Definition of Work)**

work (joules)   = W

$$W = |\mathbf{F}|\Delta s_F$$

**Equation 16.5
(from Definition of
Pressure, Eq. 16.1)**

pressure
  (newtons/m^2) = P
area (m^2)   = A

$$|\mathbf{F}| = PA$$

Equation 16.6

volume change
  (m^3)   = ΔV

$$\Delta s_F = \frac{\Delta V}{A}$$

Equation 16.7

$$W = |\mathbf{F}|\Delta s_F$$
$$= PA \times \frac{\Delta V}{A}$$
$$= P\Delta V$$

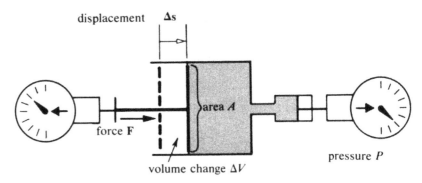

Figure 16.14 A gas in a cylinder is compressed. The force, displacement, pressure, area, and volume change are indicated. Note that the displacement is parallel to the direction of the force.

gas while it is being compressed. Work is done because a force acts to accomplish the compression, and a part of the container wall (a piston, for instance) is displaced to reduce the gas's volume (Fig. 16.14). The definition of work is given in Eq. 11.5. You can adapt the definition for use with gases by expressing the force in terms of the gas pressure (Eq. 16.5) and the displacement in terms of the volume change (Eq. 16.6). You then find that the work is equal to the gas pressure times the volume change (Eq. 16.7). The same calculation leads to the same formula for the work done by a gas while it is expanding.

To find the numerical value of the work in a particular experiment from Eq. 16.7, you must estimate an average value of the pressure, which changes according to Boyle's law during a compression or expansion process (Eq. 16.2). This can be done, but it requires mathematical techniques beyond the level of this text. Hence you may use Eq. 16.7 only for small fractional volume changes, for which the pressure remains approximately constant (Example 16.4).

EXAMPLE 16.3. A weather balloon contains 2 kilograms of helium. How much volume does it occupy at an altitude where the temperature is -50° Celsius and the pressure is 0.1 atmosphere?
Data:
M = 2 kg; T = -50° Celsius; P = 0.1 x 10^5 newtons/m^2
Solution: From Table 16.3,

$$V = \frac{2100M(T+273)}{P} = \frac{2100 \times 2 \times (-50 + 273)}{0.1 \times 10^5}$$
$$\approx \frac{2.1 \times 10^3 \times 2 \times 2.2 \times 10^2}{10^4} \approx 90 \; m^3$$

EXAMPLE 16.4. How much work is done when 1 kilogram of air is compressed from atmospheric pressure into a volume of 0.70 cubic meter at 22° Celsius?

Data:

$T = 22°$ Celsius; $M = 1$ kg; $P = 10^5$ newtons/m^2; final volume $= 0.70$ m^3

Solution: (a) Find the initial volume so the volume change can be determined. (b) Check the pressure variation to make sure it is "small." From Eq. 16·4,

(a) initial volume:

$$V = \frac{280M(T + 273)}{P} = \frac{2.8 \times 10^2 \times 1 \times 3.0 \times 10^2}{1 \times 10^5}$$

$$\approx 0.84 \text{ m}^3$$

volume change:

$$\Delta V = 0.84 \text{ m}^3 - 0.70 \text{ m}^3 = 0.14 \text{ m}^3$$

work:

$$W = P\Delta V = 1 \times 10^5 \text{ newtons/m}^2 \times 0.14 \text{ m}^3$$
$$= 1.4 \times 10^4 \text{ joules} = 3.3 \text{ Cal}$$

(b) final pressure:

$$P = \frac{280M(T + 273)}{V} = \frac{2.8 \times 10^2 \times 1 \times 3.0 \times 10^2}{0.70}$$

$$\approx 1.2 \times 10^5 \text{ newtons/m}^2$$

The pressure increased by 20% from 1×10^5 newtons per square meter to 1.2×10^5 newtons per square meter. This is a fairly small change.

Specific heat of gases. If you try to define the specific heat of a gas as the energy transfer required to change the temperature of 1 kilogram of gas by 1 degree Celsius, you encounter an interesting and important difficulty. The specific heat appears to depend on the method of measurement! The "specific heat" of air confined in a rigid cylinder, for example, is 0.18 Calorie per degree Celsius per kilogram, while the "specific heat" of air in a cylinder with a freely moving piston (where the gas expands as it is heated) is 0.25 Calorie per degree Celsius per kilogram (Fig. 16.15). In other words, 0.07 Calorie more energy is required to raise the temperature of 1 kilogram of air in the cylinder where it can expand than to raise the temperature of 1 kilogram of air in the rigid cylinder.

You can understand this experimental result if you refer to Eq. 16.7, where the volume change of a gas was related to energy transfer in the form of work. Thus, the gas in the cylinder with the sliding piston interacts with the surrounding air, which is pushed out of the way by the piston (Fig. 16.15b). The energy transferred in the form of work from

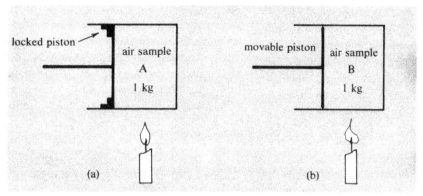

Figure 16.15 The specific heat of air.
(a) The temperature of 1 kilogram of air is raised 1 degree Celsius with the piston locked in place; 0.18 Calorie of heat is required from the candle.
(b) The temperature of 1 kilogram of air is raised 1 degree Celsius with the piston free to move; 0.25 Calorie of heat is required from the candle.
Where did the extra energy go? Answer: See top left margin on next page!

Equation 16.8 (Thermal energy of a gas)

thermal energy (Cal) = E
specific heat
  (Cal/deg C/kg) = C
temperature
  (deg C) = T
mass of gas (kg) = M

  E = CMT
(gases must be kept at constant volume)

FORMAL DEFINITION.
Isothermal processes *are processes (usually expansion or contraction of a gas) that take place at one, constant temperature. In order to keep the temperature constant, other quantities (pressure, volume,...) must usually be allowed to change in carefully controlled ways.*

the confined air to the surrounding air can be calculated from Eq. 16.7 and the ideal gas model for air (Eq. 16.4). It is indeed equal to 0.07 Calorie per kilogram, in agreement with the data (Example 16.5).

The air in the rigid cylinder (constant volume) does not gain or lose energy due to interactions other than that with the heat source. All the heat absorbed by this gas sample is stored in it in the form of thermal energy. The constant volume procedure, therefore, leads to the specific heat as it was defined in Section 10.2. The mathematical model for the thermal energy of a gas is very similar to the model for the thermal energy of solids and liquids (Eq. 16.8). An application of this model is described in Example 16.6. Specific heats of several gases are listed in Table 16.4. They do not vary appreciably as the gas pressure is changed.

Compression and expansion at a fixed temperature (isothermal).
Look again at the experiment for measuring the heat capacity (Fig. 16.15). The two air samples were in equal states at the beginning of the

TABLE 16.4 SPECIFIC HEATS OF GASES (AT CONSTANT VOLUME)

Gas	Temperature (deg Celsius)	Pressure (atm)	Specific heat (Cal/deglkg)
air	-100 to +1000	0 to 10	0.18
oxygen	-100 to +1000	0 to 10	0.16
nitrogen	-100 to +1000	0 to 10	0.18
helium	-100 to +1000	0 to 10	0.75
water	+120 to + 500	0 to 10	0.37
carbon dioxide	-50 to + 50	0 to 10	0.14
hydrogen	0 to +1000	0 to 10	2.5

experiment (equal masses, equal temperatures, equal pressures, and equal volumes) and therefore had equal energies. During the experiment, the energy stored in each sample increased by 0.18 Calorie while its temperature was being raised. Hence the two samples still had equal energies at the conclusion of the experiment, even though they were then in different states (equal masses and equal new temperatures, different pressures and different volumes. Fig. 16.16). It follows from

In Figure 16.15, the additional energy (0.07 Calories) required to raise the temperature of the air inside the cylinder at constant pressure was required because the piston moved outward, thus doing work and transferring energy to the surrounding air. When the air is heated at constant volume, the piston is locked and thus does no work.

EXAMPLE 16·5. (a) Calculate the volume change of air due to a one-degree rise in temperature from 22° to 23° Celsius.

$$V = \frac{280M(T + 273)}{P}$$

At 22°:

$$V = \frac{280M \times 295}{P}$$

At 23°:

$$V = \frac{280M \times 296}{P}$$

volume change:

$$\Delta V = \frac{280M}{P}$$

(b) Calculate the work done by 1 kilogram of air while it expands as in (a).

$$W = P\Delta V = P \times \frac{280M}{P} = 280 \times 1 = 280 \text{ joules} = 0.07 \text{ Cal}$$

Note that the result depends only on the temperature change in (a) and not on the pressure of the gas.

EXAMPLE 16·6. Find the thermal energy of one "roomful" of air (4 meters by 5 meters by 3 meters) at 22° Celsius and 1 atmosphere.

Data:

$C = 0.18$ Cal/deg/kg; $V = 60$ m³; $P = 10^5$ newtons/m²

Solution: Find the mass of air from Fig. 16·2 and the energy from Eq. 16·8.

$M = 1.2$ kg/m³ $\times$ 60 m³ $= 72$ kg

$E = CMT = 0.18$ Cal/deg/kg $\times$ 72 kg $\times$ 22 deg

$\quad = 280$ Cal

Figure 16.16 (to right) Two air samples that were in equal initial states (same values of M, T, V, P) and were heated (Fig. 16.15). (a) Sample A was heated at constant volume. The temperature, the pressure, and the energy increased. (b) Sample B was heated at constant pressure. The temperature, the volume, and the energy increased.

air sample
A

$M, T + \Delta T,$
$V, P + \Delta P$

(a)

air sample
B

$M, T + \Delta T,$
$V + \Delta V, P$

(b)

Arrows to represent energy transfer in Figs. 16·17 to 16·23.

heat

work

the law of energy conservation that the isothermal expansion of sample A to bring it to the same volume and pressure as sample B requires no change of the energy stored in the sample. The same is true for the isothermal compression of sample B to bring it to the same volume and pressure as sample A.

Interaction with a heat reservoir. If this conclusion seems abstract and unreal to you, remember this: to maintain the air samples at one fixed temperature during the isothermal expansion or compression, you have to place them in contact with an environment that has been adjusted to be at the same temperature as the samples (Fig. 16.17). The environment must also be maintained at that temperature, for example by means of an external thermostat and by actively supplying and removing heat. Such an environment is called a *heat reservoir*. By supplying

Figure 16.17 Isothermal expansion and compression of gas samples.
(a) Air sample A acts as a passive coupling element while it expands at the temperature T + ΔT. Heat input equals work output.
(b) Air sample B acts as a passive coupling element while it is being compressed at the temperature T + ΔT. Work input equals heat output.

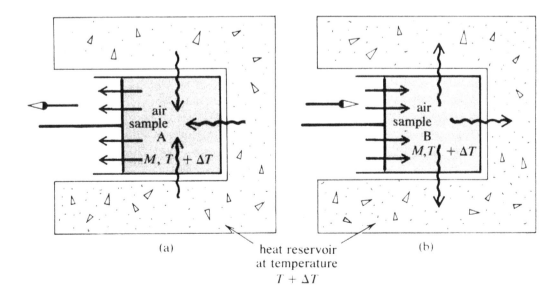

(a)

air
sample
A
$M, T + \Delta T$

air
sample
B
$M, T + \Delta T$

(b)

heat reservoir
at temperature
$T + \Delta T$

heat when the air samples cool slightly and by absorbing heat when the air samples become too warm, the heat reservoir keeps them at their original temperature. The interaction with the heat reservoir ensures heat supply or heat loss that just compensates for the work done by or on the air sample during the isothermal process.

An air sample being expanded or compressed under isothermal conditions is therefore quite different from a spring: it does not store elastic energy, even though it springs back after it is compressed. Instead of storing energy, the air sample acts only as a passive coupling element between the system acting on the piston and the heat reservoir.

Identity of "elastic" energy and thermal energy of a gas. How does a gas sample store "elastic" energy? To find the answer, you must imagine an expansion or compression process during which the gas is insulated and does not exchange heat with a heat reservoir. Then all the work done on the gas sample while it is being compressed becomes energy that is stored in the gas. Since the insulated gas cannot give off heat, its temperature rises so that its thermal energy increases by an amount equal to the work. Hence, you may conclude that the gas stored the work done to compress it in the form of thermal energy.

The reverse happens when an insulated gas sample is allowed to expand. Now the gas does work and its energy must decrease. Since it cannot absorb heat from the heat reservoir it becomes colder, so that its thermal energy decreases by an amount equal to the work. Again you may conclude that the thermal energy of a gas sample may be transferred in the form of work. In this respect also, a gas is quite different from matter in the solid or liquid phases, whose thermal energy is always transferred in the form of heat.

16.4 Heat engines and refrigerators

In the discussion of energy storage by gases in the previous section, we explained that a gas confined in a cylinder is a coupling element between a heat source and a piston. Consider now a gas confined at a constant pressure. If the heat reservoir is hotter than the gas, the gas pressure increases slightly so that the gas expands and converts its heat input to work output (Fig. 16.17a). If the reservoir is colder than the gas, the gas pressure decreases slightly, and work input is converted to heat output (Fig. 16.17b). This is the key property that makes gases so useful in heat engines.

We will now describe a working model for a cyclic heat engine. The model will illustrate the essential features of real heat engines, but will not include a correct treatment of many side effects, such as friction between the moving parts, heat losses to the surroundings, and so on.

Two basic requirements that a heat engine must satisfy are that it deliver work in proportion to the heat input and that it be able to operate over a long period of time. These requirements are met by the cyclic operation of a piston in a gas-filled cylinder. The piston moves in a repeating cycle out of and into the cylinder, as the gas is alternately

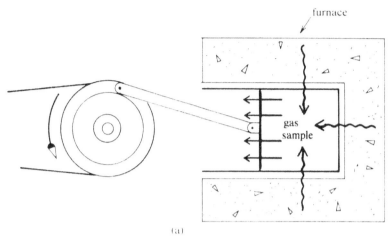

(a)

Figure 16.18 Diagram of the working model for a cyclic heat engine showing two stages in the cycle,

(a) Expansion part of the cycle. The gas is heated by the furnace,

(b) Compression part of the cycle. The gas loses heat to the cooling system.

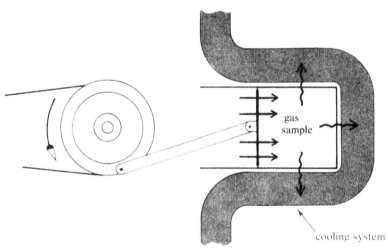

(b)

heated by being placed in a furnace and cooled by being placed in contact with a cooling system (Fig. 16.18). The piston is connected to the machinery by a shaft that converts the oscillating motion to a rotary motion. The working model for a heat engine therefore must include four major parts: a gas, a furnace, a cooling system, and a mechanical linkage (usually a piston) on which the gas can do work (Fig. 16.19).

The operating cycle. The principle of cyclic operation assures that the engine can operate for a long period, but it appears to suffer from one serious drawback. It seems that the work done during the expansion part of the cycle (Fig. 16.18a) may have to be returned when the motion of the piston is reversed during the compression part of the cycle (Fig. 16.18b). The net result would be some energy transfer from furnace to cooling system, but zero work output.

Fortunately, the properties of gases make it possible to eliminate this drawback. A gas sample at low temperature exerts a smaller pressure

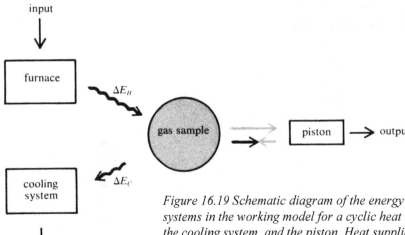

input

furnace

ΔE_H

gas sample → piston → output

cooling
system

ΔE_c

waste

Figure 16.19 Schematic diagram of the energy transfer among the four sub-systems in the working model for a cyclic heat engine: the gas, the furnace, the cooling system, and the piston. Heat supplied by the furnace at a high temperature is divided into work and heat lost to the cooling system at a low temperature.

than the same gas sample at high temperature. The work done is proportional to the pressure (Eq. 16.7). It is customary, therefore, to operate the heat engine in such a way that the gas is cold during compression and hot during expansion. Even though some work has to be returned from the machinery to compress the cold gas while it loses heat to the cooling system, this is less than the work that was done by the hot gas during the expansion part of the cycle. There is a net transfer of energy to the machinery.

The energy transfer in the operation of the working model is indicated in Fig. 16.19. Three subsystems of the engine interact by means of the gas, which is a fourth subsystem serving as coupling element among the other three. Combustion of a fuel provides energy input to the furnace. Thermal energy is transferred from the furnace, which heats the gas during the expansion stage of the cycle, while the gas is doing work on the piston. Thermal energy is transferred to the cooling system, which reduces the gas temperature during the compression stage of the cycle, while work is being done on the gas by the piston. The work returned to the gas during the compression stage is less than the work done by the gas during the expansion stage because the cold gas has a lower pressure than the hot gas. The thermal energy reaching the cooling system, however, cannot be used to heat the gas during the expansion stage, because the cooling system is at a much lower temperature than the furnace. This quantity of energy, therefore, has lost its practical usefulness in the heat engine and is indicated as "waste" in Fig 16.19. The waste of one part of the energy is the price paid for the successful conversion of another part of the energy from heat to work.

Efficiency. The *efficiency* of a heat engine is the ratio of the work output to the thermal energy input. Even though energy is conserved, the efficiency is less than 100% because the energy transferred to the cooling system is not part of the useful work output (Eq. 16.9).

Applications. Practical heat engines, such as the steam engine, internal combustion engine, and gas turbine, are all based on the same principles

Equation 16.9

heat from furnace $\quad \Delta E_H$
heat to cooling
    system $\quad \Delta E_c$
net work $\quad W$

$$\Delta E_H = W + \Delta E_c \quad (a)$$

$$efficiency = \frac{W}{\Delta E_H} \quad (b)$$

as the working model heat engine we have described (Fig 16.18). However, a heat engine built along the lines of our model inevitably wastes a substantial amount of energy because the entire cylinder must be heated and cooled for the two stages of the cycle. This was in fact the way that the first useful steam engine operated: the Newcomen engine was invented in about 1716 to pump water out of an English coal mine. It was a noisy, dirty, inefficient monster the size of a building, but it could pump water faster and at less cost than horses or humans could, and many coal mines acquired Newcomen engines during the 1700s. As you would expect, efficiency was not critical because of the abundance of fuel at a coal mine. The last working Newcomen engine was removed from active service in 1834.

James Watt's innovations. James Watt (1736-1819) addressed the problem of trying to make the steam engine more efficient and smaller in size. Watt eliminated the need to alternatively cool and heat the cylinder. He invented an engine in which, immediately after the expanding steam had pushed out the piston, a valve opened. This allowed the steam into a separate "condenser" cylinder, where it was further cooled and condensed to liquid water. This innovation meant that the main cylinder could be kept hot and the condenser kept cool at all times. This resulted in a substantial increase in the amount of work produced for every ton of coal burned. This breakthrough, supplemented by Watt's many other inventions, was the key to the vast proliferation and improvement of the steam engine.

Practical heat engines. Modern heat engines (for example, the internal combustion engine and the steam and gas turbine) differ from each other and from our ideal model in practical details of operation. In all cases, as in Watt's engine, the heating and cooling processes are separated in some way. The working fluid is heated to convert it to a gas and increase the pressure, then allowed to expand and do work, and, finally, condensed (or compressed) back to a liquid. Thus, the steam engine employs water that is first converted to steam in the boiler and then liquified during the compression part of the cycle. On the other hand, the internal combustion engine (automobile engine) uses a mixture of fuel and air as the gas in the cylinder, and the combustion takes place in the cylinder itself. This arrangement does away with the need for a separate furnace. The mixture of fuel and air is drawn into the cylinder, the fuel burns and expands, and the hot, expanded gas leaves as exhaust at a lower (atmospheric) pressure during each complete engine cycle.

In a steam turbine, high-pressure steam expands against a set of turbine blades arranged around a shaft and force the shaft to rotate. Finally, in a gas turbine there are two sets of turbine blades replacing the cylinder and piston. One set of blades acts like a fan and compresses the cold gas; the other set of blades is propelled by the pressure of the hot, exploding gas pushing against the blades. Because turbines rotate, they operate very smoothly and do not share the vibration caused by the back-and-forth motion of the piston in the steam and combustion engines. The cyclic

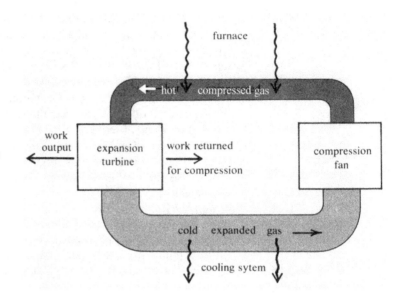

Figure 16.21 Diagram of a cyclic heat engine using turbines rather than pistons for expansion and compression of the gas. The work done on expansion is partly used for compression and partly becomes useful output.

flow of the gas in a turbine engine with a furnace is schematically shown in Fig. 16.21.

Refrigeration cycles. Refrigerators and air conditioners operate on the same principles as heat engines, but the action is reversed. A schematic diagram of a practical refrigerator is shown in Fig. 16.22. The coupling element is, for the sake of efficiency, not a gas but a liquid with a very low boiling temperature (about 0° Celsius), which can vaporize to form a gas. The liquid vaporizes in the evaporator at a low pressure and absorbs the heat of vaporization from its surroundings (the interior of the refrigerator), which are thereby cooled. The gas flows to

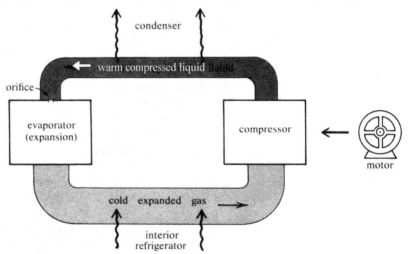

Figure 16.22 Fluid cycle in a refrigerator. Energy is transferred from the cool refrigerator interior to the warm condenser outside.

a compressor (built somewhat like a turbine) in another region of the refrigerator, where it is liquefied. The heat of vaporization, which must be removed here, is transferred in the condenser to cooling water or to air circulating around the condenser. The cooled liquid now returns to the evaporator where it expands and vaporizes once more. The net effect of the cycle is to transfer thermal energy from the environment of the place where the liquid vaporizes to the environment of the condenser where the gas liquefies. Again, three subsystems interact by means of a fluid that acts as coupling element: the inside of the refrigerator at a low temperature, the condenser at a higher temperature, and the compressor motor, which does work on the fluid (Fig. 16.23). Note that the refrigeration engine does not operate on the heat input from the refrigerator interior, but must be driven by work input to the compressor.

16.5 The kinetic theory of gases

Historical background. Newton and his contemporaries speculated about the possible micro-domain models for gases. In one model proposed by Newton, gases were composed of particles subject to a repulsive interaction-at-a-distance. This interaction spread the particles apart so they filled the entire volume of their containers and exerted a pressure on the container walls. In a second model, gases were composed of compressible particles, like fluffs of wool, touching each other. In a third model, the particles did not touch at all times, but they were in violent agitation, whirled throughout the available space within a turbulent but "subtle" fluid.

All of these particle models have some useful features, but all of them suffer shortcomings, and none of them can explain the mathematical ideal gas model of Eq. 16.4 and Table 16.3. Since the data on which the ideal gas model is based were not discovered until the end of the eighteenth century, the physicists prior to that date did not have the same reasons for eliminating their models as you have now.

Figure 16.23 Diagram of energy transfer among the four subsystems in a refrigeration cycle. Work supplied by the compressor "lifts" heat "uphill" from the cool refrigerator to the warm condenser. The fluid at higher temperature has more energy, and thus energy must be added (via work from the compressor) to raise the temperature. A heat engine (Fig. 16.19) represents the opposite process, in which heat "falls" from high to low temperature, and the engine converts some of it to external work. Thus you can think of a heat engine as similar to a waterwheel in an old-fashioned mill, where water "falls" in a controlled way, thus turning the wheel and doing useful work. The water's gravitational field energy is partially converted to other types of energy.

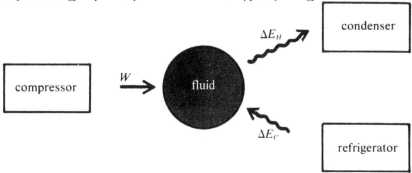

By the middle of the nineteenth century, practical knowledge and theoretical interpretations had progressed to the point where the decisive next step could be taken. The observation by Rumford that energy of motion could produce heat through friction and the measurement of the work done when a gas was heated at constant pressure led Julius Robert Mayer (1814-1878) to propose that energy is indestructible, but convertible from one type to another. James Joule thereupon constructed a micro-domain particle model for gases in which thermal energy of the gas was kinetic energy of the particles. Most significantly, Joule was able to show that his model permitted a calculation of the speed of the particles, led to Boyle's law, and gave new meaning to the concept of "absolute zero." In fact, other scientists had described such calculations during the preceding century, but their work had been ignored because it went against the accepted caloric theory (see Section 10.5).

According to Joule's model, a gas is composed of small, rapidly moving particles that do not interact with one another except for occasional collisions. Thermal energy of the gas is kinetic energy of the particles. The collisions between two particles are perfectly elastic; that is, the combined kinetic energy of the two colliding particles is the same before and after the collision. No energy is transferred to other types during a collision because the model provides for no other types of energy storage. The particles also experience elastic collisions with the container walls and exert a force on it by virtue of their impact. Each gas is composed of characteristic particles different in mass from the particles of other gases. This model is the basis for the kinetic theory of gases.

Qualitative properties of the particle model. This model explains qualitatively many phenomena:

Why do gases expand to fill the container they occupy? Because the particles travel until they collide with a container wall,

Why gases have a low density and are compressible? Because there is a great deal of empty space between the particles.

Why does gas pressure gives rise to forces on the walls? Because the particles striking and bouncing off the wall suffer changes of momentum.

Why does reducing the volume increase the gas pressure on the walls? Because the particles are crowded together so that they strike a particular segment of wall area more frequently.

Why do different gases diffuse rapidly through one another and mix completely? Because the particles' motion carries them into the empty spaces among the other particles.

The energy stored in the model gas is kinetic energy of the particles. Since the particles do not interact with one another, kinetic energy is the only type of energy storage possessed by the system of particles making up the gas. A crucial feature of the model, therefore, is the *particle velocity*, which determines both the momentum and the kinetic energy.

Derivation of Boyle's law. In the particle model, the pressure exerted by the gas is caused by the impacts of the particles with the container

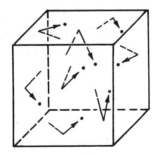

Equation 14·1b

force $\qquad$ $\mathbf{F}_{av}$
mass $\qquad$ M
momentum change $\quad$ $\boldsymbol{\Delta M}$
time interval $\qquad$ Δt

$$\mathbf{F}_{av} = \frac{\boldsymbol{\Delta M}}{\Delta t}$$

wall. As a particle strikes the wall and recoils, its momentum changes. The force exerted by the particles on the wall is equal in magnitude, and opposite in direction, to the force exerted by the wall on the particles. The net force acting on the steady stream of particles striking the wall is, according to Newton's second law, equal to the change of momentum of the particle stream divided by the time interval during which the particle stream interacted with the wall (Eq. 14.1b).

To avoid the mathematical difficulties of this calculation, we will make a still simpler particle model for gases, even though this model is in contradiction with readily observed properties of real gases. This approach, which was also used by Joule, illustrates that it is frequently more productive to use an inadequate model whose consequences can be evaluated rather than to use an adequate model that is too complicated to be exploited. Our model has the following properties.

1. The gas is in a cubical container.
2. The particles do not collide with one another at all.
3. All particles move with a speed v. Their total mass is M.
4. One sixth of the particles move at right angles toward each face of the cubical container (Fig. 16.24).

The consequences of this model for the pressure-volume relation are stated in Eq. 16.10. The force due to the particle impacts is proportional to the momentum of the particles (mass times speed) and to the frequency of the impacts (again the speed), whose combined effect gives the speed to the second power. The result is that the product of pressure times volume equals two thirds of the kinetic energy of the particles. It is therefore possible to determine the speed of the particles from the ideal gas model (Table 16.3), which was based on pressure-volume data measured by ordinary techniques (Example 16.7). The results are in excellent agreement with directly measured particle speeds.

Equation 16·10

$$PV = \tfrac{1}{3} M v^2 = \tfrac{2}{3} (\mathbf{KE})$$

Interpretation of temperature. By comparing Eq. 16.10 with the ideal gas model for air (Eq. 16.4), you can see that the total kinetic energy of the air particles is directly related to the temperature (Eq. 16.11). In fact, it is possible to show that the factor (T + 273) is directly proportional to the average kinetic energy of one gas particle, with the same

Equation 16.11

$$
\begin{aligned}
KE \ &= 1.5PV && (a)\\
&= 420M(T + 273) && (b)
\end{aligned}
$$

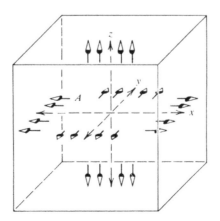

Figure 16.24 The particles move only parallel to the cube edges, and perpendicularly to the cube faces. One sixth of the particles move toward each cube face. The velocity of particle A is (—v, 0, 0).

constant of proportionality for all gases. The actual value of the constant of proportionality requires knowledge of the mass of one gas particle, which was only determined at the beginning of the twentieth century.

This, then, is the micro-domain interpretation of temperature: average kinetic energy of a gas particle. The model thereby explains what happens at a temperature of -273° Celsius: the average kinetic energy and therefore the speed of the particles is zero, so that the gas exerts zero pressure and has no "spring" at this temperature.

Thermal energy. The second principal prediction of the model concerns the specific heat of gases. If the thermal energy is equal to the kinetic energy of all the particles, as was claimed, then the mathematical model relating kinetic energy to temperature should be similar to the model for thermal energy. The result shows good agreement for helium, except that the two formulas clearly refer to different thermal reference states (Example 16.8). In the particle model, the reference state is -273° Celsius (absolute zero) where the stored energy is zero Calories. The thermal energy determined according to the operational definition (Chapter 9) is measured from the arbitrary reference state at 0° Celsius. The specific heats in both models, however, are equal to 0.75 Calorie per degree per kilogram.

For gases other than helium, the agreement is not as good as it is for helium (Example 16.9). The specific heat derived from their thermal energy is greater than the specific heat derived from their kinetic energy. This discovery led to improvements in the model and ultimately to good agreement with the data.

EXAMPLE 16.7. Find the speed of oxygen particles at 0° Celsius.

Solution: Combine Eq. 16.10 with the ideal gas model for oxygen in Table 16.3.

$$PV = \frac{1}{3}Mv^2 = 260M(T+273)$$
$$v = \sqrt{[3\times260(T+273)]} = \sqrt{(3\times2.6\times10^2\times2.7\times10^2)}$$
$$= \sqrt{21\times10^4} \approx 4.6\times10^2 = 460 \text{ m/sec}$$

This rather high speed, almost 1000 miles/hr, may seem unreasonable, but actual measurements have confirmed it! How does it compare with the speed of sound? Would gas particles moving this fast also be able to transmit sound? This result depends only on the temperature, not on the pressure of the oxygen, its volume or the mass of one particle.

EXAMPLE 16.8. Find the combined kinetic energy of all helium particles in 1 kilogram of helium gas at the temperature T and compare it with the thermal energy.

Solution:

From Eq. 16.10 and Table 16.3,

$$KE = 3100M(T + 273) \text{ joules} = 0.75M(T + 273) \text{ Cal}$$

From Eq. 16.8 and Table 16.4,

E = CMT = 0.75MT Cal

Comment. Rewrite the kinetic energy formula as follows:

KE = (205M + 0.75MT) Cal

The first term is the combined kinetic energy at 0° Celsius; the second term is the change in kinetic energy when the helium temperature is raised or lowered from 0° Celsius to the temperature T. The second term is equal to the thermal energy of a system as defined in Section 10.2.

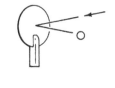

EXAMPLE 16.9. Find the combined kinetic energy of all air particles in 1 kilogram of air at the temperature T and compare it with the thermal energy.

Solution:

From Eq. 16.llb,

KE = 420M(T + 273) joules = 0.10M(T + 273) Cal

From Eq. 16.8 and Table 16.4,

Thermal Energy = E = CMT = 0.18MT Cal

Comment. Rewrite the kinetic energy formula as follows:

KE = (27M + 0.10MT) Cal

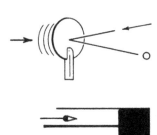

The second term here is less than the thermal energy. The first term is greater than the thermal energy so long as 0.18T is less than 27, or T is less than (27/0.18) = 150°C.

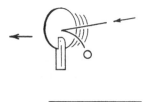

Application. With the theory of this section, you can understand qualitatively the particle mechanism whereby the gas temperature increases while a gas is being compressed. Consider the gas in a cylinder with a movable piston. The cylinder and the piston are thermally insulated and are at the same temperature as the gas. The particles strike all the walls and recoil with the same speed with which they impinged on the walls. Now, while the gas is being compressed, the piston moves forward and the particles recoil with an increased speed, just as a ping-pong ball recoils with an increased speed when it is struck by an advancing paddle. The increased speed implies an increased kinetic energy and, therefore, an increased temperature. While the gas is being compressed, therefore, the temperature rises. After the piston stops moving, incident and recoil speeds are once more equal and the temperature remains at its elevated value.

While the piston is being pulled out, the particles striking the piston recoil with a reduced speed, just as a ping-pong ball recoils with reduced speed when it strikes a receding paddle. The decreased speed implies a decreased temperature. After the piston stops moving, the temperature again remains at its reduced value. Please note that this

TABLE 16.5 MANY-INTERACTING-PARTICLES (MIP) MODEL FOR GASES

Micro domain	Macro domain
many particles	*gas*
total mass of particles	*mass of gas*
particle impacts with the walls	*pressure*
volume of container	*volume of container*
kinetic energy per particle	*temperature*
particle speed	*-*
particle collisions	*heat flow*

model includes no collisions at all between the gas particles. At thermal equilibrium, the gas temperature is related to the particle speed and not to the frequency of particle collisions with one another or with the wall. The collisions operate only to *transfer* energy, not to store it. Thus, the collisions of particles with the moving piston are important only to bring about the temperature rise during compression and the temperature drop during expansion.

Equation 16.4 (Ideal gas model for air)

pressure (newtons/m^2) = P
volume (m^2) = V
mass (kg) = M
temperature (deg C) = T

$PV = 280 M (T + 273)$
(for air)

Summary

The gas phase of matter differs from the solid and liquid phases in several significant ways. Gases have lower density, greater compressibility, and show greater thermal expansion than do solid and liquid materials. These properties are described by the mathematical ideal gas model for air (Eq. 16.4), which has the same form for many other gases as well. The ideal gas model relates the four variable factors that describe a gas sample: mass, temperature, volume, and pressure. The gas density (mass-volume ratio), the compressibility (volume-pressure relationship, Boyle's Law), and the thermal expansion (volume-temperature relationship, Charles' Law) may all be inferred from this model. A micro-domain particle model for gases can explain several of their properties (Table 16.5). The particles in the model do not interact with one another, but move at a high speed and collide (interact) with the walls of the container that confines the gas.

The most common gas is air (a mixture of oxygen and nitrogen), which surrounds the planet earth. The air is in approximate mechanical equilibrium subject to the force of gravity of the earth (downward) and the force of support exerted by the earth's surface on which the air rests (upward). The air pressure that gives rise to this force is 10^5 newtons per square meter (14.7 pounds per square inch) at the bottom of the atmosphere. This pressure, which corresponds to a mass of 10 tons for the column of air resting on 1 square meter, is known as one atmosphere, and it can support a column of mercury approximately 0.76 m (30 inches) in height.

Gas samples may exchange energy with their environment in the form of either work (the gas is compressed or expands), or heat (the

Equation 16.7

work (joules)   = W
pressure (newtons /m²) = P
volume change (m³)   = ΔV

  W = PΔV

Equation 16.8

thermal energy (Cal) = E
specific heat
  (Cal/deg C/kg)   = C
temperature (deg C) = T
mass of gas (kg)   = M

  E = CMT

gas is colder or warmer than its environment), or both at the same time. The work done by the gas is equal to the gas pressure times the volume change (Eq. 16.7). No matter how the energy transfer is accomplished (heat or work), a gas stores energy *only* in the form of thermal energy (Eq. 16.8). Cyclic heat engines and refrigerators are mechanical devices that use a gas as a coupling element to accomplish desired energy transfer. In a heat engine, thermal energy from a furnace at high temperature is transferred partly to mechanical energy of machinery and partly to thermal energy of the cooling system at low temperature. A refrigerator transfers thermal energy from the interior at low temperature to the exterior at a higher temperature, but some mechanical energy must be supplied by the motor that operates the refrigerator.

List of new terms

pressure	heat reservoir
atmosphere (unit of pressure measurement)	heat engine
barometer	efficiency
compressibility	refrigeration cycle
ideal gas model	

List of symbols

F	force	Δs$_F$	displacement component along force
P	pressure	T	temperature
A	area	C	specific heat
V	volume	M	mass
ΔV	volume change	**ΔM**	momentum change
W	work	**Δv**	velocity change
		v	speed

Problems

1. Carry out library research to trace the development of heat engines from the time of James Watt to the present.

2. Describe the impact on western civilization of the invention and development of heat engines.

3. Obtain a plastic syringe and explore the properties of the air in the syringe (Fig. 16.3).

4. Tire pressure gauges measure the excess of the pressure in the tire over the pressure of the atmosphere outside the tire.
   (a) What is the actual air pressure in an automobile tire?
   (b) The gauge shown in Fig. 16.8 measures the actual pressure. Modify the diagram so that the gauge measures the excess pressure.

5. Either the theory of atmospheric pressure or the theory that "nature abhors a vacuum" (suction) can explain many everyday phenomena.
   (a) Select three or more everyday phenomena that illustrate these theories and explain them from both points of view.
   (b) Explain the concept of suction by both theories.
   (c) Which of these theories is closer to your own common-sense approach? Explain briefly.

6. Formulate one or two operational definitions of gas pressure.

7. The ocean floor must support both the ocean water and the atmospheric air. The pressure on the ocean floor and other objects under water is therefore greater than atmospheric pressure.
   (a) Find the pressure at a depth of 30 meters (100 feet) under the surface of the water.
   (b) Find the force acting on 1 square centimeter of submarine hull 30 meters under water.
   (c) Find the force acting on 1 square centimeter of ocean bottom under 10^4 meters of water. (The bottom of the deepest ocean trench, the Marianas Trench near the Phillipines, is almost 11,000 m., or about 7 miles, below sea level. In comparison, the peak of the highest mountain, Everest, is only about 8800 m above sea level.)

8. Observe by approximately how many meters your altitude above sea level must change suddenly to create an uncomfortable feeling in your ears. Estimate the force on your eardrum that causes your discomfort, treating the eardrum as a circle 0.006 meter (1/4 inch) in diameter. (Hint: Use the data on air density, Fig. 16.2. Find the sudden change of the mass of air in a column "resting" on your eardrum.)

9. Estimate the average atmospheric pressure in Denver, which is 1600 meters above sea level.

10. Use your mathematical model for atmospheric pressure (from Problem 9) to estimate the pressure at the peak of Mount Everest and compare it with the value stated at the end of Section 16.2. Comment on any discrepancies between the two.

11. Give a critical discussion of the analogy between a gas and an elastic system.

12. Compare Boyle's law with Hooke's law and point out qualitative similarities and differences. Mention extreme examples of deformations in your discussion and compare the laws' limitations.

13. The "molecular weight" is a useful concept in chemistry and for the analysis of gases. State the ideal gas model in a form in which the mass of gas is expressed in "molecular weight units." The molecular weights of the five gases in Table 16.3 are: carbon dioxide (44); helium (4); hydrogen (2); nitrogen (28); oxygen (32).

14. Use the results of Problem 13 to estimate the "molecular weight" of air. Interpret the result in light of the knowledge that air is a mixture of oxygen and nitrogen. (The calculation has low accuracy because the numerals in Eq. 16.4 and Table 16.3 have been rounded off to two digits.)

15. (a) Use the ideal gas model to estimate the mass of 1 cubic meter of air at the peak of Mount Everest, which is about 8800 m above sea level.
(b) What are the implications of this result for the operation of internal combustion engines or of human beings at the top of Mount Everest?

16. Discuss the concept of "elastic" energy of a gas.

17. Work is done on a gas while it is being compressed at a fixed temperature. What system acts as energy receiver in this process?

18. Use the data in Table 16.4 and the ideal gas model (Table 16.3) to answer the following questions.
(a) How much heat must be supplied to 1 kilogram of hydrogen gas when it is warmed by 1 degree Celsius at constant pressure?
(b) Do the same for helium gas.

19. Estimate the increase of thermal energy stored in 1 cubic mile of air when it is heated from 50° Fahrenheit to 100° Fahrenheit. (In metric units: 4 cubic kilometers, from 10° Celsius to 38° Celsius.)

20. (a) Apply the result of Problem 19 to a 1-mile-thick air layer over the Central Valley of California (approximate dimensions: 400 miles long, 120 miles wide) to find the air's thermal energy increase on a hot summer day.
(b) Solar energy reaches the earth at the rate of approximately 1/3 Calorie per square meter per second. This numerical value is called the solar constant. How long would it take for solar energy to raise the temperature of the air over the Central Valley from 50° Fahrenheit to 100° Fahrenheit if all the solar energy is transferred to the air layer 1 mile thick? Interpret the result of your estimate.

21. Devise an experiment to measure the solar constant approximately (see Problem 20b).

22. Do you think the air being heated by the sun (Problems 19 and 20) is heated at constant pressure, at constant volume, or under still other conditions? Explain your answer.

23. When 1 kilogram of water is heated from 10° to 30° Celsius, it expands by 4×10^{-6} cubic meters. Compare the work done by water when it is heated from 10° to 30° Celsius with the work done by air when it is heated by the same amount.

24. Describe some energy sources used by man for the performance of work prior to the invention of the steam engine (other than manual and animal labor).

25. Select two cyclic heat engines from your everyday experience and identify the "furnace," the cooling system, and what happens to the "output" in each one (refer to Fig. 16.19).

26. Describe what happens to the waste heat in the following situations: automobile engine; electric power plant in your region; air conditioner.

27. Waste heat is sometimes said to give rise to "thermal pollution" in the environment of a power plant or factory. Describe some possible undesirable effects of "thermal pollution."

28. Comment on some shortcomings of the three gas models proposed in Newton's time (first paragraph, Section 16.5).

29. Discuss the three gas models described in the first paragraph of Section 16.5 with respect to their explanation of the thermal and elastic energies of gases. (Note: the conclusion of Section 16.3 about the identity of thermal and elastic energies dates to the nineteenth century.)

30. In what respect are the qualitative properties of Joule's particle model for gases superior to the models proposed in Newton's time?

31. (a) Find the speed of air "particles," hydrogen "particles," helium "particles," and carbon dioxide "particles" at room temperature.
(b) Compare the particle speeds you have calculated with the speed of sound in these gases as listed in Table 7.1.
(c) Comment on the results of the comparison in (b) by relating the particle model for gases to the wave theory of sound.

32. (a) Calculate the total kinetic energy of the "particles" in 1 kilogram of helium at room temperature.
(b) How does this energy compare with the thermal energy of 1 kilogram of helium gas, if you use absolute zero as the reference temperature for thermal energy?

(c) How does the particle model explain the fact that the thermal energy of helium gas at room temperature does not depend on the pressure or the volume of the gas?

33. Solve Problem 32 for nitrogen gas instead of helium.

34. (a) Calculate the specific heats predicted by Joule's model for hydrogen, nitrogen, and carbon dioxide.
    (b) Compare the specific heats in (a) with those listed in Table 16.4 and point out which gas has the largest and which the smallest percentage discrepancy.

35. Identify one or more explanations or discussions in this chapter that you find inadequate. Describe the general reasons for your judgment (conclusions contradict your ideas, steps in the reasoning have been omitted, words or phrases are meaningless, equations are hard to follow, . . .), and make your criticism as specific as you can.

Bibliography

M. Born, *The Restless Universe*, Dover Publications, New York, 1951.

S. G. Brush, *Kinetic Theory, Volume I: The Nature of Gases and of Heat*, Pergamon, New York, 1965.

P. W. Bridgman, *The Nature of Thermodynamics*, Harper and Row, New York, 1961.

D. Cassidy, G. Holton, J. Rutherford, *Understanding Physics*, Springer Verlag, 2002.

T. G. Cowling, *Molecules in Motion*, Hutchinson, London, 1950.

H. M. Leicester and H. S. Klickstein, *A Source Book in Chemistry*, McGraw-Hill, New York, 1952.

J. F. Sandfort, *Heat Engines*, Doubleday, Garden City, New York, 1962.

Articles from Scientific American. Some or all of these, plus many others, can be obtained on the Internet at http://www.sciamarchive.org/.

L. Bryant, "The Origin of the Automobile Engine" (March 1967).

W. Ehrenberg, "Maxwell's Demon" (November 1967).

J. F. Hogerton, "The Arrival of Nuclear Power" (February 1968).

M. B. Hall, "Robert Boyle" (August 1967).

G. A. Hoffman, "The Electric Automobile" (October 1966).

J. W. L. Kohler, "The Stirling Refrigeration Cycle" (April 1965).

Jacob Bronowski (1908-1974) was born in Poland and studied in Great Britain, receiving his doctorate in mathematics from Cambridge University. His career spanned mathematics, operations research, paleontology, science history, public service, and, especially, the philosophy, ethics and culture of science. (see the bibliography to Chapter 1). Renowned for his exceptional abilities to communicate science to the public, his last contribution was *The Ascent of Man* (1973). This extraordinary series on public TV made Dr. Bronowski a popular figure and helped millions of people understand science as a vibrant and integral, indeed essential, part of human culture.

Epilogue

This epilogue is abridged from the Damon Lecture "Science in the New Humanism" presented by Jacob Bronowski to the National Science Teachers Association 1968 Convention in Washington, D. C. It was included in the 1969 edition of this book with the permission of the National Science Teachers Association and of Dr. Bronowski. The lecture is also included in Magic, Science and Civilization *by Jacob Bronowski* © 1978, Columbia University Press. Reprinted with the permis-sion of the publisher.

The most remarkable discovery made by scientists is science itself. That is, scientists owe their growing success during the last 300 years to the way in which they have been able to turn science into a method. There are evidently two things to be asked about such a method: how it works, and why it works. Neither of these questions is simply technical; on the contrary, their answers imply a special relation and attitude of man to his environment, and that is why they are important to every thinking person.

How does the scientific method of discovery work? To answer, we must be clear in our minds that science is not a mere register of facts – and indeed, that our minds are not made (like a cash register) to tabulate a series of facts in a neutral sequence one after another. Our minds connect one fact with another, they seek for order and relationship, and in this way they arrange the facts so that they are seen to be linked by inner laws in a coherent network. Science is an organization of knowledge.

The facts are there for us to observe, but their organization is not; it has to be discovered step by step, and each step has to be probed and tested. The essence of the scientific method is the procedure of testing whether the model of the organization of nature that we have formed remains consistent with the facts when we add a new fact to those from which we began. From the known facts we form a model of how we think nature is organized – that is, of her laws; and now we ask whether the model really works as nature does, not only in those places where we already know the facts, but in places where we do not.

If nature responds to the challenge of our experiment with a fact which contradicts our prediction, then our model of her organization was wrong. That is simple and precise, as J. J. Thomson's discovery of the electron was. But when nature conforms to our prediction, the matter is not so simple. Our model now has a new fact to support it – but support is not proof. The new fact is just another fact to be added to those that we already have; it confirms the application of the model and widens its range, but it cannot be decisive – it cannot show that the model is universal. Evidently no model is universal, and the point of our experiments is inevitably to uncover a situation sooner or later in which the model fails. It is not an accident that the scientific theories of a hundred years ago look crude and primitive today, and mistaken in many of their underlying concepts. It is the fate of theories to be right up to a point in time, and thereafter wrong; and just this is the basic nature of discovery.

Thus the progress of science relies on a constant interchange between two procedures. On the one hand, there is the procedure of reasoning to find new implications of the model of nature that we have formed. And on the other hand, there is the procedure of setting up practical experiments to test these implications in new situations which are as decisive as we can make them.

Scholars in the Middle Ages (leaning on the axiomatic approach of the Greeks, and particularly of Aristotle) thought that knowledge of nature could be gained by one of these procedures alone; namely, reasoning from a basic model, whose features they believed were self-evident. Francis Bacon early in the 17th century turned this view upside down, and proposed that practical observation and experiment alone would yield a knowledge of nature. He believed that the laws of nature, and a connected model of their organization, would flow from empirical knowledge and leap to the mind as self-evident. But from the day in 1666 when Newton conceived the law of gravitation, and at once thought out a test for it by calculating the period of the moon, science has worked by coupling the two procedures in alternate steps — reasoning from a model to new implications of it, and then testing the implications empirically by practical experiment and observation. This coupling of reason and empiricism, in a constant leapfrog one over the other, was explicitly enunciated by Alfred North Whitehead 50 years ago.

In the years that have passed since then, it has become clear that there is a third leg in the progressive march of science. When an empirical test shows that our model is deficient, it is not self-evident in which of several possible ways the model is best changed; and the next model cannot be constructed by any logical process of pure reasoning. The new model with its new concepts and relations can only be conceived by an act of imagination. What Francis Bacon called induction is really imagination; and the progress of science from model to test, and from test to new model, requires the human imagination as a third active agent.

We can therefore summarize the "how" of the scientific method in a single definition:

> Science is the organization of our knowledge in such a way that it commands more of the hidden potential in nature.

The first part of the definition summarizes in the word organization the three-legged conjunction of reason, experiment, and imagination. The second part of the definition states our belief that we progress by constantly uncovering more in nature than we knew to be there.

Why does science work as a method? Why has science been successful for over 300 years in steadily enlarging man's control over his physical and biological environment? Why have we achieved a command of natural forces which is so much more effective and persuasive than our command, say, of social forces?

The answers to these questions are implicit in our definition, namely, that it is knowledge of nature that gives us command of her potential. This is the essential conception on which science rests, and it is a relatively modern conception: it was unknown in the Middle Ages. At that time, men still thought that the forces of nature could only be dominated by magic. The mystics and alchemists of those ages were not looking for the laws of nature, but, on the contrary, for spells which would turn the laws aside. Their underlying belief was that man could only turn nature to use by bewitching her so that she was compelled to run counter to her normal laws.

The essential discovery that we call science is an outright rejection of this view. Science believes that the potential of nature cannot be commanded by magic, or by exhortation, or by persuasion; it can be commanded only by knowledge. We cannot overthrow the laws of nature, or even flout them, and that is not the way to bend nature to the human will. Instead, what we have to do is to discover the laws and the organization of nature, and then think of ways in which we can use them to do for us what we want done. That is how the dynamo was invented, how radio waves, x-rays, and antibiotics were discovered, and how the jet engine and the laser beam were developed. No spell could have produced a nuclear chain reaction; it was produced from the modest finding that natural uranium contains several isotopes, and then patiently sorting out one isotope from another – nothing more.

Thus the success of science has its roots in a basic change of outlook more than 300 years ago. It was then that Renaissance thinkers came to take a different view of nature, and to see her not as an antagonist but an ally. The key to a fuller physical life, they saw, is not a secret Hermetic formula – the philosopher's stone, or the elixir of life – that will cow nature and force her to run counter to her own laws. On the contrary, the key is knowledge; an active and practical method of discovery that enters into the very processes of nature, and thereby learns to direct them to human ends.

So profound a change in outlook is not merely a technical matter, or a useful device for professionals. The scientific temper has spread through the whole community, and created a universal climate in which knowledge is indeed seen as the key to a full life – consistent intellectual life as well as a full physical life. For example, the gradual erosion of superstition by science is, we now see, something more than a happy accident; it lies at the root of our conviction that nature can be rationally understood, and can be guided to our ends by understanding. One of the major social influences of science has been this belief in the power of knowledge. This is basic to the democratic view, that every man has the right to be heard and have a voice in guiding the actions of his community. The experience of science has shown the value of the dissident voice, and this has been a powerful social influence in smoothing the way for the evolution of western society through democratic reforms.

We have a definition of science, we see how it works, and we understand why it works. Through all these there runs a strong sense that science is a method of discovery. We make good rational approximations, we link them together quite ingeniously in systems of laws or axioms, and it is wonderful that our simple models imitate nature so well. In short, we are sometimes as clever as Isaac Newton when he devised the law of gravitation, and then our discoveries look marvelously natural – what could be more natural in a space of three dimensions than a force that falls off as the second power of the distance? And we are sometimes as lucky as was Newton, whose law worked so infallibly for 200 years that it really seemed to be natural. But sooner or later our luck runs out – even Newton's luck; we have to go to work again and make another model, another imaginative discovery; and we learn that it is not in our power to be right forever.

Therefore the practice of science requires that we take a very pragmatic view of human fallibility. We cannot aspire to supernatural knowledge any more than to supernatural power. We cannot even aspire to superhuman knowledge, and we must be content to go forward as human beings do, modestly and industriously with the means that the human mind provides. That is, we must learn to work within our imperfections, for we are neither perfect gods nor perfect machines. No intellectual force (or arrogance) will preserve us from the errors of the human condition and conjure down the truth for us as an omniscient and final revelation.

The practice of science began with the basic rejection of received authority, in the time span between the Renaissance and the Puritan revolution. When men were pressed to see for themselves, by Andreas Vesalius tracing a muscle or by Isaac Newton with his prism, they seemed at first to be thrust on themselves. But that turned out to mean that they were thrust on one another, to meet and gossip and exchange speculations and experiments. From the beginning, science was not the hidden and private art of a magus, but an open, enthusiastic undertaking of friends who vied to improve on one another's tricks.

Science has continued to be a communal process, in which even the most individual genius has to build on work that other people have already done. And in order to build on previous work he must be able to rely on it: the prerequisite for personal achievement in science is communal trust. The unhesitating march of discovery, the progression from one result to the next without pausing to look suspiciously over one's shoulder, depends on our absolute assurance that we can believe what we read in a scientific paper. We know that the other man is sincerely trying to tell the truth. He may not get it right, but so far as is possible he tries.

It may seem very modest and humdrum to seek a foundation for science and humanism in simple truth to fact. Surely even science aspires to higher forms of truth than this? Yes, it does; science is the organization of our knowledge into more and more highly integrated relationships; but the hard heart of the matter is that you cannot organize what you cannot agree and rely on. Facts are the empirical foundation of our knowledge, from which our work begins and to which it constantly returns to be tested; and if we cannot rely on one another's testimony in this, then we have no common gound to build on. It is useless to talk about the higher ends of science or of humanism and then to behave as if the means were too commonplace to talk about. The basic means which a scientist uses in his work are statements of fact, by himself or by others; he and they will not always get them right, try as they may; but try to be truthful they must, in the most humble, puritanical, and merciless sense of the word. It is fatal to say to yourself that your ends are so good that you can permit yourself the expedient of bending the means a little to suit them – fatal to your status as a man as well as a scientist, fatal to the community of science and equally fatal to the human community. Here the ethic of science is simple and it is also universal:

> There is no distinction between ends and means, for there is no way to serve a great end other than by honest means.

The central tenets of humanism since the Renaissance have been its stress on the human content and meaning of experience, and a sense of our participation and immersion in the totality of nature. This is what the teaching of history, of literature, and of the classics was designed to instill and foster in the student.

The philosophy of science is the same as that of humanism, except that it reverses the sequence – it begins from a sympathy with nature, and then links the human experience to that. So there is no break here between humanism and science; on the contrary, a coherent philosophy of science gives to those who grasp it a sense of man's unique place in nature that can overcome the loss of nerve or purpose which many men now feel.

The central problem in teaching today is to establish this sense of place in the values which science derives from its active and communal search for truth. We have to show, by example rather than by precept, that the search requires an ethic of human responsibility and tolerance, an aesthetic of respect and good manners, a personal dignity, which are the essence of the humanist tradition. Above all, we have to show how science has singled out from that tradition its most powerful moral: that we are judged (and indeed formed) not by the ends we proclaim, but by the means we use day by day.

The human mind
has never invented
a labor-saving machine
equal to algebra. It is but
natural and proper that an age
like our own, characterized
by the multiplication of labor-
saving machinery, should be
distinguished by an unexampled
development of this most refined
and most beautiful of machines.

J. WILLARD GIBBS
Address to the American
Association for the Advancement
of Science,
1886

Appendix

– mathematical background

The operational definitions of physical quantities such as length, time, and mass lead to their measurement in terms of standard units. Through measurements, numbers find their way into physics. The manipulation of these numbers is based on the operations of arithmetic. The descriptions of mathematical models for relationships among the numbers may be expressed in algebraic form or in the form of line graphs. Finally, some of the quantities, such as length, distance, and relative position, refer to spatial relationships; their treatment is based on the procedures and definitions of geometry. The areas of mathematics to be reviewed in this Appendix, therefore, are arithmetic, mathematical models, and geometry.

Of paramount importance in the applications of mathematics to physics is that the results of all measurements suffer from experimental uncertainties. No numbers in physics are ever known exactly. Relationships, models, comparisons, and inferences can never be more accurate than the data on which they are based. Refinements in the technology of measurement have led to great advances in the accuracy of physical data. However, since this text is intended for beginning students, we have selected phenomena that can be observed and described successfully with modest equipment and relatively rough data. The arithmetic needed for these purposes is not complex and should always be viewed as a tool rather than as an end in itself. When you find that the mathematics is an obstacle to your understanding, seek a more approximate, more qualitative understanding first.

A.1 Arithmetic for physics

One, two, three, four, five, . . . One, ten, hundred, thousand, ten thousand, . . .

Both of these sequences of numbers have a regular pattern. The former is the way all of us have learned to count fingers, cookies, friends, and pennies. The latter enumerates the place values (ones, tens, hundreds, . . .) of the digits of a number written in the decimal number system. When you read a number and seek to grasp its magnitude, you have to determine the place value of the first digit. This step is even more important than identifying the first digit correctly. The decimal system is marvelously designed to enable you to express any number to great accuracy and to arrange any set of numbers in sequence. The two essential concepts are the digits from zero to nine in their conventional sequence and the place value of digits with its conventional meaning relative to the decimal point.

Arrange these three numbers in sequence from smallest to largest:

> *367426,*
> *36742.6,*
> *367396*

In this section we will describe the scientist's way of writing numbers and carrying out arithmetic operations. These procedures reflect the central fact that numbers are obtained by measurement and have limited accuracy. Several other considerations are important, too. One of these is that most mathematical models require the multiplication and/or division of numbers, rather than addition or subtraction. Another is the scientist's need to work with numbers throughout an enormous range of magnitudes. This range extends from the mass of an electron, which is approximately one one-thousandth of a billionth of a billionth of a billionth of 1 kilogram (0.000,000,000,000,000,000,000,000,000,-001 kilogram), to the mass of the sun, which is approximately 2000 billion billion billion kilograms (2,000,000,000,000,000,000,000,000,-000,000 kilograms). With numbers such as these, determining the place value of the first nonzero digit is quite a problem!

Place value and powers of ten. An extremely convenient notation for keeping track of very large and very small numbers makes use of the powers of ten. All the place values greater than the ones are written to the left of the decimal point and are obtained by multiplying the number one by one or more factors of ten. All place values less than the ones signify decimal fractions, are written to the right of the decimal point, and are obtained by dividing the number one by one or more factors of ten. In the powers-of-ten notation, the number of factors of ten is indicated by a superscript following the numeral 10. The superscript, called the *exponent,* is positive to indicate repeated multiplication, negative to indicate repeated division (Table A.l). The sequence of place values can also be displayed along a number line.

Multiplication and division of powers of ten can be carried out easily since the definition of the powers of ten was based on multiplication for numbers greater than one (positive exponent) and division for numbers less than one (negative exponent). You can find the net number of factors of ten in a product of powers of ten by *adding* all the exponents with due regard to their being positive or negative (Example A.l). To divide by a power of ten, change the sign of the exponent and multiply (add the resulting exponents with the changed sign, see Ex. A.1c-f).

Addition and subtraction of powers of ten is not so easy, because these operations have to be carried out digit by digit with matched

TABLE A.1 POWERS OF TEN

common name (place value)	Common notation	Significance	Scientific notation	Scientific name
one	1	1	10^0	ten to the zero power
ten	10	10	10^1	ten to the first power
hundred	100	10x10	10^2	ten to the second power
thousand	1000	10x10x10	10^3	ten to the third power
ten thousand	10,000	10x10x10x10	10^4	ten to the fourth power
hundred thousand	100,000	10x10x10x10x10	10^5	ten to the fifth power
million	1,000,000	10x10x10x10x10x10	10^6	ten to the sixth power
tenth	0.1	1/(10)	10^{-1}	ten to the negative first power
hundreth	0.01	1/(10x10)	10^{-2}	ten to the negative second power
thousandth	0.001	1/(10x10x10)	10^{-3}	ten to the negative third power
ten thousandth	0.0001	1/(10x10x10x10)	10^{-4}	ten to the negative fourth power
hundred thousandth	0.00001	1/(10x10x10x10x10)	10^{-5}	ten to the negative fifth power
millionth	0.000001	1/(10x10x10x10x10x10)	10^{-6}	ten to the negative sixth power

(margin place-value chart: ten thousands / thousands / hundreds / tens / units : 5 4 3 2 1 | tenths / hundredths / thousandths / ten thousandths : 1 2 3 4)

place values. How this is done will be described later, after the standard form of numbers is explained.

EXAMPLE A.1.
(a) $10^3 \times 10^5$ $= (10\times10\times10) \times (10\times10\times10\times10\times10)$
                                $= 10^{3+5} = 10^8$
(b) $10^6 \times 10^{-4}$ $=(10\times10\times10\times10\times10\times10) \times 1/(10\times10\times10\times10)$
                                $= 10^{6-4} = 10^2$
(c) $1/10^4$ $= 1/(10\times10\times10\times10) = 10^{0-4} = 10^{-4}$
(d) $1/10^{-4}$ $= 1/[1/(10\times10\times10\times10)] = 10^{0-(-4)} = 10^4$
(e) $10^2 \times 10^3/(10^2 \times 10^{-4})$ $= 10^2 \times 10^3 \times 10^{-2} \times 10^4 = 10^{2+3-2+4} = 10^7$
(f) $10^2 \times 10^3/(10^2 \times 10^{-4})$ $= 10^{2+3}/10^{2-4} = 10^5/10^{-2} = 10^{5-(-2)} = 10^7$

Calculators. We recommend that you use an inexpensive calculator *with* a square root key. If you wish, you can get a somewhat more expensive and powerful "scientific calculator." A scientific calculator has many built-in functions that make calculations easier and save time; *however*, learning how to use a scientific calculator involves some effort, and until you become fluent, there will be a certain number of data entry and other types of "calculator-related" errors. For a study of physics at the level of this book, doing the numerical calculations with a simple calculator and keeping track of the powers of ten on paper or in your head (as described below) will be quite adequate.

Numbers in standard form. Very large or very small numbers can be read more easily and can be multiplied and divided more easily if the place value of the first nonzero digit is written separately from the digits. The nonzero digits are usually written as a decimal number between one and ten. The place value is indicated by a power of ten (Table A.l). A number written in the form of a product, as "number between one and ten" multiplied by "power of ten," is said to be written in standard form (Example A.2, last column).

Multiplication and division of numbers in standard form. These operations are most conveniently carried out in two steps, one step for the powers of ten, the other step for the numbers between one and ten that multiply the powers of ten (Example A.3). The former was explained above. The latter is accomplished most easily with a simple calculator.

A very rough estimate of the result of any calculation can be found by keeping only the first digit of the number. By using such estimates, you can find a quick estimate for the answer you expect from a calculation, so as to catch errors in using the calculator.

EXAMPLE A.2. Exercises in the standard form
(a) three hundred twenty $= 320$ $= 3.2\times100$ $= 3.2\times10^2$

(b) seventy-three point four	= 73.4	= 7.34x10	= 7.34×10^1
(c) five million eight hundred thousand	= 5,800,000	= 5.8x1,000,000	= 5.8×10^6
(d) forty-five hundredths	= 0.45	= 4.5x1/10	= 4.5×10^{-1}
(e) sixty-two ten-thousandths	= 0.0062	= 6.2x1/1000	=
(f) one thousand three	= ____	= ____	= 1.003×10^3
(g) _____	= 0.08	= 8x1/100	=
(H) _____	= 9100	= ____	=
(I) _____	= 0.0084	= ____	=
(J) _____	= ____	= ____	= 2.8×10^{-4}

EXAMPLE A.3 (The symbol ≈ indicates approximate equality.)

$$(a) \frac{320 \times 73.4}{5,800,000} = \frac{(3.2 \times 10^2) \times (7.34 \times 10)}{(5.8 \times 10^6)} = \left(\frac{3.2 \times 7.34}{5.8}\right) \times \left(\frac{10^2 \times 10^1}{10^6}\right)$$

$$= 4.04 \times 10^{2+1-6} = 4.04 \times 10^{-3}$$

[rough estimate of $\dfrac{3.2 \times 7.34}{5.8} \approx \dfrac{3 \times 7}{6} = \dfrac{21}{6} = 3.5$]

$$(b) \frac{0.45 \times 320}{9100} = \frac{(4.5 \times 10^{-1}) \times (3.2 \times 10^2)}{(9.1 \times 10^3)} = \left(\frac{4.5 \times 3.2}{9.1}\right) \times \left(\frac{10^{-1} \times 10^2}{10^3}\right)$$

$$= 1.58 \times 10^{-1+2-3} = 1.58 \times 10^{-2}$$

[rough estimate of $\dfrac{4.5 \times 3.2}{9.1} \approx \dfrac{4 \times 3}{9} = \dfrac{12}{9} \approx 1$]

$$(c) \frac{73.4}{9100 \times 0.0062} = \frac{(7.34 \times 10^1)}{(9.1 \times 10^3) \times (6.2 \times 10^{-3})} = \left(\frac{7.34}{9.1 \times 6.2}\right) \times \left(\frac{10^1}{10^3 \times 10^{-3}}\right)$$

$$= 0.13 \times \frac{10^1}{10^0} = 1.3 \times 10^{-1} \times 10^1 = 1.3$$

[rough estimate of $\dfrac{7.34}{9.1 \times 6.2} \approx \dfrac{7}{9 \times 6} = \dfrac{7}{54} \approx 0.1$]

Supply the missing numbers to complete Example A.3d.

$$(d) \frac{0.08 \times 9100}{73.4} = \underline{\hspace{3cm}} = \underline{\hspace{3cm}} \times \left(\frac{10^{-2} \times 10^3}{10^1}\right)$$

$$= \underline{\hspace{3cm}} = \underline{\hspace{3cm}}$$

[rough estimate of $\underline{\hspace{3cm}} \approx \dfrac{8 \times 9}{7} = \dfrac{72}{7} \approx \underline{\hspace{1cm}}$]

Addition and subtraction of numbers in standard form. We have already mentioned that in addition and subtraction only digits with the same place value must be combined. The standard form is not designed for this purpose. The standard form may be adapted, however, if all the numbers to be added or subtracted are written with the same power of ten (usually the largest power occurring in any of the numbers), multiplied by a number that no longer need be between one and ten. Then the operations can be carried out in the usual way (Example A.4). Because of the need for matching place values in addition and subtraction, it is usually wisest to carry out these operations as much as possible before numbers are converted to standard form.

EXAMPLE A.4 Add these numbers: 3.2×10^2; 7.34×10^1; 4.5×10^{-1}; 1.003×10^3.

Solution: Write all numbers with ten to the third power (10^3):

$$
\begin{aligned}
3.2 \times 10^2 &= 3.2 \times (10^{-1} \times 10^3) = 0.32 \quad \times 10^3 \\
7.34 \times 10^1 &= 7.34 \times (10^{-2} \times 10^3) = 0.0734 \times 10^3 \\
4.5 \times 10^{-2} &= 4.5 \times (10^{-4} \times 10^3) = 0.00045 \times 10^3 \\
1.003 \times 10^3 &= \qquad\qquad\qquad\quad = \underline{1.003 \quad \times 10^3} \\
&\text{Sum} \qquad 1.39685 \times 10^3 \approx 1.40 \times 10^3
\end{aligned}
$$

Very small numbers may often be neglected because their contribution to the sum is no more than that of experimental uncertainties in the larger numbers.

Square roots of numbers in standard form. When you seek the square root of a number, you try to express it as a product of two equal factors; one of the two equal factors is then the square root. The square root of a number in standard form can be found easily (as shown to the left) if the standard form is modified so that the exponent is an *even* number, (2, 4, 6, 8 . . . n), that is, as a number that is the sum of two *equal* numbers ($2 = 1 + 1$, $4 = 2 + 2$, $6 = 3 + 3$, . . ., $n = n/2 + n/2$). Then the factors of ten can be split into two equal factors ($10^n = 10^{n/2}$ x $10^{n/2}$) = $(10^{n/2})^2$. The exponent of the square root is one half the original exponent (square root of $[(10^{n/2})^2] = 10^{n/2}$.

$$
\begin{aligned}
\sqrt{10^4} &= \sqrt{(10^2) \times (10^2)} \\
&= 10^2 \\
\sqrt{10^3} &= \sqrt{10 \times (10^2)} \\
&= \sqrt{10} \times \sqrt{(10^2)} \\
&\approx 3.2 \times 10^1 \\
\sqrt{5 \times 10^7} &= \sqrt{50 \times 10^6} \\
&= \sqrt{50} \times \sqrt{10^6} \\
&\approx 7 \times 10^3
\end{aligned}
$$

The number multiplying the power of ten will then be between 1 and 100, and the square root can be found easily with a calculator. You can estimate the square root of a number between 1 and 100 quickly by comparing the number with the squares of the integers from 1 to 10 (1, 4, 9, 16, 25, 36, 49, 64, 81, 100). For example, suppose you want to estimate the square root of 56; this is between 49 and 64; therefore the square root is between 7 and 8.

Rounded numbers: 20% accuracy. For the simple applications in this text, the entire capability of the decimal number system is not necessary. Numerical accuracy of about 20% is adequate for most purposes. We recommend that you use a calculator, rounding off the result to the same number of decimal places as in the numbers with which you started the calculation. We also strongly recommend that you check your answer by using rounded off numbers to do a quick calculation to estimate the approximate size of the result you expect from the calculator. This is the best way to catch the all-too-frequent errors in using a calculator (for example, entering numbers incorrectly, not fully depressing a key, and mixing up the proper order of operations). Even a simple calculator will keep track of many more digits than you will need, so you should round off the final result to as many decimal places as in the numbers with which you started the calculation. The number of figures in the answer indicates the relative accuracy and reliability.

A.2 Mathematical models

Analysis of data. A scientist who measures several variable factors associated with a phenomenon may wonder whether the data show any regular relationship. Thus, a girl may have difficulty walking on a sandy beach in shoes with high heels because the heels sink into the sand. Does the distance they sink depend on the width of the heel? How can the data in Table A.4 be interpreted? The uncertainty in the data is indicated as a "range" above and below the mean value listed. Thus, the hiking shoe's width is between 5.4 and 5.6 centimeters; the depth of the hole made by the spike heels is between 6 and 8 centimeters.

It appears from the data that the narrow heel sinks deeper than the wide heel. To detect more regularity than this qualitative inverse relationship (wide heel, shallow hole; narrow heel, deep hole), we plot the data on graph paper (Fig. A.3). For each data point we have made an oval to indicate the uncertainty of the data. The width and height of the oval represent the uncertainty in width and depth, respectively. The points on the graph show more than the inverse relationship; there is a regular progression from one point to the next. We have therefore drawn a smooth-line graph as a mathematical model in graphical form for the width-depth relationship. The line implies that the actual data

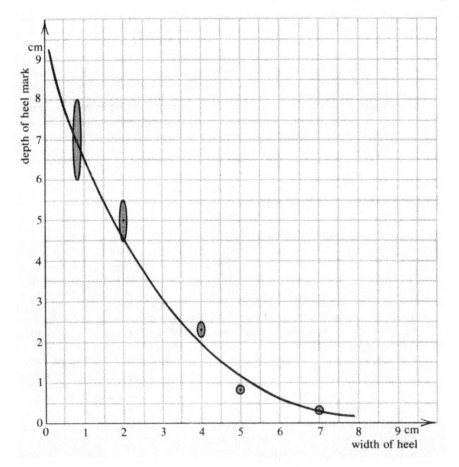

Figure A.3 Graph of the heel-mark data points (Table A.4) and a mathematical model in the form of a line graph.

points are part of a continuum of possible data points that would be obtained with heels of various widths (1.5, 3, 5 centimeters) provided other factors, such as the condition of the sand and the weight of the shoe's wearer, remain the same. Some of the predictions coming from this model are listed in Table A.5. It is difficult to predict what will happen for wider shoes than the sandals and for heels still more pointed than those used in the experiments.

The next step is to put the mathematical model into algebraic form. Two possible relations are given in Eq. A.l and drawn in Fig. A.4. You can judge their adequacy. Model I, which fits the data best, predicts that

Equation 1.2

distance traveled   = s
time elapsed   = t
speed   = v
             s = vt

TABLE A.4 DATA ON HEEL MARKS IN SAND

Shoe type (cm)	Width of heel (cm)	Depth of heel mark (cm)
sandal	7.0 ±0.1	0.3 ±0.1
hiking shoe	5.5 ±0.1	0.8 ±0.1
walking shoe	4.0 ±0.1	2.3 ± 0.2
white pump	2.0 ± 0.1	5.0 ± 0.5
spike-heeled shoe	0.8 ±0.1	7 ±1.0

TABLE A.5 PREDICTIONS FOR HEEL MARKS IN SAND — FROM FIG. A.4

Width of heel (cm)	Depth of heel mark (cm)
1.5	5.8 ± 0.5
3.	3.3 ± 0.5
5.	1.3 ± 0.2

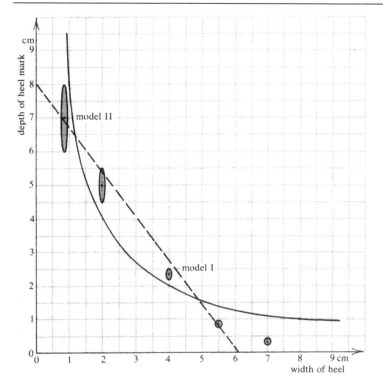

Figure A.4 Graph of the heel-mark data (Table A.4) and the mathematical models, Eq. A.l.

the girl will sink to a "depth" of -1.1 centimeters when she wears shoes with 8-centimeter-wide heels. Since a "negative depth" is actually a height above the sand level, this model predicts that on 8-centimeter-wide heels the young lady would float above the ground, which seems very unlikely. This prediction, then, is evidence of a serious shortcoming of the model. Model II does not fit the data so well as model I. Test it to determine whether it leads to any implausible predictions!

Direct proportion and ratios. One of the most useful mathematical models is the *direct proportion,* in which the two related variable factors increase in the same proportion. An example is the relationship between distance and time for the automobile trip described in Section 1.3 (Eq. 1.2 and Fig. A.5). Here the graph is a straight line that passes through the point representing zero distance in zero time. A direct proportion is characterized by a fixed ratio, such as the ratio of distance traveled to elapsed time, which is called the speed. The ratio is often given the general name "constant of proportionality," because it is constant while the variable factors change.

There are two simple tests for a direct proportion between variable factors. One is the straight line of the graph, the other is the fixed ratio between the two variables. If these tests are satisfied within the uncertainty in the data, the direct proportion is a good mathematical model for the data. If the tests are not satisfied, then a different model should be constructed.

Because the direct proportion is a simple relationship, physicists make an effort to use it whenever possible. Examples that are discussed in the text are the models relating wave speed to wavelength for a wave of particular frequency (Section 6.1), thermal energy to temperature (Section 10.2), phase energy to mass of material (Section 10.3), work to displacement for a given force (Section 11.3), Hooke's law (Section 11.6), and others. In each example, the constant of proportionality has been given a name and an interpretation.

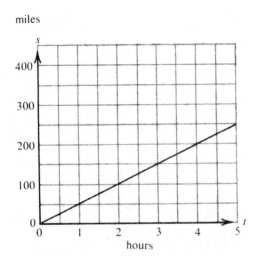

Figure A.5 Line graph for the mathematical model in Eq. 1.2.

Inverse proportion. A second common mathematical model is the *inverse proportion,* an example of which was model II for the heel marks (Eq. A.l and Fig. A.4). In an inverse proportion, one variable factor decreases as the other increases, but they change in the same ratio. That is, when one factor doubles, the other is halved; when one triples, the other decreases to one third; and so on. The graph of an inverse proportion is called a *hyperbola* (model II in Fig. A.4), which approaches but never crosses the axes of the graph on which either of the variable factors is zero. Since the factors vary inversely in the same ratio, their product does not change but remains fixed.

An arithmetic test for the inverse proportion is to calculate the product of the two variable factors and see whether it is indeed constant within the experimental uncertainty of the data. When this test is applied to the heel-mark data, it becomes clear that the model does not fit (Table A.6).

Many examples of inverse proportions are described in the text. The relation of wavelength and wave number is one example (Section 6.1), the relation of voltage and current to deliver a certain electric power is another (Section 12.4), the relation of electric current and time to transfer a specified electric charge is a third (Section 12.1).

In fact, most direct proportions also contain an inverse proportion if new variable factors are selected. In Eq. 1.2, for instance, the distance traveled to the destination is no longer variable once the route of the trip is decided. The model then becomes an inverse proportion between the speed and the travel time: the greater the speed, the shorter the travel time, and vice versa.

Power laws. In some laws of physics, the variable factors have still other relationships. Fairly common are so-called *power laws,* in which there is a direct or inverse proportion to the second, third, or higher power of a factor. The volume of a cube, for example, varies directly as the third power of the length of the edge. In the universal law of gravitation (Eq. 15.10), the force varies inversely as the second power of the distance.

TABLE A.6 TEST FOR INVERSE PROPORTION OF HEEL-MARK DATA (FROM FIG. A.3)

Width (cm)	Depth (cm)	Width times depth * (cm^2)
7.	0.3	2.1
5.5	0.8	4.4
4.0	2.3	9.2
2.0	5.0	10.0
0.8	7.0	5.6

* the values in the last column are not constant. This means that the width and depth are *not* inversely proportional.

Equation A.2
$5 + 7 = 12$	(a)
$5 = 12 - 7$	(b)
$12 - 5 = 7$	(c)
$12 = 7 + 5$	(d)

Equation A.3
$6 = 2 \times 3$	(a)
$2 = 6/3$	(b)
$6/2 = 3$	(c)
$3 \times 2 = 6$	(d)

Mathematical reasoning. The arithmetic relation among the numbers 5, 7, and 12 can be written in several alternate ways (Eq. A.2). These are mathematically equivalent but emphasize different aspects of the relation. The same is true for the relation among 2, 3, and 6 (Eq. A.3). It is trivial for you to verify from memory that each of the statements is correct.

Relations that contain letter symbols that represent numbers can also be written in several alternate ways (Examples A.7 and A.8). These are again mathematically equivalent, but now they cannot be verified from memory. Instead, you have to determine whether each one is correct in one of two ways. Either you interpret it in terms of the significance of the symbols or you derive it by mathematical reasoning from a statement you know to be correct. In each of Examples A.7 and A.8, for example, there is one relation that is incorrect. Can you find it?

EXAMPLE A.7.
total credit hours for graduation = T
credit hours in major = M
elective credit hours outside major = E
(a) T = M+E
(b) T - M = E
(c) M = T + E
(d) E + M = T

Which equation is incorrect?

EXAMPLE A.8.
semester's reading in literature course (pages) = P
weeks in semester = w
average reading for one week (pages) = p
(a) pw = P
(b) p = P/w
(c) Pw = p
(d) P = wp

Which equation is incorrect?

The rules for mathematical reasoning allow you to transform an equation in any one or several of the following ways.

1. Replace one symbol by an equivalent symbol or combination of symbols.
2. Add (or subtract) equal terms from both sides of the equation.
3. Multiply (or divide) both sides of the equation by equal factors, but do not divide by zero.

4. Perform the same operation with both sides of the equation (for example, take the square root).

Examples A.9 and A.10 further illustrate the application of these rules to instances taken from the text.

EXAMPLE A.9. Relation of wavelength and wave number (Eq. 6.1): $\lambda k = 1$

Starting point:

$\lambda k = 1$ (a)

Divide both sides of (a) by k, to get:

$\lambda = 1/k$ (b)

Or, divide both sides of (a) by λ, to get an alternative relationship:

$k = 1/\lambda$ (c)

EXAMPLE A.10. Derivation of elastic energy (Example 11.3)

Starting point:

$E = W$	assumed (energy change equals work)	
$\|\mathbf{F}\| = \frac{1}{2}\kappa\Delta s$	assumed (Hooke's Law)	(a)
$W = \|\mathbf{F}\|\Delta s$	assumed (definition of work)	(b)
$W = (\frac{1}{2}\kappa\Delta s)\Delta s$	replace $\mathbf{F}$ in (b)	(c)
$W = \frac{1}{2}\kappa(\Delta s)^2$	replace $(\Delta s)\Delta s$ with $(\Delta s)^2$	(d)
$E = \frac{1}{2}\kappa(\Delta s)^2$	replace W in (d)	(e)

A.3 Geometry and trigonometry

Very early in life everyone learns to recognize shapes. One of the most important concepts related to shape is that of a straight line. How would you check whether the type in this book had been set along straight lines? Your answer to this question describes an operational definition of the "straightness" concept. You may hold a ruler along the page to determine whether the type lines up with the ruler; then the ruler is your standard object in the definition of "straight." You may hold the book up to your eyes and sight along the type; then the light ray is your standard "object" in the definition of "straight." Or you may choose still another standard object. Some of the shapes that are formed by straight line segments are triangles, rectangles, and squares. An understanding of these figures and their properties is essential not only for the scientist, but also for the surveyor and even the archeologist.

Angles. Two intersecting straight lines form four angles that describe the direction of one of the lines relative to the other (Fig. A.6). The angles are measured in degrees of arc, with a full circle being equal to 360 degrees (360°). One fourth of a full circle (90°) is called a *right angle*. An angle smaller than a right angle is an *acute angle*. Two acute angles whose sum is a right angle are called *complementary*.

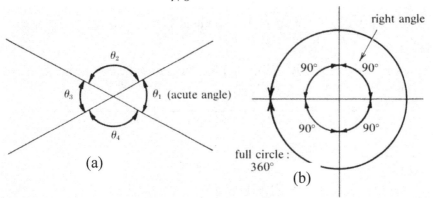

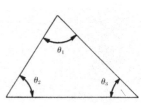

Figure A.6 Intersecting straight lines form four angles.
(a) The angles are equal in pairs, $\theta_1 = \theta_3$, $\theta_2 = \theta_4$.
(b) When all four angles are equal, each angle is a right angle.

Figure A.7 The sum of
the interior angles of a
triangle is 180 degrees
($\theta_1 + \theta_2 + \theta_3 = 180°$).

Classification of triangles. Triangles play an especially important role in physical reasoning because they are the simplest figures that include an angle. Triangles can therefore be used to make inferences about the relative directions of straight lines, such as the direction of motion of a particle, the direction of propagation of a light wave, and the direction of the surface along which two bodies are in contact. The sum of the three interior angles of a triangle is 180° (Fig. A.7). An *equilateral triangle* has all sides equal in length. An *isosceles triangle* has two equal sides. A *right triangle* contains one right angle.

Similar triangles. Two triangles are said to be *similar* if the angles of one are equal to the angles of the other (Fig. A.8). The sides of the two triangles that are opposite the equal angles are called *corresponding*

Figure A.8 Five similar triangles (OAZ, OBY, OCX, ODW, OEV) with interior angles θ_1, θ_2, θ_3 and corresponding sides in three sets: opposite θ_1: {AZ, BY, CX, DW, EV}; opposite θ_2: {OA, OB, OC, OD, OE}; opposite θ_3: {OZ, OY, OX, OW, OV}.

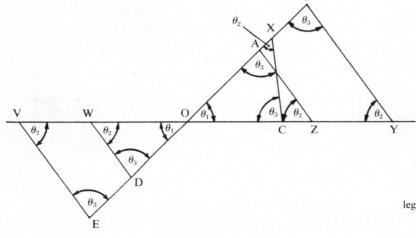

Figure A.9 (below)
Identification of the sides
of a right triangle. Angles
θ_1 and θ_2 are always acute
angles (less than 90°). Can
you see why this must be
true?

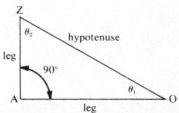

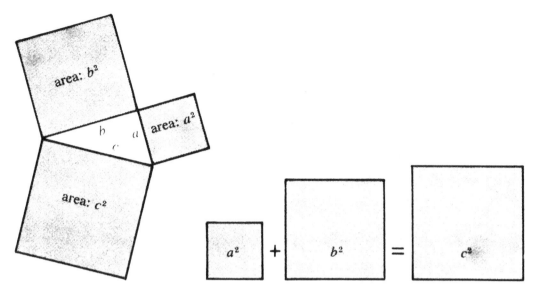

Figure A.10 Geometric significance of the Pythagorean theorem.

Equation A.4
ratios of correspond-ing sides of similar triangles (Fig. A.8):

$$\frac{AZ}{OA} = \frac{BY}{OB} = \frac{CX}{OC} = ... \text{(a)}$$

$$\frac{AZ}{OZ} = \frac{BY}{OY} = \frac{CX}{OX} = ... \text{(b)}$$

$$\frac{OA}{OZ} = \frac{OB}{OY} = \frac{OC}{OX} = ... \text{(c)}$$

sides. The ratios of corresponding sides of two similar triangles are equal (Eq. A.4). To prove that two given triangles are similar, it is sufficient to show that two angles of one are equal to two angles of the other one; the third angle of one is then automatically equal to the third angle of the other because the sum of all three angles is 180°.

Right triangles and trigonometry. In a right triangle, the two angles that are not right angles are complementary acute angles (Fig. A.9). The longest side of a right triangle, which is opposite the right angle, is called the *hypotenuse.* The other two sides are called *legs* (Fig. A.9). The Pythagorean theorem states that the square on the hypotenuse is equal in area to the sum of the squares on the legs (Fig. A.10 and Eq. A.5).

**Equation A.5
(Pythagorean theorem)**

length of
hypotenuse $= c$
lengths of legs $= a, b$

$$c^2 = a^2 + b^2$$

Trigonometric ratios. Two right triangles are similar if they have one acute angle equal (Fig. A.11). This fact leads to the most frequent identification of similar triangles in physics. In a set of similar right triangles, there are three sets of corresponding sides: the hypotenuses,

Figure A.11 Five similar right triangles with the acute angle θ. Set of hypotenuses {OZ, OY, OX, OW,OV} Set of longer legs [OA, OB, OC, OD, OE] Set of shorter legs {AZ, BY, CX, DW, EV}

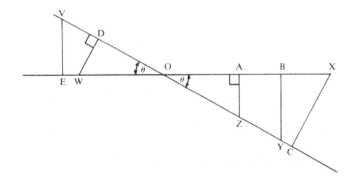

TABLE A.7 TRIGONOMETRIC RATIOS

θ	sine θ	cosine θ	tangent θ
0°	0.000	1.000	0.000
5°	0.087	0.996	0.088
10°	0.174	0.985	0.176
15°	0.259	0.966	0.286
20°	0.342	0.940	0.364
25°	0.423	0.906	0.466
30°	0.500	0.866	0.577
35°	0.574	0.819	0.700
40°	0.643	0.766	0.839
45°	0.707	0.707	1.000
50°	0.766	0.643	1.192
55°	0.819	0.574	1.428
60°	0.866	0.500	1.732
65°	0.906	0.423	2.145
70°	0.940	0.342	2.748
75°	0.966	0.259	3.732
80°	0.985	0.174	5.671
85°	0.996	0.087	11.430
90°	1.000	0.000	undefined

Equation A.6

trigonometric ratios (symbols defined in Fig. A.9):

$$sine\ \theta_1 = \frac{opposite\ leg}{hypotenuse}$$

$$= \frac{AZ}{OZ} \qquad (a)$$

$$cosine\ \theta_1 = \frac{adjacent\ leg}{hypotenuse}$$

$$= \frac{OA}{OZ} \qquad (b)$$

$$tangent\ \theta_1 = \frac{opposite\ leg}{adjacent\ leg}$$

$$= \frac{AZ}{AO} \qquad (c)$$

$$sine\ \theta_2 = \frac{OA}{OZ} \qquad (d)$$

$$cosine\ \theta_2 = \frac{AZ}{OZ} \qquad (e)$$

$$tangent\ \theta_2 = \frac{AO}{AZ} \qquad (f)$$

the longer legs, and the shorter legs. The corresponding ratios (for example, shorter legs to hypotenuses, shorter legs to longer legs, and so on) are equal for all right triangles in the set (Eq. A.4). The numerical values of these ratios depend on the size of the acute angle, but they are independent of the size of the triangle itself. They are so useful that they have been given names (Eq. A.6) and are grouped under the name *trigonometric ratios*.

The *sine* of an angle is the ratio of the leg-opposite-the-angle to the hypotenuse. The *cosine* of an angle is the ratio of the leg-adjacent-to-the-angle to the hypotenuse. The *tangent* of the angle is the ratio of the leg-opposite-the-angle to the leg-adjacent-to-the-angle.

Relations among trigonometric ratios. There are several important relations among the trigonometric ratios. One is that the tangent of an angle is equal to the ratio of the sine of the angle to the cosine of the same angle (Example A.11a). A second is that the sine of an angle is equal to the cosine of the complementary angle (Example A.11b). Both of these relations are obtained directly from the definition of the trigonometric ratios. A third relation is a consequence of the Pythagorean theorem (Example A.11c): the second power of the sine of an angle added to the second power of the cosine of the same angle always equals the number one $[(sine\ \theta)^2 + (cosine\ \theta)^2 = 1]$

Applications. By measuring the sides of a right triangle, you can find the trigonometric ratios and verify that they satisfy the relations

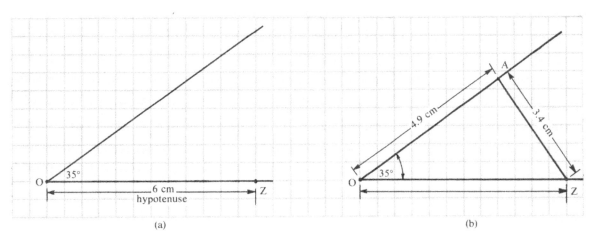

Figure A.12 Graphical construction to find the legs of a right triangle.
(a) The acute angle of 35 degrees and the hypotenuse OZ of 6-centimeter length,
(b) Drop a perpendicular from Z to the other side of the angle at A and measure
legs AZ and ZO. Does the result agree with that calculated according to Table A.7?

EXAMPLE A.11. (See Fig. A.9 for definitions)

(a) $\text{tangent } \theta_1 = \dfrac{AZ}{AO} = \dfrac{AZ/OZ}{AO/OZ} = \dfrac{\text{sine } \theta_1}{\text{cosine } \theta_1}$

(b) $\text{sine } \theta_1 = \dfrac{AZ}{OZ} = \text{cosine } \theta_2 \quad \text{if } \theta_1 + \theta_2 = 90°$

(c) $(\text{sine } \theta)^2 + (\text{cosine } \theta)^2 = \dfrac{(AZ)^2}{(OZ)^2} + \dfrac{(OA)^2}{(OZ)^2} = \dfrac{(AZ)^2 + (OA)^2}{(OZ)^2} = \dfrac{(OZ)^2}{(OZ)^2}$

$(\text{sine } \theta)^2 + (\text{cosine } \theta)^2 = 1$

Equation A.7

relationships in a right triangle (symbols defined in Fig. A.9):

$AZ = (\text{sine } \theta)OZ$
$BY = (\text{sine } \theta)OY$
$OA = (\text{cosine } \theta)OZ$
$OB = (\text{cosine } \theta)OY$

stated in Example A.11. The trigonometric ratios for angles at 5-degree intervals are listed in Table A.7.

As we explained in the section on mathematical models, a ratio can be interpreted as the constant of proportionality in direct proportion. You can apply this idea to a set of similar right triangles (Fig. A.11). As you go from one triangle to another in such a set, the length of each leg and the hypotenuse vary in the same proportion. The mathematical models for these proportions state that the side opposite one acute angle is equal to the hypotenuse times the sine of the angle, and so on (Eq. A.7). The tabulated values of the trigonometric ratios can therefore be used to calculate one leg of a right triangle if one acute angle and the hypotenuse or the other leg are known. By transforming the definitions in Eq. A.6, you can use them in many different ways. In fact, you will also recognize that Eq. A.7 is obtained by transforming the definitions in Eq. A.6.

Graphical procedures. All of the geometrical problems we have described in this section can be solved by graphical means. If you want to follow this procedure, use a ruler and a protractor to draw the diagrams carefully, and then measure the desired parts. For instance, to find the two legs of a right triangle with one acute angle of 35 degrees and a hypotenuse of 6 centimeters, you proceed as in Fig. A.12. For problems in which the lengths are too large to fit on a piece of paper, you make a scale drawing, with 1 centimeter representing 1 meter or another suitable length.

Answers

– to problems with numerical solutions

The answers to problems with numerical solutions are given here so you may compare the results of your work with them. If your answers come within 20% of the values given, you may be satisfied that you were approaching the problem correctly.

Chapter 2

5 (a) house (70°, 13 m)
      tree D (30°, 15 m)
      tree E (330°, 15 m)
6   5.1 units
7   [3, 4] =[–2, 3] + [5,1]
8   north pole
18 (a) (25°, 32 m)
  (b) [29, 14] m
19 (a) [6, 0] miles
  (b) [1.5, 2] miles
  (c) [9.5, 4.5] miles
  (d) [7.5, 2] miles
  (e) [2, 4] miles

Chapter 3

7   yes, no, yes, no

Chapter 5

13 50°, 50°
15 (a) $\theta_r = 41.7°$, $\theta_R = 30°$
    (b) $\theta_r = 59°$, $\theta_R = 40°$
19 39°
22 about 7 inches

Chapter 6

7 (a) 6°
   (b) 17°
   (c) 37°
   (d) no diffraction

8 (a) 0.05 m
   (b) 0.13 m
   (c) 0.26 m
10 (a) 1.33m,0.75/m
   (b) 0.77m, 1.3/m
   (c) 0.33m, 3.0/m

Chapter 7

2 88 m/sec
5 highest C: $f = 4100$/sec,
   $\lambda = 0.084$ m
   lowest A: $f = 28$/sec,
   $\lambda = 12.3$ m
6 B note, two octaves higher
9 0.0034 m= 3.4 mm
10 (a) approximately 65 km
19 (a) 200 m to 550 m
   (b) 2.8 m to 3.4 m

Chapter 8

5 (a) 6.6 x 10^{-10} m
  (b) 1.5 x 10^{-10} m

Chapter 9

1 $E= 11,200\ M_G$
5 (a) 4 units
  (b) –4 units
  (c) –2.5 units
  (d) 1.3 units
6 2.25 kg water and
  0.75 kg ice
11 (c) 1 unit = 9 x 10^{16} joules

Chapter 10

8 (a) 99 Cal
   (b) 75 Cal
   (c) 7 Cal
   (d) −19 Cal
9 (a) 66°
   (b) 14°
   (c) −13°
   (d) 19°
10 (a) 0.56 Cal/deg/kg
   (b) 0.16 Cal/deg/kg
   (c) 0.7 Cal/deg/kg
   (d) 0.6 Cal/deg/kg
11 (a) 57° Celsius
   (b) 2° Celsius
14 approximately 17 hours
17 5.7 kg
18 (a) 4.5 kg
   (b) 860 kg
   (c) 1.3×10^4 m (8 miles)

Chapter 11

15 2700 joules
16 3600 joules, 2750 joules
17 −1260 joules
25 one eighth degree Celsius
27 with the reference direction
   horizontal, the partial
   forces are
   on V, (45°, 425 newtons)
   on W, (45°, 390 newtons)
   on X, (45°, 640 newtons)
   on Y, (45°, 500 newtons)
   on Z, (45°, 280 newtons)

Chapter 12

10 1.2 volts
15 (a) 4, 2, 2, ?
16 4, 2, 4, ?
17 4.3×10^6 joules = 1.2 kWh
   = 1100 Cal = 13 kg ice
22 (a) 1000 volt/amp
   (b) 3×10^4 volt/amp
24 (a) 2.5 volts/amp
   (b) 40 volts/amp
   (c) 2.5, 4, 6, 7.5, 10, 13.3,
      16.7, 25, and 40
      volts/amp
27 0.031 cm
28 (a) 0.55 amp
   (b) 0.90 m
   (c) 11 amp
(d) 1200 watts

Chapter 13

2 (a) 38 m
   (b) 3.14 m/sec
   (c) with the reference direc-
      tion in the child's initial
      direction of motion,
      (90°, 2 m/sec)
   (d) (45°, 2.8 m/sec);
      (30°, 3.0 m/sec);
      (15°, 3.1 m/sec)
   (e) (0°, 3.14 m/sec)

Chapter 14

13 (a) [4000, 15000] newtons
    (b) [−4000, −15000] new-
      tons
    (c) 135 m
    (d) 540,000 joules
14 (a) 5 m/sec
    (b) 24 m/sec
    (c) 30 joules
    (d) 30 joules
19 three
31 (b) 0.4 m/sec
32 (a) 8.5×10^3 newtons
    (b) 42 m/sec
    (c) 5.3×10^6 joules
    (d) 1×10^6 joules
    (e) 5×10^7 joules
33 (a) 4×10^5 joules
35 11 m/sec
36 10 m/sec

Chapter 15

3 with the reference direction
  in the initial direction of
  motion
  (a) (−,0 m/sec)
  (b) (150°, 1.20 m/sec)
  (c) (120°, 2.6 m/sec)
  (d) (90°, 4 m/sec)
  (e) (60°, 5.2 m/sec)
  (f) (30°, 6.0 m/sec)
  (g) (15°, 6.2 m/sec)

5 (a) (−, 0 m/sec/sec)
  (b) (150°, 7.5 m/sec/sec)
  (c) (120°, 16 m/sec/sec)
  (d) (90°, 25 m/sec/sec)
  (e) (60°, 33 m/sec/sec)
  (f) (30°, 38 m/sec/sec)
  (g) (15°, 39 m/sec/sec)
7 mass of Jupiter is
  1.64×10^{27} kg
8 (a) 2 m/sec/sec
  (b) 80 newtons
18 16 m
24 3.2×10^{-2} m/sec/sec

Chapter 16

4 (a) 35 to 45 pounds per
    square inch
7 (a) 4×10^5 newtons/m^2
    (b) 40 newtons
    (c) 10^4 newtons
9 8.1×10^4 newtons/m^2
18 (a) 3.5 Cal
    (b) 1.25 Cal
19 2.4×10^{10} Cal
20 (a) 1.2×10^{15} Cal
    (b) about 8 hours
23 0.4 joules by 1 kg water
    5600 joules by 1 kg air
31 (a) 500 m/sec; 1900 m/sec;
     1400 m/sec; 400 m/sec
32 (a) 230 Cal
34 (a) hydrogen:
     1.5 Cal/deg/kg
    nitrogen:
     0.11 Cal/deg/kg
    carbon dioxide:
     0.07 Cal/deg/kg

Index

Absolute zero, 438, 452
Acceleration, 356, 375, 380, 383
  average,
    formal definition of, 357
  centripetal, 404, 410
  in circular motion, 403-404
  examples of, 358-359, 410
  of gravity, 384-385
Accelerator, 224
Aether, 194
Air, 426
Air conditioners, 448
Alhazen, 120
Alpha rays, 208, 209, 220, 221
Alternating current, 332
  applications of, 333-335
Ammeter, 315
Ampere, 315, 316
Ampère, André Marie, 310, 315
Amplitude, 140
Analogue model, 10-11, 97, 157,
    209,316-317,318,327,
    331, 342, 415, 436
Angles, 479
Aristotle, 6, 364, 368, 369, 384, 405
Astronomical unit, 405
Atmosphere, 432, 433
Atom, 91-92, 203
  Bohr's model for, 212-215
  early models for, 208-211
Atomic bomb, 4, 226
Atomic number, 221-222
Avogadro, Amadeo, 91

Bacon, Francis, 81, 462
Bacon, Roger, 1 1
Barometer, 430, 431
Beats, 115, 135, 149, 152
Becquerel, Henri, 220, 221
Beta rays, 220
Bimetallic strip, 250, 251, 252, 253
Binoculars, 132
Black, Joseph, 257, 262, 267, 274
Bohr, Niels, 212, 213
Bohr's frequency condition, 212
Bohr's theory, 2 12-213
    limitations of, 215
Boyle, Robert, 435
Boyle's law, 435, 436
Brahe, Tycho, 406
Bronowski, Jacob, 461-465

Caloric theory, 266, 267
Calorie, 237, 239, 244
Camera, 130-131
Cathode rays, 205, 206
Cause and effect, 57
Cavendish, Henry, 320, 410, 416
Cell, 322
Celsius scale, 252
Center-of-mass, 367-368
Centripetal force, 404, 408, 412
Charge, electric, 295
Charles, Alexandre, 435
Charles's law, 435-437
Chemical energy, 88, 265, 266,
    301
Children's concepts, 7, 23, 77,
    104, 105, 113, 135, 276,
    309, 344, 397
Circuit,
  closed, 313, 320
  electric *(see* Electric circuit)
  open, 313, 320
  short, 335
Circuit elements, 313, 330
  symbols for, 327
Circular motion, 402-404
  examples of, 416-419
Color, 121, 123-125, 182
  wavelength and, 183
Color filters, 125
Components, 46
Conductance, 323, 325
  example of, 330
Conductors, 319, 320
Conservation,
  of charge, 75
  of energy, 81-84
  of matter, 58, 82
  of momentum, 386
  of systems, 58-59
Control experiment, 61-62, 254
Control rods, 227
Coordinate axes, 37
Coordinate frames, 34
Coordinates,
  example of, 38
  polar, 34-35
  rectangular, 37-38
Copernicus, Nicolaus, 40, 405
Cosmic domain, 8
Coulomb, Charles Augustin de,
    295, 296, 416

Robert Karplus—A Portrait

Robert Karplus was born in Vienna, Austria in 1927 where he grew up until 11 years of age. His mother, brother, and he left Vienna in March 1938 immediately after the German Anschluss. After a 6-month stay in Switzerland, the family emigrated to the United States and settled in the Boston area. He entered Harvard in 1943 (at the age of 16) as a freshman and by 1948 (at 21) had completed his Ph. D. His thesis under E. Bright Wilson was on microwave spectroscopy and included both experimental and theoretical work. He was early recognized for his brilliance, original-ity, energy, and cheerful, positive outlook.

After completing his Ph. D., Karplus worked at the Institute for Ad-vanced Study in Princeton, where he became interested in the developing, but untested, theory of quantum electrodynamics (QED). The magnetic moment of the electron had been determined very precisely by means of a variety of experiments, but the best theoretical calculations of this im-portant quantity (based on quantum mechanics but not QED) were seri-ously at variance with the experimental results. There was great interest among physicists in knowing whether or not a calculation based on QED would agree with the experimental results, but, because of the ambigui-ties and complexity of QED, no one had so far been able to do such a calculation.

Karplus, in collaboration with Norman Kroll, used QED to calculate the value of the magnetic moment of the electron. This was an extremely difficult calculation, requiring more than a year of intense effort from both men; the agreement between their result[1] and the experimental measurements was the first, dramatic confirmation of QED (for the de-velopment of which R. P. Feynman, J. Schwinger, and S. I. Tomonaga received the Nobel Prize).

Karplus continued his work at the highest level in theoretical physics for more than 10 years, at Harvard from 1950 to '54 and then at the Uni-versity of California, Berkeley, publishing 50 research papers, mostly in QED but also in other areas of physics, including the Hall effect, Van Al-len radiation, and cosmic rays. He is best known as a theorist, but he also thoroughly enjoyed experimental work, investigating the chemistry of Land camera instant pictures and setting up an experimental germanium purification assembly line for transistors. In 1958, Karplus was promoted to the rank of Professor at Berkeley, and he was compared favorably with other theorists his age, including T. D. Lee, M. Gell-Mann, and C. N. Yang.

In 1948 Karplus married Elizabeth Frazier, whom he had met at an international folk dance group he organized while at Harvard. They had seven children born between 1950 and 1962. Bob loved camping, hiking, exploring, and playing games with his family. When the oldest child, Beverly, was 7, Karplus accepted her invitation to present a science les-son on electricity to her second-grade class, using the Wimshurst machine he had inherited from his grandfather. Unfortunately, while the children enjoyed the demonstration, the lesson was a conceptual disaster. This stimulated Karplus to think about how to teach science better, and as the other children entered school he continued to visit their classes on a "show and tell" basis with various science experiments or demonstrations. Conversing with his children and their classmates, he became increasingly interested in children's learning, reasoning, and science concept development.

Central to Karplus' approach was a genuine interest and trust in others: he really believed that people were making sense - at least to themselves - and that it was a challenge to him to discover their reasoning, to uncover what they saw as important and what might be missing. A striking example of this is in a story recounted by his wife Elizabeth about a family outing exploring a mountain near their home. They found some fossil shells, and Karplus asked the children how the shells got up there. Rick, the youngest at 1½-2 years, immediately replied "the Sun," which baffled everyone.

Karplus, characteristically, was intrigued by this response, and he immediately invited Rick to explain his thinking. Rick responded that the Sun must have dried up all the water in the lake, leaving the shells! Of course, this ignores the fact that the shells were 3000 feet up on a steep hillside where no water could have stood long enough to be dried up. But Rick's explanation, as elicited by his father, shows an unsuspected depth of thought and a particularly imaginative way of accounting for at least some of the evidence. This short dialogue is a good example of how scientific thinking actually occurs, and it clearly opens the door to discussion and to further development of how to account for the presence of the shells.

Karplus didn't seem to find an illogical answer to be particularly disappointing or disturbing; on the contrary, it would pique his unquenchable curiosity. When he encountered a response that seemed off-target, he welcomed the opportunity to figure out a new puzzle, to discuss and "play" with the ideas once more, and to discover how someone else's mind worked.

Robert G. Fuller (p. 301) tells another characteristic story (corroborated by two different physicist co-workers) about how Karplus became interested in children's thinking:

> "Robert Karplus placed the toy truck in front of a child.
> He rolled the truck slowly across the desk.
> 'Did the truck move?' he asked.
> 'No.' replied the child.
> "(... He moved the truck back to its starting position. Again, he slowly rolled the toy truck across the desk to a new location.)

'Did the truck move?' he asked again.

'No.' the child replied once again.

'Can you explain to me why you say the truck did not move?' Karplus asked.

'<u>It</u> did not move,' responded the child triumphantly, '<u>You</u> moved it.'

"In that moment of puzzlement Robert Karplus was hooked. The physics that he knew and loved had not prepared him for such an experience. He discovered the importance of one's mental state in the shaping of learning and reasoning."

Fortunately, Karplus had the necessary blend of imagination, courage, intellect, and empathy needed to appreciate and meet this challenge. He pursued this new interest in science teaching and the psychology of reasoning with the same joy in the discovery of new knowledge with which he had pursued theoretical physics.

Within a few years, Karplus had changed careers - from theoretical physics, to research on science and math learning, and then to curriculum developer. Karplus quickly mastered what was already known about the development of thinking and reasoning, studying various psychologists, especially Piaget. Characteristically, Karplus also immediately began generating his own questions about children's thinking, collecting evidence, and developing his own interpretations and explanations of what he observed. This initial research was quite informal, using his own children. With his wife Elizabeth as a close collaborator, Karplus quickly progressed to the frontiers of what was then known about intellectual development, and he successfully engaged many other collaborators in further investigations.

This research was deceptive in its directness and apparent simplicity. Individuals were presented with carefully formulated problem situations, and *both* their responses *and* their explanations were recorded in some detail. This evidence was then analyzed, classified, and interpreted, using Piaget's stages of development as a general framework. The interpretations were quite provocative, often revealing unsuspected details about what the individuals understood and what they tended to neglect about the given problems. Karplus' insights into these matters were sufficiently detailed and profound that he can, in fact, be credited with the discovery of a variety of the conceptual processes that most people use in grappling with typical math and science problems.

Karplus also extended Piaget's theory to college students and adults; Piaget's theory included four stages, and Piaget had documented children's thinking in great detail, finding that most children made the transition from the 3rd stage (concrete operations) to the 4th stage (abstract reasoning) by about 16 years of age. Karplus, however, extended Piaget's methodology to older groups and found that many of these individuals had important gaps in their ability to use abstract reasoning in solving scientific, logical, and mathematical problems. Karplus further explored and documented the details of college students' and adults' thinking as they confronted the issues involved in this critical intellectual transition, finding that many of the issues and problems that he, Piaget, and others had discovered as critical for younger students were still relevant for older individuals, particularly when they were attempting to solve a problem in a discipline that was new to them.

Most science and math teachers, and researchers, had not previously been aware of these thinking patterns, partly because no one had really looked for them. To many teachers, such conceptual processes just seemed to be mistaken or to lead to "wrong" answers, and it seemed better not to dwell on such thinking, but rather to focus strictly on the "correct" reasoning so as to arrive at the "correct" answers.

Karplus, on the other hand, like Piaget, investigated the reasoning patterns that led to the "wrong" answers. He illuminated the pathways most people naturally tended to follow when they first encountered typical science and math problems. Karplus identified common conceptual misunderstandings, and he found that dealing directly with such issues often helped students to find more productive ways of thinking about science and math.

Karplus became an expert on Piaget's theory and a pioneer exponent of how to extend and apply it effectively in science teaching and curriculum development. He gave many, many talks about this throughout the country and beyond. Karplus wrote up his research results in an important series of collaborative papers: "Intellectual Development Beyond Elementary School I-VIII" (*School Science and Mathematics*, 1970-80). In cooperation with various others, Karplus also developed and presented a series of "Workshops on Physics Teaching and the Development of Reasoning" for college and high school physics teachers.

Karplus' new passion coincided, serendipitously, with the post-Sputnik wave of efforts to upgrade US science education. Beginning in the late 50s, many other scientists also devoted themselves to science education and the schools, but Karplus was from the start the leader at the elementary level.

Initially there was substantial reluctance at the National Science Foundation (NSF) to fund science curriculum projects at the elementary level, but this was overcome in 1959, when Karplus and three colleagues received the very first of many NSF grants for science course content improvement at the elementary level. This work evolved into a monumental, 15-year effort – the Science Curriculum Improvement Study (SCIS). Under the direction of Karplus and Herbert D. Thier, SCIS became a comprehensive, fully-tested, hands-on, laboratory-based program in both physical and biological science for grades K-6.

In addition to his work on K-6 science, Karplus regularly taught science courses at Berkeley, and I was his teaching assistant in 1968-70 in two of them. I watched Karplus perform in those situations, and he was simply terrific. He organized what he wanted to teach carefully and imaginatively, preparing thoroughly and keeping the needs and preconceptions of the students in mind. Karplus was a *wonderful* lecturer, articulating his ideas powerfully and clearly, explaining physics as simply as possible (but never pretending to make it, as Einstein was reputed to have said, simpler than that!), designing demonstrations that were both entertaining and on-target, inviting questions and answering them fully, preparing clearly-written, interesting homework assignments, using humor and connecting with current events, writing excellent exams, assigning grades that were both fair and defensible, and dealing promptly with student complaints and appeals.

Those were years of great turmoil at Berkeley, with the Vietnam War, politics, riots, tear gas, and police actions exerting tremendous pressure on everyone. Karplus was a steadying and calming presence throughout, never taking the easy way out (whether to the right or the left), helping students to figure out what was important, responding to students' concerns and interests, and constantly finding ways to keep the educational dialogue going.

I particularly remember how he went about planning Physics 10 (an introductory course for non-scientists), in which we were using the hardback edition of his textbook for the first time. We were assigned one of the traditional physics lab rooms, which had been regularly used in the past for Physics 10, and we had access to a very complete set of traditional physics apparatus. Karplus asked me to meet him in the lab, so we could, as he said, "cook up" some experiments for the semester. Within roughly 15 minutes "we" had reviewed the equipment and decided on the experiments, one of which was determining the wavelength of light by measuring the angle of the fringes produced by a diffraction grating. Needless to say, the speed and (relative) thoroughness with which "we" did this was quite impressive to my half of the team!

The lab had reasonably expensive spectroscopes, which are the standard instrument for measuring optical fringes from a diffraction grating. Karplus took one look at the beautiful spectroscopes and started taking them apart; he removed the telescope (which made it possible to measure angles extremely accurately), substituting a simple wooden sight, and he quickly simplified the apparatus to make it easier for the students to mount the gratings and to see how to actually go about measuring the angle of the fringes. Within a few minutes he had converted the needlessly complex spectroscopes into much simpler instruments which measured the angle to a lower, but still quite respectable, accuracy and which had a much higher educational payoff.

I clearly remember meeting the next day with Karplus and the other teaching assistant, John, a graduate student in theoretical physics who had previously assisted another professor with Physics 10. When Karplus mentioned the grating experiment, John remarked that the students would have great difficulty with the telescope. I will never forget John's amazed look when Karplus responded that he had eliminated this problem by getting rid of the telescope! John had measured the angle of optical fringes in his own undergraduate lab exercises to high accuracy with a telescope, and he had unconsciously assumed that this was how the measuirement always had to be done. It required an effort for him to realize the value of the trade-off of accuracy for speed and ease of understanding. As the course proceeded, John assimilated, from Karplus' behavior as much as from what he said, that the primary objective was to enable non-science students to carry out the experiments and think about physics on their own rather than to simply repeat the standard experiments and to memorize and apply the classic laws and principles. By the end of the course, John had become a much more adventurous, stimulating, and engaging teacher.

I also recall Karplus' creative solution to the problem of obtaining student feedback in the 600-student lecture hall where he often lectured. He distributed two cards to each student; the cards were colored with four different colors on the four faces. In the course of the lecture, Karplus would ask questions with four possible answers and ask students to show their responses by holding up their cards. He repeatedly demonstrated an uncanny capacity, given a certain pattern of colors, for coming up, "on the fly," with just the right additional example, question or problem to help hundreds of students figure out, right then and for themselves, where and how they were going wrong. It was a given at Berkeley: if Karplus was the teacher, whether there were 600 students, or 20, or 1, he would find a way to engage them all in thinking actively about physics.

I also completed my Ph. D. at Berkeley in 1972 under Karplus (in the Graduate Group in Science and Mathematics Education, which he and other faculty had founded). Thus I had many other chance to see Karplus in action, and, in particular, to see how he used what I came to think of as his "toolkit" for effective teaching. There were 4 items in this "toolkit":

1) The learning cycle of exploration, invention, and discovery,

2) The critical interplay between autonomy and input,

3) The importance of the conceptual structure of science. Karplus understood, in a very profound sense, the extent to which science, especially physics, can be molded and shaped. He considered science and physics to be plastic, malleable structures, with logical relationships that could be organized in many alternative ways. For Karplus, the first job of

a science teacher was to come up with an organized conceptual structure that would suit the audience and that would connect and relate together the key concepts and principles. This conceptual structure should, first of all, allow the instructor to begin the discussion with language and a set of concepts that are familiar at the appropriate level of abstraction. The instructor could then use operational and formal definitions, hands-on experiments, thought experiments, examples, and explanations, plus questions to and answers from the students, to build up the rest of the structure in a logical, understandable sequence.

4) A conception of teaching as a practical, only partially understood, yet improvable, activity, where the method for improvement lay in the collection and interpretation of detailed *evidence* about what the students had learned. This evidence, in Karplus' view, had to include *both* the right and wrong answers *and* the details of the explanations for the answers.

All of these ideas are explained clearly and in depth in Karplus' published papers (see the collection edited by Fuller, *A Love of Discovery*). However, in the literature, these ideas come across as relatively established principles, and one doesn't get a picture of how flexibly he used them and how powerful they were in his hands. Karplus actually employed these ideas not so much as laws or principles but as "heuristics," that is, as practical rules of thumb which one could exploit as necessary to help work out an effective solution for a given teaching situation.

In 1977 Karplus was elected as President of the American Association of Physics Teachers (AAPT), and in 1978 the National Science Teachers Association awarded him their Citation for Distinguished Service to Science Education. In 1980 he was awarded the AAPT's highest honor, the Oersted Medal, "for his many contributions to physics teaching at all levels and especially for his work in revealing the implications for physics teaching of research in the development of reasoning." (from the presentation by Dr. James Gerhart, Past President and Chair of the AAPT Awards Committee, Fuller, p. 228).

Unfortunately, in 1982 while jogging near Seattle, Washington, Karplus suffered a severe cardiac arrest, and after an eight-year illness he died in 1990. He is survived by his wife of 42 years, Elizabeth F. Karplus (teacher, co-author with Bob of many papers, graduate of Oberlin College in physics, and holder of Master's degrees from Wellesley in physics and from St. Mary's College in special education), as well as by his seven children and many grandchildren.

Karplus' intellectual legacy is monumental, especially in science education. His many published papers, curricula, and teaching materials form an impressive and continually useful resource (see Fuller). Even more important, in my opinion, was his inspirational example for an entire generation of science teachers, college faculty, scientists, and researchers. The level of Karplus' creativity and attention to detail, the depth of his understanding of physics and of students' thinking, as well as his rigor in collecting evidence coupled with his honesty in facing it provide a true standard of excellence in science education. Anyone who spent time with Karplus went away with unforgettable first-hand experience of how much fun it was to pursue good science teaching, how worthy of one's best efforts and how dynamic and interesting it could be.

Fernand Brunschwig
New York, New York
June, 2003

Note:
[1] R. Karplus and N. Kroll, "Fourth-Order Corrections in Quantum Electrodynamics and the Magnetic Moment of the Electron." 1950. *Phys. Rev.,* 77, 536-549. As usual in science, Karplus and Kroll's breakthrough wasn't quite as clear-cut at the time. Their original calculation using QED indeed agreed substantially better with the experimental results than previous calculations, but there was still a small discrepancy that they couldn't fully explain. A few years later, other theorists discovered that Karplus and Kroll had actually made a mistake in their calculations which was responsible for much of the remaining discrepancy. It is noteworthy that the line of inquiry pursued so productively by Karplus and Kroll is still active – theorists are still calculating, and experimentalists are still measuring, the value of the magnetic moment of the electron. Of course, there are newer theories than QED to test, the experiments and calculations are now done with the help of computers, and the accuracy is much higher, but the endeavor itself, as well as much of its style and shape, has grown from Karplus and Kroll's work of 1950.

Reference: Robert G. Fuller, Editor. *A Love of Discovery: Science Education, the Second Career of Robert Karplus.* (New York, Kluwer Academic/Plenum Publishers, 2002).

Acknowledgment: I have borrowed heavily from a summary of Karplus' career written by Elizabeth F. Karplus. Mrs. Karplus also contributed many additional insights and several of the stories.

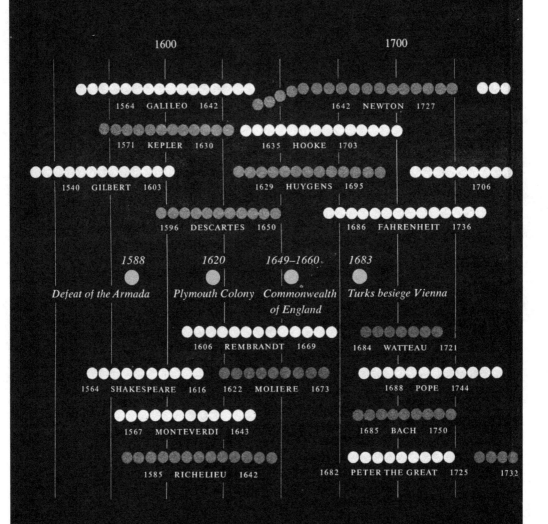

1600

1700

1564   GALILEO   1642

1642   NEWTON   1727

1571   KEPLER   1630

1635   HOOKE   1703

1540   GILBERT   1603

1629   HUYGENS   1695

1706

1596   DESCARTES   1650

1686   FAHRENHEIT   1736

1588

1620

1649–1660

1683

Defeat of the Armada   *Plymouth Colony*   *Commonwealth of England*   *Turks besiege Vienna*

1606   REMBRANDT   1669

1684   WATTEAU   1721

1564   SHAKESPEARE   1616

1622   MOLIERE   1673

1688   POPE   1744

1567   MONTEVERDI   1643

1685   BACH   1750

1585   RICHELIEU   1642

1682   PETER THE GREAT   1725   1732

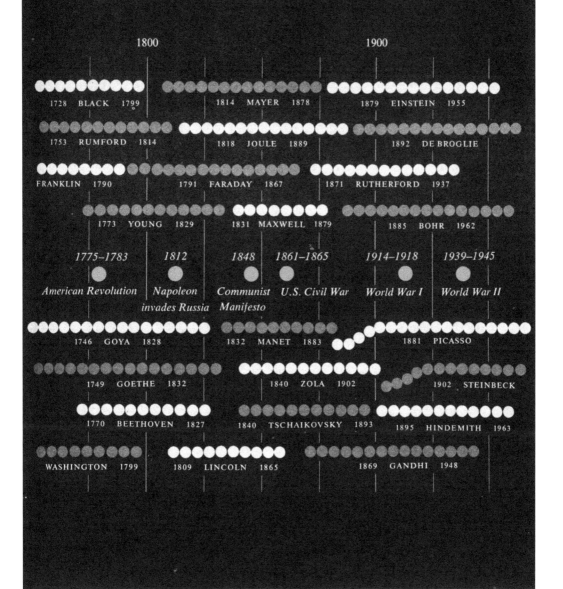

1800

1900

1728 BLACK 1799

1814 MAYER 1878

1879 EINSTEIN 1955

1753 RUMFORD 1814

1818 JOULE 1889

1892 DE BROGLIE

FRANKLIN 1790

1791 FARADAY 1867

1871 RUTHERFORD 1937

1773 YOUNG 1829

1831 MAXWELL 1879

1885 BOHR 1962

1775–1783

1812

1848

1861–1865

1914–1918

1939–1945

American Revolution

Napoleon invades Russia

Communist Manifesto

U.S. Civil War

World War I

World War II

1746 GOYA 1828

1832 MANET 1883

1881 PICASSO

1749 GOETHE 1832

1840 ZOLA 1902

1902 STEINBECK

1770 BEETHOVEN 1827

1840 TSCHAIKOVSKY 1893

1895 HINDEMITH 1963

WASHINGTON 1799

1809 LINCOLN 1865

1869 GANDHI 1948

CPSIA information can be obtained
at www.ICGtesting.com
Printed in the USA
BVHW062215190421
605296BV00007B/843